Acclaim for the first edition of

THE ANCESTOR'S TALE

"This is wonderful stuff: conversational and accessible, yet rich in scientific detail and explanation."
— *San Diego Union-Tribune*

"[*The Ancestor's Tale*] is well worth a thorough reading . . . a rewarding set of insights into life itself."
— *Seattle Times*

"A tour de force . . . It should be read by anyone who cares (positively or negatively!) about evolution."
— *New York Sun*

"One of the richest accounts of evolution ever written . . . Triumphant."
— *Financial Times*

"Intriguingly written, honest about the controversies that exist, clear about the science."
— *Washington Post Book World*

"Terrifically entertaining."
— *Slate*

"Immensely informative, sometimes caustic, always witty . . . *The Ancestor's Tale* is sure to provoke a lot of heated discussions from all sides of the evolutionary debates."
— *San Jose Mercury News*

"Marvelously lucid."
— *Science News*

"Modern and winning . . . It depicts evolution . . . as a marvelous adventure, enigmatic in origin, that teems with riddles large and small."
— *Discover*

"Brilliantly recounted."
— *Popular Science*

"An eloquent treatise on evolution."
— *Scientific American*

"The book's scope and provocativeness are truly worthy of epic treatment."
— *Library Journal*

"Erudite yet conversational . . . Should prove a natural hit with science readers."
— *Publishers Weekly,* starred review

"Lively and daring, a book certain to draw even casual readers deep into the adventure — and controversy — of science."
— *Booklist,* starred review

THE ANCESTOR'S TALE

RICHARD DAWKINS and **YAN WONG**

Revised and Expanded

A PILGRIMAGE TO

THE ANCESTOR'S TALE

THE DAWN OF EVOLUTION

Mariner Books
An Eamon Dolan Book
HOUGHTON MIFFLIN HARCOURT
BOSTON · NEW YORK
2016

Second Mariner Books edition 2016

First published in Great Britain by Weidenfeld & Nicolson,
an imprint of The Orion Publishing Group, in 2004

www.hmhco.com

Library of Congress Cataloging-in-Publication Data is available.
ISBN 978-0-544-85993-7

Printed in the United States of America
DOC 10 9 8 7 6 5 4 3 2 1

John Maynard Smith
(1920–2004)

He saw a draft and graciously accepted the dedication,
which now, sadly, must become

In Memoriam

---

*'Never mind the lectures or the "workshops"; be blowed to the
motor coach excursions to local beauty spots; forget your fancy
visual aids and radio microphones; the only thing that really
matters at a conference is that John Maynard Smith must be in
residence and there must be a spacious, convivial bar. If he can't
manage the dates you have in mind, you must just reschedule the
conference . . . He will charm and amuse the young research
workers, listen to their stories, inspire them, rekindle enthusiasms
that might be flagging, and send them back to their laboratories or
their muddy fields, enlivened and invigorated, eager to try out the
new ideas he has generously shared with them.'*

It isn't only conferences that will never be the same again.

CONTENTS

PREFACE TO THE SECOND EDITION

A decade after the first edition of this book was published, Yan Wong and I met in the fitting surroundings of the Oxford Museum of Natural History to discuss the possibility of producing a new, tenth anniversary edition. Yan, once my undergraduate pupil, had been employed as my research assistant during the writing of the original edition, before he left for his lecturing position in Leeds and his career as a television presenter. He played an enormously important part in the conception and execution of the first edition, and he was credited as joint author of several of the chapters. During the course of our discussion ten years on, we realised that much new information had come in, especially from the molecular genetics laboratories of the world. Yan undertook the bulk of the revision and I proposed to the publisher that this time he should be properly credited as joint author of the whole book.

The new research, fortunately, has not greatly changed the order of rendezvous points where pilgrims join the great trek to the origin of life which, following our Chaucerian conceit, we call Canterbury. One or two minor reversals will be discovered herein, along with a couple of additional rendezvous points and some changes of dating. In the first edition we took care to point out that different genes can be inherited through different routes. This has unexpected ramifications which are more fully explored in some new and altered tales for this edition. In particular, we provide a more specific definition of the dates quoted for each rendezvous, and take a more nuanced view of the genetic relationships between species. A single tree of life is a necessary simplification of the evolutionary process – this caution looms ever larger near our Canterbury climax, given the increasingly documented instances of horizontal gene transfer between our distant bacterial cousins.

The basic relationships between the millions of species of life on earth can now be depicted in a much more elegant fashion than in the first edition. James Rosindell, a previous collaborator of Yan's from Leeds and now at Imperial College, London, has devised a superb way to represent enormous evolutionary trees using fractals.

His marvellously explorable 'OneZoom' visualizations are ideally suited to our pilgrimage through the tree of life. Static snapshots of these fractals enhance our presentation of each rendezvous point, and are the basis for the explorable online tree of life that accompanies this book, at www.ancestorstale.net.

Some new tales have been added, and others removed or modified and handed over to other pilgrims better suited to tell them. Yan had the inspired idea of using my genome (which in 2012 had been sequenced in its entirety for television purposes) to illustrate a fascinating new technique for reconstructing human demographic history using DNA from a single person. This has been incorporated into Eve's Tale, along with similar analyses based on multiple human genomes.

The recovery of ancient genomes from fossils has radically reshaped our understanding of recent human evolution, upholding our previous surmise that Neanderthals and humans interbred, and has uncovered a previously unknown human subspecies. These 'Denisovans' take over what was the Neanderthal's Tale. To my great satisfaction, ancient DNA has also overturned the story of the elephant bird. Her moral is now recounted by the sloth, in a new tale. Additional tales have been prompted by ongoing releases of whole genome sequences from living creatures; it is extraordinary to think that such a rich source of information will be deemed commonplace by future students of nature. Here we count three new tale-tellers: chimpanzee, coelacanth, and humped bladderwort. In other cases, such as the gibbon, mouse and lamprey, increased genomic understanding has led to major revisions of previous tales, or occasionally new prologues or epilogues. Recent fossil discoveries have altered previous discussions (here we might count spectacular new *Homo, Australopithecus* and *Ardipithecus* discoveries), and have also sparked a new Lungfish's Tale. A final mention must be made of the Lava Lizard's Tale, a whimsical addition which was previously published as an article for the *Guardian*, written while I was on a small boat cruising the Galapagos archipelago.

Given the pace of new biological discoveries, it goes without saying that some of the material in this new edition may become superseded. Such is the way of science. Indeed, even in the few months before publication, several academic papers appeared which revealed major new deep branches on the tree of life. Strikingly, advances in DNA sequencing place today's naturalists in the strange position of having the full genomes of species that are otherwise almost entirely unknown. These creatures range from the human-like Denisovans at one end of our journey to various unculturable bacterial groups at the other. Who knows what future discoveries will reveal? Nevertheless, much of what we wrote in the first edition remains correct over a decade later. This surely bodes well for the vision of the natural world that we outline in these pages.

It is gratifying to see that the approach to phylogenetic questions which we tentatively adopted in the first edition – tracing the family histories of genes independently of the bodies they sit in – now underlies much of modern biology. It pervades many of the new sections in this new edition. I hope I may be forgiven for seeing it as yet another vindication of the 'gene's-eye view' which I have advocated through much of my professional life.

Any jointly authored book is placed in the awkward position of needing to establish a pronoun convention: singular or plural? 'I' or 'we'? The first edition used 'I' throughout, and was indeed written from my point of view, including personal anecdotes and caprices. The publishers rightly judged that these would not sit well with a change to 'we', and advised us, for the sake of uniformity, to stick to the same 'I' even in those chapters largely written by Yan. There are places, however, where the two of us wish to stick our necks out together, say over a particular point of theory or taxonomic technique. In such cases we have used 'we' and really mean it.

RICHARD DAWKINS
2016

ACKNOWLEDGEMENTS

I was persuaded to write this book by Anthony Cheetham, founder of Orion Books. The fact that he had moved on before the book was published reflects my unconscionable delay in finishing it. Michael Dover tolerated that delay with humour and fortitude, and always encouraged me by his swift and intelligent understanding of what I was trying to do. The best of his many good decisions was to engage Latha Menon as a freelance editor. As with *A Devil's Chaplain*, Latha's support has been beyond all estimation. Her grasp of the big picture simultaneously with the details, her encyclopaedic knowledge, her love of science and her selfless devotion to promoting it have benefited me, and this book, in more ways than I can count. Others at the publishers helped greatly, but Jennie Condell and the designer, Ken Wilson, went beyond the call of duty.

My research assistant Yan Wong has been intimately involved at every stage of the planning, researching and writing of the book. His resourcefulness and detailed familiarity with modern biology have been matched only by his green fingers with computers. If, here, I have gratefully assumed the role of apprentice, it could be said that he was my apprentice before I was his, for I was his tutor at New College. He then did his doctorate under the supervision of Alan Grafen, once my own graduate student, so I suppose Yan could be called my grandstudent as well as my student. Apprentice or master, Yan's contribution has been so great that, for certain tales, I have insisted on adding his name as joint author. When Yan left to cycle across Patagonia, the book in its final stages benefited greatly from Sam Turvey's extraordinary knowledge of zoology and his conscientious care in deploying it.

Advice and help of various kinds were willingly given by Michael Yudkin, Mark Griffith, Steve Simpson, Angela Douglas, George McGavin, Jack Pettigrew, George Barlow, Colin Blakemore, John Mollon, Henry Bennet-Clark, Robin Elisabeth Cornwell, Lindell Bromham, Mark Sutton, Bethia Thomas, Eliza Howlett, Tom Kemp, Malgosia Nowak-Kemp, Richard Fortey, Derek Siveter, Alex Freeman,

Nicky Warren, A. V. Grimstone, Alan Cooper, and especially Christine DeBlase-Ballstadt. Others are acknowledged in the Notes at the end.

I am deeply grateful to Mark Ridley and Peter Holland, who were engaged by the publishers as critical readers and gave me exactly the right kind of advice. The routine authorial claim of responsibility for the remaining shortcomings is more than usually necessary in my case.

As always, I gratefully acknowledge the imaginative generosity of Charles Simonyi. And my wife, Lalla Ward, has once again been my help and strength.

RICHARD DAWKINS
2004

Richard Dawkins approached me in late 2013, proposing to update *The Ancestor's Tale* with the intervening decade's worth of research. My main thanks must go to him for making this such an easy collaboration, justifying my enthusiastic consent. Given the scope of the book, however, my aim of finishing in 2014 was, in hindsight, wildly optimistic.

It became clear that a new edition could benefit from my previous involvement with the OneZoom project, and James Rosindell's inspirational fractals, which pack numerous species (indeed all of life) into a single diagram. This book, and popular biology in general, owes him much for his implementation. His idea to use silhouettes relied on the PhyloPic project, ably run by Mike Keesey; credits for individual silhouettes are in the endmatter. A debt of public gratitude is also owed to the team at the Open Tree of Life, one of many ambitious projects that reflect a commendable move to 'open access' in science. Similarly 'open' are the blogs assiduously updated by John Hawks and Larry Moran: I have been aided by brief communications with both authors.

I find it deeply satisfying that evolutionary biology is such a burgeoning field, but keeping up to date with research is more than a full-time job. My interpretation of the deeper parts of animal, eukaryotic and even bacterial phylogeny are due to several discussions with the ever-enthusiastic Jordi Paps. Stephan Schiffels helped enormously with the genome analysis, and Tamas David-Barrett gave excellent feedback on the human part of the story. Ryan Gregory kindly checked the new Humped Bladderwort's Tale, and Steven Balbus verified our interpretation of his theory. Of course, I take full ownership of any mistakes or misinterpretations.

Useful advice about specific research or potentially infelicitous phrasing was given by Peter Holland, Tim Lenton, Caro-Beth Stewart, Fabien Burki, David Legg and Michael Land. Thanks must also go to Rand Russell, Alex Freeman and my wife Nicky Warren, as well as Isabella Gibson and Dinah Challen for graphical assistance.

At the publishers, Bea Hemming took an admirably light hand on the tiller, while the editor Holly Harley remained resolutely positive in the face of continual scientific updates, and the designer Helen Ewing in the face of the fractal onslaught.

YAN WONG
2016

THE CONCEIT OF HINDSIGHT

History doesn't repeat itself, but it rhymes.
MARK TWAIN

History repeats itself; that's one of the things that's wrong with history.
CLARENCE DARROW

History has been described as one damn thing after another. The remark can be seen as a warning against a pair of temptations but, duly warned, I shall cautiously flirt with both. First, the historian is tempted to scour the past for patterns that repeat themselves; or at least, following Mark Twain, to seek reason and rhyme for everything. This appetite for pattern affronts those who insist that, as Mark Twain will also be found to have said, 'History is usually a random, messy affair', going nowhere and following no rules. The second connected temptation is the vanity of the present: of seeing the past as aimed at our own time, as though the characters in history's play had nothing better to do with their lives than foreshadow us.

Under names that need not trouble us, these are live issues in human history and they arise with greater force, and no greater agreement, on the longer timescale of evolution. Evolutionary history can be represented as one damn species after another. But many biologists will join me in finding this an impoverished view. Look at evolution that way and you miss most of what matters. Evolution rhymes, patterns recur. And this doesn't just happen to be so. It is so for well-understood reasons: Darwinian reasons mostly, for biology, unlike human history or even physics, already has its grand unifying theory, accepted by all informed practitioners, though in varying versions and interpretations. In writing evolutionary history I do not shrink from seeking patterns and principles, but I try to be careful about it.

What of the second temptation, the conceit of hindsight, the idea that the past works to deliver our particular present? The late Stephen Jay Gould rightly pointed out that a dominant icon of evolution in popular mythology, a caricature almost as ubiquitous as lemmings jumping over cliffs (and that myth is false too), is a shambling file of simian ancestors, rising progressively in the wake of the erect, striding, majestic figure of *Homo sapiens sapiens:* man as evolution's last word (and in this context it always is man rather than woman); man as what the whole enterprise is pointing towards; man as a magnet, drawing evolution from the past towards his eminence.

There is a physicist's version which is less obviously vainglorious and which I should mention in passing. This is the 'anthropic' notion that the very laws of physics themselves, or the fundamental constants of the universe, are a carefully tuned put-up job, calculated to bring humanity eventually into existence. It is not necessarily founded on vanity. It doesn't have to mean that the universe was deliberately made in order that we should exist. It need mean only that we are here, and we could not be in a universe that lacked the capability of producing us. As physicists have pointed out, it is no accident that we see stars in our sky, for stars are a necessary part of any universe capable of generating us. Again, this does not imply that stars exist in order to make us. It is just that without stars there would be no atoms heavier than lithium in the periodic table, and a chemistry of only three elements is too impoverished to support life. Seeing is the kind of activity that can go on only in the kind of universe where what you see is stars.

But there is more that needs to be said. Granted the trivial fact that our presence requires physical laws and constants capable of producing us, the existence of such potent ground rules may still seem tantalisingly improbable. Depending upon their assumptions, physicists may reckon that the set of possible universes vastly outnumbers that subset whose laws and constants allowed physics to mature, via stars into chemistry and via planets into biology. To some, this means that the laws and constants must have been deliberately premeditated from the start (although it baffles me why anybody regards this as an explanation for anything, given that the problem so swiftly regresses to the larger one of explaining the existence of the equally fine-tuned and improbable Premeditator).

Other physicists are less confident that the laws and constants were free to vary in the first place. When I was little it was not obvious to me why five times eight had to give the same result as eight times five. I accepted it as one of those facts that grownups assert. Only later did I understand, perhaps through visualising rectangles, why such pairs of multiplications are not free to vary independently of one another. We understand that the circumference and the diameter of a circle are not independent, otherwise we might feel tempted to postulate a plethora of possible universes, each with a different value of π. Perhaps, argue some physicists such as the Nobel Prize-winning theorist Steven Weinberg, the fundamental constants of the universe, which at present we treat as independent of one another, will in some Grand Unified fullness of time be understood to have fewer degrees of freedom than we now imagine. Maybe there is only one way for a universe to be. That would undermine the appearance of anthropic coincidence.

Other physicists, including Sir Martin Rees, the present Astronomer Royal, accept that there is a real coincidence in need of explanation, and explain it by postulating many actual universes existing in parallel, mutually incommunicado, each with its own set of laws and constants.* Obviously we, who find ourselves reflecting upon such things, must be in one of those universes, however rare, whose laws and constants are capable of evolving us.

The theoretical physicist Lee Smolin added an ingenious Darwinian spin which reduces the apparent statistical improbability of our existence. In Smolin's model, universes give birth to daughter universes, which vary in their laws and constants. Daughter universes are born in black holes produced by a parent universe, and they inherit its laws and constants but with some possibility of small random change – 'mutation'. Those daughter universes that have what it takes to reproduce (last long enough to make black holes, for instance) are, of course, the universes that pass on their laws and constants to their daughters. Stars are precursors to black holes which, in the Smolin model, are the birth events. So universes that have what it takes to make stars are favoured in this cosmic Darwinism. The properties of a universe that furnish this gift to the future are the self-same properties that incidentally lead to the manufacture of large atoms, including vital carbon atoms. Not only do we live in a universe that is capable of producing life. Successive generations of universes progressively evolve to become increasingly the sort of universe that, as a by-product, is capable of producing life.

* This 'many universes' idea is not to be confused (though it often is) with Hugh Everett's 'many worlds' interpretation of quantum theory, brilliantly championed by David Deutsch in *The Fabric of Reality*. The resemblance between the two theories is superficial and meaningless. Both theories could be true, or neither, or one, or the other. They were proposed to answer completely different problems. In the Everett theory, the different universes don't differ in their fundamental constants. But it is the entire point of the theory we are here considering that the different universes have different fundamental constants.

A colleague of mine, Andy Gardner, has recently shown that the same mathematics describes both the Smolin theory and Darwinian evolution. And the logic appeals to me, as indeed it would to anyone of imagination, but as for the physics I am not qualified to judge. I cannot find a physicist to condemn the theory as definitely wrong – the most negative thing they will say is that it is superfluous. Some, as we saw, dream of a final theory in whose light the alleged fine-tuning of the universe will turn out to be a delusion anyway. Nothing we know rules out Smolin's theory, and he claims for it the merit – which scientists rate more highly than many laymen appreciate – of testability. His book is *The Life of the Cosmos* and I recommend it.

But that was a digression about the physicist's version of the conceit of hindsight. The biologist's version is easier to dismiss since Darwin, though harder before him, and it is our concern here. Biological evolution has no privileged line of descent and no designated end. Evolution has reached many millions of interim ends (the number of surviving species at the time of observation), and there is no reason other than vanity – human vanity as it happens, since we are doing the talking – to designate any one as more privileged or climactic than any other.

This doesn't mean, as I shall continue to argue, that there is a total dearth of reasons or rhymes in evolutionary history. I believe there are recurring patterns. I also believe, though this is more controversial today than it once was, that there are senses in which evolution may be said to be directional, progressive and even predictable. But progress is emphatically not the same thing as progress towards humanity, and we must live with a weak and unflattering sense of the predictable. The historian must beware of stringing together a narrative that seems, even to the smallest degree, to be homing in on a human climax.

A book in my possession (in the main a good book, so I shall not name and shame it) provides an example. It is comparing *Homo habilis* (a human species, probably ancestral to us) with its predecessors the australopithecines.* What the book says is that *Homo habilis* was 'considerably more evolved than the Australopithecines'.

---

* The laws of zoological nomenclature follow strict precedence, and I fear there is no hope of changing the name *Australopithecus* to something less confusing to the contemporary majority who lack a classical education. It has nothing to do with Australia. No member of the genus has ever been found outside Africa. *Australo* simply means southern. Australia is the great southern continent, the Aurora australis is the southern equivalent of the Aurora borealis (boreal means northern), and *Australopithecus* was first found in *south* Africa, in the person of the Taung child.

More evolved? What can this mean but that evolution is moving in some pre-specified direction? The book leaves us in no doubt of what the presumed direction is. 'The first signs of a chin are apparent.' 'First' encourages us to expect second and third signs, towards a 'complete' human chin. 'The teeth start to resemble ours . . .' As if those teeth were the way they were, not because it suited the habiline diet but because they were embarking upon the road towards becoming our teeth. The passage ends with a telltale remark about a later species of extinct human, *Homo erectus:*

> Although their faces are still different from ours, they have a much more human look in their eyes. They are like sculptures in the making, 'unfinished' works.

In the making? Unfinished? Only with the unwisdom of hindsight. In excuse of that book it is probably true that, were we to meet a *Homo erectus* face to face, it might well look to our eyes like an unfinished sculpture in the making. But that is only because we are looking with human hindsight. A living creature is always in the business of surviving in its own environment. It is never unfinished – or, in another sense, it is always unfinished. So, presumably, are we.

The conceit of hindsight tempts us at other stages in our history. From our human point of view, the emergence of our remote fish ancestors from water to land was a momentous step, an evolutionary rite of passage. It was undertaken in the Devonian Period by lobe-finned fish a bit like modern lungfish. We look at fossils of the period with a pardonable yearning to gaze upon our forebears, and are seduced by a knowledge of what came later: drawn into seeing these Devonian fish as 'half way' towards becoming land animals; everything about them earnestly transitional, bound into an epic quest to invade the land and initiate the next big phase of evolution. That is not the way it was at the time. Those Devonian fish had a living to earn. They were not on a mission to evolve, not on a quest towards the distant future. An otherwise excellent book about vertebrate evolution contains the following sentence about fish which

> ventured out of the water on to the land at the end of the Devonian Period and jumped the gap, so to speak, from one vertebrate class to another to become the first amphibians . . .

The 'gap' comes from hindsight. There was nothing resembling a gap at the time, and the 'classes' that we now recognise were no more separate, in those days, than two species. As we shall see again, jumping gaps is not what evolution does.

It makes no more sense (and no less) to aim our historical narrative towards *Homo sapiens* than towards any other modern species – *Octopus vulgaris*, say, or *Panthera leo* or *Sequoia sempervirens*. A historically minded swift, understandably proud of flight as self-evidently the premier accomplishment of life, will regard swift-kind – those spectacular flying machines with their swept-back wings, who stay aloft for a year at a time and even copulate in free flight – as the acme of evolutionary progress. To build on a fancy of Steven Pinker, if elephants could write history they might portray tapirs, elephant shrews, elephant seals and proboscis monkeys as tentative beginners along the main trunk road of evolution, taking the first fumbling steps but each – for some reason – never quite making it: so near yet so far. Elephant astronomers might wonder whether, on some other world, there exist alien life forms that have crossed the nasal rubicon and taken the final leap to full proboscitude.

We are not swifts nor elephants, we are people. As we wander in imagination through some long-dead epoch, it is humanly natural to reserve a special warmth and curiosity for whichever otherwise ordinary species in that ancient landscape is our ancestor (it is an intriguingly unfamiliar thought that there is always one such species). It is hard to deny our human temptation to see this one species as 'on the main line' of evolution, the others as supporting cast, walk-on parts, sidelined cameos. Without succumbing to that error, there is one way to indulge a legitimate human-centrism while respecting historical propriety. That way is to do our history backwards, and it is the way of this book.

Backward chronology in search of ancestors really can sensibly aim towards a single distant target. The distant target is the grand ancestor of all life, and we can't help converging upon it no matter where we start – elephant or eagle, swift or salmonella, wellingtonia or woman. Backward chronology and forward chronology are each good for different purposes. Go backwards and, no matter where you start, you end up celebrating the unity of life. Go forwards and you extol diversity. It works on small timescales as well as large. The forward chronology of the mammals, within their large but still limited timescale, is a story of branching diversification, uncovering the richness of that group of hairy warmbloods. Backward chronology, taking any modern mammal as our starting point, will always converge upon the same unique ur-mammal: shadowy, insect-ivorous, nocturnal contemporary of the dinosaurs. This is a local convergence. A yet more local one converges on the most recent ancestor of all rodents, who lived somewhere around the time the dinosaurs went extinct. More local still is the backward convergence of all apes (including humans) on their shared ancestor, who lived about 18 million years ago. On a larger scale, there is a comparable convergence to be found if we work backwards from any vertebrate, an even larger convergence working backwards from any animal to the ancestor of all animals. The largest convergence of all takes us from any modern creature – animal, plant, fungus or bacterium – back to the universal progenitor of all surviving organisms, probably resembling some kind of bacterium.

I used 'convergence' in the last paragraph, but I really want to reserve that word for a completely different meaning in forward chronology. So for the present purpose I shall substitute 'confluence' or, for reasons that will make sense in a moment, 'rendezvous'. I could have used 'coalescence', except that, as we shall see, geneticists have already adopted it in a more precise sense, similar to my 'confluence' but concentrating on genes rather than species. In a backward chronology, the ancestors of any set of species must event-ually meet at a particular geological moment. Their point of rendez-vous is the last common ancestor that they all share, what I shall call their 'Concestor':* the focal rodent or the focal mammal or the focal vertebrate, say. The oldest concestor is the grand ancestor of all surviving life.

We can be very sure there really is a single concestor of all surviving life forms on this planet. The evidence is that all that have ever been examined share (exactly in most cases, almost exactly in the rest) the same genetic code; and the genetic code is too detailed, in arbitrary aspects of its complexity, to have been invented twice. Although not every species has been examined, we already have enough coverage to be pretty certain that no surprises – alas – await us. If we now were to discover a life form sufficiently alien to have a completely different genetic code, or one not even based on DNA, it would be the most exciting biological discovery in my adult lifetime, whether it lives on this planet or another. As things stand, it appears

---

* I am grateful to Nicky Warren for suggesting this word.

that all known life forms can be traced to a single ancestor which lived more than 3 billion years ago. If there were other, independent origins of life, they have left no descendants that we have discovered. And if new ones arose now they would swiftly be eaten, probably by bacteria.

The grand confluence of all surviving life is not the same thing as the origin of life itself. This is because all surviving species presumably share a concestor who lived after the origin of life: anything else would be an unlikely coincidence, for it would suggest that the original life form *immediately* branched and more than one of its branches survive to this day. The oldest bacterial fossils found so far date to about 3.5 billion years ago, so the origin of life must at least be earlier than that. The grand confluence – the last common ancestor of all surviving creatures – could predate the oldest fossils (it didn't fossilise) or it could have lived a billion years later (all but one of the other lineages went extinct).

Given that all backward chronologies, no matter where they start, culminate in the one grand confluence, we can legitimately indulge our human preoccupation and concentrate upon the single line of our own ancestors. Instead of treating evolution as aimed towards us, we *choose* modern *Homo sapiens* as our arbitrary, but forgivably preferred, starting point for a reverse chronology. We choose this route, out of all possible routes to the past, because we are curious about our own great grancestors. At the same time, although we need not follow them in detail, we shall not forget that there are other historians, animals and plants belonging to other species, who are independently walking backwards from their separate starting points, on separate pilgrimages to visit their own ancestors, including eventually the ones they share with us. If we retrace our own ancestral steps, we shall inevitably meet these other pilgrims and join forces with them in a definite order, the order in which their lineages rendezvous with ours, the order of ever more inclusive cousinship.

Pilgrimages? Join forces with pilgrims? Yes, why not? Pilgrimage is an apt way to think about our journey to the past. This book will be cast in the form of an epic pilgrimage from the present to the past. All roads lead to the origin of life. But because we are human, the path we shall follow will be that of our own ancestors. It will be a human pilgrimage to discover human ancestors. As we go, we shall greet other pilgrims who will join us in strict order, as we reach the common ancestors we share with them.

The first fellow pilgrims we shall greet, some 6 million years ago, deep in Africa where Stanley memorably shook hands with Living-stone, are the chimpanzees, of which there are two living species. On our backwards journey, they will have already joined forces with each other by the time we greet them.

The next pilgrims with whom we shall rendezvous as we push back along our journey are gorillas, then orang utans (quite a lot deeper into the past, and probably no longer in Africa). Next we shall greet gibbons, then Old World monkeys, then New World monkeys, then various other groups of mammals . . . and so on until eventually all the pilgrims of life are marching together in one single backward quest for the origin of life itself. As we push on back, there will come a time when it is no longer meaningful to name the continent in which a rendezvous takes place: the map of the world was so differ-ent, because of the remarkable phenomenon of plate tectonics. And further back still, all rendezvous take place in the sea.

It is a rather surprising fact that we human pilgrims pass only a little over 40 rendezvous points in all, before we hit the origin of life itself. At each of these steps we shall find one particular shared ancestor, the concestor, which will bear the same labelling number as the rendezvous. For example, Concestor 2, whom we meet at Rendezvous 2, is the most recent common ancestor of gorillas on the one hand and {humans + {chimpanzees + bonobos}} on the other. Concestor 3 is the most recent common ancestor of orang utans and {{humans + {chimpanzees + bonobos}} + gorillas}. The final con-cestor is the grand ancestor of all surviving life forms. Concestor 0 is a special case, the most recent ancestor of all surviving humans.

We shall be pilgrims, then, sharing fellowship ever more inclu-sively with other pilgrim bands, which also have been swelling on their own way to their rendezvous with us. After each meeting, we continue together on the high road back to our shared archaean goal, our 'Canterbury'. There are other literary allusions, of course, and I almost made Bunyan my model and *Pilgrim's Regress* my title. But it was to Chaucer's *Canterbury Tales* that Yan Wong and I kept returning in our discussions, and it seemed increasingly natural to think of Chaucer throughout this book.

Unlike (most of) Chaucer's pilgrims, mine do not all set out together, although they do set off at the same time, the present. These other pilgrims aim towards their ancient Canterbury from different starting points, joining our human pilgrimage at various rendezvous along the road. In this respect, my pilgrims are unlike those who gathered in London's Tabard Inn. Mine are more like the sinister canon and his understandably disloyal yeoman, who joined Chaucer's pilgrims at Boughton-under-Blee, five miles short of Canterbury. Following Chaucer's lead, my pilgrims, which are all the different species of living creature, will have the opportunity to tell tales along the way to Canterbury, which is the origin of life. It is these tales that form the main substance of this book.

Dead men tell no tales, and extinct creatures such as trilobites are deemed not to be pilgrims capable of telling them, but I shall make exceptions of two special classes. Animals such as the dodo, which survived into historical times and whose DNA is still available to us, are treated as honorary members of the modern fauna setting off on pilgrimage at the same time as us, and joining us at some particular rendezvous. Since we are responsible for their so recent extinction, it seems the least we can do. The other honorary pilgrims, exceptions to the rule that dead men tell no tales, really are men (or women). Since we human pilgrims are directly seeking our own ancestors, fossils that might plausibly be considered candidates for *being* our ancestors are deemed members of our human pilgrimage and we shall hear tales from some of these 'shadow pilgrims', for example the Handyman, *Homo habilis.*

I decided it would be twee to let my animal and plant tale-tellers speak in the first person singular, and I shall not do so. Save for occasional asides and prefatory remarks, Chaucer's pilgrims don't either. Many of Chaucer's Tales have their own Prologue, and some have an Epilogue too, all written in Chaucer's own voice as narrator of the pilgrimage. I shall occasionally follow his example. As with Chaucer, an epilogue may serve as a bridge from one tale to the next.

Before his Tales begin, Chaucer has a long General Prologue in which he sets out his cast list: the professions and in some cases the names of the pilgrims who are about to set off from the tavern. Instead, I shall introduce new pilgrims as they join us. Chaucer's jovial host offers to guide the pilgrims, and encourages them to tell their tales to while away the journey. In my role as host I shall use the General Prologue for some preparatory remarks about methods and problems of reconstructing evolutionary history, which must be faced and solved whether we do our history backwards or forwards.

Then we shall embark on our backwards history itself. Although we shall concentrate on our own ancestors, noting other creatures usually only when they join us, we shall from time to time look up from our road and remind ourselves that there are other pilgrims on their own more or less independent routes to our ultimate destination. The numbered rendezvous milestones, plus a few intermediate markers necessary to consolidate the chronology, will provide the scaffolding for our narrative. Each will mark a new chapter, where we halt to take stock of our pilgrimage, and maybe listen to a tale or two. On rare occasions, something important happens in the world around us, and then our pilgrims may pause briefly to reflect on it. But, for the most part, we shall mark our progress to the dawn of life by the measure of those 40 natural milestones, the trysts that enrich our pilgrimage.

THE GENERAL PROLOGUE

How shall we know the past, and how date it? What aids to our vision will help us peer into theatres of ancient life and reconstruct the scenes and the players, their exits and their entrances, of long ago? Conventional human history has three main methods, and we shall find their counterparts on the larger timescale of evolution. First there is archaeology, the study of bones, arrowheads, fragments of pots, oystershell middens, figurines and other relics that survive as hard evidence from the past. In evolutionary history, the most obvious hard relics are bones and teeth, and the *fossils* that they eventually become. Second, there are *renewed relics*, records that are not themselves old but which contain or embody a copy or representation of what is old. In human history these are written or spoken accounts, handed down, repeated, reprinted or otherwise duplicated from the past to the present. In evolution, I shall propose DNA as the main renewed relic, equivalent to a written and recopied record. Third, there is *triangulation*. This name comes from a method of judging distances by measuring angles. Take a bearing on a target. Now walk a measured distance sideways and take another. From the intercept of the two angles, calculate the distance of the target. Some camera rangefinders use the principle, and map surveyors traditionally relied upon it. Evolutionists can be said to 'triangulate' an ancestor by comparing two (or more) of its surviving descendants. I shall take the three kinds of evidence in order, beginning with hard relics and, in particular, fossils.

Fossils
Bodies or bones may survive for our attention, having somehow escaped that of hyenas, burying beetles and bacteria. The 'Ice Man' of the Italian Tyrol was preserved in his glacier for 5,000 years. Insects have become embalmed in amber (petrified gum from trees) for a hundred million years. Without benefit of ice or amber, hard parts like teeth, bones and shells stand the best chance of being preserved. Teeth last longest of all because, to do their job in life, they had to be harder than anything their owner was likely to eat. Bones and shells

need to be hard for different reasons, and they too can last a long time. Such hard parts and, under exceptionally lucky circumstances, soft parts too, occasionally become petrified as stone fossils that last for hundreds of millions of years. In recent years it has even become possible to scan fossil-bearing rocks, taking advantage of the sorts of technology used to scan our bodies in hospitals, opening up a whole new field of fossil analysis.

In spite of the fascination of fossils, it is surprising how much we would still know about our evolutionary past without them. If every fossil were magicked away, the comparative study of modern organisms, of how their patterns of resemblances, especially of their genetic sequences, are distributed among species, and of how species are distributed among continents and islands, would still demonstrate, beyond all sane doubt, that our history is evolutionary, and that all living creatures are cousins. Fossils are a bonus. A welcome bonus, to be sure, but not an essential one. It is worth remembering this when creationists go on (as they tediously do) about 'gaps' in the fossil record. The fossil record could be one big gap, and the evidence for evolution would still be overwhelmingly strong. At the same time, if we had *only* fossils and no other evidence, the fact of evolution would again be overwhelmingly supported. As things stand, we are blessed with both.

The word fossil is conventionally used to mean any relic dating back more than 10,000 years: not a helpful convention, for there is nothing special about a round number like 10,000. If we had fewer or more than ten fingers, we'd recognise a different set of numbers as round. When we speak of a fossil, we normally mean that the original material has been substituted or infiltrated by a mineral of a different chemical composition and therefore given, as one might say, a new lease of death. An imprint of the original form may be preserved in stone for a very long time indeed, perhaps mixed with some of the original material. There are various ways in which this can happen. I leave the details – what is technically called taphonomy – for the Ergast's Tale.

When fossils were first discovered and mapped, their ages were unknown. The most we could hope for was a rank ordering of oldness. Age ranking depends upon the assumption known as the Law of Superposition. For obvious reasons, younger strata lie atop older ones, unless the circumstances are exceptional. Such exceptions, though they sometimes cause temporary puzzlement, are usually pretty obvious. A lump of old rock, complete with fossils, may be thrown on top of a younger stratum, say by a glacier. Or a series of strata may be turned over wholesale, and its vertical ordering exactly reversed. These anomalies can be taken care of by comparing equivalent rocks in other parts of the world. Once this is done, the palaeontologist can piece together the true sequence of the whole fossil record, in a jigsaw of overlapping sequences from different parts of the world. The logic is complicated in practice, though not in principle, by the fact, documented in the Epilogue to the Sloth's Tale, that the map of the world itself changes as the ages go by.

Why is the jigsaw necessary? Why can't we just dig down as far as we like, and treat this as equivalent to digging steadily backwards through time? Well, time itself may flow smoothly, but this doesn't mean that anywhere in the world there is a single sequence of sediment deposited smoothly and continuously from start to finish through geological time. Fossil beds are laid down in fits and starts, when the conditions are right.

In any one location, at any one time, it is rather likely that no sedimentary rocks, and no fossils, are being laid down. But it is quite likely that, in *some* part of the world, fossils are being deposited at any given time. By hopping around the world, from site to site where different strata happen to be accessibly near the surface, the palaeontologist can aspire to piece together something approaching a continuous record: literally the foundation stones for the geological timescale opposite. Of course individual palaeontologists don't hop from site to site. They hop from museum to museum looking at specimens in drawers, or from journal to journal in university libraries looking at written descriptions of fossils whose site of discovery has been carefully labelled, and they use these descriptions to piece together the fragments of the puzzle from different parts of the world.

The task is eased by the fact that particular strata, with recognisably characteristic rock properties, and consistently housing the same kinds of fossils, keep turning up in different regions. For example, the Devonian period marked on the lower left of the chart owes its name to the 'Old Red Sandstone' of the beautiful county of Devon, rock which also crops up in various other parts of the British Isles, in Germany, Greenland, North America and elsewhere. Devonian rocks are recognisable as Devonian wherever they may be found, partly because of the quality of the rock but also because of the internal evidence of the fossils that they contain. This sounds like a

Simplified version of the timescale published by the International Commission on Stratigraphy, with shading corresponding to the depth of time (white is most recent, black denotes the oldest rocks). The timescale is divided into eons, eras, periods and epochs. Time is measured in 'millions of years ago' (Ma). Note that the 'Tertiary' is not officially used any more, although some geologists argue for its reintroduction, hence we have marked it here. 'Penn.' and 'Miss.' are the Pennsylvanian and Mississippian, used as an alternative to the Carboniferous period by American geologists. The lower limit of the timescale is formally undefined, though it is generally assumed to stretch back to abut 4.6 billion years ago, when the Earth and the rest of the solar system were formed.

circular argument but it really isn't: no more so than when a scholar recognises a Dead Sea Scroll, from internal evidence, as a fragment of the First Book of Samuel. Devonian rocks are reliably labelled by the presence of certain characteristic fossils.

The same goes for rocks from other geological periods, right back to the time of the earliest hard-bodied fossils. From the ancient Cambrian through to the present Quarternary, the geological periods listed in the chart we have just seen were mostly separated on the basis of changes in the fossil record. And as a result, the end of one period and the start of another is often delimited by extinctions that conspicuously interrupt the continuity of the fossils. As Stephen Jay Gould has put it, no palaeontologist has any trouble identifying whether a lump of rock lies before or after the great end-Permian mass extinction. There is almost no overlap in animal types. Indeed, fossils (especially microfossils) are so useful in labelling and dating rocks that the oil and mining industries are among their principal users.

Such 'relative dating', then, has long been possible by vertical piecing together of the jigsaw of rocks. The geological periods were named for purposes of relative dating, before absolute dating became possible. And they are still useful. But relative dating is more difficult for rocks with scarce fossils – and that includes all rocks older than the Cambrian: the first eight-ninths of Earth's history.

The ages in this book are mostly quoted in millions of years ago (inelegantly, not to say incorrectly, greco-latinised as megaannums, or simply 'Ma'). But such absolute measurements are relatively recent advances, dependent on recent developments in physics, especially the physics of radioactivity. This needs some explaining, and the details must wait for the Redwood's Tale. For now, it is enough to know that we have a range of reliable methods for putting an absolute age on fossils, or the rocks that contain or surround them. Moreover, different methods in this range provide sensitivity across the whole spectrum of ages from hundreds of years (tree rings), through thousands of years (carbon 14), millions, hundreds of millions (uranium-thorium-lead) to billions of years (potassium-argon).

Renewed Relics

Fossils, like archaeological specimens, are more-or-less direct relics of the past. We turn now to our second category of historical evidence, *renewed* relics, copied successively down the generations. For historians of human affairs this might mean eyewitness accounts, handed down by oral tradition or in written documents. We cannot ask any living witnesses what it was like to live in fourteenth-century England, but we know about it thanks to written documents, including Chaucer's. They contain information that has been copied, printed, stored in libraries, reprinted and distributed for us to read today. Once a story gets into print or, nowadays, a computer medium of some kind, copies of it have a fair chance of being perpetuated into the distant future.

Written records are more reliable than oral tradition, by a disconcerting margin. You might think that each generation of children, knowing their parents as well as most children do, would listen to their detailed reminiscences and relay them to the next generation. Five generations on, a voluminous oral tradition should, one might think, have survived. I remember my four grandparents clearly, but of my eight great-grandparents I know a handful of fragmentary anecdotes. One great-grandfather habitually sang a certain nonsense rhyme (which I can sing), but only while lacing his boots. Another was greedy for cream, and would knock the chess board over when losing. A third was a country doctor. That is about my limit. How have eight entire lives been so reduced? How, when the chain of informants connecting us back to the eyewitness seems so short, and human conversation so rich, could all those thousands of personal details that made up the lifetimes of eight human individuals be so fast forgotten?

Frustratingly, oral tradition peters out almost immediately, unless hallowed in bardic recitations like those that were eventually written down by Homer, and even then the history is far from accurate. It decays into nonsense and falsehood after amazingly few generations. Historical facts about real heroes, villains, animals and volcanoes rapidly degenerate (or blossom, depending upon your taste) into myths about demigods, devils, centaurs and fire-breathing dragons.* But oral traditions and their imperfections needn't detain us because, in any case, they have no equivalent in evolutionary history.

Writing is a huge improvement. Paper, papyrus and even stone

---

* John Reader, in his *Man on Earth*, notes that the Incas, who had no written language (unless, as has been recently suggested, their knotted strings were used for language as well as for counting), made a perhaps compensatory effort to improve the accuracy of their oral tradition. Official historians were 'obliged to memorise vast amounts of information and repeat it for the benefit of administrators as required. Not surprisingly, the role of historian passed from father to son.'

tablets may wear out or decay, but written records have the potential to be copied accurately for an indefinite number of generations, although in practice the accuracy is not total. I should explain the special sense in which I mean accuracy and, indeed, the special sense in which I mean generations. If you handwrite me a message and I copy it and pass it on to a third person (the next copying 'generation'), it will not be an exact replica, for my handwriting is different from yours. But if you write with care, and if I painstakingly match each of your squiggles with exactly one from our shared alphabet, your message has a good chance of being copied by me with total accuracy. In theory this accuracy could be preserved through an indefinite number of 'generations' of scribes. Given that there is a discrete alphabet agreed by writer and reader, copying lets a message survive the destruction of the original. This property of writing can be called 'self-normalising'. It works because letters of a true alphabet are discontinuous. The point, reminiscent of the distinction between analogue and digital codes, needs a little more explanation.

There exists a consonant sound which is intermediate between the English hard c and g (it is the French hard c in *comme*). But nobody would think of trying to represent this sound by writing a character which looked intermediate between c and g. We all understand that a written character in English must be one, and only one, member of our 26-letter alphabet. We understand that French uses the same 26 letters for sounds that are not exactly the same as ours and which may be intermediate between ours. Each language, indeed each local accent or dialect, separately uses the alphabet for self-normalising on different sounds.

Self-normalisation fights against the 'Chinese Whispers'* degrading of messages over generations. The same protection is not available to a drawing, copied and recopied along a line of imitative artists, unless the drawing style incorporates ritual conventions as its own version of 'self-normalisation'. An eyewitness record of some event, which is written down, as opposed to drawn as a picture, has a good chance of still being accurately reproduced in history books centuries later. We have what is probably an accurate account of

---

* In the game of Chinese Whispers (American children call it 'Telephone'), a number of children stand in a line. A story is whispered to the first child, who whispers it to the second, and so on until the last child, whose finally revealed version of the story turns out to be an amusingly garbled and degraded version of the original.

the destruction of Pompeii in AD 79 because a witness, Pliny the Younger, wrote down what he saw, in two epistles to the historian Tacitus, and some of Tacitus' writings survived, by successive copying and eventually printing, for us to read them today. Even in pre-Gutenberg days when documents were duplicated by scribes, writing represented a great advance in accuracy compared with memory and oral tradition.

It is only a theoretical ideal that repetitive copying retains perfect accuracy. In practice scribes are fallible, and not above massaging their copy to make it say things that they think (no doubt sincerely) the original document ought to have said. The most famous example of this, painstakingly documented by nineteenth-century German theologians, is the doctoring of New Testament history to make it conform to Old Testament prophecies. The scribes concerned were probably not wilfully mendacious. Like the gospel-makers, who themselves lived long after Jesus's death, they genuinely believed he had been the incarnation of Old Testament messianic prophecies. He 'must', therefore, have been born in Bethlehem, and descended from David. If the documents unaccountably failed to say so, it was the scribe's conscientious duty to rectify the deficiency. A sufficiently devout scribe would, I suppose, no more have regarded this as falsification than we do when we automatically correct a spelling mistake or a grammatical infelicity.

Quite apart from positive massaging, all repeated copying is subject to straightforward errors like skipping a line, or a word in a list. But in any case writing cannot take us back beyond its invention, which was only about 5,000 years ago. Identification symbols, counting-marks and pictures go back a bit further, perhaps some tens of thousands of years, but all such periods are chickenfeed compared with evolutionary time.

Fortunately, when we turn to evolution there is another kind of duplicated information which goes back an almost unimaginably large number of copying generations and which, with a little poetic licence, we can regard as the equivalent of a written text: a historical record that renews itself with astounding accuracy for hundreds of millions of generations precisely because, like our writing system, it has a self-normalising alphabet. The DNA information in all living creatures has been handed down from remote ancestors with prodigious fidelity. The individual atoms in DNA are turning over continually, but the information that they encode in the pattern of their arrangement is copied for millions, sometimes hundreds of millions, of years. We can read this record directly, using the arts of modern molecular biology to spell out the actual DNA letter sequences or, slightly more indirectly, the amino acid sequences of protein into which they are translated. Or, much more indirectly as through a glass darkly, we can read it by studying the embryological products of the DNA: the shapes of bodies and their organs and chemistries. We don't need fossils to peer back into history. Because DNA changes very slowly through the generations, history is woven into the fabric of modern animals and plants, and inscribed in its coded characters.

DNA messages are written in a true alphabet. Like the Roman, Greek and Cyrillic writing systems, the DNA alphabet is a strictly limited repertoire of symbols with no self-evident meaning. Arbitrary symbols are chosen and combined to make meaningful messages of unlimited complexity and size. Where the English alphabet has 26 letters and the Greek one 24, the DNA alphabet is a four-letter alphabet. Many vital sections of DNA spell out three-letter words from a dictionary limited to 64 words, each word called a 'codon'. Some of the codons in the dictionary are synonymous with others, which is to say that this genetic 'code' is technically 'degenerate'.*

The dictionary maps 64 code words onto 21 meanings – the 20 biological amino acids, plus one all-purpose punctuation mark. Human languages are numerous and changing, and their dictionaries contain tens of thousands of distinct words, but the 64-word DNA dictionary is universal and unchanging (with very minor variations in a few rare cases). The 20 amino acids are strung into sequences of typically a few hundred, each sequence a particular protein molecule. Whereas the number of letters is limited to four and the number of codons to 64, there is no theoretical limit to the number of proteins that can be spelled out by different sequences of codons. It is beyond all counting. A 'sentence' of codons specifying one protein molecule is an identifiable unit often called a gene. The genes are not separated

---

* 'Redundant' is sometimes mistakenly used instead of degenerate, but it means something different. The genetic code is, as it happens, redundant too, in that either strand of the double helix could be decoded to yield the same information. Only one of them is actually decoded, but the other is used for correcting errors. Engineers, too, use redundancy – repetitiousness – to correct errors. The degeneracy of the genetic code is something different, and it is what we are talking about here. A degenerate code contains synonyms and could therefore accommodate a larger range of meanings than it actually does.

from their neighbours (whether other genes or repetitive nonsense) by any delimiters apart from what can be read from their sequence. In this respect they resemble TELEGRAMS THAT LACK PUNCTU-ATION MARKS COMMA AND HAVE TO SPELL THEM OUT AS WORDS COMMA ALTHOUGH EVEN TELEGRAMS HAVE THE AD-VANTAGE OF SPACES BETWEEN WORDS COMMA WHICH DNA LACKS STOP

DNA differs from written language in that islands of sense are separated by a sea of nonsense, never transcribed. 'Whole' genes are assembled, during transcription, from meaningful 'exons' separated by meaningless 'introns' whose texts are simply discarded by the reading apparatus. And even meaningful stretches of DNA are in many cases never read – presumably they are superseded copies of once useful genes that hang around like early drafts of a chapter on a cluttered hard disk. Indeed, the image of the genome as an old hard disk, badly in need of a spring clean, is one that will serve us from time to time during the book.

It bears repeating that the DNA molecules of long-dead animals are not themselves preserved. The *information* in DNA can be preserved for ever, but only by dint of frequent re-copying. The plot of *Jurassic Park*, though not silly, falls foul of practical facts. Certainly for a short while after becoming embalmed in amber, a bloodsucking insect could contain the instructions needed to reconstruct a dino-saur, especially as fossils reveal that dinosaur red blood cells (like those of their descendants the birds but unlike ours) contained DNA. It also seems to be true that some biological molecules can survive many millions of years. For example, researchers have retrieved haemoglobin-like chemicals from the last supper of a 46-million-year-old fossilised mosquito and, almost unbelievably, collagen pro-teins from 70-million-year-old dinosaur bones. These, however, are small, robust chemicals. Long, fragile strands of DNA are another matter. Without constant maintenance, DNA starts falling to pieces. In a few years – and only days in the case of some soft tissues – it can decay to unreadable mush.

Cold, oxygen-free conditions do somewhat slow DNA's inexorable decline to illegibility. Currently, the oldest genome on record is from a 700,000-year-old horse bone preserved in Canadian permafrost. Even above freezing, a cool and stable environment can preserve DNA for hundreds of thousands of years. Bones retrieved from excavations in cool caves have provided various quantities of human

DNA, most spectacularly the entire genome of a 50,000-year-old incest-spawned Neanderthal (as we shall see). Imagine the kerfuffle if somebody managed to clone her. But long though these timespans are in human terms, they correspond to only a tiny fraction of our journey into the past. Alas, chemistry suggests that the upper limit for retaining recognisable ancient DNA is only a few million years – certainly not enough to reach back to the time of the dinosaurs.

The important point about DNA is that, as long as the chain of reproducing life is not broken, its coded *information* is copied to a new molecule before the old molecule is destroyed. In this form, DNA information far outlives its molecules. It is renewable – copied – and since the copies are literally perfect for most of its letters on any one occasion, it can potentially last an indefinitely long time. Large quantities of our ancestors' DNA information survives completely unchanged, some even from hundreds of millions of years ago, preserved in successive generations of living bodies.

Understood in this way, the DNA record is an almost unbelievably rich gift to the historian. What historian could have dared hope for a world in which every single individual of every species carries, within its body, a long and detailed text: a written document handed down through time? Moreover, it has minor random changes, which occur seldom enough not to mess up the record yet often enough to furnish distinct labels. It is even better than that. The text is not just arbitrary. In *Unweaving the Rainbow*, I made a Darwinian case for regarding an animal's DNA as a 'Genetic Book of the Dead': a descriptive record of ancestral worlds. It follows from the fact of Darwinian evolution that everything about an animal or plant, including its bodily form, its inherited behaviour and the chemistry of its cells, is a coded message about the worlds in which its ancestors survived: the food they sought; the predators they escaped; the climates they endured; the mates they beguiled. The message is ultimately scripted in the DNA that fell through the succession of sieves that is natural selection. When we learn to read it properly, the DNA of a dolphin may one day confirm what we already know from the telltale giveaways in its anatomy and physiology: that its ancestors once lived on dry land. Four hundred million years earlier, the ancestors of all land-dwelling vertebrates, including the land-dwelling ancestors of dolphins, came out of the sea where they had lived since the origin of life. Doubtless our DNA records this fact if we could read it. Everything about a modern animal, especially its DNA, but its limbs and its heart, its brain and its breeding cycle too, can be regarded as an archive, a chronicle of its past, even if that chronicle is a palimpsest, many times overwritten.

The DNA chronicle may be a gift to the historian, but it is a hard one to read, demanding deeply informed interpretation. It is made more powerful if combined with our third method of historical reconstruction, triangulation. It is to this that we now turn, and again we start with the analogous case of human history, specifically the history of languages.

Triangulation

Linguists often wish to trace languages back through history. Where written records survive it is rather easy. The historical linguist can use the second of our two methods of reconstruction, tracing back renewed relics, in this case words. Modern English goes back via Middle English to Anglo-Saxon using the continuous literary tradition, through Shakespeare, Chaucer and *Beowulf*. But speech obviously goes back long before the invention of writing, and many languages have no written form anyway. For the earlier history of dead languages, linguists resort to a version of what I am calling triangulation. They compare modern languages and group them hierarchically into families within families. Romance, Germanic, Slavic, Celtic and other European language families are in turn grouped with some Indian language families into Indo-European. Linguists believe that 'Proto-Indo-European' was an actual language, spoken by a particular tribe around 6,000 years ago. They even aspire to reconstruct many of its details by extrapolating back from the shared features of its descendants. Other language families in other parts of the world, of equivalent rank to Indo-European, have been traced back in the same way, for instance Altaic, Dravidian and Uralic-Yukaghir. Some optimistic (and controversial) linguists believe they can go back even further, uniting such major families in an even more all-embracing family of families. In this way they have persuaded themselves that they can reconstruct elements of a hypothetical ur-language which they call Nostratic, and which they believe was spoken between 12,000 and 15,000 years ago.

Many linguists, while happy about Proto-Indo-European and other ancestral languages of equivalent rank, doubt the possibility of reconstructing a language as ancient as Nostratic. Their professional scepticism reinforces my own amateur incredulity. But there is no doubt at all that equivalent triangulation methods – various techniques for comparing modern organisms – work for evolutionary history, and can be used for penetrating back hundreds of millions of years. Even if we had no fossils, a sophisticated comparison of modern animals would permit a fair and plausible reconstruction of their ancestors. Just as a linguist penetrates the past to Proto-Indo-European, triangulating from modern languages and from already reconstructed dead languages, we can do the same with modern organisms, comparing either their external characteristics or their protein or DNA sequences. As the libraries of the world accumulate long and exact DNA listings from more and more modern species, the reliability of our triangulations will increase, particularly because DNA texts have such a large range of overlaps.

Let me explain what I mean by 'range of overlaps'. Even when taken from extremely distant relations, for example humans and bacteria, large sections of DNA still unequivocally resemble each other. And very close relations, such as humans and chimpanzees, have much more DNA in common. If you choose your molecules judiciously, there is a complete spectrum of steadily increasing proportions of shared DNA, all the way in between. Molecules can be chosen which, between them, span the gamut of comparison, from remote cousins like humans and bacteria, to close cousins like two species of frogs. Resemblances between languages are harder to discern, all except close pairs of languages like German and Dutch. The chain of reasoning that leads some hopeful linguists to Nostratic is tenuous enough to make the links the subject of scepticism on the part of other linguists. Would the DNA equivalent of triangulating to Nostratic be triangulation between, say, humans and bacteria? But humans and bacteria have some genes that have hardly changed at all since the common ancestor, their equivalent of Nostratic. And the genetic code itself is virtually identical in all species and must have been the same in the shared ancestors. One could say that the resemblance between German and Dutch is comparable to that between any pair of mammals. Human and chimpanzee DNA are so similar, they are like English spoken in two slightly different accents. The resemblance between English and Japanese, or between Spanish and Basque, is so slight that no pair of living organisms can be chosen for analogy, not even humans and bacteria. Humans and bacteria have DNA sequences which are so similar that whole paragraphs are word-for-word identical.

I have been talking about using DNA sequences for triangulation. In principle it works for gross morphological characters as well but, in the absence of molecular information, distant ancestors are about as elusive as Nostratic. With morphological characters, as with DNA, we assume that features shared by many descendants of an ancestor are likely (or at least slightly more likely than not) to have been inherited from that ancestor. All vertebrates have a backbone and we assume that they inherited it (strictly inherited the genes for growing it) from a remote ancestor which lived, the fossils suggest, more than half a billion years ago and also had a backbone. It is this sort of morphological triangulation that has been used to help imagine the bodily forms of concestors in this book. For this purpose, I would have preferred to rely more heavily upon triangulation using DNA directly, but our ability to predict how a change in a gene will change the morphology of an organism is inadequate to the task.

Triangulation is even more effective if we include many species. But for this we need sophisticated methods which rely on having an accurately constructed family tree. These methods will be explained in the Gibbon's Tale. Triangulation also lends itself to a technique for calculating the date of any evolutionary branch point you like. This is the 'Molecular Clock'. Briefly, the method is to count discrepancies in molecular sequences between surviving species. Close cousins with recent common ancestors have fewer discrepancies than distant cousins, the age of the common ancestor being – or so it is hoped – proportional to the number of molecular discrepancies between their two descendants. Then we calibrate the arbitrary timescale of the molecular clock, translating it into real years, by using fossils of known date for a few key branch-points where fossils happen to be available. In practice it isn't as simple as that, and the complications, difficulties and associated controversies will occupy the Epilogue to the Velvet Worm's Tale.

Chaucer's General Prologue introduced the complete cast of his pilgrimage, one by one. My cast list is much too large for that. In any case, the narrative itself is a long sequence of introductions – at the 40 rendezvous points. But one preliminary introduction is necessary, in a way that it wasn't for Chaucer. His cast list was a set of individuals. Mine is a set of groupings. The way we group animals and plants needs introducing. At Rendezvous 11, our pilgrimage is joined by some 2,000 species of rodents, plus 90 species of rabbits, hares and pikas, collectively called Glires. Species are grouped in hierarchically inclusive ways, and each grouping has a name of its own (the grouping of mouse-like rodents is called Muridae, and of squirrel-like rodents Sciuridae). And each category of grouping has a name. Muridae is a family, so is Sciuridae. Rodentia is the name of the order to which both belong. Glires is the superorder that unites rodents with rabbits and their kind. There is a hierarchy of such category names, family and order being somewhere in the middle of the hierarchy. Species lies near the bottom of the hierarchy. We work up through genus (plural genera), family, order, class, and phylum (plural phyla), with prefixes like sub- and super- offering scope for interpolation.

Species has a particular status, as we shall learn in the course of various tales. Every species has a unique scientific binomial, consisting of its genus name with an initial capital letter, followed by its species name with no initial capital, both printed in italics. The leopard ('panther'), lion and tiger are all members of the genus *Panthera:* respectively *Panthera pardus*, *Panthera leo* and *Panthera tigris*, within the cat family, Felidae, which in turn is a member of the order Carnivora, the class Mammalia, the subphylum Vertebrata and the phylum Chordata. I shan't expatiate on the principles of taxonomy any further here, but will mention them, as necessary, during the book.

ALL HUMANKIND

It is time to set off on our pilgrimage to the past, which we can think of as a journey in a time machine in quest of our ancestors, or more strictly our ancestral genes. But *whose* are we talking about: yours or mine, a Bambuti Pygmy's or a Torres Strait Islander's?

In the long term, it doesn't matter. If we go sufficiently far back, everybody's ancestors are shared. All your ancestors are mine, whoever you are, and all mine are yours. Not just approximately but literally. This is one of those truths that turns out, on reflection, to need no new evidence. We prove it by pure reason, using the mathematician's trick of *reductio ad absurdum*. Take our imaginary time machine absurdly far back, say 100 million years, to an age when our ancestors resembled shrews or opossums. Somewhere in the world at that ancient date, at least one of my personal ancestors must have been living, or I wouldn't be here. Let us call this particular little mammal Henry (it happens to be a family name). We seek to prove that if Henry is my ancestor he must be yours too. Imagine, for a moment, the contrary: I am descended from Henry and you are not. For this to be so, your lineage and mine would have to have marched, side by side yet never touching, through 100 million years of evolution to the present, never interbreeding yet ending up at the same evolutionary destination – so alike that your relatives are still capable of interbreeding with mine. This *reductio* is clearly absurd. If Henry is my ancestor he has to be yours too. If not mine, he cannot be yours.

Without specifying how ancient is 'sufficiently', we have just proved that a sufficiently ancient individual with any human descendants at all must be an ancestor of the entire human race. Long-distance ancestry, of a particular group of descendants such as the human species, is an all-or-nothing affair. Moreover, it is perfectly possible that Henry is my ancestor (and necessarily yours, given that you arc human enough to be reading this book) while his brother William is the ancestor of, say, all the surviving wombats. Not only is it possible. It is a remarkable fact that there *must* be a moment in history when there were two animals in the same species, one of

Humankind. A stylised impression of the human family tree. It is not intended as an accurate depiction – the real tree would be unmanageably dense. Grey lines illustrate patterns of interbreeding, with lots of it within continents and occasional migration between them. The numbered circle marks Concestor 0, the most recent common ancestor of all living humans. Verify this by following routes to the right from Concestor 0: you can reach any of the modern-day-human end points.

whom became the ancestor of all humans and no wombats, while the other became the ancestor of all wombats and no humans. They may well have met, and may even have been brothers. You can cross out wombat and substitute any other modern species you like, and the statement must still be true. Think it through, and you will find that it follows from the fact that all species are cousins of one another. Bear in mind when you do so that the 'ancestor of all wombats' will also be the ancestor of lots of very different things besides wombats (in this case, all the marsupials, including kangaroos and koalas, bandicoots and bilbies and the host of other South American and Australian animals which we shall meet at Rendezvous 14).

My reasoning was constructed as a *reductio ad absurdum*. It assumed that 'Henry' lived long enough ago for it to be obvious that he begat either all living humans, or none. How long is long enough? That's a harder question. A hundred million years is more than enough to assure the conclusion we seek. If we go back only a hundred years, no individual can claim the entire human race as direct descendants. Between the obvious cases of 100 years and 100 million, what can we say about unobvious intermediates such as 10,000, 100,000 or 1 million years?

This is not simply an intriguing question. It is of relevance to this, our first rendezvous point, because we are defining Concestor o as the most recent common ancestor of all living humans. The precise calculations were beyond me when I discussed the matter in *River Out of Eden* but, happily, a Yale University statistician called Joseph T. Chang has made a start on them, and his conclusions form the basis of our first tale, which, for reasons that will become obvious, will be told by the Tasmanian.

THE TASMANIAN'S TALE

Tracing ancestors is a beguiling pastime. As with history itself, there are two methods. You can go backwards, listing your two parents, four grandparents, eight great-grandparents, and so on. Or you can pick a distant ancestor and go forwards, listing his children, grandchildren, great-grandchildren, until you end up with yourself. Amateur genealogists do both, going back and forth between generations until they have filled in the tree as far as parish registers and family Bibles allow. This tale, like the book as a whole, uses the backwards method.

Pick any two people and go backwards and, sooner or later, we hit a most recent common ancestor – MRCA. You and me, the plumber and the queen, any set of us must converge on a single concestor (or couple). But unless we pick close relatives, finding the concestor requires a vast family tree, and most of it will be unknown. This applies *a fortiori* to the concestor of all humans alive today. Dating Concestor 0, the most recent common ancestor of all living humans, is not a task that can be undertaken by a practising genealogist. It is a task in estimation: a task for a mathematician.

An applied mathematician tries to understand the real world by setting up a simplified version of it – a 'model'. The model eases thought, while not losing all power to illuminate reality. Sometimes a model gives us a baseline, departures from which elucidate the real world.

In framing a mathematical model to date the common ancestors of all surviving humans, a good simplifying assumption – a sort of toy world – is a breeding population of fixed and constant size, living on an island with no immigration or emigration. Let it be an idealised population of Tasmanian aboriginals, in happier times before they were exterminated as agricultural vermin by nineteenth-century settlers. The last pure-bred Tasmanian, Truganinni, died in 1876, soon after her friend 'King Billy', whose scrotum was made into a tobacco pouch (shades of Nazi lamps). The Tasmanian aboriginals were isolated some 13,000 years ago when land bridges to Australia were flooded by rising sea levels, and they then saw no outsiders until they saw them with a vengeance in their nineteenth-century holocaust. For our modelling purposes, we consider Tasmania to be perfectly isolated from the rest of the world for 13,000 years until 1800. Our notional 'present', for modelling purposes, will be defined as AD 1800.

The next step is to model the mating pattern. In the real world people fall in love, or into arranged marriages, but here we are modellers, ruthlessly replacing human detail by tractable mathematics. There's more than one mating model we could imagine. The random diffusion model has men and women behaving as particles diffusing outwards from their birthplace, more likely to bump into near than distant neighbours. An even simpler and less realistic model is the random mating model. Here, we forget about distance altogether and simply assume that, strictly within the island, mating between any male and any female is equally likely.

Of course neither model is remotely plausible. Random diffusion assumes that people walk in any direction from their starting point.

In reality there are paths or roads which guide their feet: narrow gene conduits through the island's forests and grasslands. The random mating model is even more unrealistic. Never mind. We set up models to see what happens under ideally simplified conditions. It can be surprising. Then we have to consider whether the real world is more surprising or less, and in which directions.

Joseph Chang, following a long tradition of mathematical geneticists, opted for random mating, and took population size as constant. He did not deal with Tasmania in particular but we shall assume, again as a calculated oversimplification, that our toy population remained constant at 5,000, which is one estimate for Tasmania's aboriginal population in 1800 before the massacres began. I must repeat that such simplifications are of the essence in mathematical modelling: not a weakness of the method but, for certain purposes, a strength. Chang of course doesn't believe people mate at random, any more than Euclid believed lines have no breadth. We follow abstract assumptions to see where they lead, and then decide whether the detailed differences from the real world matter.

So in our toy population, how many generations would we need to go back, on average, before we find the common ancestor of everyone alive in the population today? Chang's calculations reveal it to be just over 12 generations. Human family trees usually squeeze in three or four generations per century, pinning down the Tasmanian's concestor to less than four centuries ago. That may seem surprisingly recent, but there is a good reason to expect recent dates. Think of it this way. Any person has a number of ancestral lineages that increase rapidly as we go back in time. You have two parents, four grandparents, eight great-grandparents and so on, forming what mathematicians call an exponential increase: 2, 4, 8, 16, 32, 64, 128, 256, 512, 1,024, 2,048, 4,096, 8,192 . . . It is perhaps not so surprising that shared ancestors can be found only a handful of generations previously.

Following this logic, 13 generations ago you would have had 8,192 great grancestors. So too would any Tasmanian. How does that sit with our assumption of a constant Tasmanian population of only 5,000 individuals? The point is that the exponential progression counts the number of routes through your family tree, and multiple routes can overlap, tracing back to the same physical ancestor.*

---

* In fact, as we go backwards, the number of ancestral routes will continue to increase without bound, rapidly surpassing the total count of humans on earth. And by 10,000 years ago, the number far exceeds the count of atoms in the observable universe.

What about larger populations? Chang provided a general answer for how far back, on average, we have to go before we find a common ancestor. It is when the number of ancestral routes (not necessarily the number of ancestors) exactly equals the population size. How many generations ago is that? Well, it is the number of times you have to multiply 2 by itself before getting the population size. Mathematicians call this a (base 2) logarithm. For example, the logarithm (base 2) of 5,000, our idealised Tasmanian population, is about 12.3, from whence we obtained the figure of just over 12 generations to the Tasmanian's concestor.

I give the name 'Chang One' to the date of the most recent common ancestor of some specified population. Continuing backwards from Chang One, it doesn't take long before we hit the point – I shall call it 'Chang Two' – at which *everybody* in that population is either a common ancestor or has no surviving descendants. Only during the brief interregnum between Chang One and Chang Two does there exist an intermediate category of people who have some surviving descendants but are not common ancestors of everybody. A surprising and important deduction, whose rationale I won't spell out, is that at Chang Two and beyond a large number of people are universal ancestors. In fact, about 80 per cent of individuals in any generation will in theory be ancestors of everybody alive in the distant future.

As for the timing, well, the mathematics yield the result that Chang Two is approximately 1.77 times older than Chang One. 1.77 times 12.3 gives just under 22 generations, between six and seven centuries. As we ride our time machine backwards in Tasmania, therefore, around the time of Geoffrey Chaucer we enter 'all or nothing' territory. From there on backwards, to the time when Tasmania was joined to Australia and all bets are off, everyone our time machine encounters will have either the entire population as descendants or no descendants at all.

I don't know about you but, logic notwithstanding, I still find these calculated dates astonishingly recent. What's more, the conclusions don't change much if you assume a larger population. Taking a model population the size of Britain's today, 60 million, we still need to go back only 23 generations to reach Chang One and our youngest universal ancestor. If the model applied to Britain, Chang Two, when everybody is either the ancestor of all modern British people or of none, is only about 40 generations ago, or about

AD 850. If the assumptions of the model are true (of course they aren't), King Alfred the Great is the ancestor of either all today's British or none.

I must repeat the cautions with which I began. There are all sorts of differences between 'model' and 'real' populations, in Britain or Tasmania or anywhere else. Britain's population has climbed steeply in historical time to reach its present size, and that completely changes the calculations. In any real population, people don't mate at random. They favour their own tribe, language group or local area, and of course they all have individual preferences. Britain's history adds the complication that, although a geographical island, its population is far from isolated. Waves of external immigrants have swept in from Europe over the centuries: Romans, Saxons, Danes and Normans among them.

If Tasmania and Britain are islands, the world is a larger 'island' since it has no immigration or emigration. But it is imperfectly subdivided into continents and smaller islands, with not just seas but mountain ranges, rivers and deserts impeding the movement of people to varying degrees. Partial isolation of tribes plays havoc with our all-too-tidy calculations, as does any kind of non-random mating. The present population of the world is 7 billion, but it would be absurd to look up the base 2 logarithm of 7 billion and swallow the resulting medieval date for Rendezvous 0! The real date is older, if only because pockets of humanity have been separated far longer than the dates we are now calculating. If an island has been isolated for 13,000 years, as Tasmania may have been, it is impossible for the human race as a whole to have a universal ancestor younger than 13,000 years.

So we can set a lower bound on the time since Rendezvous 0 by finding when the most isolated island population in the world last bred with an outsider. But to take this lower bound seriously, isolation must be absolute. This follows from the calculated figure of 80 per cent that we met earlier. A single migrant to Tasmania, once he has been sufficiently accepted into society to reproduce normally, has an 80 per cent chance of eventually becoming a common ancestor to all Tasmanians. So even tiny amounts of migration are enough to graft the family tree of an otherwise isolated population to that of the mainland. The timing of Rendezvous 0 is likely to depend on the date at which the most isolated pocket of humanity became completely isolated from its neighbour, plus the date at which its neighbour then became completely isolated from *its* neighbour, and so on. A few island hops may be needed before we can join all the family trees together, but it is then an insignificant number of centuries back until we tumble upon Concestor 0.

I fear it is impossible to use real records of migrations to remote islands, scarce and untrustworthy as they are, to pinpoint Rendezvous 0. But it is possible to make an educated guess. In collaboration with Joseph Chang, Douglas Rohde together with Steve Olson ran computer simulations of a model Earth, with real countries, historical ports and random, infrequent movements of individuals from region to region. Although still simplistic compared to the real world, they hoped to give some sort of idea of the timescales we should be looking at. Given that there are no pure-bred Tasmanians left to push Rendezvous 0 back beyond 13,000 years, what was their answer? On average their simulations put Rendezvous 0 only 3,500 years ago!

I'm not sure I can bring myself to believe quite such an astoundingly recent date. But there are tantalising hints that it may not be far from the truth. Or at least, that our exponential swelling of ancestors overwhelms what seem otherwise strong geographical and cultural barriers to human mating. One hint comes from known genealogies. These reveal that even quite disparate people share relatively recent ancestors. These links are often found via royalty, not because the nobility are special in any way, but simply because their well-documented family trees provide fertile hunting ground for genealogists. For instance, the current US president Barack Obama can be shown to be a direct descendant of the English king Edward I. Similar royal ancestry presumably exists for virtually anyone in North America or Europe. Indeed, the Dawkins family is traceable back to King Edward III, via the Duke of Clarence who was drowned in a butt of Malmsey. With some justification, genealogists claim that anyone with even a modicum of European descent can trace their ancestry back to Charlemagne (died AD 814), since all European royal families count him amongst their ancestors. To take a non-royal equivalent, Mohammed lived only a little earlier. Given Moorish history in southern Europe, I am almost certainly Mohammed's direct descendant. More broadly, all my readers will have as ancestors not only many giants of antiquity, from Nebuchadnezzar to the first emperor of China, but also their unknown contemporaries, from the builders of Stonehenge to Queen Nerfertiti's hair stylist.

The second hint that barriers to human interbreeding are porous comes from genetics. In 2013, Peter Ralph and Graham Coop analysed genome data from over 2,000 Europeans. They were looking for long stretches of near-identical DNA, shared between different people. For reasons we shall see in Eve's Tale, long stretches like this imply recent common ancestry, and the researchers restricted their search to lengths that indicated at least one common ancestor shared within the past few thousand years. As you might expect, individuals from the same region shared more common ancestors. But even for countries as separate as the UK and Turkey, the genetic data led Ralph and Coop to conclude that 'individuals from opposite ends of Europe are expected to share millions of common genealogical ancestors over the last 1,000 years'.

With characteristic prescience, the great statistician and evolutionary geneticist Sir Ronald Fisher anticipated these startlingly recent dates. He wrote the following, in a letter dated 15 January 1929 to Major Leonard Darwin (Charles's second youngest son): 'King Solomon lived 100 generations ago, and his line may be extinct; if not, I wager he is in the ancestry of all of us, and in nearly equal proportions, however unequally his wisdom is distributed.' I don't know if by 'all of us' Fisher really did mean to include Amerindians to Zulus, and as I discuss in the epilogue to this tale I doubt his 'nearly equal proportions' claim applies to DNA. But together with Joseph Chang's calculations and the genetic evidence to boot, the conclusion for Rendezvous 0 seems clear. Concestor 0, the most recent common ancestor of all living humans, lived only thousands or possibly a few tens of thousands of years ago, no more.

As to where Rendezvous 0 took place, this is almost as surprising. You might be inclined to think of Africa, as was my initial reaction. Africa houses the deepest genetic divides within humankind, so it seems a logical place to look for a common ancestor of all living humans. It has been well said that if you wiped out sub-Saharan Africa you would lose the great majority of human genetic diversity, whereas you could wipe out everywhere except Africa and nothing much would change. Nevertheless Concestor 0 may well have lived outside Africa. Concestor 0 is the most recent common ancestor that unites the most geographically isolated population — Tasmania for the sake of argument – with the rest of the world. If we assume that populations throughout the rest of the world, including Africa, indulged in at least some interbreeding during a long period when

Tasmania was totally isolated, the logic of Chang's calculations could lead us to suspect that Concestor 0 lived outside Africa, near the take-off point for the migrants whose offspring became Tasmanian immigrants. And in fact, Rohde, Olson and Chang's computer simulations nearly always locate Concestor 0 in Eastern Asia. But that isn't to say the majority of human *genetic* lineages trace back to Asia. This seeming paradox is resolved in Eve's Tale, when we explore family trees of genes rather than of people.

The calculations concerning Chang One allow us to date our initial rendezvous point, and pinpoint Concestor 0. But our universal Chang Two is an even more compelling destination, although no one human can be singled out for attention at this slightly more distant milestone. Back at Chang Two, every person we meet is likely to be our common ancestor; and if they are not, they are no ancestor at all. It marks the point where we can stop worrying about whether it is your ancestors we trace or mine: from that milestone on, all my readers march, shoulder to shoulder, in a phalanx of pilgrims towards the past.

EPILOGUE TO THE TASMANIAN'S TALE

Our surprising conclusion is that Concestor 0 probably lived 10,000 or so years ago, and very possibly not even in Africa. Other species too may generally have quite recent common ancestors. But this is not the only part of the Tasmanian's Tale that forces us to examine biological ideas in a new light. To professional Darwinian specialists, it seems a paradox that 80 per cent of a population will become universal ancestors. Let me explain. We are used to thinking of individual organisms as striving to maximise a quantity called 'fitness'. Exactly what fitness means is disputed. One favoured approximation is 'total number of children'. Another is 'total number of grandchildren', but there is no obvious reason to stop at grandchildren, and many authorities prefer to say something like 'total number of descendants alive at some distant date in the future'. But we seem to have a problem if, in our theoretically idealised population in the absence of natural selection, 80 per cent of the population can expect to have the maximum possible 'fitness': that is, they can expect to claim the entire population as their descendants! This matters for Darwinians because they widely presume that 'fitness' is what all animals constantly struggle to maximise.

I have long argued that the only reason an organism behaves as a quasi-purposeful agent at all – an entity capable of maximising anything – is that it is built by genes that have survived through past generations. There is a temptation to personify and impute intention: to turn 'gene survival in the past' into something like 'intention to reproduce in the future'. Or 'individual intention to have lots of descendants in the future'. Such personification can also apply to genes: we are tempted to see genes as influencing individual bodies to behave in such a way as to increase the number of future copies of those same genes.

Scientists who use such language, whether at the level of the individual or the gene, know very well that it is only a figure of speech. Genes are just DNA molecules. You'd have to be barking mad to think that 'selfish' genes *really* have deliberate intentions to survive! We can always translate back into respectable language: the world becomes full of those genes that have survived in the past. Because the world has a certain stability and doesn't change capriciously, the genes that have survived in the past tend to be the ones that are going to be good at surviving in the future. That means good at programming bodies to survive and make children, grandchildren and long-distance descendants. So, we have arrived back at our individual-based definition of fitness looking into the future. But we now recognise that individuals matter only as vehicles of gene survival. Individuals having grandchildren and distant descendants is only a means to the end of gene survival. And this brings us again to our paradox: 80 per cent of reproducing individuals seem to be crammed up against the ceiling – saturated out at maximum fitness!

To resolve the paradox, we return to the theoretical bedrock: the DNA. We neutralise one paradox by erecting another, almost as if two wrongs could make a right. Think on this: an individual organism can be a universal ancestor of the entire population at some distant time in the future, yet without passing on a single bit of her DNA to that population! How can this be?

Every time an individual has a child, *exactly* half her DNA is copied into that child. Every time she has a grandchild, a quarter of her DNA *on average* goes into that grandchild. Unlike the first-generation offspring where the percentage contribution is exact, the figure for each grandchild is statistical. It could be more than a quarter, it could be less. Half your DNA comes from your father, half from your mother. In turn, when you make a child, you pass half of your DNA on to her. But which half do you give? For any piece of DNA ('gene'), you are equally likely to pass on the version you inherited from your father or that from your mother. Just by chance, you could *happen* to give all your mother's versions to your child, and none of your father's. In this case, your father would have given no DNA to his grandchild. Of course such a scenario is highly unlikely, but as we go down to more distant descendants, total non-contribution of DNA becomes more possible. On average you can expect one-eighth of your DNA to end up in each great-grandchild, one-sixteenth in each great-great-grandchild, but it could be more or it could be less. And so on until the likelihood of a literally zero contribution to a given descendant becomes significant.

Here's another way to look at it. As I mentioned, if you have any European ancestors at all, you can probably trace at least one line back to Charlemagne, 40 or so generations ago. On average, only one trillionth (2 to the power 40) of your genome will have come through that particular line. But there are only 3 billion actual letters in your genome! The average amount of DNA inherited via this route would appear to be only a tiny fraction of a single DNA letter. Under this calculation, you are vanishingly unlikely to have inherited any DNA from Charlemagne at all. In fact, go back more than a handful of generations and the norm is for you not to have inherited *any* DNA from the majority of your ancestors at that time. Well, that's not quite true, since you could – almost certainly will – be descended from any particular ancestor through many different routes. Still, on our backwards journey to the past, we are bound to meet people who, while being direct ancestors to all humanity, have not passed their DNA to anyone alive today. It is true going forwards in time too. Although approximately 80 per cent of my readers are guaranteed to become a common ancestor to all living humans a few thousand years hence, far fewer will end up with their DNA copied into posterity.

Perhaps my authorship of *The Selfish Gene* biases me, but I see this as yet another reason to return to the gene as the focus of natural selection: to think backwards about the genes that have survived up to the present, rather than forwards about individuals, or indeed genes, trying to survive into the future. The 'forward intentional' style of thought can be helpful if used carefully and not misunderstood, but it is not really necessary. 'Backwards gene' language is just as vivid when you get used to it, is closer to the truth, and is less likely to yield the wrong answer.

PROLOGUE TO THE FARMER'S TALE

We have dated our initial rendezvous to a few tens of thousands of years ago or less. As we step outside our time machine at this point, what changes will we see in our ancestors? Well, the people we meet will be no more different from us than we today are different from each other. Bear in mind that 'we today' includes Germans and Zulus, Pygmies and Chinese, Berbers and Melanesians. Even as far back as 50,000 years ago, out ancestors would have fallen within the same envelope of variability as we see around the world today.

If not biological evolution, then, what changes shall we see, as we go back through tens of millennia, as opposed to hundreds or thousands of millennia? There is an evolution-like process, orders of magnitude faster than biological evolution, which, in these early stages of our time machine's journey, has been dominating the view from the porthole. This is variously called cultural evolution, exo-somatic evolution or technological evolution. We notice it in the 'evolution' of the motor car, or of the necktie or of the English language. We mustn't overestimate its resemblance to biological evolution, and it will in any case not detain us long. We have a 4-billion-year road to run, and we shall soon have to set the time machine into a gear too high to allow us more than a fleeting glimpse of events on the scale of human history.

But while our time machine is still in bottom gear, travelling on the timescale of human rather than evolutionary history, there are two major cultural advances that we must explore. The Farmer's Tale is the story of the Agricultural Revolution, arguably the human innovation that has had the greatest repercussions for the rest of the world's organisms. And the Cro-Magnon's Tale is about the 'Great Leap Forward', that flowering of the human mind which, in a special sense, provided a new medium for the evolutionary process itself.

THE FARMER'S TALE

The agricultural revolution began at the wane of the last Ice Age, about 12,000 years ago, in the so-called Fertile Crescent between the Tigris and the Euphrates. Agriculture also arose, probably independently, in China and along the banks of the Nile, and completely independently in the New World. An interesting case can be made for yet another independent cradle of agricultural civilisation in the astonishingly isolated highland interior of New Guinea. The Agricultural Revolution dates the start of the new stone age, the Neolithic.

The transition from wandering hunter-gatherers to a settled agricultural lifestyle may represent the first time people had a concept of a home. Contemporaries of the first farmers, in other parts of the world, were unreconstructed hunter-gatherers who wandered more-or-less continuously. Indeed, the hunter-gatherer lifestyle ('hunter' can include fisher) has not died out. It is still practised in pockets around the world: by Australian Aborigines, by San and related tribes in Southern Africa (called 'bushmen'), by various Native American tribes (called 'Indians' after a navigational error), and by the Inuit of the Arctic (who prefer not to be called Eskimos). Hunter-gatherers typically do not cultivate plants and do not keep livestock. In practice all intermediates between pure hunter-gatherers and pure agriculturalists or pastoralists are found. But, earlier than about 12,000 years ago, all human populations were hunter-gatherers. Soon, probably none will be. Those not extinct will be 'civilised' – or corrupted, depending on your point of view.

Colin Tudge, in his little book *Neanderthals, Bandits and Farmers: How Agriculture Really Began*, agrees with Jared Diamond (*The Third Chimpanzee*) that the switch to agriculture from hunting and gathering was not altogether the improvement we, in our complacent hindsight, might think. The Agricultural Revolution did not, in their view, increase human happiness. Agriculture supported larger populations than the hunter-gatherer lifestyle that it superseded, but not in obviously improved health or happiness. In fact, larger populations generally harbour more vicious diseases, for sound evolutionary reasons (a parasite is less concerned to prolong the life of its present host if it can easily find new victims to infect).

Nevertheless, our situation as hunter-gatherers cannot have been a Utopia either. It has lately become fashionable to regard hunter-gatherers and primitive* agricultural societies as more 'in balance' with nature than us. This is probably a mistake. They may well have had greater knowledge of the wild, simply because they lived and survived in it. But, like us, they seem to have used their knowledge to

---

*Throughout this book I use 'primitive' in the technical sense, to mean 'more like the ancestral state'. No implication of inferiority is intended.

exploit (and often overexploit) the environment to the best of their abilities at the time. Jared Diamond emphasises overexploitation by early agriculturalists leading to ecological collapse, and the demise of their society. Far from being in balance with nature, pre-agricultural hunter-gatherers were probably responsible for widespread extinctions of many large animals around the globe. Just prior to the Agricultural Revolution, the colonisation of remote areas by hunter-gatherer peoples is suspiciously often followed in the archaeological record by the wiping out of many large (and presumably palatable) birds and mammals.

We tend to regard 'urban' as the antithesis of 'agricultural' but, in the longer perspective that this book must adopt, city dwellers should be lumped in with farmers as opposed to hunter-gatherers. Almost all the food of a town comes from owned and cultivated land – in ancient times from fields round about the town, in modern times from anywhere in the world, transported and sold on through middlemen before being consumed. The Agricultural Revolution encouraged specialisation. Potters, weavers and smiths traded their skills for food which others grew. Before the Agricultural Revolution, food was not cultivated on owned land but captured or gathered on unowned commons. Pastoralism, the herding of animals on common land, may have been an intermediate stage.

Whether it was a change for better or worse, the Agricultural Revolution was presumably not a sudden event. Husbandry was not the overnight brainwave of some genius, the neolithic equivalent of Turnip Townshend. To begin with, hunters of wild animals in open and unowned country might have guarded hunting territories against rival hunters, or guarded the herds themselves while following them about. From there it was a natural progression to herding them; then feeding them, and finally corralling and housing them. I dare say none of these changes would have seemed revolutionary when they happened.

Meanwhile the animals themselves were evolving – becoming 'domesticated' by rudimentary forms of artificial selection. The Darwinian consequences on the animals would have been gradual. Without any deliberate intention to breed 'for' domestic tractability, our ancestors inadvertently changed the selection pressures on the animals. Within the gene pools of the herds, there would no longer be a premium on fleetness or other survival skills of the wild. Successive generations of domestic animals became tamer, less able to fend for themselves, more apt to flourish and grow fat under feather-bedded domestic conditions. There are alluring parallels in the domestication, by social ants and termites, of aphid 'cattle' and fungus 'crops'. We shall hear about these in the Leaf Cutter's Tale, when the ant pilgrims join us at Rendezvous 26.

Unlike modern plant and animal breeders, our forebears of the Agricultural Revolution would not knowingly have practised artificial selection for desirable characteristics. I doubt if they realised that, in order to increase milk yield, you have to mate high-yielding cows with bulls born to other high-yielding cows, and discard the calves of low-yielders. Some idea of the accidental genetic consequences of domestication is given by some interesting Russian work on silver foxes.

D. K. Belyaev and his colleagues took captive silver foxes, *Vulpes vulpes*, and set out systematically to breed for tameness. They succeeded, dramatically. By mating together the tamest individuals of each generation, Belyaev had, within 20 years, produced foxes that behaved like Border collies, actively seeking human company and wagging their tails when approached. That is not very surprising, although the speed with which it happened may be. Less expected were the by-products of selection for tameness. These genetically tamed foxes not only behaved like collies, they looked like collies. They grew black-and-white coats, with white face patches and muzzles. Instead of the characteristic pricked ears of a wild fox, they developed 'lovable' floppy ears. Their reproductive hormone balance changed, and they assumed the habit of breeding all the year round instead of in a breeding season. Probably associated with their lowered aggression, they were found to contain higher levels of the neurally active chemical serotonin. It took only 20 years to turn foxes into 'dogs' by artificial selection.*

I put 'dogs' in inverted commas, because our domestic dogs are not descended from foxes, they are descended from wolves. Incidentally, Konrad Lorenz's well-known speculation that only some breeds of dog (his favourites such as chow chows) are derived from wolves, the rest from jackals is now known to be wrong. He supported his theory with insightful anecdotes on temperament and behaviour. But molecular taxonomy trumps human insight. Molecular evidence

---

* The Canadian archaeologist Susan Crockford has attributed such changes to changing levels of two thyroid hormones.

clearly shows that all modern breeds of dog are descended from the grey wolf, *Canis lupus.** The next closest relatives to dogs (and wolves) are coyotes, and golden 'jackals' (which it now seems should be called golden wolves). 'True' side-striped and black-backed jackals are more distantly related, although they are still placed in the genus *Canis*.

No doubt the original story of the evolution of dogs from wolves was similar to the new one simulated by Belyaev with foxes, with the difference that Belyaev was breeding for tameness deliberately. Our ancestors did it inadvertently, and it probably happened several times, independently in different parts of the world. Perhaps initially, wolves took to scavenging around human encampments. Humans may have found such scavengers a convenient means of refuse disposal, and they may also have valued them as watchdogs, and even as warm sleep comforters. If this amicable scenario sounds surprising, reflect that the medieval legend of wolves as mythic symbols of terror coming out of the forest was born of ignorance. Our wild ancestors, living in more open country, would have known better. Indeed, they evidently did know better, because they ended up domesticating the wolf, thereby making the loyal, trusted dog.

From the wolf's point of view human camps provided rich pickings for a scavenger, and the individuals most likely to benefit were those whose serotonin levels and other brain characteristics ('propensity to tameness') happened to make them feel at home with humans. Several writers have speculated, plausibly enough, about orphaned cubs being adopted as pets by children. Once tameness graduated into mutual dependency, other behaviours would have been exposed to inadvertent selection too, as ably documented by Vilmos Csányi and colleagues in Budapest. Experiments have shown that domestic dogs are better than wolves at 'reading' the expressions on human faces. At the same time we read their faces, and dog facial expressions have become more human-like than those of wolves, because of unconscious selection by humans. This is presumably why we think wolves look sinister while dogs look loving, guilty, soppy and so on.

The Russian fox experiment demonstrates the speed with which domestication can happen, and the train of incidental effects which

* Molecular evidence also shows that this probably happened before the advent of agriculture, and that various local dog breeds continued to breed with wild wolves well after their supposed domestication.

can follow in the wake of selection for tameness. It is entirely probable that cattle, pigs, horses, sheep, goats, chickens, geese, ducks and camels followed a course which was just as fast, and just as rich in unexpected side-effects. It also seems plausible that we ourselves evolved down a parallel road of domestication after the Agricultural Revolution, towards our own version of tameness and associated by-product traits.

In some cases, the story of our own domestication is clearly written in our genes. The classic example, meticulously documented by William Durham in his book *Coevolution*, is lactose tolerance. Milk is baby food, not 'intended' for adults and, originally, not good for them. Lactose, the sugar in milk, requires a particular enzyme, lactase, to digest it. (This terminological convention is worth remembering, by the way. An enzyme's name will often be constructed by adding '-ase' to the first part of the name of the substance on which it works.) Young mammals switch off the gene that produces lactase after they pass the age of normal weaning. It isn't that they lack the gene, of course. Genes needed only in childhood are not removed from the genome, not even in butterflies, which must carry large numbers of genes needed only for making caterpillars. But lactase production is switched off in human infants at the age of about four, under the influence of other, controlling genes. Fresh milk makes adults feel ill, with symptoms ranging from flatulence and intestinal cramps to diarrhoea and vomiting.

All adults? No, of course not. There are exceptions. I am one of them, and there is a good chance that you are too. My generalisation concerned the human species as a whole and, by implication, the wild *Homo sapiens* from which we are all descended. It is as if I had said 'Wolves are big, fierce carnivores that hunt in packs and bay at the moon', knowing full well that Pekineses and Yorkshire terriers belie it. The difference is that we have a separate word, dog, for domestic wolf, but not for domestic human. The genes of domestic animals have changed as a result of generations of contact with humanity, inadvertently following the same sort of course as the genes of the silver fox. The genes of (some) humans have changed as a result of generations of contact with domestic animals. Lactose tolerance has evolved in some European tribes (surprisingly recently according to recent evidence from Bronze Age DNA), and occasionally elsewhere in the world, including in the Tutsi of Rwanda (and, to a lesser extent, their traditional enemies the Hutu), the

pastoral Fulani of West Africa (though, interestingly, not the sedentary branch of the Fulani), the Sindhi of North India, the Tuareg of
West Africa and the Beja of Eastern North Africa. Significantly, what
these tribes have in common is a history of pastoralism.

At the other end of the spectrum, peoples who have retained the
normal human intolerance of lactose as adults include Chinese,
Japanese, Inuit, most Native Americans, Javanese, Fijians, Australian
Aborigines, Iranians, Lebanese, Turks, Tamils, Singhalese, Tunisians,
and many African tribes including the San, and the Tswanas, Zulus,
Xhosas and Swazis of southern Africa, the Dinkas and Nuers of
North Africa, and the Yorubas and Igbos of West Africa. In general,
these lactose-intolerant peoples do not have a history of pastoralism.
There are instructive exceptions. The traditional diet of the Masai of
East Africa consists of little else besides milk and blood, and you
might think they'd be particularly tolerant of lactose. This is not the
case, however, probably because they curdle their milk before con-
suming it. As with cheese, the lactose is largely removed by bacteria.
That's one way of getting rid of its bad effects – get rid of the stuff
itself. The other way is to change your genes. This happened in the
other pastoral tribes listed above.

Of course nobody deliberately changes their genes. Science is only
now beginning to work out how to do that. As usual, the job was
done for us by natural selection, and it happened millennia ago. I
don't know exactly by what route natural selection produced adult
lactose tolerance. Perhaps adults resorted to baby food in times of
desperation, and the individuals that were most tolerant of it
survived better. Perhaps some cultures postponed weaning, and
selection for survival of children under these conditions spilled
over gradually into adult tolerance. Whatever the details, the
change, though genetic, was culture-driven. The evolution of tameness and increasing milk yields in cattle, sheep and goats paralleled
that of lactose tolerance in the tribes that herded them. Both were
true evolutionary trends in that they were changes in gene frequencies in populations. But both were driven by non-genetic cultural
changes.

Is lactose tolerance just the tip of the iceberg? Are our genomes
riddled with evidences of domestication, affecting not just our
biochemistry but our minds? Like Belyaev's domesticated foxes, and
like the domesticated wolves that we call dogs, have we become
tamer, more lovable, with the human equivalents of floppy ears,

soppy faces and wagging tails? I leave you with the thought, and move hastily on.

While hunting was sliding into herding, gathering presumably followed a similar slide into cultivation of plants. Again, it was probably mostly inadvertent. No doubt there were moments of creative discovery, as when people first noticed that if you put seeds in the ground they make plants like those from which they came. Or when somebody first observed that it helps to water them, weed them and manure them. It was probably more difficult to work out that it might be a good idea to keep back the best seed for planting, rather than follow the obvious course of eating the best and planting the dross (my father, as a young man fresh out of college, taught agriculture to peasant farmers in central Africa in the 1940s, and he tells me that this was one of the hardest lessons to get across). But mostly the transition from gatherer to cultivator passed unnoticed by those concerned, like the transition from hunter to herder.

Many of our staple food crops, including wheat, oats, barley, rye and maize, are members of the grass family which have become greatly modified since the dawn of agriculture by inadvertent and later deliberate human selection, leaving telltale signs in their genomes. It is possible that we too have become genetically modified over the millennia to increase our tolerance of cereals, in a way parallel to our evolution of tolerance to milk. Starchy cereals such as wheat and oats cannot have featured prominently in our diets before the Agricultural Revolution. Unlike oranges and strawberries, cereal seeds do not 'want' to be eaten. Passing through an animal's digestive tract is no part of their dispersal strategy, as it is of plum and tomato seeds. On our side of the relationship, the human digestive tract is not able, unaided, to absorb much nutriment from seeds of the grass family, with their meagre starch reserves and hard, unsympathetic husks. Some aid comes from milling and cooking, but it also seems conceivable that, in parallel with the evolution of tolerance to milk, we might have evolved an increased physiological tolerance to wheat, compared to our wild ancestors. Wheat intolerance is a known problem for a substantial number of unfortunate individuals who discover, by painful experience, that they are happier if they avoid it. A comparison of the incidence of wheat intolerance in hunter-gatherers such as the San, and other peoples whose agricultural ancestors have long eaten wheat, might be revealing. If there has been a large comparative study of wheat tolerance, like the one that has been made of lactose tolerance across different tribes, I am unaware of it. A systematic comparative study of alcohol intolerance, too, would be interesting. It is known that certain genetic alleles make our livers less capable of breaking down alcohol than we might wish.

In any case, co-evolution between animals and their food plants was nothing new. Grazing animals had been exerting a kind of benevolent Darwinian selection on grasses, guiding their evolution towards mutualistic co-operation, for millions of years before we started domesticating wheat, barley, oats, rye and maize. Grasses flourish in the presence of grazers, and they probably have been doing so for most of the 20 million years since their pollens first announce them in the fossil record. It is not, of course, that individual plants actually benefit by being eaten, but that grasses can withstand being cropped better than rival plants can. My enemy's enemy is my friend, and grasses, even when grazed, thrive when herbivores eat (along with the grasses themselves) other plants that would compete for soil, sun and water. Grasses became ever more able to thrive in the presence of wild cattle, antelopes, horses and other grazers (and eventually lawnmowers), as the millions of years went by. And the herbivores became better equipped, for example with specialised teeth, and complicated digestive tracts including fermentation vats with cultures of micro-organisms, to flourish on a diet of grass.

This isn't what we ordinarily mean by domestication, but in effect it is not far from it. When, starting about 12,000 years ago, wild grasses of the genus *Triticum* were domesticated by our ancestors into what we now call wheat, it was, in a way, a continuation of what herbivores of many kinds had been doing to the ancestors of *Triticum* for 20 million years. Our ancestors accelerated the process, especially when we later switched from inadvertent, accidental domestication to deliberate, planned selective breeding (and very recently scientific hybridisation and genetically engineered mutations).

That is all I want to say about the origins of agriculture. Now, before our time machine starts moving hundreds of thousands, or even millions of years at a stretch, we briefly pause, one more time, around 50,000 years ago. Here human society, entirely consisting of hunter-gatherers, underwent what may have been an even larger revolution than the agricultural one, the 'cultural Great Leap Forward'. The tale of the Great Leap Forward will be told by Cro-Magnon Man, named after the cave in the Dordogne where fossils of this race of *Homo sapiens* were first discovered.

THE CRO-MAGNON'S TALE

Archaeology suggests that something very special began to happen to our species around 50,000 years ago. Anatomically, our ancestors who lived before this watershed date were the same as those who came later. Humans sampled earlier than the watershed would be no more different from us than they were from their own contemporaries in other parts of the world, or indeed than we are from our contemporaries. That's if you look at their anatomy. If you look at their culture, there is a huge difference. Of course there are also huge differences between the cultures of different peoples across the world today. But this wasn't true if we go back much more than 50,000 years. Something happened then – many archaeologists regard it as sudden enough to be called an 'event'. I like Jared Diamond's name for it, the Great Leap Forward.

Earlier than the Great Leap Forward, man-made artefacts had hardly changed for a million years. The ones that survive for us are almost entirely stone tools and weapons, quite crudely shaped. Doubtless wood (or, in Asia, bamboo) was a more frequently worked material, but wooden relics don't easily survive. As far as we can tell, there were no paintings, no carvings, no figurines, no grave goods, no ornamentation. After the Leap, all these things suddenly appear in the archaeological record, together with musical instruments such as bone flutes, and it wasn't long before stunning creations like the Lascaux Cave murals were created by Cro-Magnon people (see plate 1). A disinterested observer taking the long view from another planet might see our modern culture, with its computers, supersonic planes and space exploration, as an afterthought to the Great Leap Forward. On the very long geological timescale, all our modern achievements, from the Sistine Chapel to Special Relativity, from the *Goldberg Variations* to the Goldbach Conjecture, could be seen as almost contemporaneous with the Venus of Willendorf and the Lascaux Caves, all part of the same cultural revolution, all part of the blooming cultural upsurge that succeeded the long Lower Palaeolithic stagnation. Actually I'm not sure that our extraplanetary observer's uniformitarian view would stand up to much searching analysis, but it could be at least briefly defended.[*]

[*] David Lewis-Williams's *The Mind in the Cave* considers the whole question of Upper Palaeolithic cave art, and what it can tell us about the flowering of consciousness in *Homo sapiens*.

Some authorities are so impressed by the Great Leap Forward that they think it coincided with the origin of language. What else, they ask, could account for such a sudden change? It is not as silly as it sounds to suggest that language arose suddenly. Nobody thinks writing goes back more than a few thousand years, and everyone agrees that brain anatomy didn't change to coincide with anything so recent as the invention of writing. In theory, speech could be another example of the same thing. Nevertheless, my hunch, supported by the authority of linguists such as Steven Pinker, is that language is older than the Leap. We'll come back to the point a million years further into the past, when our pilgrimage reaches *Homo ergaster* (*erectus*).

If not language itself, perhaps the Great Leap Forward coincided with the sudden discovery of what we might call a new software technique: maybe a new trick of grammar, such as the conditional clause, which, at a stroke, would have enabled 'what if' imagination to flower. Or maybe early language, before the leap, could be used to talk only about things that were there, on the scene. Perhaps some forgotten genius realised the possibility of using words referentially as tokens of things that were not immediately present. It is the difference between 'That waterhole which we can both see' and 'Suppose there was a waterhole the other side of the hill'. Or perhaps representational art, which is all but unknown in the archaeological record before the Leap, was the bridge to referential language. Perhaps people learned to draw bison, before they learned to talk about bison that were not immediately visible.

Much as I would like to linger around the heady time of the Great Leap Forward, we have a long pilgrimage to accomplish and we must press on backwards. We estimated the date of Rendezvous 0 as probably thousands or tens of thousands of years ago. The next official rendezvous, our meeting with the chimpanzee pilgrims, is millions of years away, and most are hundreds of millions beyond that. To stand a chance of completing our pilgrimage, we shall need to speed up and begin the move into 'deep time'. We must accelerate past the 30 or so spectacular glaciations that typify the last 3 million years, and past other drastic events such as the drying and refilling of the Mediterranean that occurred between 4.5 and 6 million years ago. To ease this initial acceleration, I shall take the otherwise unusual liberty of stopping at a few intermediate milestones en route. At these points we shall meet individuals who are, or potentially could be, our direct ancestors. I will allow some of them to tell tales, letting these 'shadow' pilgrims satisfy our natural preoccupation with our own, human lineage.

ARCHAIC *HOMO SAPIENS*

Our first milestone on the way back to Rendezvous 1 occurs 200,000 years ago. It is a time of relatively warm temperatures, before a plunge into a period of intense glaciation, and these changes in climate must have been a major force in the unfolding of this part of the human story.

There are two reasons for choosing this way station. Firstly, it marks one of our best guesses for the date of our most recent female-line common ancestor. She is often given the title 'mitochondrial Eve', because our female line can be traced using a piece of DNA which resides in the mitochondria inside our cells and which we get exclusively from our mother. Deductions using mitochondrial DNA have played an important role in elucidating recent human history, the pitfalls of which we explore in Eve's Tale.

The second reason for stopping at this point is to look at fossil finds from beside the Omo river in Ethiopia. They were originally discovered by Richard Leakey's team in 1967, but returned to the limelight in 2008 when estimates of their ages pushed them back to as much as 195,000 years old. The most intriguing remains are two partial skulls, from opposite sides of the river, which differ slightly. In most ways, the first skull could be taken for a completely modern human. Yet the second skull isn't quite the same shape as ours: it is broader at the base and has a less domed forehead. The more complete (but at 160,000 years old, slightly more recent) finds from Herto, also in Ethiopia, include skulls with a similar 'almost modern' form. Regardless of hair-splitting distinctions between 'modern' and 'almost modern', it is clear that these skulls capture a piecemeal transition between ourselves and those predecessors that we know by the catch-all name of 'Archaic *Homo sapiens*'.

Certain authorities use the name 'Archaic *Homo sapiens*' right back to about 900,000 years ago where they grade into an earlier species, *Homo erectus*. Others prefer to give various Latin names to the bridging archaic forms. We shall sidestep the disputes by using anglicisms in the style of my colleague Jonathan Kingdon: 'Moderns', 'Archaics', 'Erects', and others that we'll mention as we come to them. We should not expect to draw a neat line between early Archaics and the Erects from whom they evolved, or between Archaics and the earliest Moderns who evolved from them. Don't be confused, incidentally, by the fact that the Erects were even more archaic (with a small a) than the Archaics (with a large A), and that all three types were erect with a small e!

As far as we know, the anatomical transition from Archaics to Moderns only happened in Africa, although fossils of the Archaic type are found all around the world, dating from various times during the last few hundred thousand years: examples are the German 'Heidelberg man', 'Rhodesian man' from Zambia (which used to be called Northern Rhodesia), and the Chinese 'Dali man'. Archaics had big brains like us, averaging 1,200 to 1,300 cubic centimetres. This is a little smaller than our average of 1,400 cubic centimetres but the range comfortably overlaps with ours. Their bodies were more robust than ours, their skulls were thicker, and they had more pronounced brow ridges and less pronounced chins. They looked more like Erects than we do, and hindsight justly sees them as intermediate. Some taxonomists recognise them as a sub-species of *Homo sapiens* called *Homo sapiens heidelbergensis* (where we would be *Homo sapiens sapiens*). Others do not recognise the Archaics as *Homo sapiens* at all, but call them *Homo heidelbergensis*. Yet others divide the Archaics into more than one species, for instance *Homo heidelbergensis, Homo rhodesiensis,* and *Homo antecessor.* If you think about it, we should be worried if there was *not* disagreement over the divisions. On the evolutionary view of life, a continuous range of intermediates is to be expected.

An unfortunate temptation, given this range of intermediate fossils, is to imagine a straight evolutionary line between them, for instance ordered by brain size. As we shall see, the true story of recent human origins requires a more sophisticated picture. Evolution is a messy historical process, unlikely to proceed in straight lines. Nowhere is this more evident than in the case of the Neanderthals, named after the Neander Valley where the first fossil of this type was discovered.*

* Pedants' Corner: *Thal* or, in modern German, *Tal,* means valley (and is the etymological ancestor of the 'thaler' or 'dollar' currency unit). When German spelling was reformed at the end of the nineteenth century, the Neander valley changed from a *Thal* to a *Tal,* but the Latin name, *Homo neanderthalensis,* was left high and dry, trapped by the laws of zoological terminology. To conform with tradition and with the Latin, we are inclined to leave the English spelling in its original form and stick with the h. Those who prefer to drop their aitches should feel free to do so.

One way to think about Neanderthals is as our parallel cousins. They represent a separate offshoot from the Archaics, evolving not in Africa, but mostly in Eurasia and the Middle East. Neanderthals resembled the Archaics more than we do in some respects, and are thought to have diverged from Archaic populations more than twice as long ago as we did, perhaps half a million years ago or more. Their separate evolutionary path is documented in the Sima de los Huesos ('cave of bones') in Northern Spain: a treasury of fossils dated to about 400,000 years ago which show some, but not all, of the characteristics we associate with the classic, later Neanderthal form.

By the time of our current rendezvous, these Eurasian humans had accumulated a distinctive enough set of features that some people prefer to give them a separate species name, *Homo neanderthalensis*. They retained some features of Archaics such as large brow ridges which Moderns did not (which is why some authorities classify them as just another type of Archaic). Adaptations to their cold environment include stockiness, short limbs and enormous noses, and they surely must have been warmly clothed, presumably in animal furs.

Fossilised Neanderthal excrement allows us to deduce their diet: mostly meat, with a small side of vegetable matter. It is often said that they had slightly bigger brains than we do, although adjusting for body weight reassures us they were less 'encephalised' (see the Handyman's Tale). Much is made of slight indications that they ceremonially buried their dead and created simple art. Nobody knows whether they could speak, and opinions differ on this important question. Archaeology hints that technological ideas may have passed both ways between Neanderthals and Moderns, but this could have been by imitation rather than by language.

This brings us to the delicate topic of interactions between Neanderthals and Moderns. As the Modern type spread out of Africa, they started overlapping with Neanderthals, firstly in the Middle East about 100,000 years ago, then later in western Asia, and finally when Moderns entered Europe around 45,000 years ago. A scant few millennia passed before Neanderthals essentially disappear from the fossil record, about 40,000 years ago. The timing leads many to argue that Moderns were responsible for their extinction, either by killing them directly or by competing with them.

The case of the disappearing Neanderthals has been rejuvenated by the astonishing extraction of DNA from a number of Neanderthal bones. One study shows a major difference between recent Neanderthals in western versus eastern Europe. Before Moderns arrived on the scene, western Neanderthals show a sudden drop in genetic diversity, coinciding with a period of extreme cold temperatures. It is evidence that Neanderthal populations fluctuated in size and location even without competition from Moderns. Other studies of ancient DNA have shed light on whether Neanderthals interbred with Moderns and so, in a way, never went extinct at all. We will come to this at the end of this chapter, in the Prologue to the Denisovan's Tale, which serves as an introduction to the tale's more startling revelation: the existence of another human cousin, in addition to the Neanderthals. This mystery 'third man' (actually the evidence comes from a young girl) has been deduced entirely from DNA, so it is to studies of genes that we first turn.

PROLOGUE TO EVE'S TALE

In the Tasmanian's Tale we talked about genealogical ancestors: historical individuals who are ancestors of modern ones in the conventional genealogist's sense: 'people ancestors'. But what you can do for people you can do for genes. Genes too have parent genes, grandparent genes, grandchild genes. Genes too have pedigrees, family trees, 'Most Recent Common Ancestors' (MRCAs). Moreover, family trees of genes have one enormous advantage over the conventional historical sort: their pattern is stamped into today's genomes. Written records of family histories peter out over centuries. Genetic histories and 'gene trees' deduced purely from living humans can illuminate our history over millions of years.

Before proceeding, I must clear up some possible confusions over the meaning of the word 'gene'. It can mean lots of things to different people, but when I talk about genes in the context of 'gene trees' I mean a distinct sequence of DNA that has been passed intact down the generations. Some biologists, especially molecular geneticists, would call this a '(genetic) variant', 'allele', or sometimes 'haplotype'. They reserve the word 'gene' to mean a particular *location* on the chromosome ('locus'), typically a region containing instructions to produce a known biochemical molecule, usually a protein. 'Allele' then denotes the different *versions* of the gene that might sit at that locus. To take an oversimplified example, the main gene for eye colour comes in different versions or alleles, including a blue allele and a brown allele. Other biologists, especially the kind to which I belong, who are sometimes called sociobiologists, behavioural ecologists or ethologists, tend to use the word gene to mean the same as allele. When we want a word for the slot in the chromosome which could be filled by any of a set of alleles, we tend to say 'locus'. People like me are apt to say, 'Imagine a gene for blue eyes, and a rival gene for brown eyes.' Not all molecular geneticists like that, but it is a well-established habit with my kind of biologist and I shall occasionally follow them.

EVE'S TALE

There's a telling difference between 'gene trees' and 'people trees'. Unlike a person who is descended from two parents, a gene has one parent only. Each one of your genes must have come from either your mother or your father, from one and only one of your four grandparents, from one and only one of your eight great-grandparents, and so on. But when whole people trace their ancestors in the conventional way, they descend equally from two parents, four grandparents, eight great-grandparents and so on. This means that a 'people genealogy' is much more mixed up than a 'gene genealogy'. In a sense, a gene takes a single path chosen from the maze of crisscrossing routes mapped by the (people) family tree. Surnames behave like genes, not like people. Your surname picks out a thin line through your full family tree. It highlights your male to male to male ancestry. DNA, with two notable exceptions which I shall come to later, is not so sexist as a surname: genes trace their ancestry through males and females with equal likelihood.

Some of the best-recorded human pedigrees are of European royal families. In the family tree of the house of Saxe-Coburg (see diagram opposite), look at the princes Alexei, Waldemar, Heinrich, and Rupert. The 'gene tree' of one of their genes is easy to trace because, unfortunately for them but fortunately for us, the gene concerned was defective. It gave the four princes, and many others of their ill-favoured family, the easily recognised blood disease haemophilia: their blood wouldn't clot properly. Haemophilia is inherited in a special manner: it is carried on the X chromosome. Males have only one X chromosome which they inherit from their mother. Females have two X chromosomes, one inherited from each parent. They suffer from the disease only if they have inherited the defective

Bloodlines in the ill-fated House of Saxe-Coburg.

version of the gene from both their mother and their father (i.e. haemophilia is 'recessive'). Males suffer from the disease if their single 'unguarded' X chromosome bears the defective gene. Very few females suffer from haemophilia, therefore, but lots of females are 'carriers'. They have one copy of the faulty gene, and a 50 per cent chance of passing it on to each child. Carrier females who are pregnant always hope for a daughter, but they still have a substantial risk of haemophiliac grandsons, regardless of their marriage choice. If a haemophiliac male lives long enough to have children, he cannot pass the gene on to a son (males never receive their X chromosome from their father), but he must pass it on to a daughter (females always receive their father's only X chromosome). Knowing these rules, and knowing which royal males had haemophilia, we can trace the faulty gene. We have marked it on the family tree: the backward path the haemophilia gene must have taken is in bold.

It seems that Queen Victoria herself was the mutant. It wasn't Albert, because his son, Prince Leopold, was haemophiliac, and sons don't get their X chromosome from their father. None of Victoria's collateral relatives suffered from haemophilia. She was the first royal individual to carry the gene. The miscopying must have occurred either in an egg of her mother, Victoria of Saxe-Coburg, or, which is more likely for reasons explained by Steve Jones in *The Language of the Genes*, 'in the august testicles of her father, Edward Duke of Kent'.

Although neither of Victoria's parents carried or suffered from haemophilia, one of them did have a gene (strictly an allele) which was the pre-mutated 'parent' of the royal haemophilia gene. We can think about (though we cannot detect) the ancestry of Victoria's haemophilia gene, back before it mutated to become a haemophilia gene. For our purposes it is irrelevant, except as a matter of diagnostic convenience, that Victoria's copy of the gene was diseased while its predecessors were not. As we trace back the family tree of the gene we ignore its effects, except in so far as they render it visible. The gene's lineage must go back before Victoria, but the visible trail goes cold when it wasn't a haemophilia gene. The lesson is that every gene has one parent gene even if, through mutation, it is not strictly identical. Similarly it has only one grandparent gene, only one great-grandparent gene, and so on. This may seem an odd way to think, but remember that we are on an ancestor-hunting pilgrimage. The present exercise is to see what an ancestor-hunting pilgrimage would look like from a gene's point of view, instead of an individual's.

In the Tasmanian's Tale we encountered the acronym MRCA (Most Recent Common Ancestor) as an alternative to 'concestor'. We would do well to reserve 'concestor' for the most recent common ancestor in a pedigree, a family tree of individual people or other creatures. So when talking about genes we shall use 'MRCA'. Two or more alleles in different individuals certainly do have an MRCA. It is the ancestral gene of which they are each a (possibly mutated) copy. The MRCA of the haemophilia genes of Princes Waldemar and Heinrich of Prussia sat on one of the two X chromosomes of their mother, Irene von Hesse und bei Rhein. When she was still a foetus, two copies of the one haemophilia gene she carried were peeled off and passed successively into two of her egg cells, the progenitors of her luckless sons. These genes in turn share an MRCA with the haemophilia gene of Tsarevitch Alexei of Russia (1904–1918), in the form of a gene carried by their grandmother, Princess Alice of Hesse. Finally, the MRCA of the haemophilia gene in all four of our chosen princes is the very one that flagged itself up for attention in the first place, the mutant gene of Victoria herself.

Geneticists often use the word 'coalescence' to describe this backwards ancestry of a gene. Looking backwards in time, two gene lineages can be said to coalesce into one at the point where, looking forwards again, a parent runs off two copies of the gene for two successive children. The point of coalescence is the MRCA. Any gene tree has many coalescence points. The haemophilia genes of Waldemar and Heinrich coalesce into the MRCA gene carried by their mother, Irene. That then coalesces with the lineage heading backwards from Tsarevitch Alexei. And, as we've seen, the grand coalescence of all the royal haemophilia genes occurs in Queen Victoria. Her genome holds the MRCA haemophilia gene for the whole dynasty.

In this example, the coalescence of the haemophilia genes of all four princes occurs in Victoria, the very individual who happens also to be their concestor, the most recent common ancestor in their pedigree. But that is just coincidence. If we were to choose another gene (say for eye colour), then the path it took through the family tree would be quite different, and the genes would coalesce in a more distant ancestor than Victoria. If we picked a gene for brown eyes in Prince Rupert and one for blue eyes in Prince Heinrich, then the coalescence must be at least as far off as the separation of an ancestral eye-colour gene into two forms, brown and blue, an event buried in prehistory. Each piece of DNA has a genealogy which may be traced in a way that is separate but parallel to the sort of genealogy where we follow surnames through records of Births, Marriages and Deaths.

We can even do this for two identical genes in the same person. Prince Charles has blue eyes, which means, since blue is recessive, that he has two blue-eyed genes.* Those two genes must coalesce somewhere in the past, but we can't tell when or where. Usually it is many millennia ago, but in the special case of Prince Charles it is possible that the two blue-eyed genes coalesce in as recent an individual as Queen Victoria. This is because, as it happens, Prince Charles is descended from Victoria twice: once via King Edward VII and once via Princess Alice of Hesse. On this hypothesis, a single blue-eyed gene of Victoria made two copies of itself at different times. These two copies of the same gene came down to the present

* Actually, there are several locations on the genome that determine eye colour, but the principle still stands.

Queen (Edward VII's great-granddaughter) and to her husband, Prince Philip (Princess Alice's great-grandson) respectively. Two copies of one Victorian gene could therefore have met again, on two different chromosomes, in Prince Charles. In fact, that almost certainly has happened for some of his genes, whether for blue eyes or not. And regardless of whether his two blue-eyed genes coalesce in Queen Victoria or in somebody further back, those two genes must have had an MRCA at some specific point in the past. It doesn't matter whether we are talking about two genes in one person (Charles) or in two people (Rupert and Heinrich): the logic is the same. Any two alleles, in different people or in the same person, are fair game for the question: When, and in whom, do these genes coalesce as we look back? And, by extension, we can ask the same question of any three genes, or any number of genes in the population, at the same genetic location ('locus').

Looking much further back still, we can ask the same question for pairs of genes at different loci, because genes give rise to genes at different loci by the process of 'gene duplication'. We shall meet this phenomenon again in the Howler Monkey's and Lamprey's Tales.

Each gene or region of DNA will have its own pattern of coalescence points, which will trace out a particular *gene tree*. Two closely related humans will appear close together in many of their gene trees. But they will also have gene trees that deliver a 'minority vote', placing them closer to otherwise more distant relatives. We can think of closeness of kinship among *people* as a kind of majority vote among genes. Some regions of your DNA vote for, say, the Queen, as a not too distant cousin. Others argue that you are closer to seemingly much more distant individuals (as we shall see, even members of other species). When quizzed, each gene has a different view of what history is all about, because each has blazed a different path through the generations. We can hope to gain a comprehensive picture only by probing many areas of the genome. But we must be wary if they are situated close to each other on a chromosome. To see why, we need to know something about the phenomenon of recombination, which happens every time a sperm or an egg is made.

In recombination, randomly chosen sections of matching DNA are swapped between chromosomes. On average, only one or two swaps are seen per human chromosome (fewer when making sperm, more when making eggs: it is not known why). Eventually, over numerous generations, many different parts of the chromosome will end up

being swapped around. Generally speaking, the nearer two pieces of
DNA are on a chromosome, the lower is the chance of a swap
occurring between them, and the more likely they are to be inherited
together.

When taking 'votes' from genes, therefore, we have to remember
that the nearer a pair of genes are to each other on a chromosome,
the more likely they are to experience the same history. And this
motivates genes which are close colleagues to back up each other's
vote. At the extreme are sections of DNA so tightly bound together
that the entire chunk has travelled through history as a single unit.
Among such caucuses within the genetic parliament, two stand out,
not because their view of history is more valid, but because they have
been extensively used to settle biological debates. Both hold sexist
views, because one has come down entirely through female bodies,
and the other has never been outside a male body. These are the two
major exceptions to unbiased gene inheritance that we previously
mentioned: Y chromosomes and mitochondria.

Like a surname, the (male-specific portion of the) Y chromosome
always passes through men only. Together with a few other genes
needed only by males, the Y chromosome contains the genetic
material that actually switches a mammal's embryo into the male
pattern of development rather than the female one. Mitochondrial
DNA, on the other hand, passes exclusively down the female line
(although in this case it is not responsible for making the embryo
develop as a female: males have mitochondria, it is just that they
don't pass them on). As we shall see in the Great Historic Rendezvous, mitochondria are tiny bodies inside cells, relics of once-free
bacteria who, probably about 2 billion years ago, took up exclusive
residence inside cells where they have been reproducing, nonsexually
by simple division, ever since. They have lost many of their bacterial
qualities and most of their DNA, but they retain enough to be useful
to geneticists. Mitochondria constitute an independent line of genetic reproduction inside our bodies, unconnected with the main
nuclear line which we think of as our 'own' genes.

Both Y chromosomes and mitochondria have been used to trace
human history. One neat study took snippets of Y-chromosome
DNA in a straight line across modern Britain. The results showed
that Anglo-Saxon Y chromosomes moved west across England from
Europe, stopping rather abruptly at the Welsh border. It is not hard
to imagine reasons why this male-carried DNA is unrepresentative of

the rest of the genome. To take a more obvious example, Viking longships carried cargoes of Y chromosomes (and other genes) and spread them among widely scattered populations. In most human societies men tend to settle and reproduce closer to their birthplace than women. Presumably Viking Y chromosomes buck the trend. We might expect them to be slightly more 'travelled' than other Viking genes, which were statistically more likely to favour home-acre over Widow-maker:

> *What is a woman that you forsake her,*
> *And the hearth-fire and the home-acre,*
> *To go with the old grey Widow-maker?*
> RUDYARD KIPLING
> 'Harp Song of the Dane Women'

Mitochondrial DNA too can be revealing. If we compare your mitochondrial DNA with mine, we can tell how long ago they shared an ancestral mitochondrion. And, since we all get our mitochondria from our mothers, and hence maternal grandmothers, maternal great-grandmothers, etc., mitochondrial comparison can tell us when our most recent female-line ancestor lived. The same can be done for Y chromosomes, to tell us when our most recent male-line ancestor lived but, for technical reasons, it is not so easy. The beauty of Y-chromosomal and mitochondrial DNA is that neither of them is contaminated by sexual mixing. This makes tracing these particular classes of ancestor easy.

The mitochondrial MRCA of all humanity, which pinpoints the 'people' common ancestor in the all-female line, is sometimes called Mitochondrial Eve – she whose tale this is. And of course the equivalent in the all-male line might as well be called Y-chromosome Adam. All human males have Adam's Y chromosome (creationists please refrain from deliberate misquotation). If surnames had always been strictly inherited by modern Western rules we'd all have Adam's surname too, which would rather lose the point of having a surname.

Eve is a great temptress to error and it is good to be forearmed. The errors are quite instructive. First, it is important to note that Mitochondrial Eve is emphatically not the most recent 'mother of us all', nor indeed is Y-Adam the 'most recent common ancestor of all men'. Variants of these terms are often used in misleading newspaper headlines, despite the fact that Eve and Adam are only two out of a multitude of male and female MRCAs that we could reach if we traced our way back through different lines. They are the special-case common ancestors that we reach if we travel up the family tree from mother to mother to mother, or father to father to father respectively. But there are many, many other ways that genes can move through the family tree: mother to father to father to mother, mother to mother to father to father, and so forth. Each of these possible pathways will have a different MRCA, all of whom unite humanity, many of them more recently than Adam or Eve.

Second, Eve and Adam were not a couple. It would be a major coincidence if they ever met, and they could well have been separated by tens of thousands of years. As a subsidiary point, there are independent reasons to expect female common ancestors to predate their male counterparts. Males are more variable in reproductive success than females. Where some females have five times as many children as other females, the most successful males could have hundreds of times as many children as unsuccessful males. That means a successful male – a prehistoric 'Genghis Khan' – can become a common ancestor relatively quickly. A successful female, since her family cannot be quite so large, needs more generations to achieve the same feat.* We find this when comparing paternal and maternal gene trees for most sets of people. But it is a statistical generalisation rather than a strict rule and, perhaps unexpectedly, recent genetic discoveries find that in the specific case of our current Y-chromosome Adam, he may have live earlier in time than our current Mitochondrial Eve.

Third, Adam and Eve are shifting honorific titles, not names of particular individuals. If, tomorrow, the last member of some outlying tribe were to die, the baton of Adam, or of Eve, could abruptly be thrown forwards several thousand years. The same is true of all the other MRCAs defined by different gene trees. To see why this is so, suppose Eve had two daughters, one of whom eventually gave rise to the Tasmanian aborigines and the other of whom spawned the rest of humanity. And suppose, entirely plausibly, that the female-line MRCA uniting 'the rest of humanity' lived 10,000 years later, all other collateral lines descending from Eve having gone extinct apart from the Tasmanians. When Truganinni, the last Tasmanian, died, the title of Eve would instantly have jumped forwards 10,000 years.

Fourth, there was nothing to single out either Adam or Eve for

---

* Rather opaquely, geneticists say that males tend to have a smaller 'effective population size' than females.

particular notice in their own times. Despite their legendary name-sakes, Mitochondrial Eve and Y-chromosome Adam were not par-ticularly lonely. Both would have had plenty of companions, and each may well have had many sexual partners, with whom they may also have surviving descendants. The only thing that singles them out is that Adam eventually turned out to be hugely endowed with descendants down the male line, and Eve with descendants down the female line. Others among their contemporaries will have left equally as many descendants all told.

The names 'Eve' and 'Adam' capture the imagination. The idea is also logically compelling. A single female-line and a single male-line ancestor *must* have existed at some point, it's simply how family trees work.* But each refers only to one changeable point: the root of an extensive gene tree. If we want to use mitochondria and Y chromo-somes to deduce human history, Eve and Adam should not be our sole focus. Far more important are all the other coalescence points in the two gene trees. And even then we must be extremely careful.

As I was writing the first edition of this book, somebody sent me a videotape of a BBC television documentary called *Motherland*, hyped as 'an incredibly poignant film', and as 'truly beautiful, a really memorable piece'. The heroes of the film were three 'black'† people whose families had immigrated to Britain from Jamaica. Their DNA was matched up against worldwide databases, in an attempt to trace the part of Africa from which their ancestors were taken as slaves. The production company then staged lachrymose 'reunions' between our heroes and their long-lost African families. They used Y-chromosomal and mitochondrial DNA because, for the reasons we have seen, it is more traceable than the rest of our genome. But unfortunately, the producers never really came clean about the limitations this imposed. In particular, no doubt for sound televisual reasons, they came close to actively deceiving these individuals, and

---

* To be scrupulously accurate, although a female-line 'Eve' is a logical certainty, the concept of *mitochondrial* Eve depends on the rule that all mitochondrial DNA always comes from the mother. That rule does not apply to some species such as mussels and conifers, and it is thought to be broken occasionally – but not enough to worry about – in mammals, including humans. Similarly, the idea of Y-chromosome Adam depends on the male-specific region of the Y chromosome being copied exclusively from fathers. Again this is unlikely to be absolutely true, since DNA sequences are very occasionally overwritten by sequences from other parts of the genome, but this is so rare that it should not affect the conclusions of this tale.

† For explanation of the inverted commas around 'black', see the Grasshopper's Tale.

also their long-lost African relatives, into becoming far more emotional about the reunions than they had any right to be.

Let me explain. When Mark, later given the tribal name Kaigama, visited the Kanuri tribe in Niger, he believed he was 'returning' to the land of 'his people'. Beaula was welcomed as a long-lost daughter by eight women of the Bubi tribe on an island off the coast of Guinea whose mitochondria matched hers. Beaula said,

> It was like blood touching blood . . . It was like family . . . I was just crying, my eyes were just filled with tears, my heart was pounding. All I just kept thinking was: 'I'm going to my motherland.'

Sentimental rubbish, and she should never have been deceived into thinking this. All that she, or Mark, were really visiting – at least as far as there was any evidence to suppose – were individuals who shared their mitochondria. As a matter of fact, Mark had already been told that his Y chromosome came from Europe (which upset him and he was later palpably relieved to discover respectable African roots for his mitochondria!). Beaula, of course, has no Y chromosome, and apparently they didn't bother to look at her father's although that would have been interesting, for she was quite light-skinned. But it was explained to neither Beaula nor Mark, nor the television audience, that DNA outside their mitochondria almost certainly came from a huge variety of 'homelands' nowhere near those identified for purposes of the documentary. If their other genes had been traced, they could have had equally emotional 'reunions' in hundreds of different sites, all over Africa, Europe and very probably Asia too. That would have spoiled the dramatic impact, of course.

Bear this in mind as we tackle a major argument about human origins, a debate in which mitochondrial and more recently Y chromosome DNA have been used as primary evidence. The 'Out of Africa' theory holds that all surviving peoples outside Africa are descended from a single exodus, between 50,000 and 100,000 years ago. At the other extreme are the 'Separate Origins' theorists or 'Multiregionalists', who believe that the races still living in, say, Asia, Australia and Europe are anciently divided, separately descended from regional populations of the earlier species, *Homo erectus*. Both names are misleading. 'Out of Africa' is unfortunate because everybody agrees that our ancestors are from Africa if you go back far enough. 'Separate Origins' is also not an ideal name because again, if you go back far enough, the separation must disappear on any theory. The disagreement concerns the date when we came out of Africa. Instead, let us call them the 'Recent African Origin' (RAO) and 'Ancient African Origin' (AAO) hypotheses. This has the added advantage of emphasising the continuum between them.

Since this is Eve's Tale, we'll start with mitochondrial evidence. As we have seen, mitochondria are copied wholly from our mothers, meaning that their ancestry forms a neat coalescing tree. Outside Africa, the indigenous mitochondria always seem to belong to one of two main branches or 'haplogroups': M (predominantly Asian) and N (found throughout Eurasia). On average, a DNA sequence belonging to group M differs from one in group N by about 30 DNA mutations. Knowing which parts of the mitochondrial DNA are important and which are freer to vary, we can use the 'molecular clock' (see the Epilogue to the Velvet Worm's Tale) to guess how long it might have taken to accumulate these mutations. Human and chimpanzee mitochondria differ by not 30, but about one and a half *thousand* mutations. Assuming the human and chimpanzee differences took 7 million years to accumulate (a controversial number which we shall examine in the Chimpanzee's Tale), the difference between M and N must have taken roughly 50,000 to 90,000 years of evolution. Calibrating mitochondrial mutation rates a different way, using DNA recovered from radiometrically dated fossil humans, produces similar dates, between 65,000 and 90,000 years ago.

More broadly, branches M and N are only two small offshoots within a much more extensive African gene tree, whose deepest coalescence (Eve) is two or three times older still. Clearly, recent mitochondria out of Africa replaced the archaic versions that until then had occupied the rest of the world. The female-only lines of Europeans, Asians, Native Americans, Australians and the rest are of Recent African Origin. With the proviso that the link between any one coalescence point and a historical event is very loose, mitochondria nevertheless support the idea of a recent African exodus. More precisely, the geographical pattern within the M and N branches hints at an exodus that left the Horn of Africa and skirted the south coast of Arabia, entering the rest of the world via Asia, just in time to explain Mungo man, a fully modern fossilised Australian.

Remember, however, that these stories are based around a single piece of DNA. To imagine that mitochondria epitomise human history is to fall into the same trap as the *Motherland* television documentary. At a minimum we must consult the Y chromosome too.

The Y chromosome contains several thousand times more DNA than the mitochondrion. That makes it a much richer source of information, although harder to study. At the moment, the evidence points to a broadly similar pattern to mitochondria. Although not necessarily involving the same groups, our Y chromosome ancestry is likewise rooted in Africa. We hesitate to give a precise estimate for the oldest coalescence, partly because of the unearthing in 2013 of a rare African lineage which implies a new 'Adam' possibly predating Eve, a story to which we shall return in this tale's epilogue. As more people in Africa have their genomes sequenced, we may well find older Adams (and possibly older Eves too).

The male lineages outside Africa occupy a few separate twigs on an otherwise Afrocentric Y-chromosome tree. Each twig originates at a broadly similar time to mitochondria (perhaps a fraction earlier), and further geographical splits in Europe and Asia roughly mirror, but do not exactly reflect, the spread of today's mitochondria. Multiple twigs could reflect more than one migration, or a single exodus that took with it a diverse set of Y chromosomes, some of whose descendants remain present in today's Eurasian men.

So far so good for the RAO theory. The problem is that both mitochondria and Y chromosomes are potentially deceptive. It's not just that they represent a tiny fraction of our genome, nor even that they are gender-biased. The problem is that any one gene tree is massively affected by both chance and (more insidiously) by natural selection. Take the spread of a Y chromosome lineage out of Africa. It could reflect a human exodus. But it could also be a result of chance or natural selection, without any exodus at all. Here's how. Imagine a populous world a few hundred thousand years ago, a world in which humans did not migrate, but merely interbred with their neighbours. Now consider the fate of a potentially beneficial new gene on the Y chromosome (say it made beards hairier). In cold climates, this gene might be favoured over alternative versions. And because all the DNA on the Y moves as one, all the genes that happened to lie alongside the advantageous one would also benefit from this 'positive selection'. As this Y chromosome and its descendants spread throughout the northern hemisphere via interbreeding, old versions would be replaced. Looking at the Y chromosome today, we could easily misinterpret its gene tree as evidence of rapid expansion and replacement of whole populations, whereas what actually happened was rapid expansion and replacement of a single 'gene'. In other words, populations may themselves stay in place, even as new genes ripple through them.

This hypothetical case illustrates an important and general principle. A single gene tree cannot distinguish between the hidden hand of natural selection and more general events such as changes in population size, migration, or division into tribes. To reconstruct these demographic features of our history, we require evidence from DNA spanning the entire genome.

Our ally in this quest is genetic recombination. Recombination, recall, is the process that splits and joins lengths of DNA. Genes that have been frequently separated by recombination – such as those widely spaced on a chromosome – can provide multiple independent witnesses to human history. In fact, choosing widely spaced genes is overly restrictive. We are starting to develop new techniques that use *all* our DNA. This must be where the future lies, but it requires us to tackle the intricacies of recombination head on.

We start by imagining what happens if we follow all the DNA from a set of individuals back through time. When two copies of a DNA sequence trace back to a single chromosome, we know that they coalesce into a common ancestor. For a large length of DNA, possibly an entire chromosome, we must add another possibility. A single copy of a DNA sequence could contain sections which trace back to *different* ancestors. It happens when, forwards in time, a new chromosome is formed by combining sections from the father with sections from the mother. As we go back, instead of two lineages merging into one, a single lineage appears to fork in two. At such a point, the histories of different portions of the chromosome diverge as they regress into the past.

The backwards pattern of coalescing and diverging lineages is known as the ancestral recombination graph (ARG for short), 'graph' in this sense meaning a network of interconnected lines. It is sometimes described as the 'holy grail' of genetic historians, for to calculate it would be to encapsulate all the genealogical information held within a set of genomes. Unfortunately, it is an impossible calculation. The relationships between even a few genomes can be represented by an interminable number of ARGs, and we can never hope to pin down which is exactly correct. Instead, geneticists use computers to average over a sample of likely possibilities. Crude perhaps, but genomes contain so much information that using this approach on just a handful of people pays generous dividends.

Astonishingly, even a single human genome can be used, because most of our DNA comes in two copies, one from our mother and one from our father. The method, due to Richard Durbin and Heng Li at the Sanger Institute in Cambridge, looks at a selection of possible ancestries (ARGs) that could reasonably explain the differences between your maternal and your paternal DNA. Coupled with the 'molecular clock', this can produce estimates of every date when a section of chromosome from your mother and from your father met ('coalesced') in a common ancestor. Since these dates are from across the genome they should give a comprehensive picture of your history, and one which is not biased by natural selection on any one gene.

We say 'your', but of course this method requires access to your personal genome sequence. One of the authors (Richard Dawkins) is fortunate enough to have had his DNA sequenced completely, for purposes of a Channel 4 television documentary (*Sex, Death and the Meaning of Life*). We will use this as a case study, but the main point is that any reader can do the same with their own genome in a few years, when sequencing becomes cheap enough. And who wouldn't want to examine the history stored in their personal genome?

Given the ancestral dates estimated in the Tasmanian's Tale, you might expect, as we did, that most of our personal coalescence dates would be recent. You would be wrong. European royal families notwithstanding, the vast majority of mother-to-father coalescences in a typical human occurred 20,000 years ago or more. In fact, a good number of the coalescence points exceed a million years in age (this does not negate the conclusion that the most recent 'person MRCA' is much more recent). Consequently, Dawkins DNA, or indeed the DNA of any single reader, can be used to triangulate right the way back to our African roots, and capture the history of much of humanity.

To see how this remarkable deduction works, understand that coalescence points echo the past size of populations. The smaller the population, the greater the probability of ancestral lines colliding, and higher the coalescence rate. So if many different parts of the genome coalesce at similar times, it indicates a small population, in extreme cases a 'bottleneck'. Conversely, a dearth of coalescences hints at a period of large population size. It follows that the frequency of Dawkins coalescences can be used to deduce an 'effective population size' at various points in the past, as shown on the plot overleaf.

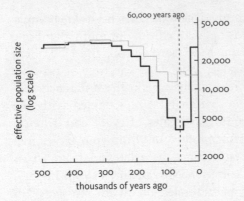

Reading the human story from your personal genome. The black line comes from dating the common ancestries (coalescences) of all sections of DNA I inherited from my (European) mother with those I inherited from my (European) father. The grey line is the equivalent for a typical Nigerian. The method, called PSMC, is due to Heng Li and Richard Durbin [247].

The pattern over time reveals a feature typical of all non-Africans: a dramatic drop in population size about 60,000 years ago. It is tempting, and probably correct, to read this as a migration of a few thousand pioneers out of Africa. While the genomes of sub-Saharan Africans show a dip too – the grey line shows a Nigerian – it is nowhere near as extreme. It is a striking fact that even if you were the last human alive on earth, much of the human story could be read from your personal genome alone.

If this is the story from one lowly individual, just imagine the power of comparing tens, hundreds, or thousands of people, and inferring full evolutionary trees for every location on their genomes. Sadly, this intensive calculation is well beyond today's computers. Nevertheless, shortcuts developed in Richard Durbin's lab, by his co-worker Stephan Schiffels, allow us to estimate something simpler: the most recent coalescence between a handful of people, for each location in their genomes. The diagram opposite shows the results of just such an analysis, based on nine indigenous people from around the globe. There are millions of coalescences within their combined genomes, which gives us a good picture of the history of human interbreeding and separation over hundreds of thousands of years. The horizontal lines on the diagram indicate periods where very few coalescence points are shared between peoples, when interbreeding must have happened, but only rarely. It is a powerful summary of where we came from.

It is this sort of full genome analysis, rather than mitochondrial or Y-chromosome trees, which is the true test of the RAO hypothesis. To emphasise this on the diagram, we have whitened-up the non-African branches. This branch of the tree does not start with a discrete split, unlike the clean separation of the Mexicans, which

A powerful summary of where we came from. Interpreting the pattern of most recent coalescences for sections of DNA across genomes from a variety of indigenous peoples. Horizontal lines indicate periods where we can trace only small amounts of shared ancestry between groups. Diagram from Schiffels and Durbin [374].

indicates the indigenous colonisation of the Americas. An African exodus did happen, perhaps 60,000 years ago, but it may not have been an instantaneous event. Limited interbreeding with other African populations may have carried on for tens of thousands of years after the initial separation. There are two additional findings which cast doubt on the pure ROA idea. The first involves the Europeans and Asians (specifically native Mexicans and Chinese), which on this diagram are shown as diverging recently, around 20,000 years ago. What the diagram doesn't show is that Schiffels and Durbin found a 100,000-year European/Asian separation in almost a tenth of those genomes. That date, well before the recent exodus, implies some sort of earlier migration too. The second finding is that a small portion of the northern European genome remains separate from Nigerians, even before 200,000 years ago. This fits in with the evidence of the next tale. Most Eurasians have a few per cent of their DNA that has been passed from Neanderthals or other ancient humans.

Fossils indicate that Modern anatomy passed to the rest of the world via recent out-of-Africa migrations. But it seems that a single, discrete African exodus is an oversimplification. This is revealed more clearly in our genes than in the fossils they left behind. True, over 95 per cent of the DNA of a typical non-African traces back to Africa within the most recent 100,000 years, but even that DNA has probably come through a few different routes. And the small remainder of the genome often conceals a more ancient and mysterious ancestry. This mix is perhaps not surprising, given that at any given time in the past each of us has a huge number of genealogical ancestors. A few thousand generations ago, the majority of them were African, regardless of who you are. But it is surely unsurprising to find a few ancestors outside Africa too? As we shall find out, those of us who come from outside Africa also have ancient Eurasian ancestors (possibly even going back to regional *H. erectus*). The evidence from Eve's Tale destroys a simple 'Adam and Eve'-like narrative of human history. Because we have so many ancestors in so many different places, it is impossible to trace human diversity back to a single population (let alone a single breeding pair!). Better to switch from people talk to gene talk: each piece of DNA traces an alternative history, and our ancestral journey is a tapestry woven by them all.

EPILOGUE TO EVE'S TALE

Eve's tale warns us not to draw general conclusions from just one stretch of DNA. It is all too easy to assume that a single gene tree can stand in for the history of a whole species. This even traps professional biologists, exemplified by the case of Albert Perry and Y-chromosome Adam. Albert Perry was an African-American former slave from South Carolina. He achieved posthumous fame in 2013 after his great-grandson's DNA was serendipitously submitted to a commercial genealogy company, and identified as a previously unknown and deeply divergent branch of the Y-chromosome tree. Albert Perry's ancient lineage was confirmed by finding the same Y chromosome in another of his male-line relatives, a great-great-grandson, through a different route. Subsequent genetic detective work at the University of Arizona also managed to locate a handful of Y chromosomes that lay on the same ancient branch. They came from a small group of modern-day men living in Southwest Cameroon, presumably the same region from where Mr Perry's unfortunate ancestors were abducted.

This recently discovered, deep branch on the Y-chromosome tree means we have to define a new 'Y-chromosome Adam'. In the time since Mr Perry's male lineage split from the rest of us, thousands of new mutations have accumulated on the various branches of the tree. In the original announcement of the find, a small subset of these mutations was used to date the split to 340,000 years ago. More recent analysis of whole Y chromosomes from around the world, including the rare Cameroonian type, places this coalescence and hence Y-Adam at around 240,000 years ago. Whichever estimate we take, it predates the appearance of modern humans, wherein lies the trap. Certain geneticists (who should have known better) have argued that this date is utterly implausible: how could a modern human gene, indeed a whole chromosome, predate the origin of modern humans themselves?

Needless to say, the argument has been seized upon by the creationist camp, eager to demonstrate that they possess an even deeper misunderstanding. Rest assured that not only are such deep divergences possible, they are to be expected. In fact, compared to other genes, these are not particularly old. An estimate made by Michael Blum and Mattias Jakobsson in 2011 found that the earliest divergence in a human gene tree typically dates to over a million years ago. That means many of our current-day genetic differences predate the origin of modern humans, often by hundreds of thousands of years. Once again, we come up against the difference between gene-level thinking and individual-level thinking.

Genetic differences can even trace back tens of millions of years. That's especially so when natural selection acts to preserve variety in the population. Here's an example. Suppose there are two blood types called A and B, which confer immunity to different diseases. Each blood type is susceptible to the disease against which the other type has immunity. Diseases flourish when the blood type that they can attack is abundant, because an epidemic can get going. So if B people, say, happen to be common in the population, the disease that hurts them will enjoy an epidemic. Consequently, B people will die until they cease to be common, and the A people increase – and vice versa. Whenever we have two types, the rarer of which is favoured because it is rare, it is a recipe for *polymorphism*: the positive maintenance of variety for variety's sake. The ABO blood group system is a famous polymorphism, long known and understood, which has probably been maintained for this kind of reason.

Some polymorphisms can be quite stable – so stable that they are maintained even while new species form. Our ABO polymorphism has been found in other apes, and even in many monkeys too. Some researchers argue that all these species have independently 'invented' the polymorphism, and for the same reason. But the most recent studies indicate that we all inherited it from a shared ancestor, and independently kept it going during millions of years of separate descent, presumably because those diseases, or something like them, have operated throughout that time. In fact, the genes responsible for the A and B blood types probably diverged before our split with the Old World monkeys, around 25 million years ago. This feature of evolution is called trans-specific polymorphism, and it shows conclusively that differences between humans can trace back to before humans themselves existed.

We can make an even more stunning deduction. For certain genes, you must be more closely related to some chimpanzees than to some humans. And I am closer to some chimpanzees than to you (or to 'your' chimpanzees). Humans as a species, as well as humans as individuals, are temporary vessels containing a mix of genes from different sources. Individuals are temporary meeting points on the crisscrossing routes that genes take through history. This is a tree-based way to express the central message of *The Selfish Gene*, my first book. As I put it there, 'When we have served our purpose we are cast aside. But genes are denizens of geological time: genes are for ever.' At the concluding banquet to a conference in America, I recited the same message in verse:

> *An itinerant selfish gene*
> *Said 'Bodies a-plenty I've seen.*
> *You think you're so clever*
> *But I'll live for ever.*
> *You're just a survival machine.'*

And, as the body's immediate reply to the gene, I parodied the very same 'Harp Song of the Dane Women' quoted previously:

> *What is a body that first you take her,*
> *Grow her up, and then forsake her,*
> *To go with the old blind watchmaker?*

This book is a natural history of DNA – a history of apparently lifeless instructions weaving different journeys through time. Prosaic though that sounds, it is anything but. Every generation, every day, every hour, these instructions are knitted into varied and exquisite life forms, which lift our pilgrimage from the pedestrian to the sublime.

PROLOGUE TO THE DENISOVAN'S TALE

Extracting DNA from ancient remains must surely rank among the most extraordinary technical feats of our age, the genetic equivalent of raising the dead. It is fraught with difficulties. DNA degrades over time, and even the fragments that remain will certainly have been chemically modified. Large numbers of microbes invade bones soon after death, mixing their genomes with the original. More insidiously, there is the ever-present problem of contamination from modern DNA which coats every surface and fills the air around us: fungal spores, drifting bacteria, droplet-borne viruses, flecks of human skin, and all the genetic detritus of the process of living. Incredibly strict and pernickety protocols have to be followed before we can reliably select, amplify, sequence and identify ancient DNA.

So, despite the appeal of cloning extinct life forms, it is not surprising that, until recently, entire genomes were rarely reassembled. Instead, most studies targeted small DNA regions. Commonly chosen, because there are thousands of copies per cell, was mitochondrial DNA, the female-line sequence that we met in Eve's Tale. Mitochondrial DNA was the first sequence to be extracted from Neanderthal bones. And not just any bones, but those of 'Neanderthal 1', the original skeleton from the Neander Valley. The results, released in 1997 by Svante Pääbo's lab in Germany,* were widely interpreted to mean that Neanderthals were an entirely separate lineage to modern humans. Their mitochondria differed from ours by an average of over 200 DNA letters, whereas modern humans differ from each other by at most about 100 letters. But remember the message of Eve's Tale. Different genes can tell different stories, and mitochondrial DNA is effectively just a single gene. This led us to strike a note of caution in the first edition of this book, when only mitochondrial DNA was available. Today we can tackle the issue comprehensively, thanks to the cool, DNA-preserving environment in which Neanderthals lived.

In 2010, Pääbo's lab released a fragmentary, draft version of the Neanderthal genome, based on bones from three Croatian specimens around 40,000 years old. Three years later, the same group released

---

* The Swedish geneticist Svante Pääbo has been the driving force behind the entire field of ancient human DNA. His popular book on the subject of this tale is entitled *Neanderthal Man: In Search of Lost Genomes*.

the full genome of a 50,000-year-old Siberian Neanderthal woman, and there are now a number of Neanderthal genomes available. These provide hundreds of thousands of times more data than just the mitochondrial DNA sequence.

Whole genomes give us a window into Neanderthal life. For example, similarities between the maternal and paternal DNA allow us to tell that the Siberian Neanderthal's parents were fairly incestuous: as closely related as (say) half-siblings or uncle and niece. Maybe they were driven to it by sheer lack of numbers? Just as we calculated ancestral population sizes from the Dawkins genome in Eve's Tale, we can also use single Neanderthal genomes to trace the gradual decline of the Neanderthals over hundreds of thousands of years. Over the same period, the population of our ancestors within Africa seems to have increased. We can also look for evidence of evolutionary changes by looking at mutations in known genes. To quote from a recent paper published by Pääbo's lab: 'we find that genes involved in skeletal morphology have changed more in the lineage leading to Neanderthals . . . whereas genes involved in behaviour and pigmentation have changed more on the modern human lineage'.

Understandably, however, popular coverage focuses on one question. Did we ever hybridise with Neanderthals? Have we inherited parts of our own genome from them? Neanderthal and modern human DNA are both just strings of letters, the vast majority of which are the same. How then can we identify any footprint of Neanderthals in our own genome? It's not as though we can simply take a modern sequence and pick out the regions that reek of Neanderthal.

There is a way, and it relies on the fact that Neanderthals lived in Eurasia. We look for new mutations in Neanderthals and living Eurasians that aren't seen in living Africans. These are indicative of recent shared ancestry. The starting point is to find DNA locations that vary among modern humans – where there are two slightly different versions of a sequence. Usually the difference is just a single letter mutation, a so-called 'SNP'.[*] There are many millions of common SNPs scattered over the human genome, on average one every 2,000 letters. They are often found in regions outside of genes, and they need not be functionally important. Here we are simply looking for neutral 'markers'.

[*] This helpful acronym is pronounced 'snip', and stands for 'Single Nucleotide Polymorphism'.

Finding single letters common to Neanderthals and (most) modern Eurasians is not quite enough to infer recent ancestry. It could be that they have the original version, and Africans a new one. To check this, we take the equivalent sequence in chimpanzees or gorillas, which usually has the original letter. This allows us to home in on new ('derived') mutations that link Eurasians to Neanderthals. The evidence is conclusive: human/Neanderthal interbreeding definitely occurred, although it wasn't an everyday event. Chunks of our genome – often tens or hundreds of thousands of DNA letters in length – show whole swathes of these shared, derived mutations.

These chunks of DNA are highly informative. For example, we can use their length to date the time of interbreeding. This relies on the fact that over time, long lengths of genome are chopped and swapped by recombination. The longer the intact sequence, the fewer generations undergone since interbreeding – the effect has been confirmed by looking at ancient DNA from Siberian contemporaries of the Neanderthals: modern-looking *Homo sapiens sapiens*, for whom Neanderthal interbreeding must have been a recent thing. These true Moderns have significantly longer Neanderthal regions in their genome. Such considerations put our Neanderthal hybridisation suggestively close to our recent African exodus, between 50,000 and 60,000 years ago.*

Another question we can ask is whether these common sections of DNA passed from Neanderthals into us or vice versa. We can do this by looking at different SNPs within the same sections of DNA. If the section originally came from a modern human, it should also contain other single letter changes – ones that accumulated in the line to modern Eurasians and Africans, but not in the ancient ancestry of Neanderthals. Very few of these are seen. Unexpectedly, practically all the DNA transfer is thought to be from Neanderthals into ancient humans. This estimate might change a little as more ancient genomes are sequenced. But if it is shown to hold more generally, might it suggest Neanderthals bred into the Moderns, with hybrid children brought up within the Modern tribe? Not necessarily. The prevailing explanation is that Neanderthal genes gained an advantage by riding

---

* New findings in this field seem to pop up every week. During the writing of this prologue, DNA from a 40,000-year-old man's jaw was analysed in the same way, showing that his ancestors had hybridised with Neanderthals only four to six generations previously – although this particular interbreeding event is thought to have been a dead end.

on the coat tails of expanding modern human populations. In contrast, Modern genes that passed into Neanderthals were probably lost as the Neanderthal population shrank.

There is another intriguing pattern too: regions of our genome are 'Neanderthal deserts', with scarcely a shared SNP in sight. The most striking case is the entire X chromosome. From hybrids between other mammal species, we know that many genes that affect fertility tend to be found on the X chromosome, and this usually affects males. It has been suggested that human/Neanderthal offspring, especially males, may have experienced fertility problems, which could also explain why we see less Neanderthal ancestry among genes that are mainly switched on in the testis. If this is true, our Neanderthal ancestry traces back mostly through hybrid women, not men.

Assuming that most of these sections of DNA have come to us from Neanderthals, rather than the other way round, how many neandergenes do we actually have? You can have your 'Neanderthal component' measured by commercial companies, but that will reveal a rather low amount. A modern European contains on average about 1.2 per cent Neanderthal DNA, a modern Asian about 1.4 per cent, and Africans have almost no neandergenes. But if we add all the Neanderthal regions together from many people, it covers about 20 per cent of the genome, and that estimate is expected to rise as we sequence more living humans. In fact, we may be carrying as much as 40 per cent of the Neanderthal genome, in various bits and pieces in different modern populations. The Neanderthals certainly didn't die out entirely. And it is hard now to think of them as a separate species to ourselves.

These deductions have all been made possible with just a single high-quality ancient Neanderthal genome. The ability to sequence ancient DNA has opened a miraculous new door into understanding our evolutionary past. It has also turned up new mysteries. Our long prologue has now equipped us to understand the most seductive mystery of them all.

THE DENISOVAN'S TALE

The Denisovan's Tale is short, as befits a set of people about whom we know so little. They are named after the Denisova Cave in the Altai mountains of Siberia, and the cave itself is named after Denis, an eighteenth-century resident hermit. Less than a decade ago, few

people would have heard of it, let alone known how to pronounce it.*
Now it is centre stage in debates surrounding recent human evolution. In 2009 Johannes Krause and Qiaomei Fu attempted to extract
DNA from one half of the tip of a 40,000-year-old finger bone,
excavated from deep under the cave floor. An archaeologist at the site
is reported to have described it as the 'most unspectacular fossil I've
ever seen'.

What did turn out to be spectacular, however, was both the degree
of DNA preservation, and the subsequent overturning of established
views. First to be sequenced was the mitochondrial DNA. This was
found to be distinct from both Moderns and Neanderthals. It lies on
a much deeper branch of the gene tree. A year or so later it was joined
by more mitochondrial DNA extracted from two molar teeth in
almost the same layer of the Denisova excavations. The teeth were
visibly larger than those of Neanderthals, more like molars in *Homo
erectus* or the earlier hominids† that we will greet further along in our
pilgrimage. Now that the fingertip has been pulverised for DNA
extraction, the two teeth constitute all the tangible evidence we have
of the Denisovans.

Although what we have described so far is titillating, it is thin
evidence for a new human subspecies. By now you should have a
sense of unease at reading too much into just one small piece of
DNA, such as that from a fossilised mitochondrion. Fortunately, in
the act of being destroyed, the Denisovan fingertip gave us something
extraordinary: an entire genome, not just the mitochondrial portion. This genome was so well-preserved that it has provided DNA
sequences as reliable as those from a living human. This certainly
makes the Denisovans worthy of a tale.

The Denisovans' whole genome is generally more Neanderthal-like
than their mitochondrial DNA, although it still supports the idea that
they were a separate subspecies. If Neanderthals split from Archaics
about 800,000 years ago, calculations put the Denisovan/Neanderthal
split at 640,000 years ago. From coalescence dating, their population
seems to have declined in the same way as Neanderthals, probably at

* de-NEE-sova and de-NEE-sovans, for the record.
† The useful word 'hominid' traditionally meant any species on or branching off the
human line – in other words, any species thought to be more closely related to us than to
chimpanzees, and this is how we use the word here. However, it has recently become
popular to call these 'hominins' and, confusingly, to use hominid more widely, to
include all the great apes and their fossil relatives.

the same time. The girl with the fingertip may have been among the last of her kind.

More surprising are comparisons with modern humans. We might expect a similar pattern to Neanderthals, who are known to have lived in Siberia too. Modern Eurasians, remember, have inherited a small but significant fraction of Neanderthal DNA. In contrast, apart from one group, modern humans show very little Denisovan DNA. The exception lies way to the south-east in . . . Oceania! Yes, thousands of miles from Siberia, and in a radically different climate, today the greatest density of Denisovan DNA is found in native Australians, New Guineans and Filipinos, and to a lesser extent in Polynesians and west Indonesian islanders. We're not just talking tiny quantities either: it can be up to 8 per cent of the genome. David Reich and colleagues, who published the Denisovan genome, also note that there are nearby East Indonesian populations with practically no Denisovan DNA, and they conclude that Denisovan DNA was not simply carried to Oceania by modern humans. Instead, so their argument goes, Denisovans themselves lived in Oceania, as well as (more provably) in Siberia. It seems that major human/Denisovan mating took place only in Oceania, but their intermingled descendants failed to return to mainland Asia, perhaps stopped by the Wallace line (an ecologically famous division caused by deep ocean straits, which we shall meet at the end of the Sloth's Tale). That's not to say there was literally no interbreeding elsewhere though. There are plausible claims that Tibetans, living between the two regions, inherited genes for high-altitude adaptation from Denisovans.

If that isn't enough in the way of surprises, let us finish with two more. The oldest *Homo* DNA recovered so far is from the Spanish Sima de Los Huesos site (we met it previously in connection with Neanderthals). The bones are ten times older than the genomes we have been talking about. Although highly fragmented, most of their DNA hints at a Neanderthal affiliation, as might be expected. Apart, that is, from the mitochondrial DNA: their mitochondrial sequences show affiliation with Denisovans instead. It hardly needs pointing out that Spain is a long way from Australia.

Most tantalising of all is the three-way comparison between Denisovans, Neanderthals and modern Africans. Remember that both Denisovan and Neanderthal genomes were calculated to be closer to each other (on average) than either were to modern Africans. But in some regions of the genome, Africans and Neanderthals are closer, and Denisovans are way out on a limb. The best explanation so far is that parts of the Denisovan genome trace back to a much older, non-African group (whisper *Homo erectus*). Can some modern South-East Asians trace ancestry directly back to Java Man, the very first ancient human fossil discovered?* Perhaps. Can we prove it? Well, in this fast-moving field, it's hard to predict what startling proofs lie around the corner. Such is the pace of change that what we write now will undoubtedly be outdated – or even overturned – in months rather than years, cautioning us to bring this tale to a close.

In the previous edition of this book, the Neanderthals told a tale. It has been replaced here by the Denisovan's Tale even though we know much less about them than we do about their Neanderthal cousins. But the Denisovans carry a pair of messages that are not specific to any one set of human ancestors. They represent our extraordinary new knowledge about our history, knowledge that can be gleaned from a mere fingertip. More importantly, they also represent our lack of knowledge. The human story is far more complex than we were previously led to believe; there are certainly many satisfying lifetimes of work ahead for students of human evolution. Almost as certain, we can expect to find other undiscovered ancient human subspecies. In hindsight, all this is unsurprising. It is the norm for animal species to exist in a continuum of groups, varieties, subspecies and species, with complex breeding patterns between them. At various stages in the past, humans are likely to have done the same. We shall see it in our close relatives the chimpanzees, in slightly more distant relatives such as the gibbons, and indeed all the way through the pilgrimage ahead.

EPILOGUE TO THE DENISOVAN'S TALE

Both tales in this chapter have focused on humans and their genes. But of course all species have family trees. All species inherit genetic material. All sexually reproducing species – and that's most of them – have their own Adam and Eve. Genes and gene trees are a ubiquitous feature of life on Earth. Cheetah DNA reveals a 12,000-year-old population bottleneck important to feline conservationists. Maize DNA has stamped upon it the unmistakable signature of its 9,000-year Mexican domestication. The coalescence patterns of HIV strains

---

* Not to mention the dwarf Indonesian 'hobbit' fossils, *Homo florensis*, whose status as modified *Homo erectus* is subject to such dispute that we will mention it no further.

can be used by epidemiologists and medical doctors to understand and contain the virus. Genes and gene trees reveal the history of the flora and fauna of the northern hemisphere: the vast migrations driven by ice ages whose waxing pushed temperate species into southern European refuges, and whose waning stranded Arctic species on isolated mountain ranges. All these events and more can be traced in the distribution of DNA around the globe.

So far, these genetic studies of other creatures have mainly looked at individual genes. But analysis of human DNA is blazing a trail of whole-genome techniques that promises to revolutionise all biology. Our children will gaze upon the evolution of life on Earth with fresh eyes. This is a certainty. Even with our own intensely studied genome, we know that our techniques are only scratching the surface. We know that there is more history contained in the shape and structure of the 'gene trees' at each point in a genome, and also in the identity of the pieces of DNA that have travelled down the generations. And we know that there are untold billions of genomes, of all manner of creatures, waiting to tell us their story.

I have already mentioned my phrase 'The Genetic Book of the Dead', which I used in *Unweaving the Rainbow*. There, I used it to dramatise the prophecy that DNA might eventually provide a detailed readout of the conditions under which our ancestors lived. The Denisovan's Tale unveils a slightly different interpretation of the same pages. Our modern reading of DNA allows us to date and place key events in the prehistory of human populations: their migrations, their partings and meetings, their bottlenecks and expansions. I am bowled over by the forensic power of today's statistical genetic techniques to unmask signals hitherto buried deep in the past. I forsee yet more stunning revelations, not just about demography and geography but also about the totality of ancestral lives, a digitally chronicled ancestor's tale.

ERGASTS

Moving deeper in time, we touch down again at 1 million years in search of ancestors. The only likely candidates of this age are of the type usually called *Homo erectus*, although some would call the African ones *Homo ergaster* and I shall follow them. In seeking an anglicised form for these creatures, I shall call them Ergasts rather than Erects, partly because I believe the majority of our genes trace back to the African form, and partly because, as I've already remarked, they were no more erect than their predecessors (*Homo habilis*) or their successors (us). Whatever name we prefer, the Ergast type persisted both in Eurasia and Africa from about 1.8 million until about a quarter of a million years ago. They are widely accepted as the immediate predecessors, and partial contemporaries, of the Archaics who are in turn the predecessors of us Moderns.

The Ergasts were noticeably different from modern *Homo sapiens*, and, unlike the Archaic *sapiens* people, they differed from us in some respects that show no overlap. Fossil finds show they lived in the Middle East and Far East including Java, and represent an ancient migration out of Africa. You may have heard them referred to by their old names of Java Man and Peking Man. In Latin, before they were admitted into the *Homo* fold, they had the generic names *Pithecanthropus* and *Sinanthropus*. They walked on two legs like us, but had smaller brains (averaging perhaps 800 cc in early specimens to over 1,000 cc in late ones), housed in lower, less domed, more 'swept-back' skulls than ours, and they had receding chins. Their jutting brow ridges made a pronounced horizontal ledge above the eyes, set in wide faces, with a pinching in of the skull behind the eyes.

At the risk of sounding obvious, we should point out that Ergasts in the same population could have markedly differed. The five oldest Ergast skulls outside Africa come from a single European cave in Dmanisi, Georgia. That implies they may well be from the same population, yet they vary drastically in size and shape, ranging from a pretty small 780 cc braincase right down to a disturbingly prognathous adult male, whose 546 cc brain is comparable to the australopithecine 'ape men' we will meet after the next chapter. Lest this variation sound extraordinary, we should point out that current humans have a range of roughly 1,000 to 1,500 cc.

Hair doesn't easily fossilise, so where to discuss the obvious fact that at some point in our evolution we lost most of our body hair, with the luxuriant exception of the tops of our heads? Barring the tantalising promise of skin and hair in hominid fossils, the best clue may come from human genetics. Researchers from the University of Utah have examined variation in the DNA letters of the melano-cortin I receptor, a gene involved in skin colour. They find a bout of natural selection in our African ancestors that dates to at least 1 million years ago. A hint, perhaps, that Ergasts had started to lose the body hair that shielded them from the hot African sun. But that doesn't necessarily mean they were as hairless as we are. And the link is tenuous enough that nobody should complain of an imaginative reconstruction as hairy as a chimpanzee, or any intermediate level of shagginess. Modern people, males at least, remain quite variable in how hairy they are. Hairiness is one of those characteristics that can increase or decrease in evolution again and again. Vestigial hairs, with their associated cellular support structures, lurk in even the barest-seeming skin, ready to evolve into a full coat of thick hair at short notice (or shrink again) should natural selection at any time call them out of retirement. Look at the woolly mammoths and woolly rhinoceroses that rapidly evolved in response to the recent ice ages in Eurasia. We shall return to the evolutionary loss of human hair in – strangely enough – the Peacock's Tale.

Subtle evidence of repeatedly used hearths suggests that at least some groups of Ergasts discovered the use of fire – with hindsight a momentous event in our history. The evidence is less conclusive than we might hope. Blackening from soot and charcoal does not survive immense timespans, but fires leave other traces that last longer. Modern experimenters have systematically constructed fires of vari-ous kinds and then examined them afterwards for their trace effects. It emerges that deliberately built campfires magnetise the soil in a way that distinguishes them from bushfires and from burnt-out tree stumps – I don't know why. But such signs provide evidence that Ergasts, both in Africa and Asia, had campfires nearly one and a half million years ago. This doesn't have to mean that they knew how to light fires. They could have begun by capturing and tending naturally occurring fire, feeding it and keeping it alive as one might look after a

Tamagotchi pet. Maybe, before they began to cook food, they used fires to scare away dangerous animals and provide light, heat, and a social focus.

The Ergasts also shaped and used stone tools, and presumably wooden and bone ones too. Nobody knows whether they could speak, and evidence is hard to come by. You might think that 'hard to come by' is an understatement, but we have now reached a point in our backward journey when fossil evidence starts to tell. Just as campfires leave traces in the soil, so the needs of speech call forth tiny changes in the skeleton: nothing so dramatic as the hollow bony box in the throat with which the howler monkeys of the South American forests amplify their stentorian voices, but still telltale signs such as one might hope to detect in a few fossils. Unfortunately, the signs that have been unearthed are not telltale enough to settle the matter, and it remains controversial.

There are two parts of the modern human brain which seem to go with speech. When in our history did these parts – Broca's Area and Wernicke's Area – enlarge? The nearest approach we have to fossil brains is endocasts, to be described in the Ergast's Tale. Unfortunately the lines dividing different regions of the brain do not fossilise very clearly, but some experts think they can say that the speech areas of the brain were already enlarged before 2 million years ago. Those who want to believe that Ergasts possessed the power of speech are encouraged by this evidence.

They are discouraged, however, when they move down the skeleton. The most complete *Homo ergaster* we know is the Turkana Boy, who died near Lake Turkana, in Kenya, about 1.5 million years ago. His ribs, and the small size of the portholes in the vertebrae through which the nerves pass, suggest that he lacked the fine control over breathing that seems to be associated with speech. Other scientists, studying the base of the skull, have concluded that even Neanderthals were speechless. The evidence is that their throat shape would not have allowed the full range of vowels that we deploy. On the other hand, as the linguist and evolutionary psychologist Steven Pinker has remarked, 'e lengeege weth e smell nember ef vewels cen remeen quete expresseve'. If written Hebrew can be intelligible without vowels, I don't see why spoken Neander or even Ergaster couldn't too. The veteran South African anthropologist Phillip Tobias suspects that language may predate even *Homo ergaster*, and he may just possibly be right. As we have seen, there are a few who go to the opposite extreme and date the origin of language to the Great Leap Forward, just a few tens of thousands of years ago.

This may be one of those disagreements that can never be resolved. All considerations of the origin of language begin by citing the Linguistic Society of Paris which, in 1866, banned discussion of the question because it was deemed unanswerable and futile. It may be difficult to answer, but it is not in principle unanswerable like some philosophical questions. Where scientific ingenuity is concerned, I am an optimist. Just as continental drift is now sewn up beyond all doubt, with multiple threads of convincing evidence, and just as DNA fingerprinting can establish the exact source of a bloodstain with a confidence that forensic experts could once only dream of, I guardedly expect that scientists will one day discover some ingenious new method of establishing when our ancestors started to speak.

Even I, however, have no hope that we shall ever know what they said to each other, or the language in which they said it. Did it begin with pure words and no grammar: the equivalent of an infant babbling nounspeak? Or did grammar come early and – which is not impossible and not even silly – suddenly? Perhaps the capacity for grammar was already deep in the brain, being used for something else like mental planning. Is it even possible that grammar, as applied to communication at least, was the sudden invention of a genius? I doubt it, but in this field I wouldn't rule anything out with confidence.

As a small step towards finding out the date at which language arose, some promising genetic evidence has appeared. A family code-named KE suffers from a strange hereditary defect. Out of approximately 30 family members spread over three generations, about half are normal, but 15 show a curious linguistic disorder, which seems to affect both speech and understanding. It has been called verbal dyspraxia, and it first shows itself as an inability to articulate clearly in childhood. Some authorities think the problem largely stems from co-ordinating rapid movements of the face and mouth. Others suspect the problem lies deeper, since it affects comprehension and written language too. What is clear is that the abnormality is genetic. Individuals either have it or they don't, and it is associated with a mutation of a gene, present in all vertebrates, called *FOXP2*. This gene encodes a 'transcription factor': a protein which binds onto other DNA and switches on and off other important genes, in this case in various parts of the brain. The particular mutation in the KE family disrupts this binding. It is fortunate that the affected individuals have a second, normal copy of *FOXP2* inherited from their other parent. Fortunate, because experiments have shown that mice which have two similarly mutant copies die very young, probably as a result of brain, lung and movement defects. This clearly shows that *FOXP2* is not simply a 'gene for language', as it is sometimes touted in the press. Nevertheless, the evidence of the KE family suggests that in humans *FOXP2* is important for the development of some part of the brain that is involved in language.

So, we naturally want to compare normal human *FOXP2* with the same gene in animals that lack language. A particularly intriguing pattern is seen when we look at the chain of amino acids that makes up the protein encoded by the *FOXP2* gene. The mouse and chimpanzee versions of the protein differ in only one amino acid out of 715 in the chain. The human version differs from both these animals in an additional two amino acids. You see what this might mean? Although humans and chimpanzees usually show rather few differences in their protein sequences, *FOXP2* is an exception. It is a gene which is active in the brain and which has evolved rapidly – and it seems uniquely – in humans in the short time since we split from chimpanzees. And one of the most important respects in which we differ from chimpanzees is that we have language and they don't. A gene that changed somewhere along the line towards us, but after the separation from chimpanzees, is exactly the sort of gene we should be looking for if we are trying to understand the evolution of language. And it is the very same gene that has mutated in the unfortunate KE family (and also, in a different way, in a wholly unrelated individual with the same kind of language defect).

We must caution that a few mutations in a single gene are unlikely to suddenly produce a new and highly adaptive function, such as the ability to learn language. There are surely other evolutionary changes in the 'downstream' genes that *FOXP2* controls.* But *FOXP2* is very likely to be one of the key players. And since the first edition of this book, we have been able to look at the Neanderthal and Denisovan

* Experiments have been done to create genetically modified mice with the 'human' version of the *FOXP2* protein, but with all other genes being the normal 'mouse' versions. Of course, these 'humanized' mice don't end up with anything like human language. But it does affect their pattern of brain development. Their squeaks end up rather different, and by some measures more complex. They also seem to be able to switch more easily between conscious and unconscious learning, as tested by the ability to solve certain types of maze.

genome sequences, and find that they too shared both our *FOXP2*
mutations – a good indication that their brains were wired in a way
that might assist the evolution of language. Whether they actually
spoke is another matter, but it looks as though they had at least a
major component of the equipment for language. Since we have their
entire genomes, perhaps in the future we will be able to understand
the genetics of brain development enough to reconstruct their
neurological capabilities and know definitively whether they spoke.
This is an amazing thought, but may not help us resolve whether
their ancestors the Ergasts could speak, unless future technology
manages to extract an Ergast genome from ancient bones (wouldn't
that be fantastic?).

There is, however, another way we can use the genetic evidence
from *FOXP2* to look at the origin of language in our ancestors. It uses
the idea of 'genetic triangulation' that we met in the General Prologue. The obvious approach would be to triangulate backwards
from variants among modern humans, and try to calculate the
antiquity of our changes to the *FOXP2* gene. But with the exception
of rare unfortunates like the members of the KE family, there is no
variation among humans in any of the *FOXP2* amino acids. So there
isn't enough variation there to triangulate from. Luckily, however,
our genes are cut up into different regions. Sections that are translated into parts of the protein ('exons') are interspersed with regions
that are never translated ('introns'). Most of the sequence in an
intron is unused, and therefore free to mutate without natural selection 'noticing': it provides a set of 'silent' code letters. The silent
letters, unlike the expressed ones, are quite variable among individual
humans, and between humans and chimpanzees. Good triangulation
fodder! We can get some understanding of the evolution of the gene if
we look at the patterns of variation in the silent areas. Even though
the silent letters are not subject to natural selection themselves, they
can be swept along by selection of neighbouring exons. Even better,
the mathematically analysed pattern of variation in the silent introns
gives a good indication of *when* the sweeps of natural selection
occurred, and where in the gene the selection was acting.

Coupled with the genetic variation seen in Neanderthals, this can
give us a powerful tool for looking at natural selection in the past. Yet
the picture is more complex than once thought. The most recent
episode of selection on the *FOXP2* gene was less than 200,000 years
ago, possibly even less than 50,000 years ago: rather difficult to

reconcile with the fact that Neanderthals have the modern human version of the *FOXP2* protein. The current suggestion is that these selective sweeps were not focused on the two amino acid changes, but on another region of the DNA, specifically within a rare, useful part of an intron – in this case a small region containing a new mutation that affects when *FOXP2* itself is switched on. If true, this opens up another can of worms, because not only do Neanderthals have the older version, but so do a minority of Africans, and presumably they can speak perfectly normally. The traditional caution that 'more research is needed' is clearly warranted. Nevertheless, these ingenious genetic techniques boost my optimism that one day science will find a way to confound the pessimists of the Linguistic Society of Paris.

Homo ergaster is the first fossil ancestor we have met on our pilgrimage who is unequivocally of a different species from ourselves. We are about to embark on a portion of the pilgrimage in which fossils provide the most important evidence, and they will continue to bulk large – though they will never overwhelm molecular evidence – until we reach extremely ancient times and relevant fossils start to peter out. It is a good moment to look in more detail at fossils, and how they are formed. The Ergast will tell the tale.

THE ERGAST'S TALE

Richard Leakey movingly describes the discovery, by his colleague Kimoya Kimeu on 22 August 1984, of the Turkana Boy (*Homo ergaster*), at 1.5 million years the oldest near-complete hominid skeleton ever found. Equally moving is Donald Johanson's description of the older, and unsurprisingly less complete, australopithecine familiarly known as Lucy. The discovery of 'Little Foot' is just as remarkable (see page 106). Whatever freak conditions blessed Lucy, 'Little Foot', and the Turkana Boy with their version of immortality, would we not wish it for ourselves when our time comes? What hurdles must we cross to achieve this ambition? How does any fossil come to be formed? This is the subject of the Ergast's Tale. To begin, we need a small digression into geology.

Rocks are built of crystals, though these are often too small for the unaided eye to see. A crystal is a single giant molecule, its atoms arranged in an orderly lattice with a regular spacing pattern repeated billions of times until, eventually, the edge of the crystal is reached. Crystals grow when atoms come out of the liquid state and build up on the expanding edge of an existing crystal. The liquid is usually water. On other occasions, it is not a solvent at all but the molten mineral itself. The shape of the crystal, and the angles at which its plane facets meet, is a direct rendition, in the large, of the atomic lattice. The lattice shape is sometimes projected very large indeed, as in a diamond or amethyst whose facets betray to the naked eye the three-dimensional geometry of the self-assembled atomic arrays. Usually, however, the crystalline units of which rocks are made are too small for the eye to detect them, which is one reason why most rocks are not transparent. Among important and common rock crystals are quartz (silicon dioxide), feldspars (mostly silicon dioxide again, but some of the silicon atoms are replaced by aluminium atoms), and calcite (calcium carbonate). Granite is a densely packed mixture of quartz, feldspar and mica, crystallised out of molten magma. Limestone is mostly calcite, sandstone mostly quartz, in both cases ground small and then compacted from sediments of sand or mud.

Igneous rocks begin as cooled lava (which in turn is molten rock). Often, as with granite, they are crystalline. Sometimes their shape may be visibly that of a glass-like solidified liquid and, with great good fortune, molten lava may sometimes be cast in a natural mould, such as a dinosaur's footprint or an empty skull. But the main usefulness of igneous rock to historians of life is in dating. As we shall see in the Redwood's Tale, the best dating methods are available for igneous rocks alone. Fossils usually cannot be precisely dated themselves, but we can look for igneous rocks in the vicinity. We then either assume that the fossil is contemporaneous, or we seek two datable igneous samples that sandwich our fossil and fix upper and lower bounds to its date. This sandwich dating is open to the slight risk that a corpse has been carried by floodwater, or by hyenas or their dinosaur equivalents, to an anachronistic site. With luck this will usually be obvious; otherwise we have to fall back on consistency with a general statistical pattern.

Sedimentary rocks such as sandstone and limestone are formed from tiny fragments, ground by wind or water from earlier rocks or other hard materials such as shells. They are carried in suspension, as sand, silt or dust, and deposited somewhere else, where they settle and compact themselves over time into new layers of rock. Most fossils lie in sedimentary beds.

It is in the nature of sedimentary rock that its materials are continually being recycled. Old mountains such as the Scottish

Highlands have been slowly ground down by wind and water, yielding materials which later settle into sediments and may ultimately push up again somewhere else as new mountains like the Alps, and the cycle resumes. In a world of such recycling, we have to curb our importunate demands for a continuous fossil record to bridge every gap in evolution. It isn't just bad luck that fossils are often missing, but an inherent consequence of the way sedimentary rocks are made. It would be positively worrying if there were no gaps in the fossil record. Old rocks, with their fossils, are actively being destroyed by the very process that goes to make new ones.

Often fossils are formed when mineral-charged water penetrates the fabric of a buried creature. In life, bone is porous and spongy, for good engineering and economic reasons. When water seeps through the interstices of a dead bone, minerals are slowly deposited as the ages pass. I say slowly almost as a ritual, but it isn't always slow. Think how fast a kettle furs up. On an Australian beach I once found a bottletop embedded in stone. But the process usually is slow. Whatever the speed, the stone of a fossil eventually takes on the shape of the original bone, and that shape is revealed to us millions of years later, even if – which doesn't always happen – every atom of the original bone has disappeared. The petrified forest in the Painted Desert of Arizona consists of trees whose tissues were slowly replaced by silica and other minerals leached out of groundwater. Two hundred million years dead, the trees are now stone through and through, but many of their microscopic cellular details can still be clearly seen in petrified form.

I've already mentioned that sometimes the original organism, or a part of it, forms a natural mould or imprint from which it is subsequently removed, or dissolved. I fondly recall two happy days in Texas in 1987 spent wading through the Paluxy River examining, and even putting my feet in, the dinosaur footprints preserved in its smooth limestone bed. A bizarre local legend grew up that some of these are giant manprints contemporary with undoubted dinosaur prints, and in consequence the nearby town of Glen Rose became home to a thriving cottage industry, artlessly faking giant manprints in blocks of cement (for sale to gullible creationists who know, all too well, that 'There were giants in the earth in those days': Genesis 6:4). The story of the real footprints has been carefully worked out, and is fascinating. The obviously dinosaurian ones are three-toed. The ones that look faintly like a human foot have no toes, and were made by dinosaurs walking on the back of the foot rather than running on their toes. Also, the viscous mud would have tended to ooze back in at the sides of the footprint, obscuring the side toes of the dinosaurs.

More poignant for us are the rare traces of real hominids. In 2013, the North Sea exposed – and then washed away – a set of nondescript footprints on the shore of Happisburgh, in the English county of Norfolk. Fortunately, in the two weeks before they were eroded to nothingness scientists photographed the footprints in detail, confirming that they we made by adults and children of a human species almost a million years ago. How astonishing that the geological record contains such captured moments, and how disappointing that most are destroyed, unnoticed. We are fortunate indeed that occasionally more permanent traces form, such as the 3.6-million-year-old tracks at Laetoli in Tanzania (see plate 2). Here, the companionable footprints of three hominids, probably *Australopithecus afarensis*, stroll together in what was then fresh volcanic ash. Who does not wonder what these individuals were to each other, whether they held hands or even talked, and what forgotten errand they shared in a Pliocene dawn?

Sometimes, as I mentioned when discussing lava, the mould may become filled with a different material, which subsequently hardens to form a cast of the original animal or organ. I am writing this on a table in the garden whose top is a six-inch thick, seven-foot square slab of Purbeck sedimentary limestone, of Jurassic age, perhaps 150 million years old.* Along with lots of fossil mollusc shells, there is an alleged (by the distinguished and eccentric sculptor who procured it for me) dinosaur footprint on the underside of the table, but it is a footprint in relief, standing out from the surface. The original footprint (if indeed it is genuine, for it looks pretty nondescript to me) must have served as a mould, into which the sediment later settled. The mould then disappeared. Much of what we know about ancient brains comes to us in the form of such casts: 'endocasts' of the insides of skulls, often imprinted with surprisingly full details of the brain surface itself.

Less frequently than shells, bones or teeth, soft parts of animals sometimes fossilise. The most famous sites are the Burgess Shale of the Canadian Rockies, and the slightly older Chengjiang in South

---

* A journalist interviewed me at this two-tonne megalith for over an hour and then described it in his newspaper as a 'white wrought-iron table': my favourite example of the fallibility of eyewitness evidence.

THE ERGAST'S TALE 91

China which we shall meet again in the Velvet Worm's Tale. At both these sites, fossils of worms and other soft, boneless and toothless creatures (as well as the usual hard ones) wonderfully record the Cambrian Period, more than half a billion years ago. We are outstandingly lucky to have Chengjiang (see plate 3) and the Burgess Shale. Indeed, as I have already remarked, we are pretty lucky to have fossils at all, anywhere. It has been estimated that 90 per cent of all species will never be known to us as fossils. If that is the figure for whole species, just think how few individuals can ever hope to achieve the ambition with which the tale began, and end up as fossils. One estimate puts the odds at one in a million among vertebrates. That sounds high to me, and the true figure must be far less among animals with no hard parts.

HABILINES

Back another million years from *Homo ergaster*, 2 million years ago there is no longer any doubt in which continent our genetic roots lie. Everyone agrees, 'multiregionalists' included, that Africa is the place. The most compelling fossil bones at this age are normally classified as *Homo habilis*. Some authorities recognise a second, very similar contemporary type, which they call *Homo rudolfensis*. Others equate it with *Kenyapithecus*, described by the Leakey team in 2001. Yet others claim that they do not really fall outside the range of variation seen in the Ergasts, and all should be called 'Early *Homo*'. As usual I shan't take a stand on names. What matters is the real flesh-and-bone creatures themselves, and I shall use 'Habilines' as an anglicism for all of them. Habiline fossils, being older, are understandably less plentiful than Ergasts. The best-preserved skull bears the reference number KNM-ER 1470 and is widely known as Fourteen Seventy. It lived about 1.9 million years ago.

The Habilines were about as different from the average Ergasts as those Ergasts from us, and, as we should expect, there were inter-mediates which are hard to classify. Habiline skulls tend to be less robust than Ergast skulls, and lack the pronounced brow ridges. In this respect, Habilines were more like us. This should cause no surprise. Robustness and brow ridges are peculiarities that, possibly like hair, hominids seem able to acquire and lose again at the drop of an evolutionary hat.

Habilines mark the place in our history where the brain, that most dramatic of human peculiarities, starts to expand. Or more accur-ately, starts to expand beyond the normal size of the already large brains of other apes. This distinction, indeed, is the rationale for placing the Habilines in the genus *Homo* at all. For many palaeontol-ogists, the large brain is the distinguishing feature of our genus. Habilines, with their brains pushing the 750 cc barrier, have crossed the rubicon and are human.

As readers may soon become tired of hearing, I am not a lover of rubicons, barriers and gaps. In particular, there is no reason to expect an early Habiline to be separated from its predecessor by a bigger gap than from its successor. It might seem tempting because the predecessor has a different generic name (*Australopithecus*) whereas the successor *(Homo ergaster)* is 'merely' another *Homo*. It is true that when we look at living species, we expect members of different genera to be less alike than members of different species within the same genus. But it can't work like that for fossils, if we have a continuous historical lineage in evolution. At the borderline between any fossil species and its immediate predecessor, there must be some individuals about whom it is absurd to argue, since the *reductio* of such an argument must be that parents of one species gave birth to a child of the other. It is even more absurd to suggest that a baby of the genus *Homo* was born to parents of a completely different genus, *Australopithecus*. These are evolutionary regions into which our zoological naming conventions were never designed to go.*

Setting names to one side frees us for a more constructive discussion about *why* the brain suddenly started to enlarge. How would we measure the enlargement of the hominid brain and plot a graph of average brain size against geological time? There is no problem about the units in which we measure time: millions of years. Brain size is harder. Fossil skulls and endocasts allow us to estimate brain size in cubic centimetres, and it is easy enough to convert this to grams. But absolute brain size is not necessarily the measure you want. An elephant has a bigger brain than a person, and it isn't just vanity that makes us think we are brainier than elephants. *Tyrannosaurus's* brain was not much smaller than ours, but all dinosaurs are regarded as small-brained, slow-witted creatures. What makes us cleverer is that we have bigger brains *for our size* than dinosaurs. But what, more precisely, does 'for our size' mean?

There are mathematical methods of correcting for absolute size, and expressing an animal's brain size as a function of how big it 'ought to be' given its body size. This is a topic worthy of a tale in its own right, and *Homo habilis*, handyman, from his uneasy vantage point straddling the brain-size 'rubicon', will tell it.

---

*The 750 cc rubicon for the definition of *Homo* was originally chosen by Sir Arthur Keith. As Richard Leakey tells us in *The Origin of Humankind*, when Louis Leakey first described *Homo habilis* his specimen had a brain capacity of 650 cc, and Leakey actually moved the rubicon to accommodate it. Later specimens of *Homo habilis* retrospectively vindicated him by turning in figures closer to 800 cc. It's not just brains. More recently a much older jawbone, from 2.8 million years ago, has been claimed as the 'earliest *Homo* species', on the basis of the relative size of the molars, even though other features of the jaw are more australopithecine. All grist to my anti-rubicon mill.

THE HANDYMAN'S TALE

We want to know whether the brain of a particular creature such as *Homo habilis* is larger or smaller than it 'ought' to be, given that animal's body size. We accept (slightly unwillingly in my case but I'll let it pass) that large animals just have to have large brains and small animals small brains. Making allowance for this, we still want to know whether some species are 'brainier' than others. So, how do we make allowance for body size? We need a reasonable basis for calculating the expected brain size of an animal from its body size, so that we can decide whether the actual brain of a particular animal is larger or smaller than expected.

In our pilgrimage to the past, we happen to have met the problem in connection with brains, but similar questions can arise with respect to any part of the body. Do some animals have larger (or smaller) hearts, or kidneys, or shoulder-blades than they 'ought' to have for their size? If so, this might suggest that their way of life makes special demands on the heart (kidney or shoulder-blade). How do we know what size any bit of an animal 'ought' to be, given that we know its total body size? Note that 'ought to be' doesn't mean 'needs to have for functional reasons'. It means 'would be expected to have, knowing what comparable animals have'. Since this is the Handyman's Tale, and since the Handyman's most surprising feature is his brain, we'll go on using brains for the sake of discussion. The lessons we learn will be more general.

We begin by making a scatter plot of brain mass against body mass for a large number of species. Each symbol in the graph on the opposite page (from my colleague the distinguished anthropologist Robert Martin) represents one species of living mammal – 309 of them, ranging from the smallest to the largest. In case you are interested, *Homo sapiens* is the point with the arrow, and the one immediately next to us is a dolphin. The heavy black line drawn through the middle of the points is the straight line that, according to statistical calculation, gives the best fit to all the points.*

A slight complication, which will make sense in a moment, is that things work better if we make the scales of both axes logarithmic, and that is how this graph was made. We plot the logarithm of

---

* It is the line that minimises the sum of the squares of the distances of the points from it.

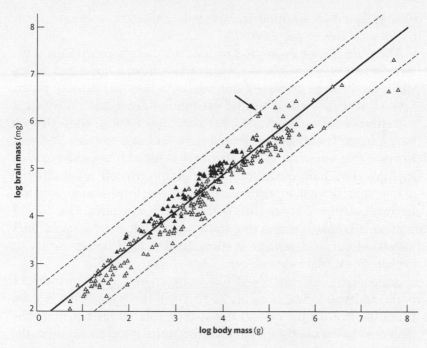

Log-log plot of brain mass against body mass for different species of placental mammal, with primates as filled triangles. Adapted from Martin [268].

an animal's brain mass against the logarithm of its body mass. Logarithmic means that equal steps along the bottom of the graph (or equal steps up the side) represent *multiplications* by some fixed number, say ten, rather than additions of a number, as in an ordinary graph. The reason ten is convenient is that we can then think of a logarithm as a count of the number of noughts. If you have to multiply a mouse's mass by a million to get an elephant's, this means you have to add six noughts to the mouse's mass: you have to add six to the logarithm of the one, to get the logarithm of the other. Half way between them on the logarithmic scale – three noughts – lies an animal that weighs a thousand times as much as a mouse, or a thousandth of an elephant: a person, perhaps. Using round numbers like a thousand and a million is just to make the explanation easy. 'Three and a half noughts' means somewhere between a thousand and ten thousand. Note that 'half way between' when we are counting noughts is a very different matter from half way between when we are counting grams. This is all taken care of automatically by looking up the logarithms of the numbers. Logarithmic scales call on a different kind of intuition from simple arithmetic scales, which is useful for different purposes.

There are at least three good reasons for using a logarithmic scale. First, it makes it possible to get a pygmy shrew, a horse and a blue whale on the same graph without needing a hundred yards of paper. Second, it makes it easy to read off multiplicative factors, which is sometimes what we want to do. We don't just want to know that we have a bigger brain than we should have for our body size. We want to know that our brain is, say, six times as big as it 'should' be. Such multiplicative judgements can be read directly off a logarithmic graph: that is what logarithmic means. The third reason for pre-ferring logarithmic scales takes a little longer to explain. One way of putting it is that it makes our scatterpoints fall along straight lines instead of curves, but there is more to it than that. Let me try to explain to my fellow dysnumerics.

Suppose you take an object like a sphere or a cube, or indeed a brain, and you inflate it evenly so it is still the same shape but ten times the size. In the case of the sphere, this means ten times the diameter. In the case of the cube, or the brain, it means ten times the width (and height and depth). In all these cases of proportionate scaling up, what will happen to the volume? It will not be ten times as great – it will be a thousand times as great! You can prove it for cubes if you imagine stacking sugar lumps. The same applies to uniformly inflating any shape you like. Multiply length by ten and, provided the shape doesn't change, you automatically multiply volume by a thou-sand. In the special case of a tenfold inflation, this is equivalent to adding three noughts. More generally, volume is proportional to the third power of length, and the logarithm is multiplied by three.

We can do the same sort of calculation for area. But area increases in proportion to the second power of length rather than the third power. Not for nothing is raising to the second power called squaring while raising to the third power is called cubing. The volume of a sugar lump determines how much sugar there is, and what it costs. But how fast it dissolves will be determined by its surface area (not a simple calculation because, as it dissolves, the remaining surface area will shrink more slowly than the volume of sugar remaining). When you uniformly inflate an object by doubling its length (width, etc.), you multiply the surface area by $2 \times 2 = 4$. Multiply its length by ten, and you multiply the surface area by $10 \times 10 = 100$ or add two noughts to the number. The logarithm of area increases as double the logarithm of length, while the logarithm of volume increases as treble the logarithm of length. A two-centimetre sugar lump will contain eight times as much sugar as a one-centimetre lump, but it will release that sugar into the tea only four times as fast (at least initially), because it is the surface of the lump that is exposed to the tea.

Now imagine that we make a scatter plot of sugar lumps of a wide range of sizes, with mass of lump (proportional to volume) along the bottom axis, and (initial) rate of dissolving up the side of the graph (assumed proportional to area). In a non-logarithmic graph, the points will fall along a *curved* line, which will be quite hard to interpret and not very helpful. But if we plot the logarithm of mass against the logarithm of initial dissolving rate, we shall see something much more informative. For every threefold increment of log mass, we shall see a doubling of log surface. On the log-log scale, the points will not fall along a curve, they will fall along a straight line. What is more, the *slope* of the straight line will mean something very precise. It will be a slope of two-thirds: for every two steps along the area axis, the line takes three steps along the volume axis. For every doubling of the logarithm of area, the logarithm of volume is tripled. Two-thirds is not the only informative slope of line we might see in a log-log plot. Plots of this kind are informative because the slope of the line gives us an intuitive feel for what is going on vis-à-vis such things as volumes and areas. And volumes and areas and the complicated relationships between them are extremely important in understanding living bodies and their parts.

I am not particularly mathematical – that's putting it mildly – but even I can see the fascination of this. And it gets better, because the same principle works for all shapes, not just tidy ones like cubes and spheres, but complicated shapes like animals and bits of animals such as kidneys and brains. All that is required is that size change should come about by simple inflation or deflation without a change of shape. This gives us a sort of null-expectation, against which to compare real measurements. If one species of animal is 10 times the length of another, its mass will be 1,000 times as great, *but only if the shapes are the same.* In fact, shape is very likely to have evolved to be systematically different as you go from small animals to large, and we can now see why.

Big animals need to be a different shape from small animals, if only because of the area/volume scaling rules we have just seen. If you turned a shrew into an elephant just by inflating it, retaining the same shape, it wouldn't survive. Because it is now about a million times heavier, a whole lot of new problems arise. Some of the problems an animal faces depend upon volume (mass). Others depend on area. Still others depend on some complicated function of the two, or on some different consideration altogether. Like a sugar lump's rate of dissolving, an animal's rate of losing heat, or of losing water through the skin, will be proportional to the area that it presents to the outside world. But its rate of generating heat is probably more related to the number of cells in the body, which is a function of volume.

A shrew scaled up to elephant size would have spindly legs that would break under the strain, and its slender muscles would be too weak to work. The strength of a muscle is proportional not to its volume but to its cross-sectional area. This is because muscular movement is the summed movement of millions of molecular fibres, sliding past each other in parallel. The number of fibres you can pack into a muscle depends upon the area of its cross-section (second power of linear size). But the task that the muscle has to perform – supporting an elephant, say – is proportional to the mass of the elephant (third power of linear size). So, the elephant *needs* proportionately more muscle fibres than a shrew, in order to support its mass. Therefore the cross-sectional area of elephant muscles needs to be larger than you'd expect from simple scaling up, and the volume of muscle in an elephant must be more than you'd expect from simple scaling up. For different particular reasons, the conclusion is similar for bones. This is why large animals like elephants have massive tree-trunk shaped legs.

Suppose an elephant-sized animal is 100 times as long as a shrew-sized animal. With no change of shape, the area of its outer skin would be 10,000 times as great as the shrew's and its volume and mass a million times as great. If touch-sensitive cells are equally spaced through the skin, the elephant will need 10,000 times as many of them, and the part of the brain that services them will perhaps need to be scaled in proportion. The total number of cells in the elephant's body will be a million times as great as in the shrew, and they'll all have to be serviced by capillary blood vessels. What does this do to the number of miles of blood vessel that we expect in a large animal, as distinct from a small one? That's a complicated calculation, and one that we'll return to in a later tale. For the moment, it is enough for us to understand that when we calculate it we cannot ignore these scaling rules for volumes and areas. And the logarithmic plot is a good method for getting intuitive clues to such things. The main conclusion is that, as animals get larger or smaller in evolution, we positively expect their shape to change in predictable directions.

We got into this through thinking about brain size. We can't just compare our brains with those of *Homo habilis, Australopithecus* or any other species without making allowance for body size. We need some index of brain size which makes allowance for body size. We can't divide brain size by body size, though that would be better than just comparing absolute brain sizes. A better way is to make use of the logarithmic plots we have just been discussing. Plot the logarithm of brain mass against the logarithm of body mass for lots of species of different sizes. The points will probably fall around a straight line, as indeed they do in the graph on page 95. If the slope of the line is $\frac{1}{1}$ (brain size exactly proportional to body size) it will suggest that each brain cell is capable of servicing some fixed number of body cells. A slope of $\frac{2}{3}$ would suggest that brains are like bones and muscles: a given volume of body (or number of body cells) demands a certain surface area of brain. Some other slope would need yet a different interpretation. So, what is the actual slope of the line?

It is neither $\frac{1}{1}$ nor $\frac{2}{3}$ but something in between. To be exact, it is a remarkably good fit to $\frac{3}{4}$. Why $\frac{3}{4}$? Well, that is a tale in itself, which will be told, as you will no doubt have guessed, by the cauliflower (well, a brain does *look* a bit like a cauliflower). Without pre-empting the Cauliflower's Tale, I will just say that the $\frac{3}{4}$ slope is not special to brains, but crops up all over the place in all sorts of living creatures, including plants like cauliflowers. Applied to brain size, and with the intuitive rationale that must wait for the Cauliflower's Tale, this observed line, with its $\frac{3}{4}$ slope, is the meaning we are going to attach to the word 'expect' as it was used in the opening paragraphs of this tale.

Although the points cluster about the 'expected' straight line of slope $\frac{3}{4}$, not all the points fall exactly on the line. A 'brainy' species is one whose point on the graph falls above the line. Its brain is larger than 'expected' for its body size. A species whose brain is smaller than 'expected' falls below the line. The distance above, or below, the line, is our measure of how *much* bigger than 'expected', or smaller, it is. A point that falls exactly on the line represents a species whose brain is exactly the size expected for its body size.

Expected on what assumption? On the assumption that it is typical of the set of species whose data contributed to calculating the line. So, if the line was calculated from a representative range of land vertebrates, from geckos to elephants, the fact that all mammals fall above the line (and all reptiles below) means that mammals have bigger brains than you would 'expect' of a typical vertebrate. If we calculate a separate line from a representative range of mammals, it will be parallel to the vertebrate line, still with a slope of $\frac{3}{4}$, but its absolute height will be higher. A separate line calculated from a representative range of primates (monkeys and apes) will be higher again, but still parallel with a slope of $\frac{3}{4}$. And *Homo sapiens* is higher than any of them.

The human brain is 'too' big, even by the standards of primates, and the average primate brain is too big by the standards of mammals generally. For that matter, the average mammal brain is too big by the standards of vertebrates. Another way to say all this is that the scatter of points in the vertebrate graph is wider than the scatter of points on the mammal graph, which is in turn wider than the primate scatter which it includes. The xenarthran scatter of points on the graph (xenarthrans are an order of South American mammals, including sloths, anteaters and armadillos) sits below the average of mammals, of which the xenarthran scatter forms a part.

Harry Jerison, the father of fossil brain size studies, proposed an index, the Encephalisation Quotient or EQ, as a measure of how much bigger, or smaller, the brain of a particular species is than it 'should' be for its size, given that it is a member of some larger grouping, such as the vertebrates or the mammals. Notice that the EQ requires us to specify the larger group which is being used as the baseline for comparison. The EQ of a species is its distance above, or below, the average line for the specified larger grouping. Jerison thought the slope of the line was $\frac{2}{3}$, whereas modern studies agree that it is $\frac{3}{4}$, so Jerison's own estimates of EQ have to be amended accordingly, as was pointed out by Robert Martin. When this is done, it turns out that the modern human brain is about six times as big as it should be, for a mammal of equivalent size (the EQ would be larger, if calculated against the standard of the vertebrates as a whole, rather than the mammals as a whole. And it would be smaller if calculated against the standard of primates as a whole).[*] A modern chimpanzee's brain is about twice the size it should be for a typical mammal, and so are the brains of australopithecines. *Homo habilis* and *Homo erectus*, the species that are probably intermediate in evolution between *Australopithecus* and ourselves, are also intermediate in brain size. Both have an EQ of about 4, meaning that their brains were about four times as big as they should have been for a mammal of equivalent size.

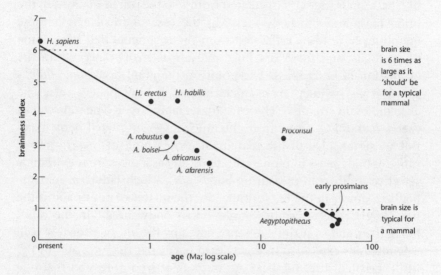

Plot of EQ or 'braininess index' for various fossil species, against time in millions of years on a log scale. The results have been corrected for a slope of ¾ for the reference baseline (see text).

The graph above shows an estimate of EQ, the 'braininess index', for various fossil primates and ape-men, as a function of the time at which they lived. With considerable pinches of salt you could read it as a rough graph of decreasing braininess as we go backwards in evolutionary time. At the top of the graph is modern *Homo sapiens* with an EQ of 6, meaning our brain is six times as heavy as it 'should' be for a typical mammal of our size. At the bottom of the graph are

[*] Much the same applies to IQ. It is *not* an absolute measure of intelligence. Rather, your IQ reflects how much more (or less) intelligent you are than the average for a particular population, that average being standardised at 100. My IQ if standardised against the background population of Oxford University would be lower than if standardised against the background population of England. Hence the joke about the politician lamenting the fact that half the population has an IQ less than 100.

fossils who might possibly represent something like Concestor 5, our common ancestor with the Old World monkeys. Their estimated EQ was about 1, meaning they had a brain which would be 'about right' for a typical mammal of their size today. Intermediate on the graph are various species of *Australopithecus* and *Homo* who might be close to our ancestral line at the time they lived. The drawn line is, once again, the straight line which best fits the points on the graph.

I advised pinches of salt, and let me raise that to ladles of salt. The EQ 'braininess index' is calculated from two measured quantities, the brain mass and the body mass. In the case of fossils, both these quantities have to be estimated from the fragments that have come down to us, and there is a huge margin of error, especially in the estimation of body mass. The point on the graph for *Homo habilis* shows it as 'brainier' than *Homo erectus*. I don't believe this. The absolute brain size of *H. erectus* is undeniably larger. The inflation of the *H. habilis* EQ comes from the much lower estimated body mass. But to get an idea of the margin of error, think of the enormous range of body mass in modern humans. EQ as a measure is extremely sensitive to error in measuring body mass, which is raised, remember, to a power in the EQ formula. So, the scatter of points about the line largely reflects erratic estimation of body mass. On the other hand, the trend over time, as represented by the line, is probably real. The methods explained in this tale, in particular the estimates of EQ in the graph at the end, bear out our subjective impression that one of the most important things that has happened during the last 3 million years of our evolution was the ballooning of our already large primate brain. The next obvious question is why. What Darwinian selection pressure drove the enlargement of the brain during the past 3 million years?

H. sapiens   H. erectus   H. habilis   A. afarensis

Ballooning of our brain. Comparison of four hominid skulls, showing clear growth in brain size beginning with *Homo habilis*. The skulls have been scaled to the same height, and are average profiles based on several fossil examples.

Because it happened after we rose up on our hind legs, some people have suggested that brain inflation was driven by the freeing of the hands and the opportunity this offered for precision-controlled manual dexterity. In a general way I find this a plausible idea, though no more than several others that have been offered. But the enlargement of the human brain looks, as evolutionary trends go, explosive. I think inflationary evolution demands a special kind of inflationary explanation. In *Unweaving the Rainbow*, in the chapter called 'The Balloon of the Mind', I developed this inflationary theme in a general theory of what I called 'software-hardware co-evolution'. The computer analogy is with software innovations and hardware innovations triggering each other in an escalating spiral. Software innovations demand an escalation in hardware, which in turn provokes an escalation in software, and so the inflation gathers pace. In the brain, my candidates for the *kind* of thing I meant by a software innovation were language, spoor tracking, throwing, and memes. One theory of brain inflation that I didn't do justice to in my earlier book was sexual selection, and it is for this reason alone that I shall give it special prominence later in this book.

Could the enlarged human brain, or rather its products such as body painting, epic poetry and ritual dances, have evolved as a kind of mental peacock's tail? I have long had a soft spot for the idea, but nobody developed it into a proper theory until Geoffrey Miller, a young American evolutionary psychologist working in England, wrote his book, *The Mating Mind*. We shall hear this idea in the Peacock's Tale, after the bird pilgrims join us at Rendezvous 16.

APE-MEN

The popular literature on human fossils is hyped up with alleged ambition to discover the 'earliest' human ancestor. This is silly. You can ask a specific question like 'Which was the earliest human ancestor to walk habitually on two legs?' Or 'Which was the first creature to be our ancestor and not the ancestor of a chimpanzee?' Or 'Which was the earliest human ancestor to have a brain volume larger than 600 cc?' Those questions at least mean something in principle, although they are hard to answer in practice and some of them suffer from the vice of erecting artificial gaps in a seamless continuum. But 'Who was the earliest human ancestor?' means nothing at all.

More insidiously, the competition to find human ancestors means that new fossil discoveries are touted as on the 'main' human line whenever remotely possible. But as the ground yields up more and more fossils it becomes increasingly clear that, during most of hominid history, Africa housed many species of bipedal apes simultaneously. This has to mean that many fossil species claimed to be ancestral will turn out to be our cousins.

For a good length of time, between about 1 and 3 million years ago, our African ancestors shared the continent with short, robust hominids, perhaps several different species of them. As usual their affinities, and the exact number of species, are hotly disputed. Names that have been attached to various of these creatures (we met them in the graph at the end of the Handyman's Tale) are *Australopithecus* (or *Paranthropus*) *robustus*, *Australopithecus* (or *Paranthropus* or *Zinjanthropus*) *boisei*, and *Australopithecus* (or *Paranthropus*) *aethiopicus*. They seem to have evolved from more 'gracile' apes (gracile being the opposite of robust). Like the robust forms, these apes tended to be about three-quarters of our height with substantially smaller brains, and palaeontologists also place them in the genus *Australopithecus*. We too almost certainly emerged from among gracile australopithecine ranks, indeed, it is often difficult to distinguish early *Homo* from gracile australopithecines – justifying my diatribe above.

The immediate ancestors of *Homo* would be classified as some kind of gracile australopithecine. Let's look at some of them. At 2 million years old, one of the latest and best preserved is *Australopithecus sediba*, from near Pretoria, South Africa. The first specimen, a skull and skeleton thought to be a nine-year-old adolescent male, was found in 2008 by (as it happens) another nine-year-old male – the son of the paleoanthropologist Lee Berger.* Alongside it are bones from what may be the rest of the family: a male, a female and three babies, who all died around the same time after falling into what was then a cave. Microscopic plant remains trapped in their tooth tartar reveal that their diet was a diverse mix of leaves, fruits, bark and certain grasses. Uniquely, these ape fossils also seem to preserve delicate traces of fossilised skin, but this has yet to be fully investigated. *Australopithecus sediba* shows rather a hotchpotch of early *Homo* and late australopithecine features, leading the discoverers to suggest they are are on the human line. Other researchers point to older fossils attributed to early *Homo*, hinting that *A. sediba*, like the robust australopithecines, was an evolutionary dead end. The predictable arguments have, of course, ensued.

From less than ten miles away, but half a million years further back in time, comes a less immediately controversial fossil, Mrs Ples. I have had special affection for her ever since the Transvaal Museum in Pretoria presented me with a beautiful cast of her skull, on the fiftieth anniversary of her discovery, when I gave the Robert Broom Memorial Lecture in honour of her discoverer. Her nickname comes from the genus *Plesianthropus*, to which she was originally assigned before people decided to incorporate her into *Australopithecus*; and from the fact that she is (probably) female. Individual fossil hominids often pick up pet names like this. 'Mr Ples', naturally, is a more recently discovered fossil from the same Sterkfontein cave system, who is in the same species as Mrs Ples, *Australopithecus africanus*. Other fossils with nicknames include 'Dear Boy', a robust australopithecine also known as 'Zinj' because he was originally named *Zinjanthropus boisei*, 'Little Foot' and 'Ardi' (both of whom we shall meet presently), and the famous Lucy, to whom we now turn.

---

* More recent tooth analysis shows the skeleton was seven and a half years old, but equivalent to a 10–13-year-old human, because they matured more quickly than we did. Incidentally, the discovery is rumoured to be overshadowed by another of Berger's excavations, in the nearby 'Rising Star Cave'. The Rising Star fossil material was announced just as this book was going to press. Named *Homo naledi*, it is indeed an exciting discovery. Its age is unknown but it has some very primitive features which make it, controversially, a possible candidate for direct ancestry of *H. sapiens*.

Lucy was found in East rather than South Africa, and we meet her as our time machine's odometer touches 3.2 million years. She is often mentioned because her species, *Australopithecus afarensis*, was for a while the main contender for a human ancestor, before the discovery of a number of similar, contemporaneous species.* Lucy still remains one of the best preserved of the gracile australopithecines. Her discoverers, Donald Johanson and his colleagues, also found fossils of 13 similar individuals in the same area, known as the 'First Family'. Other 'Lucys' have since been found between about 3 and 4 million years ago in other parts of East Africa. The 3.6-million-year-old footprints discovered by Mary Leakey at Laetoli (page 76) are attributed to *A. afarensis*. Whatever the Latin name, evidently somebody was walking bipedally at that time. More recently, paleontologists have found stone tools and bones with what are probably cut marks dating from about this time too. Lucy is not greatly different from Mrs Ples, and some people think of Lucy as an earlier version of Mrs Ples. They are anyway more like each other than either is like the robust australopithecines. Early East African Lucys are said to have a slightly smaller brain than later South African Mrs Pleses, but there isn't much in it. Their brains were no more different from each other than some modern human brains are from other modern human brains.

As we have come to expect, the more recent *afarensis* individuals such as Lucy are slightly different from the earliest 3.9-million-year-old *afarensis* forms. Differences collect over time and, by 4 million years into our journey, we find more creatures who might well be ancestral to Lucy and her kin, but who are sufficiently different, in the direction of being more chimpanzee-like, to merit a different species name. Discovered by Meave Leakey and her team, these *Australopithecus anamensis* consist of more than 80 fossils from two different sites near Lake Turkana. No intact skull has been found, but there is a splendid lower jaw which plausibly could belong to an ancestor of ours.

But perhaps the most exciting discovery from this time period is a fossil yet to be fully described in print. Affectionately known as Little Foot, this skeleton pulls us back to South Africa and the caves of Sterkfontein, around 3.67 million years into the past. Its discovery was

---

* As of 2015, these include jaws proposed to belong to two separate species: *Australopithecus bahrelghazali* and *A. deyiremeda*, an unassigned Ethiopian foot, and a controversial skull, oddly assigned to a different genus, *Kenyanthropus*.

a piece of detective work worthy of a Conan Doyle story. Bits of Little Foot's left foot were dug up from Sterkfontein in 1978, but the bones were stored away, unremarked and unlabelled, until 1994 when the palaeontologist Ronald Clarke, working under the direction of Phillip Tobias, accidentally rediscovered them in a box in the shed used by workers at the Sterkfontein cave. Three years later, Clarke chanced upon another box of bones from Sterkfontein, in a store room at Witwatersrand University. This box was labelled 'Cercopithecoids'. Clarke had an interest in this kind of monkey, so he looked in the box and was delighted to notice a hominid foot bone in amongst the monkey bones. Several foot and leg bones in the box seemed to match the bones previously found in the Sterkfontein shed. One was half a right shinbone, broken across. Clarke gave a cast of the shinbone to two African assistants, Nkwane Molefe and Stephen Motsumi, and asked them to return to Sterkfontein and look for the other half.

> The task I had set them was like looking for a needle in a haystack as the grotto is an enormous, deep, dark cavern with breccia exposed on the walls, floor and ceiling. After two days of searching with the aid of hand-held lamps, they found it on 3 July 1997.

Molefe and Motsumi's jigsaw feat was the more astonishing because the bone that fitted their cast was

> at the opposite end to where we had previously excavated. The fit was perfect, despite the bone having been blasted apart by lime workers 65 or more years previously. To the left of the exposed end of the right tibia could be seen the section of the broken-off shaft of the left tibia, to which the lower end of the left tibia with foot bones could be joined. To the left of that could be seen the broken-off shaft of the left fibula. From their positions with the lower limbs in correct anatomical relationship, it seemed that the whole skeleton had to be there, lying face downwards.

Actually, it wasn't quite there but, after pondering the geological collapses in the area, Clarke deduced where it must be and, sure enough, Motsumi's chisel found it there. Clarke and his team were indeed lucky, but here we have a first-class example of that maxim of scientists since Louis Pasteur: 'Fortune favours the prepared mind.'

The excavation of Little Foot from the solidified speleological debris in which the skeleton was entombed has taken over 20 years. It is still not fully described in print. Preliminary reports suggest a

magnificent find, rivalling Lucy in completeness, but different
enough by the (admittedly trifling) standards of palaeoanthropology
to be treated as another species, *Australopithecus prometheus*. Although more human-like than chimpanzee-like, the big toe is more
divergent than our toes. This might suggest that Little Foot grasped
tree boughs with its feet in a way that we cannot. Although it almost
certainly walked bipedally, it probably climbed too and walked with a
different gait from us. Like other australopithecines, it may have
spent time in trees, perhaps bivouacking in them at night like
modern chimpanzees.

Bipedality separates humans from the rest of the mammals so
dramatically that it deserves a tale to itself. In the first edition of this
book the tale was appropriately bestowed upon Little Foot. We did,
however, note a few promising older bones from East Africa, assigned to the species *Ardipithecus ramidus*. In 2009 this promise was
spectacularly fulfilled. Only 50 miles from where Lucy had been
found, a team led by Tim White, alumnus of both the Leakey and
Lucy schools, had been 15 years at work. Trampled into the Pliocene

**Trampled into the Pliocene mud by some
long-dead megaherbivore.** Tim White and
colleagues' meticulous reconstruction of the
skeleton of *Ardipithecus ramidus*.

mud by some long-dead megaherbivore, the ape nicknamed 'Ardi' represents an astonishing *tour de force* of palaeontological reconstruction. Although 4.4 million years old, her skull and skeleton – especially the hands and feet – are far more complete than Lucy, providing a compelling reason to call a temporary halt to our journey here. Fragmentary remains (particularly teeth) have been recovered from other individuals too, which reveal that males had relatively small canines, suggesting reduced competition between males, along with other implications for social behaviour (see the Seal's Tale). Like any ancient fossil, we must be careful not to assume Ardi was our direct ancestor, but her unexpected combination of features is highly informative. In particular, she suggests that in comparing our ancestral locomotion with chimpanzees and gorillas, we may be barking up the wrong tree.

Although chimpanzee-sized, with a brain to match, Ardi did not get around like a chimp. Or like an australopithecine. Or indeed like anything in between. Chimps and gorillas have evolved a particular way to combine tree-climbing with walking. To climb, they hang their hook-like hands round trunks and large branches in the deep forest, and pull or swing into position. On the ground, they walk on all fours, but their stiffened hands mean that they can't put the palm of their hands flat on the floor, or ball them into a fist like orang utans. Instead, they are noted for walking on extended knuckles.

Like other small woodland primates, Ardi probably spent much time in the trees: her skeleton shows an opposable big toe and other indicative features. Rather than hooking on like a chimp, her flexible wrists and shorter arms would have meant grasping and carefully clambering were the order of the day. But on the ground (and perhaps sometimes in the trees), Ardi was fully bipedal, as evidenced by the placement of the skull, and features of the foot and pelvis. She had no clear specialisations for walking on the ground on all fours, either with knuckles or otherwise. Even without being our direct ancestor, Ardi suggests it is possible to evolve from a tree-based to a bipedal lifestyle without necessarily going through a four-legged, knuckle-walking intermediary. It is possible that none of our ancestors ever walked like a chimp, and that gorillas and chimps have independently converged upon the same style of locomotion, something also hinted upon by certain anatomical studies.

Reminding people that we didn't evolve from chimpanzees can be tediously repetitive. It is actually rather a relief to find evidence indicating that, in a few important ways, our common ancestor need not have resembled either a chimp or a gorilla. That may be Ardi's legacy, and, given her impact, she rather than Little Foot will now take up the whys and wherefores of our bipedality.

ARDI'S TALE

It isn't particularly helpful to dream up reasons why walking on two legs might be generally a good thing. If it were, the chimps would do it too, to say nothing of other mammals. There is no obvious reason for saying that either bipedal or quadrupedal running is faster or more efficient than the other. Galloping mammals can be astonishingly fleet, using the up-and-down flexibility of the backbone to achieve – among other benefits – a lengthened effective stride. But ostriches show that a man-like bipedal gait can be a match for a quadrupedal horse. Indeed a top human sprinter, though noticeably slower than a horse or dog (or ostrich or kangaroo, for that matter), is not disgracefully slow. Quadrupedal monkeys and apes are generally undistinguished runners, perhaps because their bodily designs have to compromise with the needs of a climber. Even baboons, which normally forage and run on the ground, resort to the trees to sleep and as a defence against predators, but baboons can run fast when they need to.

So, when we ask why our ancestors rose up on their hind legs, and when we imagine the quadrupedal alternative that we forsook, it is unfair to 'think cheetah', or anything like it. When our ancestors first stood up, there was no overwhelmingly strong advantage in efficiency or speed. We should look elsewhere for the natural selection pressure which drove us to this revolutionary change in gait.

Like some other quadrupeds, chimpanzees can be trained to walk bipedally, and they often do it anyway over short distances. So it probably wouldn't be insuperably difficult for them to make the switch if there were strong benefits to doing so. Orang utans are even better at it. Wild gibbons, whose fastest method of locomotion is brachiation – swinging under the boughs by their arms – also run across clearings on their hind legs. Some monkeys rise upright, to peer over long grass or to wade through water. A lemur, Verreaux's sifaka, although it lives mainly in trees where it is a spectacular acrobat, 'dances' across the ground between trees on its hind legs, the arms held up with balletic grace.

Doctors sometimes ask us to run on the spot in a mask, so they can measure our oxygen consumption and other metabolic indices when we are exerting ourselves. In 1973 two American biologists, C. R. Taylor and V. J. Rowntree, did this with trained chimpanzees and capuchin monkeys, running on a treadmill. By making the animals run the treadmill either on four legs or on two (they were given something to hold on to), the researchers could compare the oxygen consumption and efficiency of the two gaits. They expected that quadrupedal running would be more efficient. This, after all, is what both species naturally do, and it is what their anatomy fits them for. Maybe bipedalism was helped by the fact that they had something to hold on to. In any case, the result was otherwise. There was no significant difference between the oxygen consumption of the two gaits. Taylor and Rowntree concluded that:

> The relative energy cost of bipedal versus quadrupedal running should not be used in arguments about the evolution of bipedal locomotion in man.

Even if this is an exaggeration, it should at least encourage us to look elsewhere for possible benefits of our unusual gait. It arouses the suspicion that, whatever non-locomotor benefits of bipedality we might propose as drivers of its evolution, they probably did not have to fight against strong locomotor costs.

What might a non-locomotor benefit look like? A stimulating suggestion is the sexual selection theory of Maxine Sheets-Johnstone, of the University of Oregon. She thinks we rose on our hind legs as a means of showing off our penises. Those of us that have penises, that is. Females, in her view, were doing it for the opposite reason: concealing their genitals which, in primates, are more prominently displayed on all fours. This is an appealing idea but I don't carry a torch for it. I mention it only as an example of the *kind* of thing I mean by a non-locomotor theory. As with so many of these theories, we are left wondering why it would apply to our lineage and not to other apes or monkeys.

A different set of theories stresses the freeing of the hands as the really important advantage of bipedality. Perhaps we rose on our hind legs, not because that is a good way of getting about, but because of what we were then able to do with our hands – carry food, for instance. Many apes and monkeys feed on plant matter that is widely available but not particularly rich or concentrated, so you must eat as you go, more or less continuously like a cow. Other kinds of food such as meat or large underground tubers are harder to acquire but, when you do find them, they are valuable – worth carrying home in greater quantity than you can eat. When a leopard makes a kill, the first thing it normally does is drag it up a tree and hang it over a branch, where it will be relatively safe from marauding scavengers and can be revisited for meals. The leopard uses its powerful jaws to hold the carcass, needing all four legs to climb the tree. Having much smaller and weaker jaws than a leopard, did our ancestors benefit from the skill of walking on two legs because it freed their hands for carrying food – perhaps back to a mate or children, or to trade favours with other companions, or to keep in a larder for future needs?

Incidentally the latter two possibilities may be closer to each other than they appear. The idea (I attribute this inspired way of expressing it to Steven Pinker) is that before the invention of the freezer the best larder for meat was a companion's belly. How so? The meat itself is no longer available, of course, but the goodwill it buys is safe in long-term storage in a companion's brain. Your companion will remember the favour and repay it when fortunes are reversed.* Chimpanzees are known to share meat for favours. In historic times, this kind of i.o.u. became tokenised as money.

A particular version of the 'carrying food home' theory is that of the American anthropologist Owen Lovejoy, who has recently re-applied it to *Ardipithecus*. He suggests that females would often have been hampered in their foraging by nursing infants, therefore unable to travel far and wide looking for food. The consequent poor nutrition and poor milk production would have delayed weaning. Suckling females are infertile. Any male who feeds a nursing female accelerates the weaning of her current child and brings her into receptiveness earlier. When this happens, she might make her receptiveness especially available to the male whose provisioning accelerated it. So, a male who can bring lots of food home might gain a direct reproductive advantage over a rival male who just eats where he finds. Hence the evolution of bipedalism to free the hands for carrying.

Other hypotheses of bipedal evolution invoke the benefits of height, perhaps standing upright to look over the long grass; or to keep the head above water while wading. This last is the imaginative 'aquatic

* There is a well-developed theory of reciprocal altruism in Darwinism, beginning with the pioneering work of Robert Trivers and continuing with the modelling of Robert Axelrod and others. Trading favours, with delayed repayment, really works. My own exposition of it is in *The Selfish Gene*, especially the second edition.

ape' theory of Alister Hardy, ably championed by Elaine Morgan. Another theory, favoured by John Reader in his fascinating biography of Africa, suggests that upright posture minimises exposure to the sun, limiting it to the top of the head which is consequently furnished with protective hair. Moreover, when the body is not hunched close to the ground, it can lose heat more rapidly.

My colleague the distinguished artist and zoologist Jonathan Kingdon has centred a whole book, *Lowly Origin*, around the question of the evolution of human bipedality. After a lively review of 13 more-or-less distinct hypotheses, including the ones I have mentioned, Kingdon advances his own sophisticated and multifaceted theory. Rather than seek an immediate benefit of walking upright, Kingdon expounds a complex of quantitative anatomical shifts which arose for some other reason, but which then made it easier to become bipedal (the technical term for this kind of thing is pre-adaptation). The pre-adaptation that Kingdon proposes is what he calls squat feeding. Squat feeding is familiar from baboons in open country, and Kingdon visualises something similar in our ape ancestors in the forest, turning over stones or leaf litter for insects, worms, snails and other nutritious morsels. To do this effectively they would have had to undo some of their adaptations to living up trees. Their feet, previously hand-like for gripping branches, would have become flatter, forming a stable platform for squatting on the haunches. You will already be getting a glimmering of where the argument is going. Flatter, less hand-like feet for squatting are later going to serve as pre-adaptations for upright walking. And you will, as usual, understand that this apparently purposeful way of writing – they had to 'undo' their tree-swinging adaptations, etc. – is a shorthand which is easily translated into Darwinian terms. Those individuals whose genes happened to make their feet more suitable for squat feeding survived to pass on those genes because squat feeding was efficient and aided their survival. I shall continue to employ the shorthand because it chimes with the way humans naturally think.

A tree-swinging, 'brachiating' ape could fancifully be said to walk upside down under the branches – run and leap in the case of an athletic gibbon – using the arms as its 'legs' and the shoulder girdle as its 'pelvis'. Our ancestors probably passed through a brachiating phase, and the true pelvis consequently became rather inflexibly bound to the trunk by long blades of bone, which form a substantial part of a rigid trunk that can be swung as a single unit. Much of this, according to Kingdon, would have needed to change, to make an efficient squat feeder out of an ancestral brachiator. Not all, however. The arms could have remained long. Indeed, long brachiating arms would have been a positively beneficial 'pre-adaptation', increasing the reach of the squat feeder and decreasing the frequency with which it had to shuffle to a new squatting position. But the massive, inflexible, top-heavy ape trunk would have been a disadvantage in a squat feeder. The pelvis would have needed to free itself and become less rigidly tied to the trunk, and its blades would have shrunk – to more human proportions. This, to anticipate the later stages of the argument again (you might say that anticipation is what a pre-adaptation argument is all about) just happens to make a better pelvis for bipedal walking. The waist became more flexible, and the spine was held more vertically, to allow the squat-feeding animal to search all around with its arms, turning on the platform of the flat feet and the squatting haunches. The shoulders became lighter and the body less top-heavy. And the point is that these subtle quantitative changes, and the balancing and compensating shifts that went with them, incidentally had the effect of 'preparing' the body for bipedal walking.

Not for a moment is Kingdon proposing any kind of anticipation of the future. It is just that an ape whose ancestors were tree-swingers, but which has switched to squat feeding on the forest floor, now has a body which feels relatively *comfortable* walking on its hind feet. And it would have begun to do this while squat feeding, shuffling to a new squatting position as the old one became depleted. Without realising what was happening, squat feeders were, over the generations, preparing their bodies to feel more comfortable when upright and on two legs; to feel more awkward on four. I use the word comfortable deliberately. It is not a trivial consideration. We are capable of walking on all fours like a typical mammal, but it is uncomfortable: hard work, because of our altered body proportions. Those proportional changes which now make us feel comfortable on two legs originally came about, Kingdon suggests, in the service of a minor shift in food habits – to squat feeding.

There is much more in Jonathan Kingdon's subtle and complex theory, but I will now recommend his book, *Lowly Origin*, and move on. My own slightly wayout theory of bipedality is very different but not incompatible with his. Indeed, most of the theories of human bipedality are mutually compatible, with the potential to assist rather than oppose one another. As in the case of the enlargement of the human brain, my tentative suggestion is that bipedality may have evolved through sexual selection, so again I postpone the matter to the Peacock's Tale.

Whatever theory we believe about the evolutionary origins of human bipedality, it subsequently turned out to be an extremely important event. In former times it was possible to believe, as respected anthropologists did up to the 1960s, that the decisive evolutionary event that first separated us from the other apes was the enlargement of the brain. Rising up on the hind legs was secondary, driven by the benefits of freeing the hands to do the kind of skilled work which the enlarged brain was now capable of controlling and exploiting. Recent fossil finds point decisively towards the reverse sequence. Bipedality came first. Lucy, who lived long after Rendezvous 1, was bipedal, yet her brain was approximately the same size as a chimpanzee's. The enlargement of the brain could still have been associated with the freeing of the hands, but the sequence of events was reversed. If anything it would be the freeing of the hands by bipedal walking that drove the enlargement of the brain. The manual hardware came first, then the controlling brainware evolved to take advantage of it, rather than the other way around.

EPILOGUE TO ARDI'S TALE

Whatever the reason for the evolution of bipedality, recent fossil discoveries seem to indicate that hominids were already bipedal at a date which is pushing disconcertingly close to Rendezvous 1, the fork between ourselves and chimpanzees (disconcerting because it seems to leave little time for bipedality to evolve). In the year 2000, a French team led by Brigitte Senut and Martin Pickford announced a new fossil from the Tugen Hills, east of Lake Victoria in Kenya. Dubbed 'Millennium Man', dated at 6 million years and given yet another new generic name, Orrorin tugenensis was also, according to its discoverers, bipedal. Indeed, they claim that the top of its femur, near the hip joint, was more human-like than that of Australopithecus. This evidence, supplemented by fragments of skull bones, suggested to Senut and Pickford that orrorins are ancestral to later hominids and that Lucys are not. These French workers go further and suggest that Ardipithecus might be ancestral to modern chimpanzees rather than to us. Clearly we need more fossils to settle these arguments. Other scientists are sceptical of these French claims, and some doubt that there is enough evidence to show whether *Orrorin* was or was not bipedal. If it was, and Rendezvous 1 took place between 5 and 7 million years ago, this raises difficult questions about the speed with which bipedality must have arisen.

If a bipedal *Orrorin* pushes back alarmingly close to Rendezvous 1, a skull discovered a year later from Chad in southern Sahara, found by another French team led by Michel Brunet, is even more disturbing to accepted ideas. Nicknamed Toumai (Hope of Life in the local Goran language) its official name is *Sahelanthropus tchadensis*, after the Sahel region of the Sahara in Chad where it was found. It is an intriguing skull, looking rather human from in front (lacking the protruding face of a chimpanzee or gorilla) but chimpanzee-like from behind, with a chimpanzee-sized braincase (see plate 4). It has an extremely well-developed brow-ridge, even thicker than a gorilla's, which is the main reason for thinking Toumai was male. The teeth are rather human-like, especially the thickness of the enamel which is intermediate between a chimpanzee's and our own. The foramen magnum (the big hole through which the spinal cord passes) is placed further forward than in a chimpanzee or gorilla, suggesting to Brunet himself, though not to some others, that Toumai was bipedal. Ideally, this should be confirmed by pelvis and leg bones but, unfortunately, nothing but a skull has so far been described.

There are no volcanic remains in the area to provide radiometric dates, and Brunet's team had to use other fossils in the area as an indirect clock. These are compared with already known faunas from other parts of Africa which can be dated absolutely. The comparison yields a date for Toumai of between 6 and 7 million years. Brunet and his colleagues claim it as older than *Orrorin*, which has predictably elicited indignant ripostes from *Orrorin*'s discoverers. One of them, Brigitte Senut, of the Natural History Museum in Paris, has said that Toumai is 'a female gorilla', while her colleague Martin Pickford described Toumai's canine teeth as typical 'of a large female monkey'. These were the two, remember, who (perhaps rightly) wrote off the human credentials of *Ardipithecus*, another threat to the priority of their own baby, *Orrorin*. Other authorities have hailed Toumai more generously: 'Astonishing.' 'Amazing.' 'This will have the impact of a small nuclear bomb.'

If their discoverers are right that *Orrorin* and Toumai were bipedal, this poses problems to any tidy view of human origins. The naive expectation is that evolutionary change spreads itself uniformly to fill the time available for it. If 6 million years elapsed between Rendezvous 1 and modern *Homo sapiens*, the quantity of change ought to be spun out, *pro rata* one might naively think, through the 6 million years. But *Orrorin* and Toumai both lived very close to the date identified from molecular evidence as that of Concestor 1, the split between our line and that of chimpanzees. These fossils even predate Concestor 1 according to some reckonings.

Assuming that the molecular and fossil dates are correct, there seem to be four ways (or some combination from among the four) in which we might respond to *Orrorin* and Toumai.

1. *Orrorin* and/or Toumai were not bipedal after all. This is quite possible, but the remaining three possibilities assume, for the sake of argument, that it is wrong. If we accept option 1, the problem just goes away.
2. An extremely rapid burst of evolution occurred immediately after Concestor 1, which itself moved around like a normal ape. The more humanoid Toumai and *Orrorin* evolved their bipedality so swiftly after Concestor 1 that the separation in dates cannot easily be resolved.
3. Humanoid features such as bipedality have evolved more than once, maybe many times. *Orrorin* and Toumai could represent earlier occasions when African apes experimented with bipedality, and perhaps other human features too. On this hypothesis, they could indeed predate Concestor 1 while being bipedal, and our own lineage would constitute a later foray into bipedality.
4. Chimpanzees and gorillas descend from more human-like, even bipedal ancestors, and have reverted to all fours more recently. On this hypothesis, Toumai, say, could actually be Concestor 1.

The last three hypotheses all have difficulties, and many authorities are driven to doubt either the dating, or the supposed bipedality, of Toumai and *Orrorin*. But if we accept these for the moment and look at the three hypotheses that assume ancient bipedality, there is no strong theoretical reason to favour or disfavour any particular one of them. We shall learn from the Galapagos Finch's Tale and the Coelacanth's Tale that evolution can be extremely rapid or can be extremely slow. So Theory 2 is not implausible. The Marsupial Mole's Tale will teach us that evolution can follow the same path, or strikingly parallel paths, on more than one occasion. There's nothing particularly implausible, then, about Theory 3. Theory 4, at first sight, seems the most surprising. We are so used to the idea that we have risen 'up' from the apes that Theory 4 seems to put the cart before the horse, and may even insult human dignity into the bargain (often good for a laugh in my experience). Also there is a so-called law, Dollo's Law, which states that evolution never reverses itself, and it might seem that Theory 4 violates it.

The Blind Cave Fish's Tale, which is about Dollo's Law, will reassure us that this last is not the case. There is nothing in principle wrong with Theory 4. Chimpanzees really could have passed through a more humanoid, bipedal stage before reverting to quadrupedal apehood. As it happens, this very suggestion has been revived by John Gribbin and Jeremy Cherfas, in their two books, *The Monkey Puzzle* and *The First Chimpanzee*. Although it is not a strictly necessary part of their theory, they even suggest that chimpanzees might be descended from gracile australopithecines (like Lucy), and gorillas from robust australopithecines (like 'Dear Boy'). For such an in-your-face radical suggestion, they make a surprisingly good case. It centres on an interpretation of human evolution which has long been widely accepted, although not without controversy: people are juvenile apes who have become sexually mature. Or, putting it another way, we are like chimpanzees who have never grown up.

The Axolotl's Tale explains the theory, which is known as neoteny. To summarise, the axolotl is an overgrown larva, a tadpole with sex organs. In a classic experiment by Vilém Laufberger in Germany, hormone injections persuaded an axolotl to grow into a fully adult salamander of a species that nobody had ever seen. More famously in the English-speaking world, Julian Huxley later repeated the experiment, not knowing it had already been done. In the evolution of the axolotl, the adult stage had been chopped off the end of the life cycle. Under the influence of experimentally injected hormone, the axolotl finally grew up, and an adult salamander was recreated, presumably never before seen. The missing last stage of the life cycle was restored.

The lesson was not lost on Julian's younger brother, the novelist Aldous Huxley. His *After Many a Summer** was one of my favourite novels when I was a teenager. It is about a rich man, Jo Stoyte, who resembles William Randolph Hearst and collects *objets d'art* with the same voracious indifference. His strict religious upbringing has left

---

* The American edition rounds off the Tennyson quotation: 'Dies the Swan'.

him with a terror of death, and he employs and equips a brilliant but cynical biologist, Dr Sigismund Obispo, to research how to prolong life in general and Jo Stoyte's life in particular. Jeremy Pordage, a (very) British scholar, has been hired to catalogue some eighteenth-century manuscripts recently acquired as a job lot for Mr Stoyte's library. In an old diary kept by the Fifth Earl of Gonister, Jeremy makes a sensational discovery which he imparts to Dr Obispo. The old Earl was hot on the trail of everlasting life (you have to eat raw fish guts), and there is no evidence that he ever died. Obispo takes the increasingly fretful Stoyte to England in quest of the Fifth Earl's remains . . . and finds him still alive at 200. The catch is that he has finally matured from the juvenile ape which all the rest of us are into a fully adult ape: quadrupedal, hairy, repellent, urinating on the floor while humming a grotesquely distorted vestige of a Mozart aria. The diabolical Dr Obispo, beside himself with gleeful laughter and evidently acquainted with Julian Huxley's work, tells Stoyte he can start on the fish guts tomorrow.

Gribbin and Cherfas are in effect suggesting that modern chimpanzees and gorillas are like the Earl of Gonister. They are humans (or australopithecines, orrorins or sahelanthropes) who have grown up and become quadrupedal apes again, like their, and our, more distant ancestors. I never thought the Gribbin/Cherfas theory was obviously silly. The new findings of very ancient hominids like *Orrorin* and Toumai, whose dates push up against our split with chimpanzees, could almost justify them in a *sotto voce* 'We told you so'.

Even if we accept *Orrorin* and Toumai as bipedal, I would not choose with confidence between Theories 2, 3 and 4. And we mustn't forget the distinct possibility of Theory 1, that they weren't actually bipedal, in which case the problem goes away. But of course these different theories make predictions about Concestor 1, our next stopping point. Theories 1, 2, and 3 agree in assuming that Concestor 1 walked on all fours, using both hands and feet, with only the occasional and uncomfortable two-legged walk. Theory 4 by contrast differs in assuming a more humanoid Concestor 1. In narrating Rendezvous 1, I have been forced to make a decision between the theories. Somewhat reluctantly, I'll go with the majority, and assume a four-legged concestor. On to meet it.

CHIMPANZEES

Between 5 and 7 million years ago, somewhere in Africa, we human pilgrims enjoy a momentous encounter. It is Rendezvous 1, our first meeting with pilgrims from another species. Two other species to be precise, for the band that joins us here consists of two living representatives: common chimpanzees and pygmy chimpanzees, or bonobos. By the time we meet them, these pilgrims have already joined forces: their rendezvous with each other took place perhaps 2 million years ago, far more recently than their rendezvous with us. The common ancestor we share with them, Concestor 1, is our 250,000-greats-grandparent – an approximate guess this, of course, like the comparable estimates that I shall be making for other concestors.

As we approach Rendezvous 1, then, the chimpanzee pilgrims are approaching the same point from another direction. Unfortunately we know very little about that other direction. Although Africa has yielded up some thousands of hominid fossils or fragments of fossils, the only fossils which can definitely be regarded as along the chimpanzee line of descent from Concestor 1 are a few recent teeth from Kenya. This may be because chimpanzees are forest animals, and the leaf litter of forest floors is not friendly to fossils. Whatever the reason, it means that the chimpanzee pilgrims are effectively searching blind. Their equivalent contemporaries of the Turkana Boy, of 1470, of Mrs Ples, Lucy, Little Foot, Dear Boy and the rest of 'our' fossils – have never been found.

Nevertheless, in our fantasy the chimpanzee pilgrims meet us in some Miocene forest clearing, and their dark brown eyes, like our less predictable ones, are fixed upon Concestor 1: their ancestor as well as ours. In trying to imagine the shared ancestor, an obvious question to ask is, is it more like modern chimpanzees or modern humans, is it intermediate, or completely different from either?

Notwithstanding the pleasing speculation that ended the previous section – which I would by no means rule out – the prudent answer is that Concestor 1 was more like a chimpanzee, if only because chimpanzees are more like the rest of the apes than humans are. Humans are the odd ones out among apes, both living and fossil. Which is only

Chimpanzees join. There are over 2 million species on earth. To show them all on a single evolutionary tree, we use the fractal representation pioneered by James Rosindell [364]. This zoomed-in region of the tree shows the branch containing ourselves and our closest relatives, the two species of chimpanzee. The dark line from humans back up the branch marks our journey into the past. Note that the length and width of the branches and the size of the circles for each species have no specific meaning: they are simply chosen to fit into the available space.

Dates on this tree, in millions of years ago (Ma), are indicated by contour lines. These show that the concestor of common chimpanzees and bonobos is thought to have lived slightly less than 2 million years ago. Concestor 1, the most recent common ancestor of humans and chimpanzees, probably lived around 6 million years ago. This is Rendezvous 1, marked by a numbered circle with a date.

Branches in the tree appear solid, but are better imagined as an enormous genealogical network, as in the schematic illustration at Rendezvous 0. Within this pedigree, most genes unite somewhat further back in time than the rendezvous. This results in some genes showing relationships at odds with those shown here, for example tying humans more closely to common chimpanzees than to bonobos. This is the topic of the Bonobo's Tale.

to say that more visible change has occurred along the human line of descent from the common ancestor, than along the lines leading to the chimpanzees. We must not assume, as many laymen do, that our ancestors *were* chimpanzees, or like them in all respects. The very phrase 'missing link' is suggestive of this misunderstanding. You still hear people saying things like, 'Well, if we are descended from chimpanzees, why are there still chimpanzees around?'

So, when we and the chimpanzee/bonobo pilgrims meet at the rendezvous point, the likelihood is that the shared ancestor that we greet in that Miocene clearing was hairy like a chimpanzee, and had a chimpanzee-sized brain. Reluctantly to set aside the speculations which concluded the previous chapter, it probably moved around using its hands, as well as its feet, although maybe not exactly as chimpanzees do today. It probably spent some time up trees, but also lots of time on the ground, maybe squat feeding as Jonathan Kingdon would say. All available evidence suggests that it lived in Africa, and only in Africa. It probably used and made tools, following local traditions as modern chimpanzees still do. It probably was omnivorous, sometimes hunting, but with a preference for fruit.

Bonobos have been seen to kill duikers, but hunting is more frequently documented for common chimpanzees, including highly co-ordinated group pursuits of colobus monkeys. But meat is only a supplement to fruit, which is the main diet of both species. Jane Goodall, who first discovered hunting and intergroup warfare in chimpanzees, was also the first to report their now famous habit of termite fishing, using tools of their own construction. Bonobos have not been seen to do this, but that may be because they have been studied less. Common chimpanzees in different parts of Africa develop local traditions of tool use. Where Jane Goodall's animals on the east side of the range fish for termites, other groups to the west have developed local traditions of cracking nuts using stone or wood hammers and anvils. Some skill is required. You have to hit hard enough to break the kernel but not so hard as to pulp the nut itself.

Although often spoken of as a new and exciting discovery, by the way, nut cracking was mentioned by Darwin in Chapter 3 of *The Descent of Man* (1871):

> It has often been said that no animal uses any tool; but the chimpanzee in a state of nature cracks a native fruit, somewhat like a walnut, with a stone.

The evidence cited by Darwin (a report by a missionary in Liberia in the 1843 issue of the *Boston Journal of Natural History*) is brief and non-specific. It simply states that 'the *Troglodytes niger*, or Black Orang of Africa' is fond of a species of unidentified nut, which 'they crack with stones precisely in the manner of human beings'.

The especially interesting thing about nut cracking, termite fishing and other such chimpanzee habits is that local groups have local customs, handed down locally. This is true culture. Local cultures extend to social habits and manners. For example, one local group in the Mahale Mountains in Tanzania has a particular style of social grooming known as the grooming hand clasp. The same gesture has been seen in another population in the Kibale forest in Uganda. But it has never been seen in Jane Goodall's intensively studied population at Gombe Stream. Interestingly, this gesture also spontaneously arose and spread among a captive group of chimpanzees.

If both species of modern chimpanzee used tools in the wild as we do, this would encourage us to think that Concestor 1 probably did too. I think it probably did – even though bonobos have not been seen using tools in the wild, they are adept tool-users in captivity. The fact that common chimpanzees use different tools in different areas, following local traditions, suggests to me that lack of such a tradition in a particular area should not be taken as negative evidence. After all, Jane Goodall's Gombe Stream chimpanzees haven't been seen to crack nuts. Presumably they would, if the West African nut-cracking tradition were introduced to them. I suspect that the same might be true of bonobos. Maybe they just haven't been studied enough in the wild. In any case, I think the indications are strong enough that Concestor 1 made and used tools. This idea is strengthened by the fact that tool use also occurs in wild orang utans, local populations again differing in ways that suggest local traditions. And tool use is widespread among mammals and birds, as Jane Goodall herself (among others) has documented.

The present-day representatives of the chimpanzee lineage are both forest apes, whereas we are savannah apes, more like baboons except, of course, that baboons are not apes at all but monkeys. Bonobos today are confined to the forests south of the great curve of the River Congo and north of its tributary the Kasai. Common chimpanzees inhabit a wider belt of the continent, north of the Congo, westward to the coast, and extending as far as the Rift Valley in the east.

As we shall see in the Cichlid's Tale, current Darwinian orthodoxy suggests that usually, in order for an ancestral species to split into two daughter species, there is an initial, accidental separation between them, often geographical. Without the geographical barrier, sexual mixing of the two gene pools keeps them together. It is plausible that the great Congo river provided the barrier to gene flow which assisted the evolutionary divergence of the two chimpanzee species from each other, a few million years ago.

Geography also divides each of the other great ape lineages into two species: the western and eastern gorillas, and the Bornean and Sumatran orang utans. This might lead us to suspect that some sort of geographical separation was responsible for dividing the population that was to become humans from that which was to become chimpanzees. One candidate is the emergence of the Great Rift Valley and the east-west climatic differences that it generated. Recent discoveries have rendered this idea unfashionable. Half-a-million-year-old chimpanzee teeth from Kenya show that chimps are not purely a West African species, as once thought. Neither are early hominids solely East African any more: the 'Toumai' skull comes from Chad, thousands of miles to the west of the Rift Valley, as does the younger and more poorly known *Australopithecus bahrelghazali*.

Whether or not it was the Great Rift Valley, or another feature of geography, some ancestral population of apes must have split in two, leaving as evidence humans on one side and the chimpanzees (including bonobos) on the other. The traces of this split are strewn across our genomes, and the details will occupy our next two tales.

PROLOGUE TO THE CHIMPANZEE'S TALE

Chimpanzees fascinate us because they bear comparison to ourselves. The Chimpanzee's Tale compares human and chimpanzee genomes, which can help us to date Rendezvous 1. Before that, it is worth tackling a couple of oft-asked questions that crop up when carrying out such comparisons.

The most common question sounds simple: by what percentage does the chimp genome differ from that of humans? It turns out to be rather tricky to answer, largely because of problems of definition. A reasonable analogy might be made with two different edits of a book. Edits involve deleting sentences and changing words, but can also encompass all sorts of other copy-and-paste changes such as moving paragraphs or even whole chapters around. That makes it hard to put an exact value on the degree of difference between the two end products. Similarly, large chunks of DNA move around the genome during evolution. For example, there is incontrovertible evidence that our chromosome 2 was formed by pasting together two great ape chromosomes, which explains why humans have 23 pairs of chromosomes and the rest of the great apes have 24. It is not a clear choice whether we should count this as a very large mutation, or (given that the genes themselves have been kept) as a rather minor one.

Smaller sections of DNA can be duplicated or excised too, and it has been estimated that these 'indels' (insertion/deletion differences) number about 5 million, covering an estimated 3 per cent of the genome. But it would be misleading to equate this to a 3 per cent difference between our genomes, because in one sense a duplicated section is not different at all. Would a duplicated page in this book count as a difference of (say) 400 words from the original? Or would it just be a single 'difference'? To set aside that semantic issue, we can match fragments of DNA from humans and chimps, then focus on small, single-letter differences. Compared to indels, there are far more of these 'single nucleotide changes', around 35 million, but they cover a smaller fraction of the genome: just over 1 per cent. In this very limited sense, we might say that chimps and humans are roughly 98–99 per cent similar. Bear in mind, however, that this not only omits indels, but also regions which are so different that human and chimp sequences can't be matched, common in certain gene-poor regions, in particular the Y chromosome.

Another common question is where these genetic differences lie. Again, this turns out to require qualification, because about half of our DNA is parasitic 'junk', and much of the rest is probably unused. Conventional 'genes' that code for proteins only account for between 1 and 2 per cent of human (and chimp) DNA. Perhaps another 8 per cent is useful in other ways (some is used for turning genes on or off, other sequences have an as yet unknown function). In other words, only about 10 per cent of our DNA consists of actively useful sequences. These regions experience rather little change over evolutionary time. It is the remaining 90 per cent of the genome where we differ most from chimpanzees, since this is where any mutation can happen without much effect – where the accumulated changes are invisible to natural selection.

A more incisive question is where the *significant* genetic differences between humans and chimps lie: those that are responsible for our contrasting forms and behaviour. A knotty problem. It involves working out what our various DNA sequences actually do: a major (if not *the* major) task for biologists of the future. Preliminary studies indicate that many important differences affect not so much the protein-coding genes, but the DNA sequences that switch these genes on and off. This intriguing topic is the focus of the Mouse's Tale. Here in the Chimpanzee's Tale, we are not interested in the few significant mutations, but in the many that have no effect, and are free to accumulate over time. It is these that can be used to date our current rendezvous.

THE CHIMPANZEE'S TALE

The word 'mutation' conjures up images of grotesquely distorted creatures, perhaps generated by unscrupulous experimenters, or springing up as a consequence of some radioactive catastrophe. The truth is somewhat different. We are all mutants. The DNA passed on to us from our parents contains novel changes – mutations – which were not present in the DNA that they inherited from their parents. Fortunately so, for mutations provide the raw material, sculpted over millennia by natural selection, used to build the bodies of all the pilgrims on our journey. How many new mutations do we each inherit? That is the topic of the Chimpanzee's Tale, of particular interest because this 'mutation rate' is used to calibrate the timescale of human evolution.

There are many types of mutation, and there are many different mutation rates. In this tale we are only concerned with single letter mutations, which are mostly due to mistakes made when copying DNA. How many DNA letters can be copied, on average, before an uncorrected typo slips in? We don't know for sure, but biochemical observations suggest a range of once every 1 billion to once every 100 billion letters copied. The problem is that this error rate varies, and it is tricky to measure such a tiny number with any accuracy. To do so requires examining huge numbers of copying events.

We have a few options. One possibility is to look at a small section of DNA over many millions of copying events. Indirectly, that is what we are doing when we compare a gene between a human and a chimpanzee. Another possibility is to focus on one gene in many millions of individuals, as in a particular study that counted all the novel instances of haemophilia mutations in the USA. Finally, with modern DNA sequencing technology, we can directly count the number of mutations we pass on to our offspring, by sequencing the 6 billion DNA letters in a child and comparing those to both its parents. This direct method is currently threatening to double many previously established dates in human evolution. To understand why requires some background detail.

Before the molecular era, fossil discoveries suggested that the ancestors of chimps and humans split apart more than 15 million years ago. In a now famous paper from 1967, Vincent Sarich and Allan Wilson from Berkeley threw down a major challenge to these dates, claiming that molecular similarities argued for closer to 5 million years. In lieu of being able to read DNA directly, or even using protein sequences encoded by genes, Sarich and Wilson took an indirect measure: the strength of antibody reactions against a particular protein, in their case a blood protein called albumin. Any animal with an immune system would serve: they used rabbits. It's an ingenious technique. Rabbits are injected with albumin from (say) a human, and their antibodies collected. These anti-human versions react strongly against human albumin of course, but also more weakly against chimpanzee, gorilla and monkey albumin. The strength of reaction is used to measure the similarity between albumin proteins of different species.

Sarich and Wilson's rabbits revealed that humans, chimpanzees and gorillas were far more similar at the molecular level than expected, their differences accounting for only about a sixth of that between apes and Old World monkeys. As it happened, the split from Old World monkeys could be dated with the hard evidence of fossils. It took place around 30 million years ago, so Sarich and Wilson were able to use this to calibrate their estimate of the split between human and chimpanzee albumins: they put it at 5 million years ago. This is a classic example of using the 'molecular clock', something we will meet in earnest in the Epilogue to the Velvet Worm's Tale. As we shall discuss there, the molecular clock can have problems when the species in question are very different creatures, with radically different rates of molecular evolution. And there can be issues with problematic fossil 'calibration' points. Neither problem should apply strongly in the case of the apes and monkeys, however. And indeed the Sarich and Wilson estimate became widely accepted. As geneticists developed more sophisticated techniques for reading and analysing DNA sequences, similar studies tended to agree, more or less, with the original estimates. Until recently, the average date of divergence of chimpanzee and human genes was placed at 6 or perhaps 7 million years ago.

If we take these divergence dates at face value, we can use them to calculate the yearly mutation rate, as long as we discount mutations that have been subject to natural selection. We simply divide the number of mutations that differ between humans and chimps by the number of years separating the sequences, remembering that any two modern species are separated by twice the number of years since their common ancestor (mutations accumulate down both forks from the common ancestor). The percentage of single letters that differ is 1.23 per cent, and the vast majority are thought to be invisible to natural selection. For ease of calculation, the Sarich/Wilson dates of just over 6 million years can be equated to 12.3 million years of total separation. The percentage difference and separation time then cancel, giving a mutation rate of exactly one in a billion DNA letters every year.

So much for the indirect, Sarich and Wilson method for measuring mutation rates. How about the direct one? Does it agree? A typical example is a recent study by Augustine Kong and colleagues from Reykjavik. Sequencing a large number of Icelandic families, they saw that children differed from their parents by on average around one DNA letter in every 80 million.* If we take an average human generation time of 25 years (also the figure found in wild chimpanzees), that corresponds to one mutation in every 2 billion letters per year: half the Sarich/Wilson estimate. If instead we assume an average human generation time of 30 years, a figure more in line with historical European records and studies of hunter-gatherer populations, the per-year mutation rate which emerges is yet smaller still.

We need to treat these figures with caution. Genome sequencing itself is error-prone, and we must take care not to mix up true mutations with mistakes made by sequencing machines. But after taking precautions to avoid such errors, most studies are converging on approximately the same answer: one in 2 billion DNA letters per

---

* Incidentally, since we have billions of letters in our genome, that means that you and I clearly must be mutants. In fact it implies that our 6 billion DNA letters experience, on average, around 80 new copying errors each generation. With the world population in the billions, this also means that in a single generation, almost every DNA letter in the human genome will be miscopied, somewhere in the world. But of course, the overwhelming majority of these mutations end up being lost by chance.

year. Moreover this Reykjavik figure is roughly the same mutation rate obtained from a similar study of a wild chimpanzee population, and also from studies of disease-causing mutations in large populations. To top it all, it is backed up by analysis of ancient DNA. Qiaomei Fu and colleagues from Svante Pääbo's group in Leipzig compared DNA in modern humans with ancient DNA from a human thigh bone, eroded out of a cold Siberian river bank. We have had 45,000 more years to accumulate new mutations than had their ancient human. The difference in the number of mutations between the genomes – the 'missing mutations' – can therefore be used to estimate the mutation rate. Again, they found it to be around one in 2 billion letters per year. This slower mutation rate, about half as great as the previous molecular clock calculations, is evidently not just a recent thing. On the face of it we have a problem, though not an insuperable one, as we shall see.

Should we be worried about the twofold discrepancy in calculated mutation rates? In many respects, not. Given the difficulty of the calculation, the two rates are not all that far apart. And both are compatible with the number of copying errors per cell division, as estimated from biochemistry. Nevertheless, for students of human prehistory, a halving of the mutation rate makes a great difference. If this mutation rate applies all the way back to our common ancestor with chimpanzees, then dates of the Sarich/Wilson school are inaccurate, and should be doubled. No longer would we propose a 6-million-year-split between humans and chimpanzees. Instead the split should be more like 12 million years. Other dates are pushed back too. For example, it was originally thought that Neanderthals split from Archaics around 350,000 years ago. The new estimates of mutation rate suggest more like 700,000 years.

There is a way to reconcile both estimates. In the distant past, our mutation rates could have been higher, then gradually become slower on the line leading to humans, and perhaps to chimps and gorillas too. A version of this idea, the 'hominoid slowdown', was proposed as far back as 1985 by the distinguished molecular anthropologist Morris Goodman. Goodman's slowdown was not an *ad hoc* device to explain the very dilemma we are now facing. He originally suggested it in order to explain why primates in general (and humans in particular) showed slower rates of molecular change compared to other animals such as cows. A slowdown is not unreasonable. It could happen, for example, if our common ancestors with chimps had roughly the same mutation rate per generation, but modern chimps and ourselves have longer generation times. Whatever the explanation, some sort of molecular slowdown is now seeming quite likely. That would push the divergence between chimp and human genes somewhat further back in time than the Sarich estimate, although perhaps not twice as far. Our current guess is that the average genetic divergence occurred around 10 million years ago. We have gone down the same route when quoting dates in the previous chapters, such as for our recent out-of-Africa exodus. Regardless of whether our figures are exactly right, it is uncontroversial to say that human mutation rates are lower – and our date estimates more variable – than we once thought.

EPILOGUE TO THE CHIMPANZEE'S TALE

In the Chimpanzee's Tale we argued that the average divergence between human and chimpanzee genes traces back further than we once thought, to 10 million years ago or more. Yet at the start of this chapter we quoted a date for Rendezvous 1 of between 5 and 7 million years ago. This apparent contradiction is straightforward to explain. We have already seen that gene copies within a single interbreeding population (in fact even within your own body) can coalesce millions of years into the past. When a species splits into two, the genes in each population will carry with them a pre-existing history. That means splits between genes can predate splits between species by millions of years.

Making a few simplifying assumptions, we can estimate the time between the divergence of species and the divergence of genes. In a recent analysis of the gorilla genome, Aylwyn Scally and colleagues observed that the average time since human and gorilla genes split was about 12 million years. According to their calculations, this corresponds to an ancestral split into separate populations about 9 million years ago, if we assume that split was clean and instantaneous. Similarly their average human-chimp genetic divergence time, 10 million years, could be produced by an ancestral split as recent as 6 million years ago, depending on the population sizes involved. In general, average gene divergence dates are likely to predate the splitting of populations by perhaps a few million years. Although that is a relatively large difference for these initial rendezvous points, it is of decreasing importance as our rendezvous start to be measured in many tens or hundreds of millions of years.

So our contradiction in dates is rather easy to resolve. But there is another implication of genetic history which also throws up apparent contradictions. It involves the relationships between different species. The bonobo is well placed to take up the tale.

THE BONOBO'S TALE

The bonobo, *Pan paniscus*, looks pretty much like a common chimpanzee, *Pan troglodytes*, and before 1929 they were not recognised as separate species. The bonobo, despite its other name of pygmy chimpanzee, which should be abandoned, is not noticeably smaller than the common chimpanzee. Its body proportions are slightly different, and so are its habits, and that is the cue for its tale. The primatologist Frans de Waal put it neatly: 'The chimpanzee resolves sexual issues with power; the bonobo resolves power issues with sex . . .' Bonobos use sex as a currency of social interaction, somewhat as we use money. They use copulation, or copulatory gestures, to appease, to assert dominance, to cement bonds with other troop members of any age or sex, including small infants. Paedophilia is not a hang-up with bonobos; all kinds of philia seem fine to them. De Waal describes how, in a group of captive bonobos that he watched, the males would develop erections as soon as a keeper approached at feeding time. He speculates that this is in preparation for sexually mediated food-sharing. Female bonobos pair off to practise so-called GG (genital–genital) rubbing.

> One female facing another clings with arms and legs to a partner that, standing on both hands and feet, lifts her off the ground. The two females then rub their genital swellings laterally together, emitting grins and squeals that probably reflect orgasmic experiences.

The 'Haight-Ashbury' image of free-loving bonobos has led to a piece of wishful thinking among nice people, who perhaps came of age in the 1960s – or maybe they are of the 'medieval bestiary' school of thought, in which animals exist only to point moral lessons to us. The wishful thinking is that we are more closely related to bonobos than to common chimpanzees. The Margaret Mead in us feels closer to this gentle role-model than to the patriarchal, monkey-butchering chimpanzee. Unfortunately, like it or not, this hope conflicts with a standard tenet of zoological dogma. Common chimpanzees and bonobos share a common ancestor with each other 2 million years ago or less – much more recently than Concestor 1. This is usually taken as proof that *Pan paniscus* and *P. troglodytes* are exactly equally related to us. It is a general and commonly assumed principle, and was the moral of this tale in the first edition of this book. But there are two subtle twists to the Bonobo's Tale, and we must admit to having overlooked them in the first edition. Both have the potential to undermine zoological dogma, and we give the bonobo the pleasant task of explaining our oversight.

The first twist comes when we think about family trees of individuals: pedigrees. In our own pedigrees we tend to classify relatives in terms of their nearest link to us. This is slightly misleading. Take, for example, your third cousins (you almost certainly have some, even if you don't know who they are). Contrary to what you might think, your third cousins aren't all *exactly* equal relatives. That's because they don't just share a single ancestor with you. In fact, every one of their ancestors must be shared with you, to different degrees. One of your third cousins might well also be your sixth cousin through a different route. Another might be your eighth cousin through three different routes, and so on. The exact details don't matter, the point is that the true degree of relationship in a pedigree is not a discrete measure but an average, a blurry continuum.

Genealogists mostly choose to ignore this, and for practical reasons look at only the closest links. The fact that your third cousin is also your sixth cousin twice removed is ignored. But for very distant relations, the closest link – what we have named the concestor – may not be a good measure of the relationship in general. Hybridisation between two species is the extreme example. Take, for example, the following *reductio ad absurdum*. What if, almost 2 million years ago, a particularly adventurous proto-bonobo managed to mate with one of our australopithecine ancestors, producing a single fertile daughter who interbred into the pre-human population.* She would be one

---

*We say 'daughter' rather than son because of 'Haldane's rule'. The polymath J. B. S. Haldane, whom we shall meet at several points in this book, pointed out that in interspecies hybrids, if one gender is infertile, it is invariably the one with two different sex chromosomes. Thus in mammals, where the male has X and Y sex chromosomes, hybrids such as male mules are always infertile, whereas female mules (XX) can sometimes have normal babies. In birds and butterflies this pattern is reversed, because males are defined by their two Z chromosomes, whereas a W and a Z define a female. This pattern has long intrigued evolutionary biologists, particularly those interested in the origin of new species.

hybrid among maybe hundreds of thousands of 'normal' ancestors. Chances are that would make very little difference to the species today. We might not even have inherited any DNA from her. Nevertheless, if we based the evolutionary tree on the most recent common ancestor, it would make humans the sister group to bonobos, leaving the common chimpanzee as a more distant cousin to both. That would make a nonsense of the patent similarities linking bonobos to chimps. Even under this *reductio* we would surely still want to treat *Pan paniscus* and *P. troglodytes* as joint 'closest' relatives, despite humans sharing a recent one-off tie with bonobos. While this is a deliberately absurd example, hybridisation certainly does occur between many species. Even chimpanzees and bonobos, considered true species by all zoologists today, can hybridise in captivity, and may have occasionally done so in the recent past, although hard evidence is difficult to come by.

This discussion of hybrids and pedigrees is admittedly rather theoretical, especially as we stand no chance of knowing all the historical lines of interbreeding that link our three species. That is why our preference is to turn to DNA, and the outward features of bodies which dance to its molecular tune. It is here that we hit the second twist. We previously saw that different genes can show different relationships. For example, historical interbreeding means that from the point of view of some genes, Europeans are closer cousins to Neanderthals than Africans. Here we are making a wider point. From the genes' point of view, patterns of conflicting relationships often extend across species, even ones as different as bonobos, chimps, gorillas and humans. Moreover, this does not require interbreeding or hybridisation. It is down to an important process called 'incomplete lineage sorting'. The effect is easiest to describe when different branches ('lineages') of a gene tree can be easily labelled. Consider for instance the distinct 'A' and 'B' branches in the tree of the ABO blood group gene that we met in the Epilogue to Eve's Tale. We previously saw that from the gene's eye view, an 'A' is a closer relative of another 'A' than to any 'B', whatever species the genes come from. For example, a human 'A' and a bonobo 'A' are closer to each other than a chimpanzee 'B'. What if, hypothetically, humans and bonobos lost all blood groups except A, and chimps lost all except B. That would mean from the point of view of the ABO gene (and the blood transfusion service) humans and bonobos would be closest relatives, to the exclusion of all chimpanzees. Contrary to outward appearance, this particular aspect of our biology would ally humans with bonobos.

In the case of blood group genes, such very particular losses are unlikely, because both the A and B lineages of the tree are kept going within each species by natural selection. That is unusual. In most genes, new lineages continually spring into being, and frequently persist for some time until eventually one wins out, and the others go extinct. As long as our ancestral population contained multiple genetic lineages – which it did – chance extinctions of the sort we have just discussed can sometimes create these unexpected relationships.

Now that we can peer into the genome, we see this all over the place, even in genes previously thought to be uncontroversial. Well-established relationships, such as between humans, chimps and gorillas, still show incomplete lineage sorting at the molecular level. In fact, almost a third of our genome argues against humans and chimps being each others' closest relatives, either linking us more closely to gorillas, or placing chimps and gorillas together. Although rarely acknowledged by traditional taxonomists, this is bound to mean that our external features show conflicting evolutionary relationships too. More relevant to our tale is a calculation made by Kay Prüfer and colleagues, when they sequenced the bonobo genome in 2012. They found 1.6 per cent of human DNA is more closely related to bonobos than to common chimpanzees. And here's the killer: 1.7 per cent of our DNA shows a closer relationship to chimps than bonobos. If anything, at the biological level we are (marginally) closer relatives of common chimpanzees than bonobos.

This subversive point is a gratifying example of why the evolutionarily important unit is not the individual or the species but the 'gene'. Or to be more precise, the partial fragment of DNA sequence, for even within a single gene different letters can show different patterns of inheritance. In fact, Prüfer and his colleagues estimate that 'about 25 per cent of human genes contain parts that are more closely related to one of the two apes than the other'.

That leaves one uneasy question, qualm even. Don't both the pedigree and (especially) the genetic arguments subvert the whole premise of an evolutionary tree of life? Don't they undermine our concept of rendezvous points? We will tackle this in the Epilogue to the Gibbon's Tale. For the moment, we simply note that the conflict between relationships at the genetic level usually involves only a minority of the genome. Over 70 per cent of our genes argue that chimps and bonobos are our closest relatives, not gorillas. And over 90 per cent of our genome affirms that common chimpanzees and bonobos are our equal cousins. We can build a tree of life based on the majority opinion among genes. Or perhaps more accurately, we can build a tree of life that is maximally informative, in the true statistical sense of the word. A tree that tells us what to expect when approaching an organism for the first time. A tree that informs us *a priori* not to expect any particular feature – either genetic or morphological – to link us more closely to bonobos than common chimpanzees.

By the way, it does not follow from this genetic argument that we *resemble* common chimpanzees and bonobos equally. If chimpanzees have changed more than bonobos since the shared ancestor, Concestor 1, we might be more like bonobos than chimpanzees, or vice versa – and we shall probably find different things in common with both our *Pan* cousins, perhaps in roughly equal measure. But as a rule of thumb, we should not expect to be closer to one than the other, either in kinship or in appearance, and this applies within each pilgrim band as we journey into the past.

Our pilgrim's progress towards our prehistoric Canterbury requires a final qualification. We have been arguing that we must view our pilgrims in terms of their DNA and its physical manifestation, treating the relationships between creatures as averages among gene trees. Yet this is slightly at odds with the idea of concestors and rendezvous points. The most recent common ancestor of a set of individuals – the concestor – is a feature of a pedigree: a family tree of individuals, not a tree of genes. Why use concestors to pace our pilgrimage?

Although in this book we are using gene trees to map our relationship to the rest of life, using them to date our journey is problematic, because different parts of the genome coalesce at radically different times. Gene divergences vary by millions or even tens of millions of years. True, we could take an average, but this is not linked to any meaningful point in history. Alternatively, we could place each redezvous point at the time of speciation. Yet this is also an ill-defined date, because isolation seldom happens instantaneously. As we saw, even 'true' species like chimps and bonobos retain the ability to hybridise. Seldom is there a specific moment at which we can claim that one species has definitively become two.

Defining an exact date of speciation is to impose an arbitrary value on what is a biological continuum. If we wish to pick a single definitive point, a single milestone, we are left with one option: the instant in the past when two current-day species last shared a common genealogical ancestor. This will be more recent than the average genetic divergence, and (because of hybridisation) also later than most estimates of the 'time of speciation'. But because our concestors define precise and indisputable moments in history, using them to pace out our pilgrimage is the most principled approach we can take.

GORILLAS

The molecular clock tells us that Rendezvous 2, where the gorillas join us, again in Africa, is only a few million years further into our pilgrimage than Rendezvous 1. Eight million years ago, North and South America were not joined, the Andes had not undergone their major uplift and the Himalayas only just so. Nevertheless the continents would have looked pretty much as now and the African climate, while less seasonal and slightly wetter, would have been similar. Africa was more thoroughly forested then than now – even the Sahara would have been wooded savannah at the time.

Unfortunately there are no fossils to bridge the gap between Concestors 2 and 1, nothing to guide us in deciding whether Concestor 2, which is perhaps our 300,000-greats-grandparent, was more like a gorilla or more like a chimpanzee or, indeed, more like a human. My guess would be chimpanzee, but this is only because the huge gorilla seems more extreme, and less like the generality of apes. Don't let's exaggerate the unusualness of gorillas, however. They are not the largest apes that have ever lived. The Asian ape *Gigantopithecus*, a sort of giant orang utan, would have stood head and massive shoulders over the largest gorilla. It lived in China, and went extinct only recently, about half a million years ago, overlapping with *Homo erectus* and archaic *Homo sapiens*. This is so recent that some enterprising fantasists have gone so far as to suggest that the Yeti or Abominable Snowman of the Himalayas . . . but I digress. *Gigantopithecus* presumably walked like a gorilla, maybe on the knuckles of its hands and the soles of its feet as gorillas and chimpanzees do, and as orang utans, committed as they are to life up trees, do not.

As we have previously discussed, it is unclear whether Concestor 2 was a knuckle-walker. Either way, it seems probable that, like chimpanzees, it spent time in trees as well, especially at night. Natural selection under a tropical sun favours dark pigmentation as protection against ultraviolet rays, so if we had to guess at Concestor 2's colour we would presumably say black or dark brown. All apes except humans are hairy, so it would be surprising if Concestors 1 and 2 were

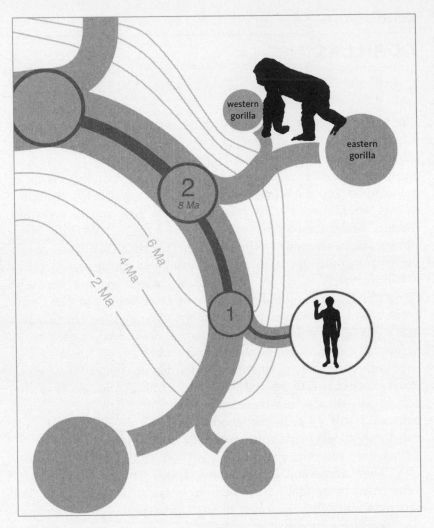

Gorillas join. We have zoomed out a little from the fractal depicted at the previous rendezvous. Chimpanzees and humans can still be seen in the same positions (although chimpanzees are no longer distinguished by a silhouette), but we can also see the next band of pilgrims – the gorillas – joining at Rendezvous 2. The position of humans is marked by a silhouette within a circle, which will go on to represent our approximate place in the tree of life as we travel further backwards in time. Note that humans occupy no special place in this tree: the direction of branching is determined only by the number of living species on each branch. The branch with more species forks to the right, and the branch with fewer species to the left.

Gorillas are currently thought to comprise two species and four subspecies. This silhouette depicts the western gorilla (*Gorilla gorilla*), which lives in the lower Congo basin, near the coast. The slightly larger eastern species (*Gorilla beringei*) lives inland, in lowland and mountainous forested areas in Rwanda, Uganda, and the Democratic Republic of the Congo.

not. Since chimpanzees, bonobos and gorillas are inhabitants of deep forest, it is plausible to locate Rendezvous 2 in a forest, in Africa, but there is no strong reason to guess any particular part of Africa.

Gorillas are not just giant chimpanzees, they are different in other respects which we need to think about in trying to reconstruct Concestor 2. Gorillas are entirely vegetarian. The males have harems of females. Chimpanzees are more promiscuous, and the differences in breeding systems have interesting consequences on the size of their testes as we shall learn from the Seal's Tale. I suspect that breeding systems are evolutionarily labile, meaning easily changed. I don't see any obvious way to guess where Concestor 2 stood in this respect. Indeed, the fact that different human cultures today show a large range of breeding systems, from faithful monogamy to potentially very large harems, reinforces my reluctance to speculate about such matters for Concestor 2, and persuades me to bring my speculations as to its nature to a swift end.

Apes, perhaps especially gorillas, have long been potent generators – and victims – of human myths. The Gorilla's Tale considers our changing attitudes to our closest cousins.

THE GORILLA'S TALE

The rise of Darwinism in the nineteenth century polarised attitudes towards the apes. Opponents who might have stomached evolution itself balked with visceral horror at cousinship with what they perceived as low and revolting brutes, and desperately tried to inflate our differences from them. This was nowhere more true than with gorillas. Apes were 'animals'; we were set apart. Worse, where other animals such as cats or deer could be seen as beautiful in their own way, gorillas and other apes, precisely because of their similarity to ourselves, seemed like caricatures, distortions, grotesques.

Darwin never missed an opportunity to put the other side, sometimes in little asides such as his charming observation in *The Descent of Man* that monkeys 'smoke tobacco with pleasure'. T. H. Huxley, Darwin's formidable ally, had a robust exchange with Sir Richard Owen, the leading anatomist of the day, who claimed (wrongly as Huxley showed) that the 'hippocampus minor' was uniquely diagnostic of the human brain. Nowadays, scientists not only think we resemble apes. We include ourselves within the apes, specifically the African apes. We emphasise, by contrast, the distinctness of apes, including humans, from monkeys. To call a gorilla or a chimpanzee a monkey is a solecism.*

It has not always been so. In former times, apes were frequently lumped with monkeys, and some of the early descriptions confused apes with baboons, or with Barbary macaques, which indeed are still known as Barbary apes. More surprisingly, long before people thought in terms of evolution at all, and before apes were clearly distinguished from each other or from monkeys, great apes were often confused with humans. Agreeable as it would be to approve this apparent prescience of evolution, it unfortunately may owe more to racism. Early white explorers in Africa saw chimpanzees and gorillas as close kin only to black humans, not to themselves. Interestingly, tribes in both South East Asia and Africa have traditional legends suggesting a reversal of evolution as conventionally seen: their local great apes are regarded as humans who fell from grace. Orang utan means 'man of the woods' in Malay.

A picture of an 'Ourang Outang' by the Dutch doctor Bontius in 1658 is, in T. H. Huxley's words, 'nothing but a very hairy woman of rather comely aspect and with proportions and feet wholly human'. Hairy she is except, oddly, in one of the few places where a real woman is: her pubic region is conspicuously naked. Also very human are the pictures made, a century later, by Linnaeus's pupil Hoppius (1763). One of his creatures has a tail, but is otherwise wholly human, bipedal, and carries a walking stick. Pliny the Elder says that 'the tailed species have even been known to play at draughts' (American 'checkers').

One might have thought such a mythology would have prepared our civilisation for the idea of evolution when it arrived in the nineteenth century, and might even have accelerated its discovery. Apparently not. Instead, the picture is one of confusion between apes, monkeys and humans. This makes it hard to date the scientific discovery of each species of great ape, and it is often unclear which one is being discovered. The exception is the gorilla, which became known to science the most recently.

In 1847 an American missionary, Dr Thomas Savage, saw in the house of another missionary on the Gaboon river 'a skull represented by the natives to be a monkey-like animal, remarkable for its size, ferocity and habits'. The unjust reputation for ferocity, later to be

---

* This may be one of the only points where the authors disagree. Yan Wong thinks the word 'monkey' should be redefined so that it includes apes (and therefore humans) too. But that is not currently the norm.

Confusion between apes, monkeys and humans. Engraving by E. Hoppius, in Linnaeus' *Amoenitates Academicae, VI. Anthropomorpha,* ('Human forms'), published in 1763. The characters are labelled Troglodytes, Lucifer, Satyr, and 'Pygmie'.

hyperbolised in the story of King Kong, comes through loud and clear in an article about the gorilla in the *Illustrated London News* published in the same year as the *Origin of Species.* This piece is replete with falsehoods of a quantity and magnitude that try even the high standards set by travellers' tales of the time:

> . . . a close inspection is almost an impossibility, especially as the moment it sees a man it attacks him. The strength of the adult male being prodigious, and the teeth heavy and powerful, it is said to watch, concealed in the thick branches of the forest trees, the approach of any of the human species, and, as they pass under the tree, let down its terrible hind feet, furnished with an enormous thumb, grasp its victim round the throat, lift him from the earth, and, finally, drop him on the ground dead. Sheer malignity prompts the animal to this course, for it does not eat the dead man's flesh, but finds a fiendish gratification in the mere act of killing.

Savage believed the skull in the missionary's possession belonged 'to a new species of Orang'. He later decided that his new species was none other than the 'Pongo' of earlier travellers' tales in Africa. In naming it formally, Savage, with his anatomist colleague Professor Wyman, avoided *Pongo* and revived *Gorilla*, the name used by an ancient Carthaginian admiral for a race of wild hairy people which he claimed to have found on an island off the African coast. Gorilla has survived as both the Latin and common name for Savage's animal, while *Pongo* is now the Latin name of the orang utan of Asia.

Judging from its location, Savage's species must have been the western gorilla, *Gorilla gorilla.* Savage and Wyman put it in the same genus as the chimpanzees, and called it *Troglodytes gorilla.* By the rules of zoological nomenclature, *Troglodytes* had to be relinquished by both chimpanzee and gorilla because it had already been used for – of all things – the tiny wren. It survived as the specific name of the common chimpanzee, *Pan troglodytes,* while the former specific name of Savage's gorilla was promoted to become its generic name, *Gorilla.* The 'mountain gorilla' was 'discovered' – he shot it! – by the German Robert von Beringe as late as 1902. As we shall see, it is now regarded as a subspecies of the eastern gorilla, and the whole eastern species now – unfairly, one might think – bears his name: *Gorilla beringei.*

Savage did not believe his gorillas really were the race of islanders reported by the Carthaginian sailor. But the 'pygmies', originally mentioned by Homer and Herodotus as a legendary race of very small humans, were later assumed by seventeenth- and eighteenth-century explorers to be none other than the chimpanzees then being discovered in Africa. Tyson (1699) shows a drawing of a 'Pygmie' which, as Huxley says, is plainly a young chimpanzee although it, too, is depicted walking upright and carrying a walking stick. Now, of course, we use the word pygmy for small humans again.

This leads us back to the racism which, until relatively late in the twentieth century, was endemic in our culture. Early explorers often assigned the native peoples of the forests a closer affinity with chimpanzees, gorillas or orangs than with the explorers themselves. In the nineteenth century, after Darwin, evolutionists often regarded African peoples as intermediate between apes and Europeans, on the upward path to white supremacy. This is, of course, complete nonsense. As the Bonobo's Tale explains, such ideas are not only factually wrong. They violate a fundamental principle of evolution. Racism and speciesism, and our perennial confusion over how inclusively we wish to cast our moral and ethical net, are brought into sharp and sometimes uncomfortable focus in the history of our attitudes to our fellow humans, and our attitudes to apes – our *fellow* apes.*

* The Great Ape Project, dreamed up by the distinguished moral philosopher Peter Singer, goes to the heart of the matter by proposing that great apes should be granted, as far as is practically possible, the same moral status as humans. My own contribution to the book *The Great Ape Project* is one of the essays reprinted in *A Devil's Chaplain.*

ORANG UTANS

Molecular evidence puts Rendezvous 3 – where our ancestral pil-grimage is joined by the orang utans – at 14 million years ago, right in the middle of the Miocene Epoch. Although the world was starting to enter its current cool phase, the climate was warmer and the sea levels higher than at present. Coupled with minor differences in the positions of the continents, this led to the land between Asia and Africa, as well as much of south-east Europe, being intermittently submerged by sea. This bears, as we shall see, on our calculation of where Concestor 3, perhaps our two-thirds-of-a-million-greats-grandparent, might have lived. Did it live in Africa like 1 and 2, or Asia? As the common ancestor of ourselves and an Asian ape, we should be prepared to find it in either continent, and partisans of both are not hard to find. In favour of Asia is its richness of plausible fossils from around the right time, the mid-to-late Miocene. Africa, on the other hand, seems to be where the apes originated, before the beginning of the Miocene. Africa witnessed a great flowering of ape life in the early Miocene, in the form of proconsulids (several species of the early ape genus *Proconsul*) and others such as *Afropithecus* and *Kenyapithecus*. Our closest living relatives today, and all our post-Miocene fossils, are African.

But our special relationship to chimpanzees and gorillas has been known only for a few decades. Before that, most anthropologists thought we were the sister group to all the apes, and therefore equally close to African and Asian apes. The consensus favoured Asia as the home of our late Miocene ancestors, and some authorities even picked out a particular fossil 'ancestor', *Ramapithecus*. This animal is now thought to be the same as one previously called *Sivapithecus* which therefore, by the laws of zoological nomenclature, takes pre-cedence. *Ramapithecus* should no longer be used – a pity because the name had become familiar. Whatever one feels about *Sivapithecus/ Ramapithecus* as a human ancestor, many authorities agree that it is close to the line that gave rise to the orang utan and might even be the orang utan's direct ancestor. *Gigantopithecus* could be regarded as a kind of giant, ground-dwelling version of *Sivapithecus*. Several

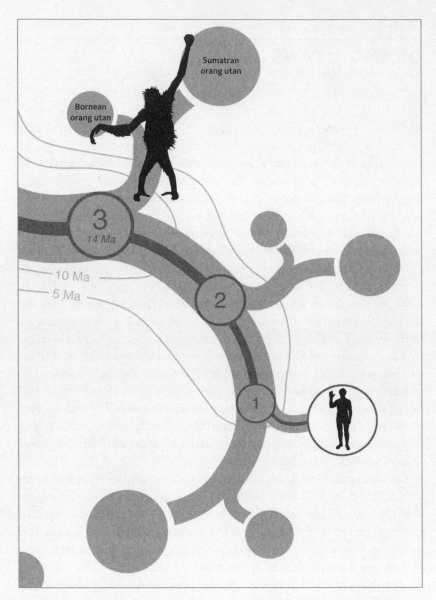

Orang utans join. The two species of Asian orang utan, Bornean (*Pongo pygmaeus*) and Sumatran (*Pongo abelii*), are generally accepted to join our pilgrimage at Rendezvous 3. Detailed notes for this and further fractal diagrams are given in the endnotes, on page 703.

other Asian fossils occur from about the right time. *Ouranopithecus* and *Dryopithecus* seem almost to be jostling for the title of most plausible human ancestor of the Miocene. If only, it is tempting to remark, they were in the right continent. As we shall see, this 'if only' just might turn out to be true.

If only the late Miocene apes were in Africa instead of Asia, we'd have a smooth series of plausible fossils linking the modern African apes all the way back to the early Miocene and the rich proconsulid ape fauna of Africa. When molecular evidence established beyond any doubt our affinities with the African chimpanzees and gorillas, rather than with the Asian orangs, seekers of human ancestors reluctantly turned their backs on Asia. They assumed, in spite of the plausibility of the Asian apes themselves, that our ancestral line must lie in Africa right through the Miocene and concluded that, for some reason, our African ancestors had not fossilised after the early burgeoning of proconsulid apes in the early Miocene.

That's where things stood until 1998, when an ingenious piece of lateral thinking appeared in a paper called 'Primate evolution – in and out of Africa' by Caro-Beth Stewart and Todd R. Disotell. This tale, of back and forth traffic between Africa and Asia, will be told by the orang utan. Its conclusion will be that Concestor 3 probably lived in Asia after all.

But never mind, for the moment, where it lived. What did Concestor 3 look like? It is the common ancestor of the orang utans and all today's African apes, so it might resemble either or both of them. Which fossils might give us helpful clues? Well, looking at the family tree, the fossils known as *Lufengpithecus, Oreopithecus, Sivapithecus, Dryopithecus* and *Ouranopithecus* all lived around the right time or slightly later. Our best-guess reconstruction of Concestor 3 (see plate 5) might combine elements of all five of these Asian fossil genera – but it would help if we could accept Asia as the location of the concestor. Let's listen to the Orang Utan's Tale and see what we think.

THE ORANG UTAN'S TALE

Perhaps we have been too ready to assume that our links with Africa go back a very long way. What if, instead, our ancestral lineage hopped sideways out of Africa around 20 million years ago, flourished in Asia until around 10 million years ago, and then hopped back to Africa?

On this view, all the surviving apes, including the ones that ended up in Africa, are descended from a lineage that migrated out of Africa into Asia. Gibbons and orang utans are descendants of these migrants who stayed in Asia. Later descendants of the migrants returned to Africa, where the earlier Miocene apes had gone extinct. Back in their old ancestral home of Africa, these migrants then gave rise to gorillas, chimpanzees and bonobos, and us.

The known facts about the drifting of the continents and the fluctuations of sea levels are compatible. There were land bridges available across Arabia at the right times. The positive evidence in favour of the theory depends upon 'parsimony': an economy of assumptions. A good theory is one that needs to postulate little, in order to explain lots. (By this criterion, as I have often remarked elsewhere, Darwin's theory of natural selection may be the best theory of all time.) Here we are talking about minimising our assumptions about migration events. The theory that our ancestors stayed in Africa all along (no migrations) seemed, on the face of it, more economical with its assumptions than the theory that our ancestors moved from Africa into Asia (a first migration) and later moved back to Africa (a second migration).

But that parsimony calculation was too narrow. It concentrated on our own lineage and neglected all the other apes, especially the many fossil species. Stewart and Disotell did a recount of the migration events, but they counted those that would be needed to explain the distribution of all the apes including fossils. In order to do this, you first have to construct a family tree on which you mark all the species about which you have sufficient information. The next step is to indicate, for each species on the family tree, whether it lived in Africa or Asia. In the diagram on the opposite page, which is taken from Stewart and Disotell's paper, Asian fossils are highlighted in black, African ones are in white. Not all the known fossils are there, but Stewart and Disotell did include all whose position on the family tree could be clearly worked out. They also drew in the Old World monkeys, who diverged from the apes around 25 million years ago (the most obvious difference between monkeys and apes, as we shall see, is that the monkeys retained their tails). Migration events are indicated by arrows.

Taking into account the fossils, the 'hop to Asia and back again' theory is now more parsimonious than the 'our ancestors were in Africa all along' theory. Leaving out the monkeys which, on both

In and out of Africa. Stewart and Disotell's family tree of African and Asian apes. Swollen areas represent dates known from fossils, while the lines linking these to the tree are inferred from parsimony analysis. Arrows represent inferred migration events. Adapted from Stewart and Disotell [403].

theories, demand two migration events from Africa to Asia, it need postulate only two ape migrations, as follows:

1. A population of apes migrated from Africa to Asia around 20 million years ago and became all the Asian apes including the living gibbons and orang utans.
2. A population of apes migrated back from Asia to Africa and became today's African apes including us.

Conversely, the 'our ancestors were in Africa all along' theory demands six migration events to account for ape distributions, all from Africa to Asia, by ancestors of the following:

1. Gibbons, around 18 million years ago
2. *Oreopithecus*, around 16 million years ago
3. *Lufengpithecus*, around 15 million years ago

4. *Sivapithecus* and orang utans, around 14 million years ago
5. *Dryopithecus*, around 13 million years ago
6. *Ouranopithecus*, around 12 million years ago

Of course all these migration counts are valid only if Stewart and Disotell have got the family tree right, based on anatomical comparisons. They think, for example, that among the fossil apes, *Ouranopithecus* is the closest cousin to the modern African apes (its branch is the last to come off the family tree in the diagram before the African apes). The next closest cousins, according to their anatomical assessments, are all Asian (*Dryopithecus*, *Sivapithecus*, etc.). If they have got the anatomy all wrong: if, for instance, the African fossil *Kenyapithecus* is actually closest to the modern African apes, then the migration counts would have to be done all over again.

The family tree was itself constructed on grounds of parsimony. But it is a different kind of parsimony. Instead of trying to minimise the number of geographical migration events we need to postulate, we forget about geography and try to minimise the number of anatomical coincidences (convergent evolution) we need to postulate. Having got our family tree without regard to geography, we then superimpose the geographical information (the black and white coding on the diagram) to count migration events. And we conclude that it is most likely that the 'recent' African apes, that is gorillas, chimpanzees and humans, arrived from Asia.

Now here's an interesting little fact. A leading textbook of human evolution, by Richard G. Klein of Stanford University, gives a fine description of what is known of the anatomy of the main fossils. At one point Klein compares the Asian *Ouranopithecus* and the African *Kenyapithecus* and asks which most resembles our own close cousin (or ancestor) *Australopithecus*. Klein concludes that *Australopithecus* resembles *Ouranopithecus* more than it resembles *Kenyapithecus*. He goes on to say that, if only *Ouranopithecus* had lived in Africa, it might even make a plausible human ancestor. 'On combined geographic-morphologic grounds', however, *Kenyapithecus* is a better candidate. You see what is going on here? Klein is making the tacit assumption that African apes are unlikely to be descended from an Asian ancestor, even if the anatomical evidence suggests that they were. Geographical parsimony is being subconsciously allowed to pull rank over anatomical parsimony. Anatomical parsimony suggests that *Ouranopithecus* is a closer cousin to us than *Kenya-*

pithecus is. But, without being explicitly so called, geographical parsimony is assumed to trump anatomical parsimony. Stewart and Disotell argue that, when you take into account the geography of *all* the fossils, anatomical and geographical parsimony *agree* with each other. Geography turns out to agree with Klein's initial anatomical judgement that *Ouranopithecus* is closer to *Australopithecus* than *Kenyapithecus* is.

This argument may not be settled yet. It is a complicated business juggling anatomical and geographical parsimony. When it was first published, Stewart and Disotell's paper unleashed a flourishing correspondence in the scientific journals, both for and against. Seventeen years later, Caro-Beth Stewart writes to us that 'I have seen nothing published since our paper that refutes it', and on balance the 'hop to Asia and back' theory of ape evolution still seems to fit with the available evidence. Two migration events is more parsimonious than six. And there really do seem to be some telling resemblances between the late Miocene apes in Asia and our own line of African apes such as *Australopithecus* and chimpanzees. It is only a preference 'on balance', but it leads us to locate Rendezvous 3 (and Rendezvous 4) in Asia rather than Africa.

The moral of the Orang Utan's Tale is twofold. Parsimony is always in the forefront of a scientist's mind when choosing between theories, but it isn't always obvious how to judge it. And possessing a good family tree is often an essential first prerequisite to powerful further reasoning in evolutionary theory. But building a good family tree is a demanding exercise in itself. The ins and outs of it will be the concern of the gibbons, in the tale that they will tell us in melodious chorus after they join our pilgrimage at Rendezvous 4.

GIBBONS

Rendezvous 4, where we are joined by the gibbons, occurs around 18 million years ago, probably in Asia, in the warmer and more wooded world of the early Miocene. Depending on which authority you consult, there are up to 18 modern species of gibbons. All live in South East Asia, including Indonesia and Borneo. They used to be placed in the genus *Hylobates*, with the larger siamang separated off on account of its distinctive voice-amplifying throat sac. With the realisation that they divide into four groups, not two, the need to refer to 'gibbons and siamangs' has become outdated. I shall call them all gibbons.

Gibbons are small apes, and perhaps the finest arboreal acrobats that have ever lived. In the Miocene there were lots of small apes. Getting smaller and getting larger are easy changes to achieve in evolution. Just as *Gigantopithecus* and *Gorilla* got large independently of each other, plenty of apes, in the Miocene golden age of apes, got small. The pliopithecids, for instance, were small apes which flourished in Europe in the early Miocene and probably lived in a similar way to gibbons, without being ancestral to them. I suppose, for example, that they 'brachiated'.

Brachia is the Latin for 'arm'. Brachiation means using your arms rather than your legs to get about, and gibbons are spectacularly good at it. Their big grasping hands and powerful wrists are like upside-down seven-league boots, spring-loaded to slingshot the gibbon from branch to branch and from tree to tree. A gibbon's long arms, perfectly in tune with the physics of pendulums, are capable of hurling it across a sheer ten-metre gap in the canopy. My imagination finds high-speed brachiation more exciting even than flying, and I like to dream of my ancestors enjoying what must surely have been one of the great experiences life could offer. Unfortunately, current thinking doubts that our ancestry ever went through a fully gibbon-like stage, but it is reasonable to conjecture that Concestor 4, approximately our 1-million-greats-grandparent, was a small tree-dwelling ape with at least some proficiency in brachiation.

Among the apes, gibbons are also second only to humans in the

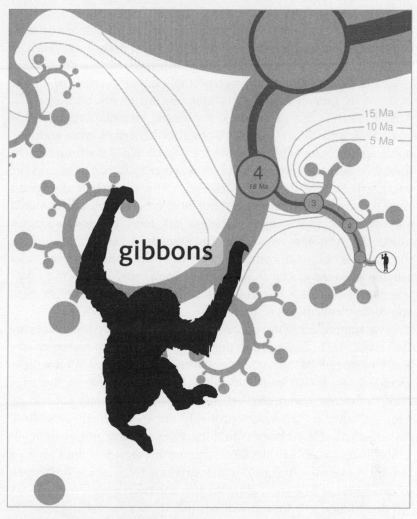

Gibbons join. There are roughly 16 or so species of gibbon, now recognised as falling into four groups, as described on page 155. In this fractal phylogeny, the relationships between these four groups have been taken from whole genome studies, which show an initial division between the *Hylobates* gibbons and all the others, but this is much disputed. Whatever arrangement used, it will merely summarise the majority vote among gene trees, as discussed in the Epilogue to the Gibbon's Tale. From this point on there is not enough room to show names of individual species: readers who wish to identify individual species, or explore the tree in more detail, are encouraged to visit the Ancestor's Tale website (www.ancestorstale.net) or the OneZoom website, from which it is derived (www.onezoom.org).

difficult art of walking upright. Using its hands only to steady itself, a gibbon will use bipedal walking to travel along the length of a branch, whereas it uses brachiation to travel across from branch to branch. If Concestor 4 practised the same art and passed it on to its gibbon descendants, could some vestige of the skill have persisted in the brain of its human descendants too, waiting to resurface again in Africa? That is no more than a pleasing speculation, but it is consistent with the recent reconstruction of *Ardipithecus* and certainly apes in general have a tendency to walk bipedally from time to time. We can also only speculate on whether Concestor 4 shared the vocal virtuosity of its gibbon descendants, and whether this might have presaged the unique versatility of the human voice, in speech and in music. Then again, gibbons are faithfully monogamous, unlike the great apes which are our closer relatives. Unlike, indeed, the majority of human cultures, in which custom and in several cases religion encourages (or at least allows) polygyny. We do not know whether Concestor 4 resembled its gibbon descendants, or its great ape descendants in this respect.*

Let's summarise what we can guess about Concestor 4, making the usual weak assumption that it had a good number of the features shared by all its descendants, which means all the apes including us. It was probably more dedicated to life in the trees than Concestor 3, and smaller. If, as I suspect, it hung and swung from its arms, its arms were probably not so extremely specialised for brachiation as those of modern gibbons, and not so long. It probably had a gibbon-like face, with a short snout. It didn't have a tail. Or, to be more precise, its tail vertebrae were, as in all the apes, joined together in a short internal tail, the coccyx.

I don't know why we apes lost our tail. It is a subject that biologists discuss surprisingly little. An exception is Jonathan Kingdon in *Lowly Origin*, but even he reaches no satisfactory closure. Zoologists faced with this kind of conundrum often think comparatively. Look around the mammals, note where taillessness (or a very short tail) has independently cropped up, and try to make sense of it. I don't

---

* Perhaps the good old-fashioned family values of the gibbons, and the pious hope that our evolutionary ancestors once shared them, should be drawn to the attention of the right-wing 'moral majority', whose ignorant and single-minded opposition to the teaching of evolution endangers educational standards in several backward North American States. Of course, to draw any moral would be to commit the 'naturalistic fallacy' but fallacies are what these people do best.

think anyone has done this systematically, and it would be a nice thing to undertake. Apart from apes, tail loss is found in moles, hedgehogs, the tailless tenrec *Tenrec ecaudatus*, guinea pigs, hamsters, bears, bats, koalas, sloths, agoutis and several others. Perhaps most interesting for our purposes, there are tailless monkeys, or monkeys with a tail so short it might as well not be there, as in a Manx cat. Manx cats have a single gene that makes them tailless. It is lethal when homozygous (present twice) so is unlikely to spread in evolution. But it has crossed my mind to wonder whether the first apes were 'Manx monkeys'. My normal bias is against such 'hopeful monster' theories of evolution, but could this be an exception? It would be interesting to examine the skeleton of tailless mutants of normally tailed 'Manx' mammals, to see whether they 'do' taillessness in the same kind of way as apes.

The Barbary macaque *Macaca sylvanus* is a tailless monkey and, perhaps in consequence, is often miscalled the Barbary ape. The 'Celebes ape' *Macaca nigra* is another tailless monkey. Jonathan Kingdon tells me it looks and walks just like a miniature chimpanzee. Madagascar has some tailless lemurs, such as the indri, and several extinct species including 'koala lemurs' (*Megaladapis*) and 'sloth lemurs', some of which were gorilla-sized.

Any organ which is not used will, other things being equal, shrink for reasons of economy if nothing else. Tails are used for a surprisingly wide variety of purposes among mammals. Sheep keep a fat reserve in the tail. Beavers use it as a paddle. The spider monkey tail has a horny gripping pad and is used as a 'fifth limb' in the treetops of South America. The massive tail of a kangaroo is spring-loaded to assist bounding. Hoofed animals use the tail as a fly whisk. Wolves and many other mammals use it for signalling, but this is likely to be secondary 'opportunism' on natural selection's part.

But here we must be especially concerned with animals who live up trees. Squirrel tails catch the air, so a 'leap' is almost like flying. Tree-dwellers often have long tails as counterweights, or as rudders for leaping. Lorises and pottos, whom we shall meet at Rendezvous 8, creep about the trees, slowly stalking their prey, and they have extremely short tails. Their relatives the bushbabies, on the other hand, are energetic leapers, and they have long feathery tails. Tree sloths are tailless, like the marsupial koalas who might be regarded as their Australian equivalents, and both move slowly in the trees like lorises.

In Borneo and Sumatra, the long-tailed macaque lives up trees, while the closely related pig-tailed macaque lives on the ground and has a short tail. Monkeys that are active in trees usually have long tails. They run along the branches on all fours, using the tail for balance. They leap from branch to branch with the body in a horizontal position and the tail held out as a balancing rudder behind. Why, then, do gibbons, who are as active in trees as any monkey, have no tail? Maybe the answer lies in the very different way in which they move. All apes, as we have seen, are occasionally bipedal, and gibbons, when not brachiating, run along branches on their hind legs, using their long arms to steady themselves. It is easy to imagine a tail being a nuisance for a bipedal walker. My colleague Desmond Morris tells me that spider monkeys sometimes walk bipedally, and the long tail is obviously a major encumbrance. And when a gibbon projects itself to a distant branch it does so from a vertically hanging position, unlike the monkey's horizontal leaping posture. Far from being a steadying rudder streaming out behind, a tail would be a positive drag for a vertical brachiator like a gibbon or, presumably, Concestor 4.

That is the best I can do. I think zoologists need to give more attention to the puzzle of why we apes lost our tail. The *a posteriori* counterfactual engenders pleasing speculations. How would the tail have sat with our habit of wearing clothes, especially trousers? It gives a different urgency to the classic tailor's question, 'Does Sir hang to the left or to the right?'

THE GIBBON'S TALE

Rendezvous 4 is the first time we greet a pilgrim band of more than two species. Any more than that, and there can be problems with deducing relationships. These problems will become worse as our pilgrimage advances. How to solve them is the topic of the Gibbon's Tale.*

---

* The subject matter of this tale inevitably makes it tougher than other parts of the book. Readers should either don thinking caps for the next 13 pages, or skip now to page 171 and return to the tale when they want their neurons exercised. Incidentally, I have often wondered what a 'thinking cap' actually is. I wish I had one. My benefactor Charles Simonyi, one of the world's greatest computer programmers, is said to wear a special 'debugging suit' which may help to account for his formidable success.

We have mentioned that the 16 or so species of gibbon fall into four groups. Each group has a different number of chromosomes and nowadays they are given separate genus names. They are *Hylobates* (seven species, of which the best-known is the white-handed gibbon *Hylobates lar*), *Hoolock* (the two Bengali hoolock gibbons, renamed in 2005), *Symphalangus* (the siamang), and *Nomascus* (six species of 'crested' gibbons). This tale explains how to build an evolutionary relationship, or phylogeny, relating the four groups.

Family trees can be 'rooted' or 'unrooted'. When we draw a rooted tree, we know where the ancestor is. Most of the tree diagrams in this book are rooted. But if we don't have any idea where the ancestor lies on a tree, we have to draw it as unrooted. Unrooted trees have no sense of direction, no arrow of time, and they are often shown in the form of star diagrams. Above are three examples, which exhaust the possibilities for relating four entities.

At every fork in a tree, it makes no difference which is the left and which the right branch. And so far (though that will change later in the tale) no information is conveyed by the lengths of the branches. A tree diagram whose branch lengths are meaningless is known as a cladogram (an unrooted cladogram in this case). The order of branching is the only information conveyed by a cladogram: the rest is cosmetic. For example, swivelling the position of *Hoolock* and *Hylobates* in tree (a) above makes no difference to the relationships.

These three unrooted cladograms represent the only possible ways of connecting four species, as long as we restrict ourselves to connections via branches that only ever split in two (dichotomies). As with rooted trees, it is conventional to discount three-way splits (trichotomies) or more (polytomies) as temporary admissions of ignorance – 'unresolved'.

Any unrooted cladogram turns into a rooted one the moment we specify the oldest point (the 'root') of the tree. It is not always an easy

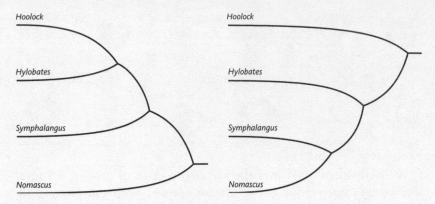

decision to make, something which will come back to haunt us at the end of our pilgrimage. And unfortunately the order of branching can change rather dramatically if we change the position of the root. For example, if we take the unrooted tree (a), and dangle the root between *Nomascus* and the other three gibbons, we obtain the rooted tree shown on the left above. Yet if we take the same unrooted tree (a), but this time place the root between *Hoolock* and the other three, we obtain the tree on the right. Both these trees have been suggested by gibbon researchers, and to the uninitiated the pattern of relationships looks very different. Nevertheless, they differ only in where the root has been placed.

How do we 'root' a tree? The usual method is to extend the tree to include at least one – and preferably more than one – 'outgroup': a species that is universally agreed beforehand to be reasonably distantly related to all the others. In the gibbon tree, for example, orang utans or gorillas – or indeed elephants or kangaroos – could do duty as the outgroup. However uncertain we may be about relationships *among* gibbons, we know that the common ancestor of *any* gibbon with a great ape or an elephant is older than the common ancestor of any gibbon with any other gibbon: it is uncontroversial to place the root of a tree that includes the gibbons and the great apes somewhere between the two.

These preliminaries out of the way, we can now get down to choosing which of the possible unrooted trees is likely to be correct. With our four gibbons this comes down to choosing between three unrooted trees. If we had five animals to classify we would need to consider 15 possible trees. But don't even try to count the number of possible trees for, say, 20 animals. It is up in the hundreds of millions of millions of millions.* As the count of trees rises steeply with the number of groups to be classified, even the fastest computer can take for ever. In principle, however, our task is simple. Of all possible trees we must choose that which best explains the similarities and differences between our groups.

How do we judge 'best explains'? Infinitely rich similarities and differences present themselves when we look at a set of animals. But they are harder to count than you might think. Often one 'feature' is an inextricable part of another. If you count them as separate, you've really counted the same one twice. As an extreme example suppose there are four millipede species, A, B, Y and Z. A and B resemble each other in all respects except that A has red legs and B has blue legs. Y and Z are the same as each other and very different from A and B, except that Y has red legs while Z has blue legs. If we count leg colour as a single 'feature' we correctly group AB apart from YZ. But if we naively count each of 100 legs as separate, their colours will give a hundredfold boost to the number of features supporting the alternative grouping of AY as against BZ. Everyone would agree that we have spuriously counted the same feature 100 times. It is 'really' only one feature, because a single embryological 'decision' determined the colour of all 100 legs simultaneously.

The same goes for left-right symmetry: embryology works in such a way that, with few exceptions, each side of an animal is a mirror image of the other. No zoologist would count each mirrored feature twice in making a cladogram, but non-independence isn't always so obvious. A pigeon needs a deep breastbone to attach the flight muscles. A flightless bird like a kiwi does not. Do we count deep breastbone and flapping wings as two separate features by which pigeons differ from kiwis? Or do we count them as only a single feature, on the grounds that the state of one character determines the other, or at least reduces its freedom to vary? In the case of the millipedes and the mirroring, the sensible answer is pretty obvious. In the case of the breastbones it isn't. Reasonable people can be found arguing on opposite sides.

That was all about visible resemblances and differences. But visible features evolve only if they are manifestations of DNA sequences. Nowadays we can compare DNA sequences directly. As an added

---

* The actual number is $(3 \times 2 - 5) \times (4 \times 2 - 5) \times (5 \times 2 - 5) \times \ldots \times (n \times 2 - 5)$ where n is the number of groups.

benefit, being long strings, DNA texts provide a lot more items to count and compare. Problems of the wing-and-breastbone variety are likely to be drowned out in the flood of data. Even better, many DNA differences will be invisible to natural selection and so provide a 'purer' signal of ancestry. As an extreme example, some DNA codes are synonymous: they specify exactly the same amino acid. A mutation that changes a DNA word to one of its synonyms is invisible to natural selection. But to a geneticist, such a mutation is no less visible than any other. The same goes for 'pseudogenes' (usually accidental duplicates of real genes) and for many other 'junk DNA' sequences, which sit in the chromosome but whose information is never actually used. Freedom from natural selection leaves DNA free to mutate in ways that leave highly informative traces for taxonomists. By the way, none of this alters the fact that some mutations do have real and important effects. Even if these are only the tips of icebergs, it is those tips that are visible to natural selection and account for all the visible and familiar beauties and complexities of life.

DNA is not a complete panacea. It can evolve in unexpected ways, fooling the unwary. As we shall see in the Humped Bladderwort's Tale, over half of our DNA consists of viruses or viral-like parasites that hijack the machinery of DNA replication to spread themselves about the genome. It would be misleading to group together two organisms on the basis of a virus that one passed on to the other! Even DNA that confines itself to spreading within the genome of a single organism can give us multiple counting problems, like the millipede's legs. A more insidious problem arises because organisms contain similar duplicated DNA sequences, such as the various haemoglobin genes we will meet in the Lamprey's Tale. An ancestral chromosome containing both haemoglobin alpha and haemoglobin beta genes could produce one group of descendants that have lost the alpha, and another group the beta. Comparing the two groups, we are likely to end up comparing the alpha gene in one with the beta gene in another – apples to oranges. For this reason, when comparing different species we must ensure we are using the same 'orthologous' genetic sequence, to use the technical term.

There are also cases where extensive regions of DNA show enigmatic resemblances between comparatively unrelated creatures. Nobody doubts that birds are more closely related to turtles, lizards, snakes and crocodiles than to mammals (see Rendezvous 16). Nevertheless, the DNA sequences of birds and mammals have resemblances greater than one might expect given their distant relationship. Both have regions of their DNA which show an excess of G-C pairings, especially near genes. This seems to be a result of certain features of their DNA repair mechanisms. So across the genome, DNA sequences from mammals and birds have a slight tendency to accumulate Gs and Cs in the same places. In the early days of genetic studies, this led to birds and mammals being grouped together. We now know that these seemingly multiple similarities are not independent of each other: they are due to a single coincidental change in the genetic machinery. DNA seems to promise a Utopia for biological systematists, but we must be aware of such dangers: there is a lot that we still don't understand about genomes.

So, having taken the necessary invocation of caution, how can we use the information present in DNA? Fascinatingly, literary scholars use the same techniques as evolutionary biologists in tracing the ancestries of texts. And – almost too good to be true – one of the best examples happens to be the work of the *Canterbury Tales* Project. Members of this international syndicate of literary scholars have used the tools of evolutionary biology to trace the history of 85 different manuscript versions of *The Canterbury Tales*. These ancient manuscripts, hand-copied before the advent of printing, are our best hope of reconstructing Chaucer's lost original. As with DNA, Chaucer's text has survived through repeated copyings, with accidental changes perpetuated in the copies. By meticulously scoring the accumulated differences, scholars can reconstruct the history of copying, the evolutionary tree – for it really is an evolutionary process, consisting of a gradual accumulation of errors over successive generations. So similar are the techniques and difficulties in DNA evolution and literary text evolution, that each can be used to illustrate the other.

So, let's temporarily turn from our gibbons to Chaucer, and in particular four of the 85 manuscript versions of *The Canterbury Tales:* the 'British Library', 'Christ Church', 'Egerton', and 'Hengwrt' versions.* Here are the first two lines of the General Prologue:

---

* The 'British Library' manuscript belonged to Henry Dene, Archbishop of Canterbury in 1501, and, together with the Egerton manuscript and others, is now kept at the British Library in London. The 'Christ Church' manuscript now resides close to where this book is being written, in the library of Christ Church, Oxford. The earliest record of the 'Hengwrt' manuscript shows it belonging to Fulke Dutton in 1537. Damaged by rats gnawing at the sheepskin on which it is written, it is now in the National Library of Wales.

BRITISH LIBRARY: Whan that Aprylle / wyth hys showres soote
                 The drowhte of Marche / hath pcede to the rote
CHRIST CHURCH: Whan that Auerell w^t his shoures soote
                 The droght of Marche hath pced to the roote
     EGERTON: Whan that Aprille with his showres soote
                 The drowte of marche hath pced to the roote
     HENGWRT: Whan that Aueryll w^t his shoures soote
                 The droghte of March / hath pced to the roote

The first thing that we must do with either DNA or literary texts is to locate the similarities and differences. For this we have to 'align' them – not always an easy task, for texts can be fragmentary or jumbled and of unequal length. A computer is a great help when the going gets tough, but we don't need it to align the first two lines of Chaucer's General Prologue, which we have shaded at the 15 points where the scripts disagree (see opposite page).

Having compiled a list of differences, we now work out which tree best explains them. A quick and dirty way is to group the texts on the basis of overall similarity using some variant on the following method. First we locate the pair of texts that are the most similar. We then treat this pair as a single averaged text, and put it alongside the remaining texts while we look for the next most similar pair. And so on, forming successive, nested groups until a tree of relationships is built up. These sorts of techniques – one of the most common is known as 'neighbour-joining' – are fast because they don't require us to churn through all the possible trees. But they do not incorporate the logic of the evolutionary process. They are purely measures of similarity. For this reason, the 'cladist' school of taxonomy, which is deeply evolutionary in its rationale (although not all its members realise this) prefers other methods, of which the earliest to be devised was the parsimony method.

Parsimony, as we saw in the Orang Utan's Tale, means economy of explanation. In evolution, whether of animals or manuscripts, the most parsimonious explanation is the one that postulates the least quantity of evolutionary change. If two texts share a common feature, the parsimonious explanation is that they have jointly inherited it from a shared ancestor rather than that each evolved it independently. It is very far from an invariable rule, but it is at least more likely to be true than the opposite. The method of parsimony – at least in principle – looks over all possible trees and chooses the one that minimises the quantity of change.

Bl: Whan that Aprylle / wyth hys showres soote   The drowhte of Marche / hath pcede to the rote
Ch: Whan that Auerell   wt his shoures soote   The droght of Marche hath pced to the roote
En: Whan that Aprille with his showres soote   The drow te of marche hath pced to the roote
Hg: Whan that Auerylll   wt his shoures soote   The droghte of March / hath pced to the roote

When we are choosing trees for their parsimony, certain types of difference can't help us. Differences that are unique to a single manuscript, or a single species of animal, are *uninformative*. The neighbour-joining method uses them, but the method of parsimony ignores them completely. Parsimony relies upon *informative* changes: ones that are shared by more than one manuscript. The preferred tree is the one that uses shared ancestry to explain as many inform- ative differences as possible. In our Chaucerian lines, nine of the differences are uninformative and can be ignored. The six inform- ative differences are highlighted above, and you can check that the first five neatly divide the manuscripts into *Cc* plus *Hg* on the one hand, versus *Bl* plus *Eg* on the other. The remaining difference (the last 'virgule', or diagonal line) splits the texts differently, grouping *Bl* with *Hg* and *Cc* with *Eg*. This last difference conflicts with the others. No unrooted tree can explain the scriptural differences purely through unique, inherited copying mistakes. Two scribes must have made identical errors at some point.

Parsimony tells us to choose the tree with the least change: that which only requires a single coincidental error, such as two monks independently inserting a virgule in the same place. That tree is the following:

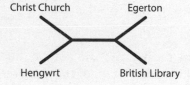

Christ Church   Egerton

Hengwrt   British Library

The two other trees would require five or six identical errors to have been made in transcription – much less likely. Still, it's not impossible, and we should treat our Chaucerian conclusion with caution, especially because convergences or reversions are common in texts. A medieval scribe might have little compunction in changing a spelling, and even less in inserting or removing a punctuation mark such as a virgule. Better indicators of relationship would be changes such as the reordering of words. The genetic equivalents are 'rare genomic changes': events such as large insertions, deletions, or duplications of DNA. We can explicitly acknowledge these by giving more or less weight to different types of change. Changes known to be common or unreliable are downweighted when counting up extra changes. Changes known to be rare, or reliable indicators of kinship, are given increased weighting. Heavy weighting to a change means we especially don't want to count it twice. The most parsimonious tree, then, is the one with the lowest overall weight.

The parsimony method is much used to find evolutionary trees. But if convergences or reversions are common – as with many DNA sequences and also in our Chaucerian texts – parsimony can be misleading. It is the notorious bugbear known as 'long branch attraction'. Here's what this means.

Cladograms, whether rooted or unrooted, convey only the order of branching. *Phylograms*, or phylogenetic trees (Greek *phylon* = race/tribe/class), are similar but also use the length of branches to convey information. Typically branch lengths represent evolutionary distance: long branches represent a lot of change, short ones little change. For example, the relationships between our four Canterbury Tales extracts could be pictured like this:

In this phylogram, the branches are not too different in length. But imagine what would happen if two of the manuscripts changed a lot, compared to the other two. The branches leading to these two would be drawn very long. And a proportion of the changes would not be unique. They would just happen to be identical to changes elsewhere on the tree, but (and now here is the point) *especially* to those on the other long branch. This is because long branches are where the most changes are anyway. With enough evolutionary changes, the ones that spuriously link the two long branches will drown out the true signal. Based upon a simple count of the number of changes, parsimony erroneously groups together the termini of especially long branches. The method of parsimony makes long branches spuriously 'attract' one another.

The problem of long branch attraction is an important headache for biological taxonomists. It rears its head whenever convergences and reversions are common, and unfortunately we cannot hope to avoid it by looking at more text. On the contrary, the more text we look at, the more erroneous similarities we find, and the stronger our conviction in the wrong answer.* Unfortunately, DNA data are particularly vulnerable to long branch attraction. The main reason is that there are only four letters in the DNA code. If the majority of differences are single letter changes, independent mutation to the same letter by accident is extremely likely. This sets up a minefield of long branch attraction. Clearly we need an alternative to parsimony in these cases. It comes in the form of a technique known as likelihood analysis, which is increasingly favoured in biological taxonomy.

Likelihood analysis burns even more computer power than parsimony, because now the lengths of the branches matter. So we have vastly more to contend with. In addition to all the branching patterns, we must also look at the different possible branch lengths and mutation rates – a Herculean task. We can only hope to do this by using clever approximations and shortcuts, which is an active area for computational biologists.

'Likelihood' is not a vague term. On the contrary, it has a precise meaning. Think about it like this: take a guess at the probabilities of each type of change (the probabilities of swapping one letter for another, of missing out a letter entirely, and so on). At the same time, also take a guess at an evolutionary tree (remembering to include branch lengths). Pretend those guesses are correct and we can calculate the exact chance they would produce the DNA sequences that we actually see. This is the 'likelihood' of those guesses (it may be a very small value). If we take another set of guesses, we will get a different 'likelihood' value, which can be compared with the first. And so on, for as many guesses – as many trees and probabilities – as we can possibly make.

Using likelihood, there are various methods of defining the 'best guess' tree. The simplest is to assume it is the tree that has the highest

* Such trees are said to lie in the scary-sounding 'Felsenstein zone', named after the distinguished American biologist Joe Felsenstein, whose recent book *Inferring Phylogenies* is considered the bible in this field.

likelihood value. Not unreasonably, this goes under the name 'maximum likelihood', but just because it is the single most likely tree doesn't mean that other possible trees aren't *almost* as likely. Perhaps, instead of believing in a single most likely tree, we should look at all possible trees, but give proportionally more credence to the more likely ones. This approach is known as Bayesian phylogenetics, and part of a recent move towards a 'Bayesian' approach for all sorts of probability calculations (one example is internet spam filters). With evolutionary trees, it provides two benefits. It produces numerical probabilities for each branch point (although experience indicates the numbers can be overly optimistic). More importantly, it provides scope for adjusting the rate of evolution along each branch, so that lengths of the branches become an estimate of actual time, rather than the amount of change. This essentially means that the changes are being used as a 'molecular clock', of the sort that is used to produce many of the dates in this book, and to which we will return in the Epilogue to the Velvet Worm's Tale. Of course, just as in maximum likelihood, Bayesian analysis can't examine all possible trees, but there are computational shortcuts and they work pretty well.

Our *confidence* in the tree we finally choose will depend on our certainty that its various branches are correct, and it is common to place measures of this beside each branch point. Probabilities are automatically calculated when using the Bayesian method, but for others such as parsimony or maximum likelihood, we need alternative measures. Commonly used is the 'bootstrap' method, which resamples different parts of the data repeatedly to see how much difference it makes to the final tree – how robust the tree is, in other words, to error. The nearer the 'bootstrap' value to 100 per cent, the more robust the branch point, but even experts struggle to interpret exactly what a particular bootstrap value tells us. Similar methods are the 'jackknife', and the 'decay index'. All are measures of how much we should believe each branch point on the tree.

Before we leave literature and return to biology, opposite is a summary diagram of the evolutionary relationships between the first 250 lines of 24 Chaucer manuscripts. It is a phylogram, in which not just the branching pattern but the lengths of the lines are meaningful. You can immediately read off which manuscripts are minor variants of each other, which are aberrant outliers. It is unrooted – it doesn't commit itself as to which of the 24 manuscripts is closest to the 'original'. And note with satisfaction that the four manuscripts we

'By me was nothyng added ne mynusshyd' (Caxton's Preface). Unrooted phylogenetic tree of the first 250 lines of 24 different manuscript versions of *The Canterbury Tales*. This represents a subset of the manuscripts studied by the *Canterbury Tales* Project, whose abbreviations for the manuscripts are used here. The tree was constructed by parsimony analysis, and bootstrap values are shown on the branches. The four versions discussed are named in full.

discussed (those with names in brackets) show exactly the relationship that we calculated from only their first two lines.

It's time to return to our gibbons. Over the years, many people have tried to work out gibbon relationships. Parsimony suggested four groups of gibbons. What follows is a rooted cladogram based on physical characteristics.

The cladogram overleaf shows reasonably convincingly that gibbons fall into the four known genera, with *Nomascus* species grouped together (100 per cent bootstrap), and *Hylobates* species probably so (80 per cent bootstrap). But most other relationships are poorly resolved. And even though *Hylobates* and *Hoolock* are grouped together, the bootstrap value, 63, is unconvincing to those trained to read such runes. Physical features are not enough to work out the relationships between the gibbon genera.

For this reason, researchers have increasingly turned to molecular genetics. Eve's Tale introduced mitochondrial DNA, a sequence inherited down the female line which is commonly used in genetic studies. The leading authority in the case of gibbons is Christian Roos, who has pointed us to a recent analysis of their complete mitochondrial DNA.

Rooted cladogram of gibbons, based on morphology. Adapted from Geissmann [148].

Sequences from a number of individual gibbons, as well as from seven outgroups, were matched letter-for-letter (as we did for the Chaucer texts) with a few badly matching sections removed. Then, maximum likelihood and Bayesian analyses were performed, the latter allowing evolutionary rates to vary. The resulting mitochondrial DNA tree (opposite) has the end points lined up, with the branch lengths providing an estimate of geological time. In this diagram, branch points lacking numbers are trustworthy, with bootstrap values of 100 per cent and Bayesian probabilities which round to 1.0. This tree is therefore better resolved than the one based on physical characteristics.

Unfortunately, two slightly less certain branch points, with bootstrap values marked as 87 and 77 per cent, still affect our gibbon genera. One thing is certain: *Nomascus* mitochondria branch off first. The subsequent divergence of the *Hylobates* gibbons (87 per cent bootstrap) has reasonable support. But the 77 per cent bootstrap value linking the siamang and the hoolock gibbons is disappointingly low. Given that the tree is based on almost 16,000 DNA letters, this is perhaps surprising. One problem is the lengths of the branches linking the groups. These basal gibbon divergences happened in rather short succession, providing little time to accumulate evolutionary changes. Fortunately for our pilgrimage, the intermediate branches in the human part of the tree should be less problematic. Apart from one short branch from 6 to 8 million years ago, humans

Phylogeny of gibbons using full mitochondrial DNA sequences, bootstrap percentages lower than 100 per cent and Bayesian probabilities less than 1.00 are marked. The Bayesian approach allows branch points to be dated using human/chimp, human/orang utan, and human/macaque calibration points of 6-7 Ma, 14 Ma, and 24-29 Ma respectively, with bars on each branch point showing the 95 per cent range of probable dates. Adapted from Carbone *et al* [56].

are connected by branches which extend for 5 million years or longer, enough time for many informative evolutionary changes to accumulate.

Short branches in a tree – a paucity of evolutionary changes – can be problematic. But so can overly long branches – very large numbers of changes. As more and more differences accumulate between DNA sequences, even the sophisticated techniques of maximum likelihood and Bayesian analysis start to fail us. There can come a point where an unacceptably large proportion of similarities are coincidental. The DNA differences are then said to be saturated. No fancy techniques can recover the signal of ancestry, because any vestiges of relationship have been overwritten by the ravages of time. The problem is especially acute with neutral DNA differences. Strong natural selection keeps genes on the straight and narrow. In extreme cases, important functional genes can stay literally identical over hundreds of millions of years. But, for a pseudogene that never does anything, such lengths of time are enough to lead to hopeless saturation. In such cases, we need different data. The most promising idea is to use the rare genomic changes that we mentioned before – changes that involve DNA reorganisation rather than single letter changes. These being rare, indeed usually unique, coincidental resemblance is much less of a problem. And once found, they can reveal remarkable relationships, as we shall learn when our swelling pilgrim band is joined by the hippo, and we are bowled over by its whale of a surprising tale.

EPILOGUE TO THE GIBBON'S TALE

The Gibbon's Tale explained how to use physical features or genetic sequences to deduce an evolutionary tree. For gibbon mitochondrial DNA that gives a reasonable answer. But we must qualify the tale with an important afterthought; we might call it the Gibbon's decline and fall of the species tree. Mitochondrial DNA is passed on through females only, so we know it reflects a strictly branching tree. The rest of our genome is more problematic. As we have argued in several tales so far, species are composites of DNA from many different sources. Each gene, in fact each letter in a genetic sequence, takes its own path through history. Each piece of DNA, and each aspect of an organism, can have a different evolutionary tree, which means there may not be a neat and simple relationship between species. We have previously touched upon one example of this: the ABO blood groups. There is a more obvious case, which we overlook only because of its familiarity. A Martian taxonomist shown only the genitals of a male human, a female human and a male gibbon would have no hesitation in classifying the two males as more closely related to each other than either is to the female. Indeed, the gene determining maleness (called *SRY*) has never been in a female body, at least since long before we and the gibbons diverged. Traditionally, morphologists plead a special case for sexual characteristics, to avoid 'nonsensical' classifications. But as we look deeper into the genome, scientists are finding that this problem is much more extensive than previously assumed.

Opposite is a representation, by Lucia Carbone and colleagues, of the genetic relationships among gibbons, not just deduced from the mitochondria, but from the entire genome. It is clearly a mix of different evolutionary trees. The most commonly seen tree, and the

Nomascus leucogenys · Hoolock leuconedys · Symphalangus syndactylus · Hylobates pileatus · Hylobates moloch

13.2%

15.4%

10.9%

one supported by 15 per cent of the genome, shows the two *Hylobates* species branching first. Another tree, showing *Nomascus* branching first, is supported by 13 per cent of the genome. A third set of relationships, supported by 11 per cent of the genome, also puts *Hylobates* on the earliest splitting branch, but has a different branching order among the remaining three genera. The analysis picks up other less supported trees too, although they have been omitted from the diagram for clarity. Unlike our Chaucerian example, these conflicting trees aren't a result of accidental convergences. We know this because irreconcilable evolutionary histories are also exposed when analysing rare insertions of larger DNA chunks.

We can trace this problem to two effects. In the gibbon's case both are likely to contribute. First is the possibility of hybridisation. Gibbons from different genera can produce hybrids in zoos, and hybrids between wild species presumably occurred in the past, which would have passed various genes between species. Second is a possibly we met in the Bonobo's Tale: 'incomplete lineage sorting'. This happens because none of the gibbon species trace back to a single breeding pair. Ancestral gibbon populations have always numbered in the thousands or tens of thousands. In large populations, each gene inevitably shows some diversity: its gene tree will consist of many pre-existing lineages. The larger the population, the longer these ancestral lineages are likely to persist. When large ancestral populations are coupled with closely timed speciation events, different pieces of DNA can retain conflicting affiliations, simply by chance.

This book, and our journey to the dawn of life, is based around the concept of a single evolutionary tree, something that these two effects throw into question. Fortunately, we are rescued by the sweep of geological time. As populations evolve in different directions, the chance of hybrids decreases and (perhaps more importantly) ancestral genetic lineages are lost. So when speciation events are separated by many millions of years, most genes end up agreeing on a single tree. With widely spaced branch points, conflicts are rare, restricted to unusual cases such as sex or blood group genes. Gibbons show this too. Although the lower branches of the gibbon tree are entangled, the rest is more clear-cut. For example, the two *Hylobates* gibbons are consistently grouped together, and that reflects the fact that they split from each other 4 or 5 million years after their initial divergence from the other gibbon genera.

As we regress in time, we will find that most of our rendezvous points are separated by 5 million years or more. That justifies our picture of a universal tree of life. It is only when successive speciation events are close in time that this idea is undermined, and we shall tackle these cases at Rendezvous 9, 10 and 13.

OLD WORLD MONKEYS

As we near this rendezvous and prepare to greet Concestor 5 – approximately our 1.5-million-greats-grandparent – we cross a momentous (if somewhat arbitrary) boundary. We leave one geological period, the Neogene, to enter an earlier one, the Palaeogene. The next time we do this will be to burst into the Cretaceous world of the dinosaurs. Rendezvous 5 is scheduled at about 25 million years ago, in the Palaeogene. More specifically it is in the Oligocene Epoch of that period, the last stop on our backward journey when the climate and vegetation of the world are recognisably similar to today's. Much further back, and we shall not find any evidence of the open grasslands that so typify our own Neogene Period, or the wandering herds of grazers that accompanied their spread. Twenty-five million years ago, Africa was completely isolated from the rest of the world, separated from the nearest piece of land – Spain – by a sea as wide as that which separates it from Madagascar today. It is on that gigantic island of Africa that our pilgrimage is about to be invigorated by a new influx of spirited and resourceful recruits, the Old World monkeys – the first pilgrims to arrive bearing tails.

Today, the Old World monkeys number around 130 species, some of which have migrated out of their mother continent into Asia (see the Orang Utan's Tale). They are divided into two main groups: on the one hand are the colobus monkeys of Africa together with the langurs and proboscis monkeys of Asia; on the other hand are the mostly Asian macaques plus the baboons and guenons, etc. of Africa.

The last common ancestor of all surviving Old World monkeys lived some 11 million years later than Concestor 5, probably around 14 million years ago. The most helpful fossil genus for illuminating the period is *Victoriapithecus*, which is now known from more than a thousand fragments, including a splendid skull, from Maboko Island in Lake Victoria. All the Old World monkey pilgrims join hands around 14 million years ago to greet their own concestor, perhaps *Victoriapithecus* itself, or something like it. They then march on backwards to join the ape pilgrims at our own Concestor 5, 25 million years ago.

And what was Concestor 5 like? Perhaps a bit like the fossil genus *Aegyptopithecus*, which actually lived about 7 million years earlier. Concestor 5 itself, according to our usual rule of thumb, is more likely than not to have had the characteristics shared by its descendants, the catarrhines, defined as consisting of the apes and the Old World monkeys. For example (it's the feature that gives the catarrhines their name) Concestor 5 probably had a narrow septum between close-set downwards-facing nostrils, unlike the rounded, widely separated and sideways-facing nostrils of the New World monkeys, the platyrrhines. The females probably showed full menstruation, as is common among apes and Old World monkeys but not New World monkeys. It probably had an ear tube formed by the tympanic bone, unlike New World monkeys whose ear lacks a bony tube.

Did it have a tail? Almost certainly yes. Given that the most obvious difference between apes and monkeys is the presence or absence of the tail, we are tempted by the *non sequitur* that the divide of 25 million years ago corresponds to the moment at which the tail was lost. In fact, Concestor 5 was presumably tailed like virtually all other mammals, and Concestor 4 was tailless like all its descendants the modern apes. But we don't know at what point along the road leading from Concestor 5 to Concestor 4 the tail was lost. Nor is there any particular reason for us suddenly to start using the word 'ape' to signify the loss of the tail. The African fossil genus *Proconsul*, for example, can be called an ape rather than a monkey, because it lies on the ape side of the fork at Rendezvous 5. But the fact that it lies on the ape side of the fork tells us nothing about whether it had a tail. As it happens, the balance of the evidence suggests that, to quote the title of an authoritative scientific paper, '*Proconsul* did not have a tail.' But that in no way follows from the fact that it is on the ape side of the rendezvous divide.

What, then, should we call the intermediates between Concestor 5 and *Proconsul* before they lost their tail? A strict cladist would call them apes, because they lie on the ape side of the fork. A different kind of taxonomist would call them monkeys because they were tailed. Not for the first time, I say it is silly to become too worked up over names.

The Old World monkeys, Cercopithecidae, are a true clade, a group that includes all descendants of a single common ancestor. However 'monkeys' as a whole are not, because they include the New World monkeys, Platyrrhini. The Old World monkeys are closer

baboons, drills, & mangabeys

macaques

vervets, guenons etc.

langurs, surilis etc.

colobus monkeys

5
25 Ma

20 Ma
10 Ma

4

Old World monkeys join. Here we show the widely accepted division of the 130 or so species of Old World monkey into two subfamilies. The three groups at the top constitute one subfamily, the Cercopithecinae, containing baboons, vervets and macaques. These are primarily African animals. The other subfamily, the Colobinae, shown here at the bottom left, consists of the langurs (Asian) and the colobus monkeys (African). This arrangement is supported by comprehensive molecular analysis.

cousins to apes, with whom they are united in the Catarrhini, than to New World monkeys. All apes and monkeys together constitute a natural clade, the Anthropoidea. 'Monkeys' constitutes an artificial (technically 'paraphyletic') grouping because it includes all the platyrrhines plus some of the catarrhines but excluding the ape portion of the catarrhines. It might be better to call the Old World monkeys tailed apes. Catarrhine, as I mentioned earlier, means 'down nose': the nostrils face downwards – in this respect we are ideal catarrhines. Voltaire's Dr Pangloss observed that 'the nose is formed for spectacles, therefore we come to wear spectacles'. He could have added that our catarrhine nostrils are beautifully directed to keep out the rain. Platyrrhine means flat or broad nose. It is not the only diagnostic difference between these two great groups of primates, but it is the one that gives them their names. Let's press on to Rendezvous 6, and meet the platyrrhines.

NEW WORLD MONKEYS

Rendezvous 6, where the New World platyrrhine 'monkeys' meet us and our approximately 3-million-greats-grandparent, Concestor 6, the first anthropoid, is some 40 million years ago. It was a time of lush tropical forests – even Antarctica was at least partly green in those days. Although all platyrrhine monkeys now live in South or Central America, the rendezvous itself almost certainly did not take place there. My guess is that Rendezvous 6 occurred somewhere in Africa. A little later, a group of African primates somehow managed, in the form of a small founding population, to get across to South America. It can't have been much later, because by 36 million years ago squirrel-sized monkeys existed in what is now Amazonia. We know this from the unearthing in 2015 of fossilised monkey teeth in Peru. Moreover, these teeth are gratifyingly similar to monkey teeth in slightly older fossil deposits from North Africa. The two continents were closer to each other than they are now, and sea levels were low, perhaps exposing a chain of islands across the gap from West Africa, convenient for island-hopping. The monkeys probably rafted across, perhaps on fragments of mangrove swamps that could support life as floating islands for a short while. Currents were in the right direction for inadvertent rafting. Another major group of animals, the hystricognath rodents, probably arrived in South America around the same time. Again probably they came from Africa, and indeed they are named after the African porcupine, *Hystrix*. Probably the monkeys rafted across the same island chain as the rodents, using the same favourable currents, though presumably not the same rafts.

Are all the New World primates descended from a single immigrant? Or was the island-hopping corridor used* more than once by primates? What would constitute positive evidence for a double immigration? In the case of the rodents, there are still hystricognath rodents in Africa, including African porcupines, mole rats, dassie rats

---

* 'Used' is, of course, unfortunate if it implies anything more than inadvertence. As we shall see in the Dodo's Tale, no animal ever tries to colonise brand new territory. But when it accidentally happens, the evolutionary consequences can be momentous.

and cane rats. If it turned out that some of the South American rodents were close cousins of some African ones (say porcupines) while other South American rodents were closer cousins to other African ones (say mole rats) this would be good evidence that rodents more than once drifted to South America. That this is not the case is compatible with the view that rodents dispersed to South America only once, though it is not strong evidence. The South American primates, too, are all closer cousins to each other than they are to any African primate. Again this is compatible with the hypothesis of a single dispersal event, but again the evidence is not strong.

This is a good moment to repeat that the improbability of a rafting event is very far from being a reason for doubting that it happened. This sounds surprising. Usually, in everyday life, massive improbability is a good reason for thinking that something won't happen. The point about intercontinental rafting of monkeys, or rodents or anything else, is that it only had to happen once, and the time available for it to happen, in order to have momentous consequences, is way outside what we can grasp intuitively. The odds against a floating mangrove bearing a pregnant female monkey and reaching landfall in any one year may be ten thousand to one against. That sounds tantamount to impossible by the lights of human experience. But given millions of years it becomes almost inevitable. Once it happened, the rest was easy. The lucky female gave birth to a family, which eventually became a dynasty, which eventually branched to become all the species of New World monkeys. It only had to happen once: great things then grew from small beginnings.

In any case, accidental rafting is not nearly so rare as you might think. Small animals are often seen on flotsam. And the animals aren't always small. The green iguana is typically a metre long and can be up to two metres. I quote from a note to *Nature* by Ellen J. Censky and others:

> On 4 October 1995, at least 15 individuals of the green iguana, *Iguana iguana*, appeared on the eastern beaches of Anguilla in the Caribbean. This species did not previously occur on the island. They arrived on a mat of logs and uprooted trees, some of which were more than 30 feet long and had large root masses. Local fishermen say the mat was extensive and took two days to pile up on shore. They reported seeing iguanas on both the beach and on logs in the bay.

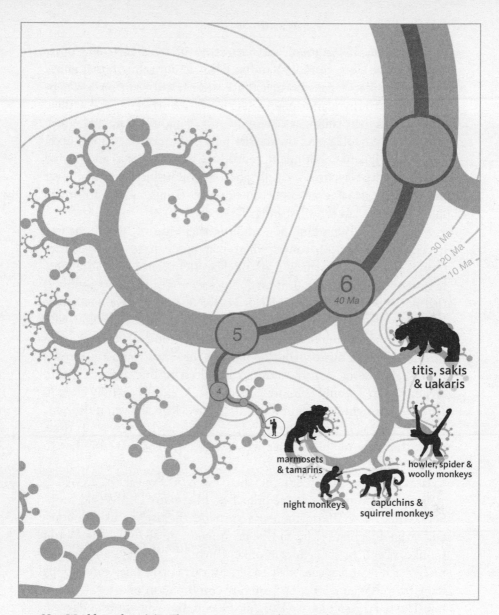

New World monkeys join. The 100 or so species of New World monkey are usually classified into the five families labelled here. Clockwise from the top, they go under the scientific names of Pitheciidae, Atelidae, Cebidae, Aotidae and Callitrichidae. The higher-level relationships between them are somewhat disputed: the position of the night monkeys is particularly problematic. This probably reflects rapid speciation early in their evolutionary history.

The iguanas were presumably roosting in trees on some other island, which were uprooted and sent to sea by a hurricane: either Luis, which had raged through the Eastern Caribbean on 4–5 September, or Marilyn, a fortnight later. Neither hurricane hit Anguilla. Censky and her colleagues subsequently caught or sighted green iguanas on Anguilla, and on an islet half a kilometre offshore. As of 2014, green iguanas continue to reproduce there. Iguanas and related lizards, by the way, are especially good at colonising islands, all over the world. Iguanas even occur on Fiji and Tonga, which are much more remote than the West Indian islands.

I can't resist remarking how chilling this kind of 'it only had to happen once' logic becomes when you apply it to contingencies nearer home. The principle of nuclear deterrence, and the only remotely defensible justification for possessing nuclear weapons, is that nobody will dare risk a first strike, for fear of massive retaliation. What are the odds against a mistaken missile launch: a dictator who goes mad; a computer system that malfunctions; an escalation of threats that gets out of hand? What are the odds against a terrible mistake, initiating Armageddon? A hundred to one against, within any one year? I would be more pessimistic. We came awfully close in 1963. What might happen in Kashmir? Israel? Korea? Even if the odds per year are as low as one in a hundred, a century is a very short time, given the scale of the disaster we are talking about. It only has to happen once.

Let's return to a happier topic, the New World monkeys. As well as walking quadrupedally above branches, like many Old World monkeys, some New World monkeys suspend themselves like gibbons, and even brachiate. The tail is prominent in all the New World monkeys, and in the spider monkeys, woolly monkeys and howler monkeys it is prehensile, wielded like an extra arm. They can happily hang from the tail alone, or from any combination of arms, legs and tail. The tail doesn't have a hand at the end, but you almost believe it has, when you watch a spider monkey.*

New World monkeys also include some spectacularly acrobatic

---

* Prehensile tails are also found in several other South American groups, including kinkajous (Carnivores), porcupines (rodents), tree-anteaters (xenarthrans), opossums (marsupials), and even the salamander *Bolitoglossa*. Is there something special about South America? But prehensile tails also occur in pangolins, some tree rats, some skinks and chameleons not from South America!

leapers, as well as the only nocturnal anthropoids, the owl monkeys. Like owls and cats, owl monkeys have large eyes – the largest eyes of all the monkeys or apes. Pygmy marmosets are the size of a dormouse, smaller than any other anthropoid. The largest howler monkeys, however, are only about as big as a large gibbon. Howlers resemble gibbons, too, in being good at hanging and swinging from their arms, and in being very noisy – but where gibbons sound like New York police sirens in full cry, a troop of howler monkeys, with their resonating hollow bony voice boxes, remind me more of a ghost squadron of jet planes, roaring eerily through the treetops. As it happens, howler monkeys have a particular tale to tell us Old World monkeys – about the way we see colour, for they have independently arrived at the same solution.

THE HOWLER MONKEY'S TALE

New genes aren't added to the genome out of thin air. They originate as duplicates of older genes. Then, over evolutionary time, they go their separate ways by mutation, selection and drift. We don't usually see this happening but, like detectives arriving on the scene after a crime, we can piece together what must have happened from the evidence that remains. The genes involved in colour vision provide a striking example. For reasons that will emerge, the howler monkey is especially well placed to tell the tale.

During their formative megayears, mammals were creatures of the night. The day belonged to the dinosaurs, who probably, if their modern relatives are any guide, had superb colour vision. So, we may plausibly imagine, did the mammals' remote ancestors, the mammal-like reptiles, who filled the days before the rise of the dinosaurs. But during the mammals' long nocturnal exile, their eyes needed to snap up whatever photons were available, regardless of colour. Not surprisingly, for reasons of the kind that we shall examine in the Blind Cave Fish's Tale, colour discrimination degenerated. To this day most mammals, even those who have returned to live in the daylight, have rather poor colour vision, with only a two-colour system ('dichromatic'). This refers to the number of different classes of colour-sensitive cells – 'cones' – in the retina. We catarrhine apes and Old World monkeys have three: red, green and blue, and are therefore trichromatic, but the evidence suggests that we *regained* a third class of cone, after our nocturnal ancestors lost it. Most other vertebrates, such as fish and reptiles but *not* mammals, have three-cone ('trichromatic') or four-cone ('tetrachromatic') vision, and birds and turtles can be even more sophisticated. We'll come to the very special situation in the New World monkeys, and the even more special situation in the howler monkey, in a moment.

Interestingly, there is evidence that Australian marsupials differ from most mammals in having good trichromatic colour vision. Catherine Arrese and her colleagues, who discovered this in honey possums and dunnarts (it has also been demonstrated in wallabies), suggest that Australian (but not American) marsupials kept an ancestral reptilian visual pigment that the rest of the mammals lost. But mammals in general probably have the poorest colour vision among vertebrates. Most mammals see colour, if at all, only as well as a colourblind man. The notable exceptions are to be found among primates, and it is no accident that they, more than any other group of mammals, make use of bright colours in sexual display.

Unlike the Australian marsupials who perhaps never lost it, we can tell by looking at our relatives among the mammals that we primates did not retain trichromatic vision from our reptilian ancestors but rediscovered it – not once, but twice independently: first in the Old World monkeys and apes; and second in the New World howler monkeys, although not among the New World monkeys generally. Howler monkey colour vision is like that of apes, but different enough to betray its independent origin.

Why would good colour vision be so important that trichromacy evolved independently in New and Old World monkeys? A favoured suggestion is that it has to do with eating fruit. In a predominantly green forest, fruits stand out by their colours. This, in turn, is no accident. Fruits have probably evolved bright colours to attract frugivores, such as monkeys, who play the vital role of spreading and manuring their seeds. Trichromatic vision also assists in the detection of younger, more succulent leaves (often pale green, sometimes even red), against a background of darker green – but that is presumably not to the advantage of the plants.

Colour dazzles our awareness. Colour words are among the first adjectives that infants learn, and the ones they most eagerly tie to any noun that's going. It is hard to remember that the hues we perceive are labels for electromagnetic radiations of only slightly differing wavelengths. Red light has a wavelength around 700 billionths of a metre, violet around 420 billionths of a metre, but the whole gamut of visible electromagnetic radiation that lies between these bounds is an almost ludicrously narrow window, a tiny fraction of the total spectrum whose wavelengths range from kilometres (some radio waves) down to fractions of a nanometre (gamma rays).

All eyes on our planet are set up in such a way as to exploit the wavelengths of electromagnetic radiation in which our local star shines brightest, and which pass through the window of our atmosphere. For an eye that has committed itself to biochemical techniques suitable for this loosely bounded range of wavelengths, the laws of physics impose sharper bounds to the portion of the electromagnetic spectrum that can be seen using those techniques. No animal can see far into the infrared. Those that come closest are pit vipers, who have pits in the head which, while in no sense focusing a proper image with infrared rays, allow these snakes to achieve some directional sensitivity to the heat generated by their prey. And no animal can see far into the ultraviolet although some, bees for instance, can see a bit further than we can. But on the other hand, bees can't see our red: they might call it 'infrayellow'. All animals agree that 'light' is a narrow band of electromagnetic wavelengths lying somewhere between ultraviolet at the short end and infrared at the long end. Bees, people and snakes differ only slightly in where they draw the lines at each end of 'light'.

An even narrower view is taken by each of the different kinds of light-sensitive cells within a retina. Some cones are slightly more sensitive towards the red end of the spectrum, others towards the blue. It is the comparison between cones that makes colour vision possible, and the quality of colour vision depends largely on how many different classes of cones there are to compare. Dichromatic animals have only two populations of cones interspersed with one another. Trichromats have three, tetrachromats four. Each cone has a graph of sensitivity, which peaks somewhere in the spectrum and fades away, not particularly symmetrically, on either side of the peak. Out beyond the edges of its sensitivity graph, the cell may be said to be blind.

Suppose a cone's sensitivity peaks in the green part of the spectrum. Does this mean, if that cell is firing impulses towards the brain, that it is looking at a green object like grass or a billiard table? Emphatically not. It is just that the cell would need more red light (say) to achieve the same firing rate as a given amount of green light. Such a cell would behave identically towards bright red light or dimmer green light.* The nervous system can tell the colour of an object only by *comparing* the simultaneous firing rates of (at least) two cells that favour different colours. Each one serves as a 'control' for the other. You can get an even better idea of the colour of an object by comparing the firing rate of three cells, all with different sensitivity graphs.

Colour television and computer screens, doubtless because they are designed for our trichromatic eyes, also work on a three-colour system. On a normal computer monitor, each 'pixel' consists of three dots placed too close together for the eye to resolve. Each dot always glows with the same colour – if you look at the screen at sufficient magnification you always see only the same three colours, usually red, green and blue although other combinations can do the job. Flesh tones, subtle shades – any hue that humans might wish to see – can be achieved by manipulating the intensities with which these three primary colours glow. But tetrachromatic turtles, for example, might be disappointed by the unrealistic (to them) pictures on our television and cinema screens.

By comparing the firing rates from just three kinds of cones, our brains can perceive a huge range of hues. But most placental mammals, as already stated, are not trichromats but dichromats, with only two populations of cones in their retinas. One class peaks in the violet (or in some cases the ultraviolet), the other class peaks somewhere between green and red. In us trichromats, the short wavelength cones peak between violet and blue, and they are normally called blue cones. Our other two classes of cones can be called green cones and red cones. Confusingly, even the 'red' cones peak at a wavelength that is actually yellowish. But their sensitivity curve as a whole stretches into the red end of the spectrum. Even if they peak in the yellow, they still fire strongly in response to red light. This means that, if you subtract the firing rate of a 'green' cone from that of a 'red' cone, you'll get an especially high result when looking at red light. From now on I shall forget about peak sensitivities (violet, green and yellow) and refer to the three classes of cones as blue, green and red. In addition to cones, there are also rods: light-sensitive cells

---

*This raises an intriguing possibility. Imagine that a neurobiologist inserts a tiny probe into, say, a green cone and stimulates it electrically. The green cell will now report 'light' while all other cells are silent. Will the brain 'see' a 'super green' hue such as could not possibly be achieved by any real light? Real light, no matter how pure, would always stimulate all three classes of cones to differing extents.

of a different shape from cones, which are especially useful at night and which are not used in colour vision at all. They'll play no further part in our story.

The chemistry and the genetics of colour vision are rather well understood. The main molecular actors in the story are opsins: protein molecules which serve as visual pigments sitting in the cones (and rods). Each opsin molecule works by attaching to, and encasing, a single molecule of retinal: a chemical derived from vitamin A.[*] The retinal molecule has been forcibly kinked beforehand to fit it into the opsin. When hit by a single photon of light of an appropriate colour, the kink straightens out. This is the signal to the cell to fire a nervous impulse, which says to the brain 'my kind of light *here*'. The opsin molecule is then recharged with another kinked retinal molecule, from a store in the cell.

Now, the important point is that not all opsin molecules are the same. Opsins, like all proteins, are made under the influence of genes. DNA differences result in opsins that are sensitive to light of different wavelengths, and this is the genetic basis of the two-colour or three-colour systems we have been talking about. Of course, since all genes are present in all cells, the difference between a red cone and a blue cone is not which genes they possess, but which genes they turn on. And there is some kind of rule that says that any one cone *only* turns on one class of gene.

The genes that make our green and red opsins are very similar to each other, and they are on the X chromosome (the sex chromosome of which females have two copies and males only one). The gene that makes the blue opsin is a bit different, and lies not on a sex chromosome but on one of the ordinary non-sex chromosomes called autosomes (in our case it is chromosome 7). Our green and red cells have clearly been derived from a recent gene duplication event, and much longer ago they must have diverged from the blue opsin gene in another duplication event. Whether an individual has dichromatic or trichromatic vision depends on how many distinct opsin genes it has in its genome. If it has, say, blue- and green-sensitive opsins but not red, it will be a dichromat.

That's the background to how colour vision works in general.

---

[*] Carrots are rich in beta-carotene from which vitamin A can be made: hence the rumour about seeing in the dark, started by World War II strategists to avoid revealing the secret of radar. Beta-carotene does contribute to healthy eyesight, but it's unlikely to *improve* your night vision.

Now, before we come to the special case of the howler monkey itself and how it became trichromatic, we need to understand the strange dichromatic system of the rest of the New World monkeys (some lemurs have it too, by the way, and not all New World monkeys do – for example, nocturnal owl monkeys have monochromatic vision). For the purposes of this discussion, 'New World monkey' temporarily excludes howler monkeys and other exceptional species. We'll come to the howler monkeys later.

First, set aside the blue gene as an unvarying fixture on an autosome, present in all individuals whether male or female. The red and green genes, on the X chromosome, are more complicated and will occupy our attention. Each X chromosome has only one locus where a red or a green* allele might sit. Since a female has two X chromosomes, she has two opportunities for a red or green gene. But a male, with only one X chromosome, has *either* a red or a green gene but *not* both. So a typical male New World monkey has to be dichromatic. He has only two kinds of cones: blue plus *either* red or green. By our standards, all males are colourblind, but they are colourblind in two different ways; some males within a population lack green opsins, others lack red opsins. All have blue.

Females are potentially more fortunate. Having two X chromosomes, they could be lucky enough to have a red gene on one and a green gene on the other (plus the blue which again goes without saying). Such a female would be a trichromat.† But an unlucky female might have two reds, or two greens, and would therefore be a dichromat. By our standards such females are colourblind, and in two ways, just like males.

A population of New World monkeys such as tamarins or squirrel monkeys, therefore, is an oddly complicated mixture. All males, and some females, are dichromats: colourblind by our standards but in two alternative ways. Some females, but no males, are trichromats, with true colour vision which is presumably similar to ours.

* Actually, red and green are only two out of a range of possibilities at this locus, but we have enough complications to be going on with. For the purposes of this tale they will be firmly 'red' and 'green'.
† As for ensuring that, in any one cone, only the red or the green opsin gene, but not both, is turned on, this happens to be easy for females. They already have a mechanism for turning the whole of one X chromosome off in any cell. A random half of the cells deactivates one of the two X chromosomes, the other half the other one. This is important, because all the genes on an X chromosome are set up to work if only one is active – necessary because males only have one X chromosome.

Experimental evidence with tamarins searching for food in camou-flaged boxes showed that trichromatic individuals were more successful than dichromats. Perhaps foraging bands of New World monkeys rely on their lucky trichromat females to find food that most of them would otherwise miss. On the other hand, there is a possibility that the dichromats, either alone or in collusion with dichromats of the other kind, might have strange advantages. There are anecdotes of bomber crews in the Second World War deliberately recruiting one colourblind member because he could spot certain types of camouflage better than his otherwise more fortunate trichromat comrades. Experimental evidence confirms that human dichromats can indeed break certain forms of camouflage that fool trichromats. Is it possible that a troop of monkeys consisting of trichromats and two kinds of dichromats might collectively find a greater variety of fruits than a troop of pure trichromats? This might sound far-fetched, but it is not silly.

The red and the green opsin genes in New World monkeys constitute an example of a 'polymorphism'. Polymorphism is the simultaneous existence, in a population, of two or more alternative versions of a gene, where neither is rare enough to be just a recent mutant. It is a well-established principle of evolutionary genetics that visible polymorphisms like this don't just happen without good reason. Unless something very special is going on, monkeys with the red gene will be either better off, or worse off, than monkeys with the green gene. We don't know which, but it is highly unlikely that they would be exactly equally good. And the inferior kind should go extinct.

A stable polymorphism in a population, then, indicates that something special is going on. What sort of thing? Two main suggestions have been made for polymorphisms in general, and either might apply to this case: frequency-dependent selection, and heterozygous advantage. Frequency-dependent selection happens when the rarer type is at an advantage, simply by virtue of being rarer. So, as the type which we had thought was 'inferior' starts to go extinct, it ceases to be inferior and bounces back. How could this be? Well, suppose 'red' monkeys are especially good at seeing red fruits while 'green' monkeys are especially good at seeing green fruits. In a population dominated by red monkeys, most of the red fruits will be already taken, and a lone green monkey, able to see green fruits, might be at an advantage – and vice versa. Even if that is not especially plausible, it is an example of the *kind* of special cir-cumstance that can maintain both types in a population, without one of them going extinct. It is not hard to see that something along the lines of our 'bomber crew' theory might be the kind of special circumstance that maintains a polymorphism.

Turning now to heterozygous advantage, the classic example – cliché almost – is sickle-cell anaemia in humans. The sickling gene is bad, in that individuals with two copies of it (homozygotes) have damaged blood corpuscles that look like sickles, and suffer from debilitating anaemia. But it is good in that individuals with only one copy (heterozygotes) are protected against malaria. In areas where malaria is a problem, the good outweighs the bad, and the sickling gene tends to spread through the population, in spite of the adverse effects on individuals unlucky enough to be homozygotes.* Professor John Mollon and his colleagues, whose research is mainly responsible for uncovering the polymorphic system of colour vision in New World monkeys, propose that the heterozygous advantage enjoyed by the trichromatic females is enough to favour the coexistence of the red and green genes in the population. But the howler monkey does it better, and this brings us to the teller of the tale itself.

Howler monkeys have managed to enjoy the virtues of both sides of the polymorphism, by combining them in one chromosome. They have done this by means of a lucky translocation. Translocation is a special kind of mutation. A chunk of chromosome somehow gets pasted into a different chromosome by mistake, or into a different place on the same chromosome. This seems to have happened to a lucky mutant ancestor of the howler monkeys, which consequently ended up with *both* a red gene and a green gene next door to one another on a single X chromosome. This monkey would have been well on its evolutionary way towards becoming a true trichromat, even if it was a male. The mutant X chromosome spread through the population until, now, all howler monkeys have it.

It was easy for howler monkeys to perform this evolutionary trick, because the three opsin genes were already knocking around the population in New World monkeys: it is just that, with the exception of a few lucky females, any one individual monkey had only two of

---

* This sadly affects many African-Americans, who no longer live in a malarial country but inherit the genes of ancestors who did. Another example is the debilitating disease cystic fibrosis whose gene, in the heterozygous condition, seems to confer protection against cholera.

them. When we apes and Old World monkeys independently did the same kind of thing, we did it differently. The dichromats from which we sprang were dichromats in only one way: there wasn't a polymorphism to take off from. Evidence suggests that the doubling up of the opsin gene on the X chromosome in our ancestry was a true duplication. The original mutant found itself with two tandem copies of an identical gene, say two greens next door to each other on the chromosome, and it therefore was not a near-instant trichromat like the ancestral howler monkey mutant. It was a dichromat, with a blue and two green genes. The Old World monkeys became trichromats gradually in subsequent evolution, as natural selection favoured a divergence of the colour sensitivities of the two X opsin genes, towards green and red respectively.

When a translocation happens, it isn't just the gene of interest that is seen to move. Sometimes its travelling companions – its neighbours on the original chromosome who move with it to the new chromosome – can tell us something. And so it is in this case. The gene called *Alu* is well known as a 'transposable element': a short, virus-like piece of DNA that replicates itself around the genome, as a sort of parasite, by subverting the cell's DNA replication machinery. Was *Alu* responsible for moving the opsin? It seems so. We find the 'smoking gun' when we look at the details. There are *Alu* genes at both ends of the duplicated region. Probably the duplication was an unintended by-product of parasitic reproduction. In some long-forgotten monkey of the Eocene Epoch, a genomic parasite near to the opsin gene tried to reproduce, accidentally replicated a much larger chunk of DNA than intended, and set us on the road to three-colour vision. Beware, by the way, of the temptation – it is all too common – to think that, because a genomic parasite seems, with hindsight, to have done us a favour, genomes therefore harbour parasites in the hope of future favours. That isn't how natural selection works.

Whether engineered by *Alu* or not, mistakes of this kind still sometimes happen. When two X chromosomes line up, prior to crossing over, it is possible for them to line up incorrectly. Instead of lining the red gene on one chromosome with the corresponding red on the other, the similarity of the genes can confuse the lining-up process so that a red is lined up with a green. If crossing over then happens it is 'unequal': one chromosome could end up with an extra green (say) while the other X chromosome gets no green gene at all.

Even if crossing over doesn't happen, a process called 'gene conversion' can take place, where a short sequence of one chromosome is converted to the matching sequence in the other. With misaligned chromosomes, a part of the red gene may be replaced by the equivalent part of the green gene, or vice versa. Both unequal crossing over and misaligned gene conversion can lead to red–green colourblindness.

Men suffer more frequently from red–green colourblindness than women (the suffering is not great, but it is still a nuisance and they presumably are deprived of aesthetic experiences enjoyed by the rest of us) because if they inherit one faulty X chromosome they do not have another to serve as a backup. Nobody knows whether they see blood and grass in the way the rest of us see blood, or in the way the rest of us see grass, or whether they see both in some completely different way. Indeed, it may vary from person to person. All we know is that people who are red–green colourblind think grass-like things are pretty much the same colour as blood-like things. In humans, dichromatic colourblindness afflicts about 2 per cent of males. Don't be confused, incidentally, by the fact that other kinds of red–green colourblindness are more common (affecting about 8 per cent of males). These individuals are called anomalous trichromats: genetically they are trichromats, but one of their three kinds of opsins doesn't work.*

Unequal crossing over doesn't always make things worse. Some X chromosomes end up with more than two opsin genes. The extra ones nearly always seem to be green rather than red. The record number is a staggering 12 extra green genes, arrayed in tandem. But there is no evidence that individuals with extra green genes can see any better. Nevertheless, the high mutation rate along this part of the X chromosome means that not all 'green' genes in the population are exactly the same as each other. So it is theoretically possible for a female, with her two X chromosomes, to have not trichromatic vision but vision which is tetrachromatic (or even pentachromatic, if her red genes also differ). I don't know that anybody has tested this.

---

* Mark Ridley, in *Mendel's Demon* (retitled *The Cooperative Gene* in America), points out that the 8 per cent (or higher) figure applies to Europeans, and others with a history of good medicine. Hunter-gatherers, and other 'traditional' societies closer to the cutting edge of natural selection, show a lower percentage. Ridley suggests that a relaxation of natural selection has allowed colourblindness to increase. The whole business of colourblindness is treated, in characteristically original fashion, by Oliver Sacks in *The Island of the Colour-Blind*.

It is possible that an uneasy thought has occurred to you. I have talked as though the acquisition, by mutation, of a new opsin automatically confers enhanced colour vision. But of course differences between the colour sensitivities of cones are no earthly use unless the brain has some means of knowing which kind of cone is sending it messages. If it were achieved by genetic hard wiring – this brain cell is hooked up to a red cone, that nerve cell is hooked up to a green cone – the system would work, but it couldn't cope with mutations in the retina. How could it? How could brain cells be expected to 'know' that a new opsin, sensitive to a different colour, has suddenly become available and that a particular set of cones, in the huge population of cones in the retina, have turned on the gene for making the new opsin?

It seems that the most likely answer is that the brain learns. Presumably it compares the firing rates that originate in the population of cone cells in the retina and 'notices' that one sub-population of cells fires strongly when tomatoes and strawberries are seen; another sub-population when looking at the sky; another when looking at grass. This is a 'toy' speculation, but I suppose something like it enables the nervous system swiftly to accommodate a genetic change in the retina. My colleague Colin Blakemore, with whom I raised the matter, sees this problem as one of a family of similar problems that arise whenever the central nervous system has to adjust itself to a change in the periphery.*

The final lesson of the Howler Monkey's Tale is the importance of gene duplication. The red and the green opsin genes are clearly derived from a single ancestral gene that xeroxed itself to a different part of the X chromosome. Further back in time, we may be sure, it was a similar duplication that separated the blue† autosomal gene from what was to become the red/green X-chromosomal gene. It is common for genes on completely different chromosomes to belong to the same 'gene family'. Gene families have arisen by ancient DNA duplications followed by divergence of function. Various studies have found that a typical human gene has an average probability of duplication of about 0.1 to 1 per cent per million years. DNA

---

* I expect that some such learning must be used by birds and reptiles, who enhance their range of colour sensitivities by planting tiny coloured oil droplets over the surface of the retina.

† Or ultraviolet or whatever it was in those days. The exact colour sensitivities of all these classes of opsin will have been modified over the evolutionary years anyway.

duplication can be a piecemeal affair, or it can happen in bursts, for example when a newly virulent DNA parasite like *Alu* spreads throughout the genome, or when a genome is duplicated wholesale. (Entire-genome duplication is common in plants, and as we shall see in the Lamprey's Tale, is postulated to have happened at least twice in our ancestry, during the origination of the vertebrates.) Regardless of when or how it happens, accidental DNA duplication is one of the major sources of new genes. Over evolutionary time, it isn't only genes that change, within genomes. Genomes themselves change.

TARSIERS

We anthropoid pilgrims have arrived at Rendezvous 7, 60 million years ago in the dense and varied forests of the Palaeocene Epoch. There we greet a little evolutionary trickle of cousins, the tarsiers. We need a name for the clade that unites anthropoids and tarsiers, and it is haplorhines. The haplorhines consist of Concestor 7, perhaps our 6-million-greats-grandparent, and all its descendants: tarsiers, 'monkeys' and apes.

The first thing you notice about a tarsier is its eyes. Looking at the skull, it is almost the only thing there is to notice: a pair of eyes on legs pretty well sums up a tarsier. Each one of its eyes is as large as its entire brain, and the pupils open very wide too. The skull seen head-on seems to be wearing a pair of fashionably outsize, not to say giant, spectacles. Their huge size makes the eyes hard to rotate in their sockets but tarsiers, like some owls, are equal to the challenge. They rotate the whole head, on an extremely flexible neck, through nearly 360 degrees. The reason for their huge eyes is the same as in owls and night monkeys – tarsiers are nocturnal. They rely on moonlight, starlight and twilight, and need to sweep up every last photon they can.

Other nocturnal mammals have a tapetum lucidum – a reflecting layer behind the retina, which turns photons back in their tracks, so giving the retinal pigments a second chance to intercept them. It is the tapetum that makes it easy to spot cats and other animals at night.* Shine a torch all around you. It will catch the attention of any animals in the vicinity, and they'll look straight at your light out of curiosity. The beam will be reflected back off the tapetum. Sometimes you can locate dozens of pairs of eyes with a single sweep of the torch. If electric light beams had been a feature of the environment in which animals evolved, they might well not have evolved a tapetum lucidum, as it is such a giveaway.

Tarsiers, surprisingly, have no tapetum lucidum. It has been suggested that their ancestors, along with other primates, passed

---

* Most nocturnal birds, too, have reflecting eyes, but not the owlet nightjars (Aegothelidae) of Australasia, nor the Galapagos swallowtailed gull *Creagrus furcatus*, the only nocturnal gull in the world.

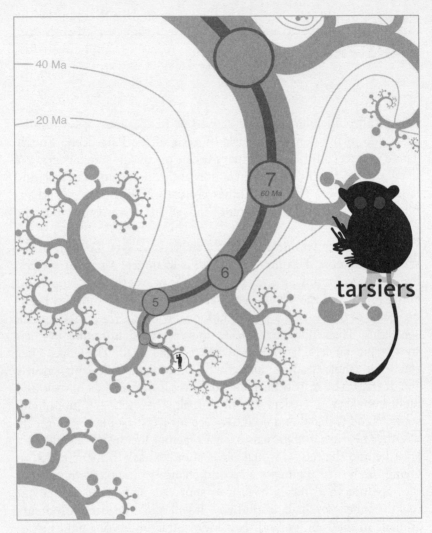

40 Ma

20 Ma

7
60 Ma

6

5

tarsiers

Tarsiers join. Living tarsiers fall into three groups, all from South East Asia. The earliest divergence splits the family east/west. To the west live Horsfield's tarsier (*Cephalopachus bancanus*) and the Philippine tarsier (*Carlito syrichta*). To the east are a handful of species in the genus *Tarsius*, all of which live on the Indonesian island of Sulawesi. The exact number of living species is disputed, but there are certainly more 'cryptic' Eastern species than the five shown in the tree above (four additional *Tarsius* species have been described, some of which are known only from museum specimens and have not been observed alive).

through a diurnal phase and lost the tapetum. This is supported by the recently discovered mouse-sized fossil *Archicebus*, which is currently the earliest known branch of the tarsier line and had smaller, diurnal eye sockets. It is telling too that modern tarsiers have the same weird system of colour vision as most of the New World monkeys. Several groups of mammals that were nocturnal in the time of the dinosaurs became diurnal when the death of the dinosaurs made it safe to do so. The suggestion is that the tarsiers subsequently returned to the night, but for some reason the evolutionary avenue of regrowing the tapetum was blocked to them. So they achieved the same result, of capturing as many photons as possible, by making their eyes very big indeed.* On this theory, if the tarsiers had managed to regrow a tapetum they wouldn't need such huge eyes and this would presumably have been a good thing.

The other descendants of Concestor 7, the 'monkeys' and apes, also lack a tapetum lucidum, not surprisingly given that they are all diurnal except the owl monkeys of South America. And the owl monkeys, like the tarsiers, have compensated by growing very large eyes – although not quite so large, in proportion to the head, as those of the tarsiers. We can make a good guess that, like all its descendants, Concestor 7 lacked a tapetum lucidum but unlike tarsiers was probably diurnal.

Eyes apart, what is to be said about tarsiers? They are outstanding leapers, with long legs like frogs or grasshoppers. Although smaller than the size of your fist, a tarsier can jump more then 3 metres horizontally and 1.5 metres vertically. They have been called furry frogs. It is probably no accident that they resemble frogs too in uniting the two bones of the lower legs, the tibia and the fibula, to make a single strong bone, the tibiofibula.

In this respect today's tarsiers seem to be, dare we say, far more jumped-up than the earlier forms of tarsier. *Archicebus*, for example,

---

* The absolutely largest eyes in the entire animal kingdom are those of the giant squid, nearly a foot in diameter. They too have to cope with very low light levels, this time not because they are nocturnal but because so little light penetrates to the great depths of ocean that they inhabit.

had long legs but lacked fused bones and its heel bone was more monkey-like, hence its full Latin (or should that be Greek) name, *Archicebus achilles*. It was a tree-dwelling insectivore, as are tarsiers today, a lifestyle which was also common in the other tarsier relatives known as omomyids. Like virtually all primates, *Archicebus* had nails in place of claws, so would have clutched branches, rather than clawing into them (curiously tarsiers do have 'grooming claws' on their second and third toes). It seems a reasonable guess that Concestor 7 ran, hopped and scrambled in the trees, spending its small life in pursuit of insects. Of its two descendant lineages, one stayed in the light and blossomed into the anthropoid monkeys and apes. The other reverted to the darkness and became the modern tarsiers.

We can't guess with any certainty where Rendezvous 7 takes place. But contrary to our speculation in the first edition of this book, *Archicebus* was unearthed in China. Some of the early omomyids seem to have originated in Asia too, hopping with remarkable rapidity through the forests of Europe and into North America, which in those days firmly joined to Eurasia via what is now Greenland. Perhaps Concestor 7 was Chinese.

LEMURS, BUSHBABIES AND THEIR KIN

Gathering the little leaping tarsiers into our pilgrimage, we head off back towards Rendezvous 8, where we are to be joined by the rest of the primates traditionally called prosimians: the lemurs, pottos, bushbabies and lorises. We need a name for those 'prosimians' that are not tarsiers. 'Strepsirrhines' has become customary. It means 'split nostril' (literally twisted nose). It is a slightly confusing name. All it means is that the nostril is shaped like a dog's. The rest of the primates, including us, are haplorhines (simple nose: our nostrils are each just a simple hole).

We haplorhine pilgrims, then, greet our strepsirrhine cousins, of which the great majority are lemurs, at Rendezvous 8. Various dates have been suggested for this point. I have taken it as 65 million years into the past, a commonly accepted date and one just this side of the cataclysmic events which, as we shall see in the next chapter, mark the end of the Cretaceous – and the dinosaurs – and herald the start of the modern era. Bear in mind, however, that a few researchers place this rendezvous even further back in time, during the Cretaceous itself. At 65 million years ago, most of the continents were in a similar position to those of today (see plate 8), and the Earth's vegetation and climate were starting to rebound from the 'Great Cretaceous Catastrophe'. The world was largely wet and forested, with at least the northern continents covered in a relatively restricted mix of deciduous conifers, and a scattering of flowering plant species.

Perhaps in the branches of a tree, we encounter Concestor 8, seeking fruit or maybe an insect (see plate 6). This most recent common ancestor of all surviving primates is approximately our 7-million-greats-grandparent. Fossils that might help us reconstruct what Concestor 8 was like include the large group called plesiadapiforms. They lived about the right time, and they have many of the qualities you would expect of the grand ancestor of all the primates. Not all of them, however, which makes their supposed position close to the primate ancestor controversial.

Of the living strepsirrhines, the majority are lemurs, living exclusively in Madagascar, and we'll come to them in the tale that follows.

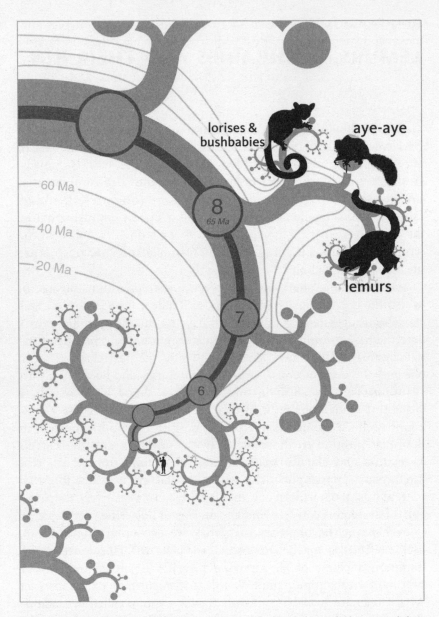

Lemurs and their kin join. The living primates can be divided into the lemurs and their kin, and the rest. The time of this divergence is debated – some experts place it as much as 20 million years earlier, with a consequent increase in the age of Concestors 9, 10, and 11. The five Madagascan lemur families (100 or so species), together with the loris and bushbaby families (32 species) are known as 'strepsirrhines'. The order of branching within lemurs remains controversial. In particular, the position of the aye-aye is the subject of continued research.

The others divide into two main groups, the leaping bushbabies and the creeping lorises and pottos. When I was a child of three in Nyasaland (now Malawi) we had a pet bushbaby. Percy was brought in by a local African, and was probably an orphaned juvenile. He was tiny: small enough to perch on the rim of a glass of whisky, into which he would dip his hand and drink with evident enjoyment. He slept during the day, clasping the underside of a beam in the bathroom. When his 'morning' came (in the evening), if my parents failed to catch him in time (which was often, because he was extremely agile and a terrific leaper) he would race to the top of my mosquito net and urinate on me from above. When leaping, for example onto a person, he did not exhibit the common bushbaby habit of urinating on his hands first. On the theory that 'urine washing' is for scent-marking, this would make sense given that he was not an adult. On the alternative theory that the urine improves grip, it is less clear why he didn't do it.

I shall never know to which of the 21 species of bushbaby Percy belonged, but he was most certainly a leaper, not a creeper. The creepers are the pottos of Africa and the lorises of Asia. They move much more slowly – especially the 'slow loris' of the Far East, which is a stealth hunter, inching along a branch until within reach of prey, whereupon it lunges with great speed.

Bushbabies and pottos remind us that a forest is a three-dimensional world like the sea. Seen from above the canopy, the green waves at its surface billow towards the horizon. Dive down into the darker green world beneath, and you pass through distinct layers, again as in the sea. The animals of the forest, like fish in the sea, find it as easy to move up and down as horizontally. But, also as in the sea, each species in practice specialises in making its living at a particular level. In the West African forests by night, the surface canopy is the province of the pygmy bushbabies hunting insects, and the fruit-eating pottos. Below the level of the canopy, the trunks of the trees are separated by gaps, and this is the domain of the needle-clawed bushbaby whose eponymous equipment enables it to cling to the trunks after leaping the gaps between them. Deeper still, in the understorey, the golden potto and the closely related angwantibo hunt caterpillars. At dawn, the nocturnal bushbabies and pottos give place to day-hunting monkeys, who parcel up the forest into similarly stratified layers. The same kind of stratification goes on in the South American forests, where as many as seven species of (marsupial) possum can be found, each at its own level.

The lemurs are descended from those early primates who were marooned on Madagascar during the time when monkeys were evolving in Africa. Madagascar is a large enough island to serve as a laboratory for natural experiments in evolution. The tale of Madagascar will be told by one of the lemurs, by no means the most typical of them, the aye-aye *Daubentonia*. I don't remember much from the discourse on lemurs that Harold Pusey – wise and learned warhorse of the lecture hall – gave to my generation of Oxford zoologists, but I do remember the haunting refrain with which he concluded almost every sentence about lemurs: 'Except *Daubentonia*.' 'EXCEPT *Daubentonia!*' Despite appearances, *Daubentonia*, the aye-aye, is a perfectly respectable lemur, and lemurs are the most famous inhabitants of the great island of Madagascar. The Aye-Aye's Tale is about Madagascar, textbook showcase of biogeographic natural experiments, a tale not just of lemurs but of all of Madagascar's peculiar – in the original sense of the word – fauna and flora.

THE AYE-AYE'S TALE

A British politician once described a rival (who later went on to become their party leader) as having 'something of the night' about him. The aye-aye conveys a similar impression, and indeed it is wholly nocturnal – the largest primate to be so. It has disconcertingly wide-set eyes in a ghostly pale face. The fingers are absurdly long: the fingers of an Arthur Rackham witch. 'Absurd' only by human standards, however, for we may be sure those fingers are long for a good reason: aye-ayes with shorter fingers would be penalised by natural selection, even if we don't know why. Natural selection is a strong enough theory to be predictive in this fashion, now that science no longer needs convincing of its truth.

One finger, the middle finger, is unique. Hugely long and thin, even by aye-aye standards, it is used specifically to make holes in dead wood and lever out grubs. Aye-ayes detect prey in wood by drumming with the same long finger, and listening for the changes in tone that betray an insect underneath.* That isn't quite all they use

---

* The same habit, with the same long finger (except it is the fourth finger instead of the third), has convergently evolved in a group of New Guinean marsupials, the striped possum and the trioks (*Dactylopsila*). These marsupials seem to be champion convergers, by the way. They are striped in the same pattern as skunks. And like skunks, they emit a powerful smell in defence.

the long middle finger for. At Duke University, which surely has the
largest collection of lemurs outside Madagascar, I have seen an aye-
aye, with great delicacy and precision, insert the long middle finger
up its own nostril – in quest of what, I don't know. The late Douglas
Adams wrote a wonderful chapter about the aye-aye in *Last Chance
to See*, his travel book about his journeys with the zoologist Mark
Carwardine.

> The aye-aye is a nocturnal lemur. It is a very strange-looking creature
> that seems to have been assembled from bits of other animals. It looks a
> little like a large cat with a bat's ears, a beaver's teeth, a tail like a large
> ostrich feather, a middle finger like a long dead twig and enormous
> eyes that seem to peer past you into a totally different world which
> exists just over your left shoulder . . . Like virtually everything that
> lives on Madagascar, it does not exist anywhere else on earth.

What wonderfully pithy writing, how sadly missed its author. Adams
and Carwardine's purpose in *Last Chance to See* was to call attention
to the plight of endangered species. The 100 or so surviving species of
lemur seems a huge number for a single island (and some authorities
would recognise only half that number). Either way, they are relics of
a much larger fauna that survived up until Madagascar was invaded
by destructive humans about 2,000 years ago.

Madagascar is a fragment of Gondwana (see the Sloth's Tale)
which became separated from what is now Africa about 165 million
years ago, and finally separated from what became India about 90
million years ago. This order of events may seem surprising but, as
we shall see, once India had shaken itself free of Madagascar it moved
away unusually fast by the sub-lorisoid standards of plate tectonics.

Setting aside bats (which presumably flew in) and human introductions, Madagascar's terrestrial inhabitants are descendants either
of the ancient Gondwana fauna and flora, or of rare immigrants
rafted in with improbable good luck from elsewhere. It is a natural
botanical and zoological garden, which houses about 5 per cent of
all the plant and animal land species in the world, more than 80 per
cent of them being found nowhere else. Yet, notwithstanding this
astonishing richness of species, it is also remarkable for the number
of major groups that are totally absent. Unlike Africa or Asia,
Madagascar has no native antelopes, no horses or zebras, no giraffes,
no elephants, no rabbits, no elephant shrews, no members of the cat

or the dog family: none of the expected African fauna at all, although fossil remains suggest that several species of hippo survived until recent times. There are bushpigs which seem to have arrived quite recently, perhaps introduced by humans. (We shall return to the aye-aye and the other lemurs at the end of the tale.)

The only Carnivores on Madagascar are about ten members of the mongoose family, which are clearly related to each other and must have arrived in the form of a single founder species from Africa, and subsequently branched. Of these, the most famous is the fossa, a sort of giant mongoose the size of a beagle but with a very long tail. Its smaller relatives are the falanouc, and the fanaloka whose Latin name, confusingly, is *Fossa fossana*. The fossa's own Latin name is *Cryptoprocta* ('hidden anus'), referring to the fact that its anus is concealed by a pouch, thought to help in scent marking.

There is a group of peculiarly Madagascan rodents, nine genera in all, united in one subfamily, the Nesomyinae. These include a burrowing giant rat-like form, a tree-climber, a tufted-tail 'marsh rat' and a jumping jerboa-like form. It has long been controversial whether these peculiarly Madagascan rodents result from one immigration event, or several. Like the Madagascan Carnivores, DNA evidence now unites these rodents in an exclusive group, implying that a single founding population diversified to fill all these different rodent niches: a very Madagascan story. Fascinatingly the evidence also produces almost identical dates (20 to 25 Ma) for the colonisation of Madagascar by Carnivores and rodents. Was one chasing the other?

Six out of the eight species of baobab tree are unique to Madagascar, and its count of 130 species of palm trees dwarfs the number found in the whole of Africa (see plate 7). Some authorities think chameleons originated there. Certainly, two-thirds of the world's species of chameleons are native to Madagascar. And there is a peculiarly Madagascan family of shrew-like animals, the tenrecs. Once classified in the Insectivora order, they are nowadays placed with the Afrotheria whom we shall meet at Rendezvous 13. Like the Carnivores and rodents, they probably arrived on Madagascar as a single founder population from Africa, a few tens of millions of years ago. They have now diversified into 27 species, including some that resemble hedgehogs, some that resemble shrews, and one that lives largely underwater like a water shrew. The resemblances are convergent – independently evolved, in typical Madagascan fashion.

Madagascar being isolated, there were no 'true' hedgehogs and no 'true' water shrews. So tenrecs, who had the good fortune to be on the spot, evolved to become the local equivalents of hedgehogs and water shrews.

Madagascar has no monkeys or apes at all, and their absence set the scene for the lemurs themselves, which on DNA evidence are the oldest surviving mammals on the island. By lucky chance, perhaps 50–60 million years ago, a founder population of early strepsirrhine primates accidentally found their way to Madagascar. As usual, we have no idea how this happened, although helpfully the sea currents at the time are now thought to have been from Africa to Madagascar, rather than the reverse as they are today. However the lemurs' ancestors got there, it must have been after Rendezvous 8 (at 65 Ma). As you can see from plate 8, that's far more recent than Madagascar's geographical separation from Africa (165 Ma) and India (88 Ma), so they certainly weren't Gondwanan residents sitting there all along. In several places in this book I have used 'rafting' as a kind of shortened code for 'fluke sea-crossing by some means unknown, of great statistical improbability, which only had to happen once, and which we know must have happened at least once because we see the later consequences'. I should add that 'great statistical improbability' is in there for form's sake. The evidence, as we saw at Rendezvous 6, is actually that 'rafting' in this general sense is commoner than intuition would expect. The classic example is the swift recolonisation of the remnants of Krakatoa after it was abruptly destroyed by a catastrophic volcanic event. E. O. Wilson's *The Diversity of Life* has a beautiful account.

In Madagascar, the consequences of the lucky rafting were dramatic and delightful: lemurs great and small, ranging in size from the pygmy mouse lemur, smaller than a hamster, to the recently extinct *Archaeoindris*, which was heavier than a large silverback gorilla and looked like a bear; familiar lemurs like the ringtailed, with their long, striped, hairy-caterpillar tails wafting in the air as the troop runs along the ground; or the indri, or the dancing sifaka which may be the most bipedally accomplished primate after ourselves.

And of course there is the aye-aye, teller of this tale. The world will be a sadder place when the aye-aye goes extinct, as I fear it may. But a world without Madagascar would be not just sadder – it would be impoverished. If you wiped out Madagascar, you would destroy only about a thousandth of the world's total land area, but fully 4 per cent of all species of animals and plants.

For a biologist, Madagascar is the Island of the Blest. Along our pilgrim voyage, it is the first of five large – in some cases very large indeed – islands, whose isolation, at crucial junctures in Earth history, radically structured the diversity of mammals. And not just mammals. Something similar happens with insects, birds, plants and fish, and when we are eventually joined by more distant pilgrims we shall find other islands playing the same role – not all of them dry land islands. The Cichlid's Tale will persuade us that each of the great African lakes is its own watery Madagascar, and cichlid fishes are its lemurs.

The islands or island continents that have shaped the evolution of mammals are, in the order we shall visit them, Madagascar, Laurasia (the great northern continent which was once isolated from its southern counterpart, Gondwana), South America, Africa, and Australia. Gondwana itself might be added to the list, for, as we shall discover at Rendezvous 15, it too bred its own unique fauna, before it broke up into all our Southern Hemisphere continents. The Aye-Aye's Tale has shown us the faunistic and floristic extravagance of Madagascar. Laurasia is the ancient home, and Darwinian proving-ground, of the huge influx of pilgrims we shall meet at Rendezvous 12, the laurasiatheres. At Rendezvous 13 we shall be joined by two strange bands of pilgrims, the xenarthrans, who served their evolutionary apprenticeship on the then island continent of South America, and the afrotheres, another hugely varied group of mammals, whose diversity was honed on the island continent of Africa. Then, at Rendezvous 14 it is the turn of Australia and the marsupials. Madagascar is the microcosm which sets the pattern – large enough to follow it, small enough to display it in exemplary clarity.

THE GREAT CRETACEOUS CATASTROPHE

Rendezvous 8, where our pilgrims meet the lemurs 65 million years ago, is the oldest rendezvous that takes place this side of the 66-million-year barrier, the so-called K/T boundary, which separates the Age of Mammals from the much longer Age of Dinosaurs that preceded it.* The K/T was a watershed in the fortunes of the mammals. They had been small, shrew-like creatures, nocturnal insectivores, their evolutionary exuberance held down under the weight of reptilian hegemony for more than 100 million years. Suddenly the pressure was released and, in a geologically very short time, the descendants of those shrews expanded to fill the ecological spaces left by the dinosaurs.

What caused the catastrophe itself? A controversial question. At the time there was extensive volcanic activity in India, spewing out lava flows covering well over a million square kilometres (the 'Deccan Traps') which must have had a radical effect on the climate. In favour of this explanation are the 'Siberian Traps', over 5 times larger, which are the main suspects in an even worse extinction – indeed the biggest of all time – at the end of the Permian, about a quarter of a billion years ago. In the Cretaceous, however, a variety of evidence is building a consensus that the final deathblow was more sudden and more drastic. It seems that a projectile from space – a large meteorite or comet – hit Earth. Detectives proverbially reconstruct events from cigar ash and footprints. The ash in this case is a worldwide layer of the element iridium at just the right place in the geological strata. Iridium is normally rare in the Earth's crust but common in meteorites. The sort of impact we are talking about would have pulverised the incoming bolide, and scattered its remains as dust throughout the atmosphere, from which it would eventually

* K/T stands for Cretaceous–Tertiary, with 'K' rather than 'C' because 'C' had already been granted by geologists to the Carboniferous Period. Cretaceous comes from *creta*, the Latin for chalk, and the German for chalk is Kreide, hence the K. The 'Tertiary' was part of a mostly defunct system of nomenclature, which we have nevertheless indicated on the geological timescale on page 15. The boundary is now officially called Cretaceous–Palaeogene. Nevertheless, the abbreviation 'K/T' remains in common use, and we will use it here.

have rained down all over the Earth's surface. The footprint – 100 miles wide and 30 miles deep – is a titanic impact crater, Chicxulub, at the tip of the Yucatan peninsula in Mexico (see plate 8).

Space is full of moving objects, travelling in random directions and at a great variety of speeds relative to one another. There are many more ways in which objects can be travelling at high speeds relative to us than low speeds. So, most of the objects that hit our planet are travelling very fast indeed. Fortunately, most of them are small and burn up in our atmosphere as 'shooting stars'. A few are large enough to retain some solid mass all the way to the planet's surface. And, once in a few tens of millions of years, a very large one catastrophically collides with us. Because of their high velocity relative to Earth, these massive objects release an unimaginably large quantity of energy when they collide. A gunshot wound is hot because of the velocity of the bullet. A colliding meteorite or comet is likely to be travelling even faster than a high-velocity rifle bullet. And where the rifle bullet weighs only ounces, the mass of the celestial projectile that ended the Cretaceous and slew the dinosaurs was measured in gigatons. The noise of the impact, thundering round the planet at a thousand kilometres per hour, probably deafened every living creature not burned by the blast, suffocated by the wind-shock, drowned by the 150-metre tsunami that raced around the literally boiling sea, or pulverised by an earthquake a thousand times more violent than the largest ever dealt by the San Andreas fault. And that was just the immediate cataclysm. Then there was the aftermath – the global forest fires, the smoke and dust and ash which blotted out the sun in a two-year nuclear winter that killed off most of the plants and stopped dead the world's food chains.

No wonder all the dinosaurs, with the notable exception of the birds, perished – and not just the dinosaurs, but about half of all other species too, particularly the marine ones.[*] The wonder is that any life at all survives these cataclysmic visitations. We are uneasily aware that a similar catastrophe could hit us at any moment. Unlike the dinosaurs in the Cretaceous, astronomers would give us several years' warning, or at least months. But this would not be a blessing for, at least with present-day technology, there is nothing we could do to prevent it. Fortunately, the odds that this will happen in any particular

---

[*] It is tempting to see the catastrophe as strangely selective. The deep sea Foraminifera (protozoa in tiny shells which fossilise in enormous numbers and are therefore much used by geologists as indicator species) were almost entirely spared.

person's lifetime are, by normal actuarial standards, negligible. At the same time, the odds that it will happen in *some* unfortunate individuals' lifetime are near certainty. Insurance companies are just not used to thinking that far ahead. And the unfortunate individuals concerned will probably not be human, for the statistical likelihood is that we shall be extinct before then anyway.

A rational case can be mounted that humanity should start research into defensive measures now, to bring the technology up to the point where, if a credible warning were sounded, there would be time to put measures into effect. Present-day technology could only minimise the impact, by storing a suitable balance of seeds, domestic animals, machines including computers and databases full of accumulated cultural wisdom, in underground bunkers with privileged humans (now *there's* a political problem). Better would be to develop so far only dreamed-of technologies to avert the catastrophe by diverting or destroying the intruder. Politicians who invent external threats from foreign powers, in order to scare up economic or voter support for themselves, might find that a potentially colliding meteor answers their ignoble purpose just as well as an Evil Empire, an Axis of Evil, or the more nebulous abstraction 'Terror', with the added benefit of encouraging international co-operation rather than divisiveness. The technology itself is similar to the most advanced 'star wars' weapons systems, and to that of space exploration itself. The mass realisation that humanity as a whole shares common enemies could have incalculable benefits in drawing us together rather than, as at present, apart.

Evidently, since we exist, our ancestors survived the Permian extinction, and later the Cretaceous extinction. Both catastrophes, and the others that have also occurred, must have been extremely unpleasant for them, and they may have only survived by the skin of their teeth, but still just capable of reproducing, otherwise we wouldn't be here. Then, in the fullness of evolutionary time, they reaped the benefits. In the case of the Cretaceous survivors, there were now no dinosaurs to eat them, no dinosaurs to compete with them. You might think there was a down side: no dinosaurs for them to eat. But few mammals were large enough, and few dinosaurs small enough, to make that much of a loss. There can be no doubt that the mammals flowered massively after the K/T, but the form of the flowering and how it relates to our rendezvous points is debatable. Three 'models' have been suggested, and now is the time to discuss them. The three shade into each other, and I shall present them in their extreme forms only for simplicity. For reasons of clarity, as I believe, I shall change their usual names to the Big Bang Model, the Delayed Explosion Model, and the Non-explosive Model. There are parallels in the controversy over the so-called Cambrian Explosion, to be discussed in the Velvet Worm's Tale.

1. The Big Bang Model, in its extreme form, sees a single mammal species surviving the K/T catastrophe, a sort of Palaeocene Noah. Immediately after the catastrophe, the descendants of this Noah started proliferating and diverging. On the Big Bang Model, most of the rendezvous points occurred in a bunch, just this side of the K/T boundary – the backwards way of viewing the rapidly divergent branching of the Noah's descendants.

2. The Delayed Explosion Model acknowledges that there was a major explosion of mammal diversity after the K/T boundary. But the mammals of the explosion were not descended from a single Noah, and most of the rendezvous points between mammal pilgrims predate the K/T boundary. When the dinosaurs suddenly left the scene, there were lots of little shrew-like lineages who survived to step into their shoes. One 'shrew' evolved into Carnivores, a second 'shrew' evolved into primates, and so on. These different 'shrews', although probably quite similar to each other, traced their separate ancestry deep into the past, eventually to unite way back in the Age of Dinosaurs. Those ancestors followed, in parallel, their long fuses into the future through the Age of Dinosaurs to the K/T boundary. Then they all exploded in diversity, more or less simultaneously, when the dinosaurs disappeared. The consequence is that the concestors of modern mammals long predate the K/T boundary, although they only started diverging from each other in appearance and way of life after the death of the dinosaurs.

3. The Non-explosive Model doesn't see the K/T boundary as marking any kind of sharp discontinuity in the evolution of mammalian diversity at all. Mammals just branched and branched, and this process went on before the K/T boundary in much the same way as it went on after it. As with the Delayed Explosion Model, the concestors of modern mammals predate the K/T boundary. But in this model they had already diverged considerably by the time the dinosaurs disappeared.

Of the three models, the evidence, especially molecular evidence but increasingly fossil evidence too, seems to favour the Delayed Explosion Model. Most of the major splits in the mammal family tree go way back, deep into dinosaur times. But most of those mammals that coexisted with dinosaurs were pretty similar to each other, and remained so until the removal of the dinosaurs freed them to explode into the Age of Mammals. A few members of those major lineages haven't changed much since those early times, and they consequently resemble each other, even though the common ancestors that they share are extremely ancient. Eurasian shrews and tenrec shrews, for example, are very similar to each other, probably not because they have converged from different starting points but because they haven't changed much since primitive times. Their shared ancestor, Concestor 13, may have lived over 90 million years ago, half as long before the K/T boundary as the K/T is before the present.

COLUGOS AND TREE SHREWS

In this chapter we find ourselves 70 million years into the past, still in the time of the dinosaurs and before the flowering of mammalian diversity properly began. Actually, the flowering of flowers themselves had only just begun. Flowering plants, while diverse, had been previously restricted to disturbed habitats such as those uprooted by elephantine dinosaurs or ravaged by fire, but by now had gradually evolved to include a range of forest-canopy trees and understorey bushes. Here our pilgrimage is joined by two groups of South East Asian mammals, the 20 species of squirrel-like tree shrews and the four colugos (more like flying squirrels).

The tree shrews are all very similar to each other, and are placed in the family Tupaiidae. Most live like squirrels, in trees, and some species resemble squirrels even down to having long, fluffy tails. The resemblance, however, is superficial. Squirrels are rodents. Tree shrews are certainly not rodents. As to what they are, well, that is partly what the next tale will be about. Are they shrews, as their common name would suggest? Are they primates, as certain authorities have long thought? Or are they something else altogether? The pragmatic solution has been to place them in their own, uncertainly placed, mammalian order, the Scandentia (Latin *scandere*, to climb).

Colugos have long been known as flying lemurs, prompting the obvious put-down: they neither fly nor are lemurs. Recent evidence suggests that they are closer to lemurs than was realised even by those responsible for the misnomer. And, while they don't have powered flight like a bat or a bird, they are adept gliders. Traditionally, two species have been recognised, *Cynocephalus volans*, the Philippine colugo, and *Galeopterus variegatus*, the Sunda or Malayan colugo (which lately has been split into three species, Javanese, Bornean and mainland Malay). They have a whole order to themselves, the Dermoptera. It means 'wings of skin'. Like the flying squirrels of America and Eurasia, the more distantly related flying scaly-tailed squirrels of Africa, and the marsupial gliders of Australia and New Guinea, colugos have a single large flap of skin, the patagium, which works a bit like a controlled parachute. Unlike that of the other

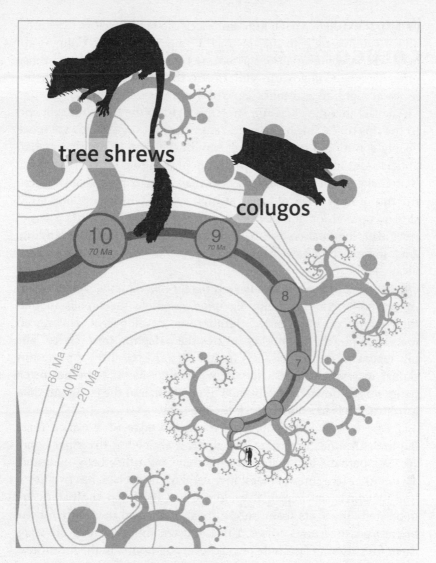

tree shrews

colugos

10
70 Ma

9
70 Ma

8

7

60 Ma

40 Ma

20 Ma

Tree shrews and colugos join. This is one of the most uncertain phylogenies in the book (see the Colugo's Tale). The two disputed groups are the colugos (four species) and the tree shrews (20 species). To emphasise the uncertainty, we have placed them in the same chapter, but joining at different rendezvous, 9 and 10 respectively.

A study from 2007 finds insertion/deletion mutations (indels) in eight genes, seven of which support the arrangement shown here (the other indel suggests that tree shrews are our closest relatives instead). In contrast, another study identifies a single large chromosomal rearrangement that unites tree shrews and colugos into a single group, an arrangement loosely supported by fossils. If these rendezvous points occured only a few million years apart, different parts of the genome really could have different evolutionary trees (see the Epilogue to the Gibbon's Tale). The dates of this and the next rendezvous are not well established.

gliders, the colugo's patagium embraces the tail as well as the limbs, and it extends right to the tips of the fingers and toes. Colugos are also, with a 'wing' span of 70 centimetres, larger than any of those other gliders. Colugos can glide more than 70 metres through the forest at night, to a distant tree, with little loss of height.

The fact that the patagium stretches right to the tip of the tail, and to the tips of the fingers and toes, suggests that the colugos are more deeply committed to the gliding way of life than other mammalian gliders. And indeed, they are pretty inept on the ground. They more than make up for it in the air, where their huge parachute gives them the run of large areas of forest at high speed. This necessitates good stereoscopic vision for steering accurately at night towards a target tree, avoiding fatal collisions, and then making a precision landing. And indeed they have large stereoscopic eyes, excellent for night vision.

Colugos and tree shrews have unusual reproductive systems, but in very different directions. Colugos resemble marsupials in that their young are born early in embryonic development. Having no marsupial pouch, the mother presses the patagium into service. The tail region of the patagium is folded forwards to form a makeshift pouch in which the (usually single) young sits. The mother often hangs upside down from a branch like a sloth, and the patagium then looks and feels like a hammock for the baby.

To be a baby colugo peeping over the edge of a warm, furry hammock sounds appealing. A baby tree shrew, on the other hand, receives perhaps less maternal care than any other baby mammal. The mother tree shrew, at least in several of the species, has two nests, one in which she herself lives, the other in which the babies are deposited. She visits them only to feed them, and then only for the briefest possible time, between five and ten minutes. And she visits them for this brief feed only once in every 48 hours. In the meantime, with no mother to keep them warm as any other baby mammal would have, the little tree shrews need to heat themselves from their food. To this end, the mother's milk is exceptionally rich.

Uniquely in our pilgrimage, we have placed two separately numbered rendezvous points in the same chapter. This highlights the fact that the affinities of the tree shrews and the colugos, to each other and to the rest of the mammals, are subject to dispute and uncertainty. There is a lesson in that very fact, and it is the lesson of the Colugo's Tale.

THE COLUGO'S TALE

The colugo could tell a tale of nocturnal gliding through the forests of South East Asia. But for the purposes of our pilgrimage it has a more down-to-earth tale to tell, whose moral is a warning. It is the warning that our apparently tidy story of concestors, rendezvous points, and the sequence in which pilgrims join us is heavily subject to disagreement and revision as new research is done. The phylogeny diagram at the start of this chapter is based on a study of insertions and deletions of small chunks of DNA code – the sort of rare and reliable changes discussed at the end of the Gibbon's Tale. These suggest that we primates are slightly more closely related to colugos than we are to tree shrews. The previous edition of this book had the colugos and tree shrews joining us as a single band, on the basis of earlier genetic studies. And prior to any molecular evidence, ortho-dox taxonomy would have had the tree shrews alone joining the primates at this rendezvous: the colugos would have joined us further down the road, not even very close.

There is no guarantee that our present picture, and the numbers we give to rendezvous points, will stay settled. Other recent evidence, from fossils and from a single large-scale chromosome change, argues for resurrecting our previous view. Future developments may even prompt a completely different arrangement, and change our numbering system yet again. This continuing uncertainty is one of the reasons why we have not given colugos and trees shrews their own separate chapters. Doubt and uncertainty may seem rather unsatisfac-tory as the moral for a tale, but it is an important lesson that must be taken on board before our pilgrimage to the past proceeds much further. The lesson will apply to many other rendezvous.

We could have signalled our uncertainty by having multi-way splits ('polytomies': see the Gibbon's Tale) in our phylogenetic trees. This is the solution adopted by certain authors, notably Colin Tudge in his masterly phylogenetic summary of all life on Earth, *The Variety Of Life*. But having polytomies on some branches risks giving false confidence in the others. As we shall see at Rendezvous 13, placental mammals are now thought to divide into four main groups, a major taxonomic transformation that happened in 2001, after Tudge's book was published. These areas of his classification, previously considered reasonably resolved, are now radically changed. The same could happen with this book, and it isn't just the colugos and tree shrews. We are even more unsure about the joining order of the two super-orders of mammals, afrotheres and xenarthrans, whom we shall meet at Rendezvous 13. At Rendezvous 18 and 19, the ordering of lungfish and coelacanths is still occasionally disputed. Fossil and genetic evidence disagrees on the placement of turtles (Rendezvous 16). The recent sequencing of the ctenophore genome (Rendezvous 29) has thrown their position into doubt. And the branching of the major groups of complex cellular life (Rendezvous 38) provides perhaps the most uncertain example of all.

Other rendezvous, such as that with the orang utans, are as near certain as it is possible to be, and there are a growing number in that happy category. There are also some borderline cases. So, rather than make what comes close to a subjective judgement about which groups deserve fully resolved trees and which do not, we have nailed our more or less uncertain colours to the mast in 2015, explaining the doubts in the text whenever possible (apart from a single rendezvous, number 36, where the order is so unsure that even the experts are not willing to hazard a guess). In the fullness of time, it is likely that some (but relatively few, we hope) of our rendezvous points and their phylogenies will turn out to be wrong, in the light of new evidence.[*]

Readers who find uncertainty unsettling can draw comfort from three observations. First, the educated guesses in the first edition of this book have largely been upheld, rather than overturned, by a decade of new data – these include increased consensus about the placement of tarsiers (Rendezvous 7), and the grouping of lampreys and hagfish (Rendezvous 22). Second, we know our areas of uncertainty. With a single exception (swapping numbers 23 and 24), changes to the rendezvous in this book have happened in areas that we earlier flagged up as controversial. Third, there are cases where we positively expect to see conflicting studies. In particular, as discussed in the Epilogue to the Gibbon's Tale, when two rendezvous points occur close in time, different regions of the genome are predicted to show different evolutionary trees. That may be why a few of the colugo's genes still disagree with the order of branching shown in this rendezvous. The shrew-like mammals which eventually evolved into colugos, tree shrews, and primates simply speciated within a few

___

[*] Creationist misquotation alert: creationists, please do not quote this as indicating that 'the evolutionists can't agree about anything' with the implication that the whole massive underlying theory can therefore be thrown out.

million years of each other. This effect probably explains conflicting studies at a few other points in our journey (in particular at Rendezvous 13). It leaves a dwindling list of cases where controversy is caused by a true misunderstanding of evolutionary history.

Earlier systems of taxonomy that were not tied to the evolution-standard might be controversial, in the way that matters of taste or judgement are controversial. A taxonomist might argue that, for reasons of convenience in exhibiting museum specimens, tree shrews should be grouped with shrews and colugos with flying squirrels. In such judgements there is no absolutely right answer. The phyletic taxonomy adopted in this book is different. There is still room for human judgement, but it is judgement about what will eventually turn out to be the undisputable truth. It is only because we haven't looked at enough details yet, especially molecular details, that we are still unsure what that truth is. The truth really is hanging up there waiting to be discovered. The same cannot be said for judgements of taste or of museum convenience.

RODENTS AND RABBITKIND

Rendezvous 11 occurs 75 million years into our journey.[*] It is here that our pilgrims are joined – overwhelmed, rather – by a teeming, scurrying, gnawing, whisker-quivering plague of rodents. For good measure, we also greet at this point the rabbits, including the very similar hares and jack-rabbits, and the rather more distant pikas. Rabbits were once classified as rodents, because they also have very prominent gnawing teeth at the front – indeed they outpoint the rodents, with an extra pair. They were then separated off, and are still placed in their own order, Lagomorpha, as opposed to Rodentia. But modern authorities group the lagomorphs together with the rodents in a 'cohort' called Glires. In other words, the lagomorph pilgrims and the rodent pilgrims joined up with each other more recently than the whole lot of them joined our pilgrimage. Concestor 11 is approximately our 15-million-greats-grandparent. It is the latest ancestor we share with a mouse, but the mouse is connected to it through a very much larger number of greats, because of short generation times.

Rodents are one of the great success stories of mammaldom. More than 40 per cent of all mammal species are rodents, and there are said to be more individual rodents in the world than all other mammals combined. Rats and mice have been the hidden beneficiaries of our own Agricultural Revolution, and they have travelled with us across the seas to every land in the world. They destroy our granaries and our health. For over 1,000 years, rats and their cargo of fleas have shipped bubonic plague to Europe, from its reservoir in the gerbils and ground squirrels of central Asia. The effects have been devastating: not just the 'Great' plagues of the seventeenth century but, as genetic evidence now reveals, the far more deadly Justinian plague of the sixth and seventh centuries and the worst pandemic of them all, the Black Death, which wiped out over half the population of medieval Europe. When even the four horsemen are laid low by the apocalypse, it will be rats that scavenge their remains, rats that will

---

[*] Note that in the first edition of this book this rendezvous point and the next were labelled numbers 10 and 11 respectively.

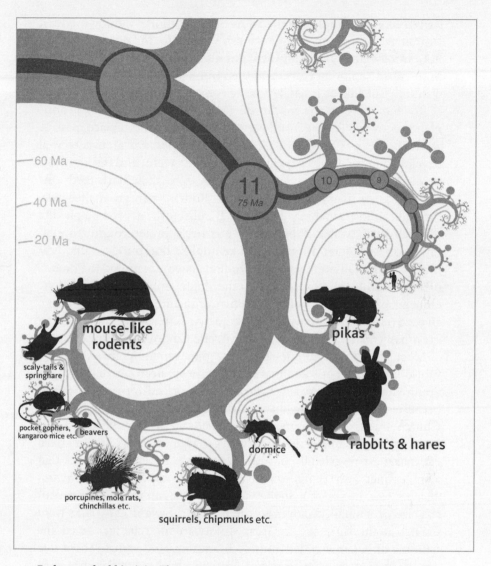

Rodents and rabbits join. There are 90 or so species of rabbits, hares, and pikas, together known as the lagomorphs. With the 2,300 rodent species (two-thirds of which are in the mouse family), they form a group named 'Glires', which molecular studies now convincingly place as close relatives to primates, colugos and tree shrews. Together, we constitute one of the four superorders of placental mammals. Neither of the titles for this superorder roll easily off the tongue, although 'supraprimates' is marginally more pronounceable than the more commonly used name, 'euarchontoglires'.

swarm like lemmings over the ruins of civilisation. And, by the way, lemmings are rodents, too – northern voles who, for reasons that are not entirely clear, build up their populations to plague proportions in so-called 'lemming years', and then indulge in frantic – though not wantonly suicidal as is falsely alleged – mass migrations.

Rodents are gnawing machines. They have a pair of very prominent incisor teeth at the front, perpetually growing to replace massive wear and tear. The gnawing masseter muscles are especially well developed in rodents. They don't have canine teeth, and the large gap or diastema that separates their incisors from their back teeth improves the efficiency of their gnawing. Rodents can gnaw their way through almost anything. Beavers fell substantial trees by gnawing through their trunks. Mole rats live entirely underground, tunnelling, not with their front paws like moles, but purely with their incisor teeth.* Different species of rodents have penetrated the deserts of the world (gundis, gerbils), the high mountains (marmots, chinchillas), the forest canopy (squirrels, including flying squirrels), rivers (water voles, beavers, capybaras), rainforest floor (agoutis), savannah (maras, springhares), and Arctic tundra (lemmings).

Most rodents are mouse-sized, but they range up through marmots, beavers, agoutis and maras to the sheep-sized capybaras of the South American waterways. Capybaras are prized for meat, not just because of their large size but because, bizarrely, the Roman Catholic Church traditionally deemed them honorary fish for Fridays, presumably because they live in water. Large as they are, modern capybaras are dwarfed by various giant South American rodents that went extinct only quite recently. The giant capybara, *Protohydrochoerus*, was the size of a donkey. *Telicomys* was an even larger rodent the size of a small rhinoceros which, like the giant capybara, went extinct at the time of the Great American Interchange, when the Isthmus of Panama ended South America's island status. These two groups of giant rodents were not particularly closely related to each other, and seem to have evolved their gigantism independently.

A world without rodents would be a very different world. It is less

---

* With the exception of one of the 15 species, they gnaw their way through the earth. The extreme troglodytes among mole rats, the naked mole rats, form tandem trains to mass-produce burrows, kicking back up the line the soil gnawed from the raw earth face by the lead worker. I use 'worker' advisedly, for the naked mole rats are further remarkable in being the nearest approach to social insects that the mammal world has to offer. They even *look* a bit like oversized termites – extraordinarily ugly by our standards, but they themselves are blind so they presumably don't care.

likely to come to pass than a world dominated by rodents and free of people. If nuclear war destroys humanity and most of the rest of life, a good bet for survival in the short term, and for evolutionary ancestry in the long term, is rats. I have a post-Armageddon vision. We and all other large animals are gone. Rodents emerge as the ultimate post-human scavengers. They gnaw their way through New York, London and Tokyo, digesting spilled larders, ghost super-markets and human corpses and turning them into new generations of rats and mice, whose racing populations explode out of the cities and into the countryside. When all the relics of human profligacy are eaten, populations crash again, and the rodents turn on each other, and on the cockroaches scavenging with them. In a period of intense competition, short generations perhaps with radioactively enhanced mutation-rates boost rapid evolution. With human ships and planes gone, islands become islands again, with local populations isolated save for occasional lucky raftings: ideal conditions for evolutionary divergence. Within 5 million years, a whole range of new species replace the ones we know. Herds of giant grazing rats are stalked by sabretoothed predatory rats.* Given enough time, will a species of intelligent, cultivated rats emerge? Will rodent historians and scien-tists eventually organise careful archaeological digs (gnaws?) through the strata of our long-compacted cities, and reconstruct the peculiar and temporarily tragic circumstances that gave ratkind its big break?

THE MOUSE'S TALE

Of all the thousands of rodents, the house mouse, *Mus musculus*, has a special tale to tell because it has become the second most intensively studied mammal species after our own. Much more than the proverbial guinea pig, the mouse is a main staple of medical, physiological and genetic laboratories the world over. In particular, the mouse was the second mammal, after ourselves, to have its genome completely sequenced.†

---

* Dougal Dixon long ago foresaw this, and he had the talent to paint it, in his imaginative book *After Man: A Zoology of the Future*.
† Although a common phrase in genetics, 'completely sequenced' is a bit of a misnomer. At the time of writing, in 2016, the 'complete' human genome is still missing about 5 per cent of the letters. Mostly these lie in areas that do not contain any genes, but help maintain the structure of chromosomes. Their sequence continues to elude us because it is highly repetitive, difficult to match up in the right order, and hence almost impossible to string together.

Two things about these genomes have sparked unwarranted surprise. The first is that mammal genomes contain rather few genes: recent estimates count a little under 20,000 in humans. And the second is that they are so similar to each other. Human dignity seemed to demand that we should have far more genes than a tiny mouse. And shouldn't it be absolutely larger than 20,000 anyway?

This last expectation has led people, including some who should know better, to deduce that the 'environment' must be more important than we thought, because there aren't enough genes to specify a body. That really is a breathtakingly naive piece of logic. By what standard do we *decide* how many genes you need to specify a body? This kind of thinking is based on a subconscious assumption which is wrong: the assumption that the genome is a kind of blueprint, with each gene specifying its own little piece of body. As the Fruit Fly's Tale will tell us, it is not a blueprint, but something more like a recipe, a computer program, or a manual of instructions for assembly.

If you think of the genome as a blueprint, you might expect a big, complicated animal like yourself to have more genes than a little mouse, with fewer cells and a less sophisticated brain. But, as I said, that isn't the way genes work. Even the recipe or instruction-book model can be misleading unless it is properly understood. My colleague Matt Ridley develops a different analogy which I find beautifully clear, in his book *Nature via Nurture*. Most of the genome that we sequence is not the book of instructions, or master computer program, for building a human or a mouse, although parts of it are. If it were, we might indeed expect our program to be larger than the mouse's. But our genes are more like the dictionary of words available for writing the book of instructions – or, we shall soon see, the set of subroutines that are called by the master program. As Ridley says, the list of words in *David Copperfield* is almost the same as the list of words in *The Catcher in the Rye*. Both draw upon the vocabulary of an educated native speaker of English. What is completely different about the two books is the order in which those words are strung together.

When a person is made, or when a mouse is made, both embryologies draw upon the same dictionary of genes: the normal vocabulary of mammal embryologies. The difference between a person and a mouse comes out of the different orders with which the genes, drawn from that shared mammalian vocabulary, are deployed, the different places in the body where this happens, and its timing. All this is controlled by a bewildering array of mechanisms, some of which we are only just starting to understand. Much lies outside the protein-coding sections of our DNA. We now know that the sequences which specify the proteins that build and run our bodies – and which are often thought of as the genes proper – make up only just over 1 per cent of our genome. About 8 per cent seems to carry out other important functions, and it's thought most of them help to switch genes on and off. These regulatory regions, often adjacent to the genes that they control, are critical for shaping the organism. A common *modus operandi* involves proteins called transcription factors which attach to the regulatory regions, turning their associated genes on or off. Some of these genes may themselves encode transcription factors, leading to complicated and exquisitely timed cascades and feedback loops. With such an intricate system, it only needs a few subtle tweaks at the DNA level to change the order in which different genes are eventually used.

Don't misunderstand 'order' as meaning the order in which the genes are strung out along the chromosomes. With notable exceptions, which we shall meet in the Fruit Fly's Tale, the order of genes along a chromosome is as arbitrary as the order in which words are listed in a vocabulary – usually alphabetical but, especially in phrase books for foreign travel, sometimes an order of convenience: words useful in airports; words useful when visiting the doctor; words useful for shopping, and so on. The order in which genes are stored on chromosomes is unimportant. What matters is that the cellular machinery finds the right gene when it needs it, and it does this using methods that are becoming increasingly understood. In the Fruit Fly's Tale, we'll return to those few cases, very interesting ones, where the order of genes arranged on the chromosome is non-arbitrary in something like the foreign phrase-book sense. For now, the important point is that what distinguishes a mouse from a man is mostly not the genes themselves, nor the order in which they are stored in the chromosomal 'phrase-book', but the order in which they are turned on: the equivalent of Dickens or Salinger choosing words from the vocabulary of English and arranging them in sentences.

In one respect the analogy of words is misleading. Words are shorter than genes, and some writers have likened each gene to a sentence. But sentences aren't a good analogy, for a different reason. Different books are not put together by permuting a fixed repertoire of sentences. Most sentences are unique. Genes, like words but unlike sentences, are used over and over again in different contexts. A better analogy for a gene than either a word or a sentence is a toolbox subroutine in a computer.

The computer I happen to be familiar with is the Macintosh, and it is some years since I did any programming so I am certainly out of date with the details. Never mind – the principle remains, and it is true of other computers too. The Mac has a toolbox of routines which are available for any program to use. There are thousands of these toolbox routines, each one doing a particular operation, which is likely to be needed, over and over again, in slightly different ways, in different programs. For example the toolbox routine called ObscureCursor hides the cursor from the screen until the next time the mouse is moved. Unseen to you, the ObscureCursor 'gene' is called every time you start typing and the mouse cursor vanishes. Toolbox routines lie behind the familiar features shared by all programs on the Mac (and their imitated equivalents on Windows machines): pulldown menus, scrollbars, shrinkable windows that you can drag around the screen with the mouse, and many others.

The reason all Mac programs have the same 'look and feel' (that very similarity famously became the subject of litigation) is precisely that all Mac programs, whether written by Apple, or by Microsoft, or by anybody else, call the same toolbox routines. If you are a programmer who wishes to move a whole region of the screen in some direction, say following a mouse drag, you would be wasting your time if you didn't invoke the ScrollRect toolbox routine. Or if you want to place a check mark by a pulldown menu item, you would be mad to write your own code to do it. Just write a call of CheckItem into your program, and the job is done for you. If you look at the text of a Mac program, whoever wrote it, in whatever programming language and for whatever purpose, the main thing you'll notice is that it consists largely of invocations of familiar, built-in toolbox routines. The same repertoire of routines is available to all programmers. Different programs string calls of these routines together in different combinations and sequences.

The genome, sitting in the nucleus of every cell, contains the toolbox of DNA routines available for performing standard biochemical functions. Different cells, for example liver cells, bone cells and muscle cells, string 'calls' of these routines together in different orders and combinations when performing particular cell functions including growing, dividing, or secreting hormones. Mouse bone cells are more similar to human bone cells than they are to mouse liver cells – they perform very similar operations and need to call the same repertoire of toolbox routines in order to do so. This is the kind of reason why all mammals have roughly the same number of genes – they all need the same toolbox.

Nevertheless, mouse bone cells do behave differently from human bone cells; and this too will be reflected in different calls to the toolbox in the nucleus. The toolbox itself is not identical in mouse and man, but it might as well be identical without in principle jeopardising the main differences between the two species. For the purpose of building mice differently from humans, what matters is differences in the calling of toolbox routines, more than differences in the toolbox routines themselves.

EPILOGUE TO THE MOUSE'S TALE

We saw in the Mouse's Tale that the main differences between mice and men reside in the pattern of choosing genetic routines from the available toolbox. It is of even greater importance in the case of differentiated cells in a single multicellular body. What distinguishes your nerve cells from your liver cells, your muscle cells from your skin cells, is not the DNA instruction set, but which instructions have been switched on or off. Unlike the case of mice and men, this is not a genetic difference, for your cells have essentially identical DNA. The difference is 'epigenetic' – outside of the genome. This has been known since before the discovery of DNA – the word was coined by Conrad Waddington in 1942 – but it has become disappointingly popular to trumpet this as something unexpected, even as a threat to our conventional understanding of genetics. Newspapers state with ill-informed astonishment that 'new' research shows the environment can modify which genes are turned on and off. Well of course it can! Embryology depends on it. And it's not just when switching between different types of cell. Compared to a couch potato, we would expect a body builder to have a different pattern of genes being switched on in his muscles. Responding to environmental cues – be those hormones in embryonic development or external influences in an adult – is exactly what genes do. This has been known since Jacob and Monod's investigations of bacteria in the early 1960s. And wasn't it obvious even before?

There is an extension to epigenetics which is more controversial. This is the idea that the pattern of gene use can be passed on to future generations: epigenetic *inheritance*. We are fêted with stories of characteristics being passed on from parent to offspring, in a modern resurgence of the Larmarckian idea of a blacksmith passing on his strong muscles to his children. There seems to be something in the human psyche to which this disastrous concept appeals – disastrous because it would also mean the blacksmith's child inheriting his father's gammy leg, scarred face and political attitudes (of whatever persuasion). Most of us would consider it fortunate that we do not inherit all the acquired characteristics of our parents. There is a lot to be said for starting from scratch.

Nevertheless, if you define epigenetic inheritance to include cellular inheritance within the body, it is undoubtedly true. Liver cells beget liver cells, muscle cells muscle cells, although the DNA is the same in all. Would it be too surprising if the same epigenetic inheritance carried over to a new body, that of the offspring? There is some suggestive evidence that mothers pass on the effects of starvation to their offspring and even their grandchildren. It's perhaps not too surprising that chemicals in the egg should carry an epigenetic whiff of the mother. But some stories have the whiff of implausibility. Recent experiments claim that male mice, conditioned to fear certain smells, pass on the fear through their sperm. This result may be solid, but 'extraordinary claims require extraordinary evidence'.

We should make the point that in order for such effects to be of evolutionary significance in the same sense as true mutations, they would have to be passed on not just to the grandchild generation but indefinitely through future generations. In fact, to the extent that they occur at all, they seem to die away over generational time. This is why we expect all adaptive evolution to be caused by Darwinian natural selection on the DNA sequence. In fact, all epigenetic inheritance we know of is under the control of the genome, the most important case being the cellular inheritance required to build a multicellular body. In contrast, pseudo-Lamarckian examples of epigenetic inheritance from parent to child are insignificant, and threaten to distract us from a much more important thought. We said 'all' adaptive evolution is based on DNA, but that is not true for humans. We pass on acquired characteristics in the form of ideas, a case of non-genetic inheritance if ever there was one. This alternative form of information transmission is the root of our culture, and in many senses distinguishes humanity from the rest of life on Earth.

THE BEAVER'S TALE 223

THE BEAVER'S TALE

A 'phenotype' is that which is influenced by genes. That pretty much means everything about a body. But there is a subtlety of emphasis which flows from the word's etymology. *Phaino* is Greek for 'show', 'bring to light', 'make appear', 'exhibit', 'uncover', 'disclose', 'manifest'. The phenotype is the external and visible manifestation of the hidden genotype. The *Oxford English Dictionary* defines it as 'the sum total of the observable features of an individual, regarded as the consequence of the interaction of its genotype with its environment' but it precedes this definition by a subtler one: 'A type of organism distinguishable from others by observable features.'

Darwin saw natural selection as the survival and reproduction of certain types of organism at the expense of rival types of organism. 'Types' here doesn't mean groups or races or species. In the subtitle of *The Origin of Species*, the much misunderstood phrase 'preservation of favoured races' most emphatically does not mean races in the normal sense. Darwin was writing before genes were named or properly understood, but in modern terms what he meant by 'favoured races' was 'possessors of favoured genes'.

Selection drives evolution only to the extent that the alternative types owe their differences to genes: if the differences are not inherited, differential survival has no impact on future generations. For a Darwinian, phenotypes are the manifestations by which genes are judged by selection. When we say that a beaver's tail is flattened to serve as a paddle, we mean that genes whose phenotypic expression included a flattening of the tail survived by virtue of that phenotype. Individual beavers with the flat-tailed phenotype survived as a consequence of being better swimmers; the responsible genes survived inside them, and were passed on to new generations of flat-tailed beavers.

At the same time, genes that expressed themselves in huge, sharp incisor teeth capable of gnawing through wood also survived. Individual beavers are built by permutations of genes in the beaver gene pool. Genes have survived through generations of ancestral beavers because they have proved good at collaborating with other genes in the beaver gene pool, to produce phenotypes that flourish in the beaver way of life.

At the same time again, alternative co-operatives of genes are surviving in other gene pools, making bodies that survive by prosecuting other life trades: the tiger co-operative, the camel co-operative, the cockroach co-operative, the carrot co-operative. My first book, *The Selfish Gene*, could equally have been called *The Co-operative Gene* without a word of the book itself needing to be changed. Indeed, this might have saved some misunderstanding (some of a book's most vocal critics are content to read the book by title only). Selfishness and co-operation are two sides of a Darwinian coin. Each gene promotes its own selfish welfare, by co-operating with the other genes in the sexually stirred gene pool which is that gene's environment, to build shared bodies.

But beaver genes have special phenotypes quite unlike those of tigers, camels or carrots. Beavers have lake phenotypes, caused by dam phenotypes. A lake is an *extended phenotype*. The extended phenotype is a special kind of phenotype, and it is the subject of the rest of this tale, which is a brief summary of my book of that title. It is interesting not only in its own right but because it helps us to understand how conventional phenotypes develop. It will turn out that there is no great difference of principle between an extended phenotype like a beaver lake or dam, and a conventional phenotype like a flattened beaver tail.

How can it possibly be right to use the same word, phenotype, on the one hand for a tail of flesh, bone and blood, and on the other hand for a body of still water, stemmed in a valley by a dam? The answer is that both are manifestations of beaver genes; both have evolved to become better and better at preserving those genes; both are linked to the genes they express by a similar chain of embryological causal links. Let me explain.

The embryological processes by which beaver genes shape beaver tails are not known in detail, but we know the kind of thing that goes on. Genes in every cell of a beaver behave as if they 'know' what kind of cell they are in. Skin cells have the same genes as bone cells, but different genes are switched on in the two tissues. We saw this in the Mouse's Tale. Genes, in each of the different kinds of cells in a beaver's tail, behave as if they 'know' where they are. They cause their respective cells to interact with each other in such a way that the whole tail assumes its characteristically hairless flattened form. There are formidable difficulties in working out how they 'know' which part of the tail they are in, but we understand in principle how these difficulties are overcome; and the solutions, like the difficulties themselves, will be of the same general kind when we turn to the development of tiger feet, camel humps and carrot leaves.

They are also of the same general kind in the development of the neuronal and neurochemical mechanisms that drive behaviour. Copulatory behaviour in beavers is instinctive. A male beaver's brain orchestrates, via hormonal secretions into the blood, and via nerves controlling muscles tugging on artfully hinged bones, a symphony of movements. The result is precise co-ordination with a female, who herself is moving harmoniously in her own symphony of move-ments, equally carefully orchestrated to facilitate the union. You may be sure that such exquisite neuromuscular music has been honed and perfected by generations of natural selection. And that means selec-tion of genes. In beaver gene pools, genes survived whose phenotypic effects on the brains, the nerves, the muscles, the glands, the bones, and the sense organs of generations of ancestral beavers improved the chances of those very genes passing through those very generations to arrive in the present.

Genes 'for' behaviour survive in the same kind of way as genes 'for' bones, and skin. Do you protest that there aren't 'really' any genes for behaviour; only genes for the nerves and muscles that make the behaviour? You are still wrecked among heathen dreams. Anatomical structures have no special status over behavioural ones, where 'direct' effects of genes are concerned. Genes are 'really' or 'directly' respons-ible only for proteins or other immediate biochemical effects. All other effects, whether on anatomical or behavioural phenotypes, are indirect. But the distinction between direct and indirect is vacuous. What matters in the Darwinian sense is that *differences* between genes are rendered as *differences* in phenotypes. It is only differences that natural selection cares about. And, in very much the same way, it is differences that geneticists care about.

Remember the 'subtler' definition of phenotype in the *Oxford English Dictionary:* 'A type of organism distinguishable from others by observable features.' The key word is distinguishable. A gene 'for' brown eyes is not a gene that directly codes the synthesis of a brown pigment. Well, it might happen to be, but that is not the point. The point about a gene 'for' brown eyes is that its possession makes a *difference* to eye colour *when compared* with some alternative version of the gene – an 'allele'. The chains of causation that culminate in the difference between one phenotype and another, say between brown and blue eyes, are usually long and tortuous. The gene makes a protein which is different from the protein made by the alternative gene. The protein has an enzymatic effect on cellular chemistry, which affects X which affects Y which affects Z which affects . . . a long chain of intermediate causes which affects . . . the phenotype of interest. The allele makes the *difference* when its phenotype is compared with the corresponding phenotype, at the end of the correspondingly long chain of causation that proceeds from the alternative allele. Gene differences cause phenotypic differences. Gene changes cause phenotypic changes. In Darwinian evolution alleles are selected, vis à vis alternative alleles, by virtue of the differences in their effects on phenotypes.

The beaver's point is that this comparison between phenotypes can happen anywhere along the chain of causation. All intermediate links along the chain are true phenotypes, and any one of them could constitute the phenotypic effect by which a gene is selected: it only has to be 'visible' to natural selection, nobody cares whether it is visible to us. There is no such thing as the 'ultimate' link in the chain: no final, definitive phenotype. Any consequence of a change in alleles, anywhere in the world, however indirect and however long the chain of causation, is fair game for natural selection, so long as it impinges on the survival of the responsible allele, relative to its rivals.

Now, let's look at the embryological chain of causation leading to dam-building in beavers. Dam-building behaviour is a complicated stereotypy, built into the brain like a fine-tuned clockwork mechanism. Or, as if to follow the history of clocks into the electronic age, dam-building is hard wired in the brain. I have seen a remarkable film of captive beavers imprisoned in a bare, unfurnished cage, with no water and no wood. The beavers enacted, 'in a vacuum', all the stereotyped movements normally seen in natural building behaviour when there is real wood and real water. They seem to be placing virtual wood into a virtual dam wall, pathetically trying to build a ghost wall with ghost sticks, all on the hard, dry, flat floor of their prison. One feels sorry for them: it is as if they are desperate to exercise their frustrated dam-building clockwork.

Only beavers have this kind of brain clockwork. Other species have clockwork for copulation, scratching and fighting, and so do beavers. But only beavers have brain clockwork for dam-building, and it must have evolved by slow degrees in ancestral beavers. It evolved because the lakes produced by dams are useful. It is not totally clear what they are useful for, but they must have been useful for the beavers who built them, not just any old beavers. The best guess seems to be that a lake provides a beaver with a safe place to build its lodge, out of reach for most predators, and a safe conduit for transporting food. Whatever the advantage it must be a substantial one, or beavers would not devote so much time and effort to building dams. Once again, note that natural selection is a predictive theory. The Darwinian can make the confident prediction that, if dams were a useless waste of time, rival beavers who refrained from building them would survive better and pass on genetic tendencies not to build. The fact that beavers are so anxious to build dams is very strong evidence that it benefited their ancestors to do so.

Like any other useful adaptation, the dam-building clockwork in the brain must have evolved by Darwinian selection of genes. There must have been genetic variations in the wiring of the brain which affected dam-building. Those genetic variants that resulted in improved dams were more likely to survive in beaver gene pools. It is the same story as for all Darwinian adaptations. But which is the phenotype? At which link in the chain of causal links shall we say the genetic difference exerts its effect? The answer, to repeat it, is all links where a difference is seen. In the wiring diagram of the brain? Yes, almost certainly. In the cellular chemistry that, in embryonic development, leads to that wiring? Of course. But also *behaviour* – the symphony of muscular contractions that is behaviour – this too is a perfectly respectable phenotype. Differences in building behaviour are without doubt manifestations of differences in genes. And, by the same token, the *consequences* of that behaviour are also entirely allowable as phenotypes of genes. What consequences? Dams, of course. And lakes, for these are consequences of dams (see plate 9). Differences between lakes are influenced by differences between dams, just as differences between dams are influenced by differences between behaviour patterns, which in turn are consequences of differences between genes. We may say that the characteristics of a dam, or of a lake, are true phenotypic effects of genes, using exactly the logic we use to say that the characteristics of a tail are phenotypic effects of genes.

Conventionally, biologists see the phenotypic effects of a gene as confined within the skin of the individual bearing that gene. The Beaver's Tale shows that this is unnecessary. The phenotype of a gene, in the true sense of the word, may extend outside the skin of the individual. Birds' nests are extended phenotypes. Their shape and size, their complicated funnels and tubes where these exist, all are Darwinian adaptations, and so must have evolved by the differential survival of alternative genes. Genes for building behaviour? Yes. Genes for wiring up the brain so it is good at building nests of the right shape and size? Yes. Genes for nests of the right shape and size? Yes, by the same token, yes. Nests are made of grass or sticks or mud, not bird cells. But the point is irrelevant to the question of whether differences between nests are influenced by differences between genes. If they are, nests are proper phenotypes of genes. And nest differences surely must be influenced by gene differences, for how else could they have been improved by natural selection?

Artefacts like nests and dams (and lakes) are easily understood examples of extended phenotypes. There are others where the logic is a little more . . . well, extended. For example, parasite genes can be said to have phenotypic expression in the bodies of their hosts. This can be true even where, as in the case of cuckoos, they don't live inside their hosts. And many examples of animal communication – as when a male canary sings to a female and her ovaries grow – can be rewritten in the language of the extended phenotype. But that would take us too far from the beaver, whose tale will conclude with one final observation. Under favourable conditions the lake of a beaver can span several miles, which may make it the largest phenotype of any gene in the world.

LAURASIATHERES

Eighty-five million years ago, in the hot-house world of the Upper Cretaceous, we greet Concestor 12, approximately our 25-million-greats-grandparent. Here we are joined by a much more diverse band of pilgrims than the rodents and rabbits who swelled our party at Rendezvous 11. Zealous taxonomists recognise their shared ancestry by giving them a name, Laurasiatheria, but it is seldom used because, in truth, this is a miscellaneous bunch. The rodents are all built to the same toothy design and have proliferated and diversified, presumably because it works so well. 'Rodents' therefore really means something strong; it unites animals that have much in common. 'Laurasiatheria' is as awkward as it sounds. It unites highly disparate mammals which have only one thing in common: their pilgrims all join up with each other more recently than the point at which they join us. And they all hail, originally, from the old northern continent of Laurasia.

And what a diverse crew these laurasian pilgrims are, some of them flying, some of them swimming, many of them galloping, half of them nervously looking over their shoulder for fear of being eaten by the other half. They belong to seven different orders, the Pholidota (pangolins), Carnivora (dogs, cats, hyenas, bears, weasels, seals, etc.), the Perissodactyla (horses, tapirs and rhinos), Cetartiodactyla (antelopes, deer, cattle, camels, pigs, hippos and . . . well, we'll come to the surprise member of this group later), Microchiroptera and Megachiroptera (respectively small and big bats) and Insectivora (moles, hedgehogs and shrews, but NOT elephant shrews or tenrecs: we have to wait for Rendezvous 13 to meet them).

Carnivora is an irritating name because, after all, it simply means meat-eater, and meat-eating has been invented literally hundreds of times independently in the animal kingdom. Not all carnivores are Carnivora (spiders are carnivores and so was the hoofed *Andrewsarchus*, the largest meat-eater since the end of the dinosaurs) and not all Carnivora are carnivores (think of the gentle giant panda, eating almost nothing but bamboo). Within the mammals the order Carnivora does appear to be a genuinely monophyletic clade: that is, a group of animals, all descended from a single concestor who would have been

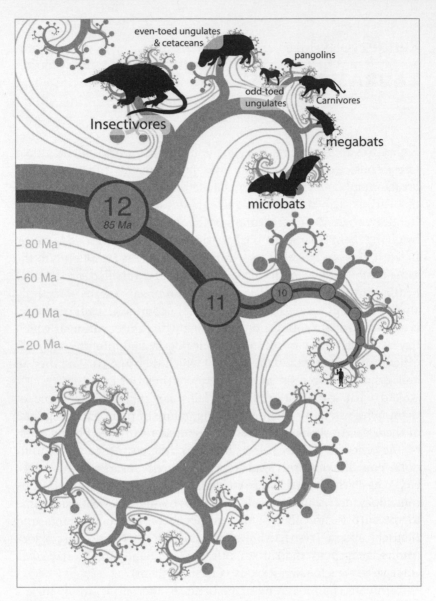

Laurasiatheres join. Molecular studies consistently find that the closest relatives of primates and rodents are a major group of mammals: the superorder Laurasiatheria. Numbering over 2,000 species, laurasiatheres range from lions to bats, hedgehogs to whales. Although now generally accepted as a natural grouping, there is still considerable debate about the relationships between the major contingents shown, especially among the Carnivores and the 'ungulate' groups, but also in the exact placement of the two major groups of bat: the large, often fruit-eating 'megabats' and the smaller echolocating and insect-hunting 'microbats'. The arrangement shown here may well be incorrect in places: our uncertainty probably reflects a rapid pattern of branching early in laurasiathere evolution.

classified as one of them. Cats (including lions, cheetahs and sabretooths), dogs (including wolves, jackals and Cape hunting dogs),
weasels and their kind, mongooses and their kind, bears (including
pandas), hyenas, wolverines, seals, sea lions and walruses, all are
members of the laurasiatherian order Carnivora, and all are descended
from a concestor which would have been placed in the same order.

Carnivores and their prey need to outrun each other, and it is not
surprising that the demands of fleetness have pushed them in similar
evolutionary directions. You need long legs for running, and the
great laurasiatherian herbivores and carnivores have, independently
and in different ways, added extra length to their legs by com-
mandeering bones which, in us, are inconspicuously buried within
the hands (metacarpals) or feet (metatarsals). The 'cannon-bone' of a
horse is the enlarged third metacarpal (or metatarsal) fused together
with two tiny 'splint' bones that are vestiges of the second and fourth
metacarpal (metatarsal). In antelopes and other even-toed ungulates,
the cannon-bone is a fusion of the third and fourth metacarpal
(metatarsal). Carnivores, too, have elongated their metacarpals and
metatarsals, but these five bones have stayed separate instead of
fusing together or disappearing altogether, as in horses, cattle and
the rest of the so-called ungulates.

Unguis is Latin for nail, and ungulates are animals that walk on
their nails – hooves. But the ungulate way of walking has been
invented several times and ungulate is a descriptive term rather than
a respectable taxonomic name. Horses, rhinos and tapirs are odd-
toed ungulates. Horses walk on a single toe, the middle one. Rhinos
and tapirs walk on the middle three, as did early horses and some
atavistically mutant horses today. Even-toed or cloven-hoofed ungulates walk on two toes, the third and fourth. The convergent
resemblances between the two-toed cattle family and the one-toed
horse family are modest compared to the convergent resemblances of
both, separately, to certain extinct South American herbivores. A
group called the litopterns independently, and earlier, 'discovered'
the horse habit of walking on a single middle toe. Their leg skeletons
are almost identical to those of horses. Other South American
herbivores, among the so-called notoungulates, independently discovered the cattle/antelope habit of walking on toes three and four.
Such stunning resemblances really did fool a senior Argentinian
zoologist in the nineteenth century, who thought that South America
was the evolutionary nursery of many of our great groups of

mammals. In particular, he believed that litopterns were early relatives of the true horse (perhaps with a little national pride that his
country might have been the cradle of that noble animal).

The laurasiathere pilgrims now joining us include small animals as
well as the large ungulates and Carnivores. Bats are remarkable for all
sorts of reasons. They are the only surviving vertebrates to put up any
sort of competition to birds in flight, and very impressive aerobats
they are. With nearly a thousand species, they far outnumber all
other orders of mammals except rodents. And bats have perfected
sonar (the sound equivalent of radar) to a higher degree than any
other group of animals, including human submarine designers.*

The other main group of small laurasiatheres are the so-called
insectivores. The order Insectivora includes shrews, moles, hedgehogs and other small, snouty creatures which eat insects and small
terrestrial invertebrates like worms, slugs and centipedes. As with
Carnivora, I shall use a capital letter to denote the taxonomic group,
Insectivora, as opposed to insectivore with a small i, which means
just anything that eats insects. So, a pangolin (or scaly anteater) is an
insectivore but not an Insectivore. A mole is an Insectivore, which
actually eats insects. As I have already remarked, it is a pity the early
taxonomists used names like Insectivora and Carnivora, which are
only loosely correlated to the descriptions of preferred diet with
which they are so readily confused.

Related to the other Carnivores like dogs, cats and bears are the seals,
sea lions and walruses. We shall soon hear the Seal's Tale, which is about
mating systems. I find seals interesting for another reason, too: they
have moved into the water, and have modified themselves in that
direction to about half the extent that dugongs have, or whales have.
And that reminds me – there is one other major group of laurasiatheres
that we haven't dealt with. On to the Hippo's Tale, for a real surprise.

THE HIPPO'S TALE

When I was a schoolboy studying Greek, I learned that *hippos* meant
'horse' and *potamos* 'river'. Hippopotamuses were river horses (see
plate 10). Later, when I gave up Greek and read Zoology, I was not too
disconcerted to learn that hippopotamuses weren't close to horses

---

* I would have liked, at this point, to insert the Bat's Tale, but it would be pretty much
the same as a chapter in another of my books, so I won't. Incidentally, I had to exercise
similar restraint over 'the Spider's Tale', 'the Fig Tree's Tale' and half a dozen others.

after all. Instead, they were classified firmly with pigs, in the middle of the even-toed ungulates or artiodactyls. I have now learned something so shocking that I am still reluctant to believe it, but it looks as though I am going to have to. Hippos' closest living relatives are whales. The even-toed ungulates include whales! Whales, needless to say, don't have hooves at all, whether odd- or even-toed. Indeed, they don't have toes, so it might be less confusing if we adopt the scientific name, artiodactyls (which is actually just the Greek for even-toed, so the change doesn't help much). For completeness, I should add that the equivalent name for the horse order is Perissodactyla (Greek for uneven-toed). Whales, it would now seem from strong molecular evidence, are artiodactyls. But since they previously had been placed in the order Cetacea, and since Artiodactyla was also a well-established name, a new composite has been coined: Cetartiodactyla.

Whales are wonders of the world. They include the largest organisms that have ever moved. They swim with up-and-down movements of the spine derived from the mammalian gallop, as opposed to the side to side wave motion of the spine of a swimming fish or a running lizard. Or presumably a swimming ichthyosaur which, in other respects, looked rather like a dolphin, except for its telltale vertical tail: the dolphin's is horizontal for galloping through the sea. A whale's front limbs are used for steering and stabilising. There are no externally visible hind limbs at all, but some whales have small vestigial pelvic and leg bones buried deep in their bodies.

It would not be too hard to believe that whales are closer cousins to even-toed ungulates than they are to any other mammals. A bit strange, perhaps, but not shocking to accept that some remote ancestor branched to the left and went to sea to give rise to the whales, while it branched to the right to give all the even-toed ungulates. What is shocking is that, according to the molecular evidence, whales are deeply embedded *within* the even-toed ungulates. Hippos are closer cousins to whales than hippos are to anything else including other even-toed ungulates such as pigs.* On their backward journey, the hippo pilgrims and the whale pilgrims unite with each other more recently than the two of them join the ruminants, and it seems that the other even-toed ungulates such as pigs and camels join them yet deeper in time. Whales are the surprise inclusion that I coyly

---

* By the way, we even had it wrong when we classified hippos most closely with pigs within the artiodactyls. The molecules suggest that the sister group to the hippo–whale clade is the ruminants: the cows, sheep and antelopes. Pigs lie outside them all.

referred to when I introduced the cetartiodactyls at the start of this rendezvous. It is known as the Whippo Hypothesis.

All this supposes that we believe the testimony of the molecules.* What do the fossils say? To my initial surprise, the new theory fits quite nicely. Most of the great orders of mammals (though not the subdivisions within them) go well back into the age of dinosaurs, as we saw in connection with the Great Cretaceous Catastrophe. Rendezvous 11 (with the rodents and rabbits) and Rendezvous 12 (the one we have just reached) both took place during the Cretaceous Period at the height of the dinosaur regime. But mammals in those days were all rather small, shrew-like creatures, whether their respective descendants were destined to become mice or hippos. The real growth of mammal diversity started suddenly after the dinosaurs went extinct 66 million years ago. It was then that the mammals were able to blossom into all economic trades vacated by the dinosaurs. Large body size was just one thing that became possible for mammals when the dinosaurs were gone. The process of divergent evolution was swift, and a huge range of mammals, of all sizes and shapes, roamed the land within 5 million years of 'liberation'. Five to ten million years later, in the late Palaeocene to early Eocene Epoch, there are abundant fossils of even-toed ungulates.

Another 5 million years later, in the early to middle Eocene, we find a group called the archaeocetes. The name means 'old whales', and most authorities accept that among these animals are to be found the ancestors of modern whales. An early one of these, *Pakicetus* from Pakistan, seems to have spent at least some of its time on land. Later ones include the unfortunately named *Basilosaurus* (unfortunate not because of Basil but because saurus means lizard: when first discovered, *Basilosaurus* was thought to be a marine reptile, and the rules of naming rigidly enforce priority, even though we now know better†). It had an immensely long body, and would have been a good candidate for the giant sea serpent of legend, if only it were not long extinct. Around the time that whales were rep-

---

* The molecular evidence for this radical view is what I have referred to in the Gibbon's Tale as a Rare Genomic Change (RGC). Transposable element genes that are highly recognisable are found in particular places in the genome, and are presumably inherited from the hippo–whale ancestor. While this is very strong testimony, it is nevertheless prudent to look at the fossils too.

† The celebrated Victorian anatomist Richard Owen tried to get the name changed to *Zeuglodon,* and Haeckel followed suit in his phylogeny, reproduced in the epilogue to this tale. But we are stuck with *Basilosaurus.*

resented by the likes of *Basilosaurus*, the contemporary hippo ancestors may have been members of a group called the anthracotheres,
some reconstructions of which make them look quite like hippos.

Returning to the whales, what of the antecedents of the archaeocetes, before they re-invaded the water? If the molecules are right
that whales' closest affinities are with hippos, we might be tempted to
seek their ancestors among fossils which show some evidence of
herbivory. On the other hand, no modern whale or dolphin is
herbivorous. The completely unrelated dugongs and manatees, by
the way, show that it is perfectly possible for a purely marine
mammal to have a purely herbivorous diet. But whales eat either
planktonic crustacea (baleen whales), fish or squid (dolphins and
most toothed whales), or large prey such as seals (killer whales). This
has led people to look for whale ancestors among carnivorous land
mammals, beginning with Darwin's own speculation, sometimes
ridiculed though I have never understood why:

> In North America the black bear was seen by Hearne swimming for
> hours with widely open mouth, thus catching, like a whale, insects in
> the water. Even in so extreme a case as this, if the supply of insects were
> constant, and if better adapted competitors did not already exist in the
> country, I can see no difficulty in a race of bears being rendered, by
> natural selection, more and more aquatic in their structure and habits,
> with larger and larger mouths, till a creature was produced as mon-
> strous as a whale (*Origin of Species*, 1859, p. 184).

As an aside, this suggestion of Darwin illustrates an important general
point about evolution. The bear seen by Hearne was evidently an
enterprising individual, feeding in an unusual way for its species. I
suspect that major new departures in evolution often start in just such
a way, with a piece of lateral thinking by an individual who discovers
a new and useful trick, and learns to perfect it. If the habit is then
imitated by others, including perhaps the individual's own children,
there will be a new selection pressure set up. Natural selection will
favour genetic predispositions to be good at learning the new trick,
and much will follow. I suspect that something like this is how
'instinctive' feeding habits such as tree-hammering in woodpeckers,
and mollusc-smashing in thrushes and sea otters, got their start.*

* This idea has a name, the Baldwin Effect, although the idea was independently proposed
by Lloyd Morgan in the same year, and by Douglas Spalding earlier. I follow Alister
Hardy's way of developing it in *The Living Stream*. For some reason it is a favourite of
mystics and obscurantists.

For a long time, people looking over the available fossils for a plausible antecedent to the archaeocetes have favoured the mesony-chids, a large group of land mammals that flourished in the Palaeo-cene Epoch, just after the extinction of the dinosaurs. The mesonychids seem to have been largely carnivorous, or omnivorous like Darwin's bear, and they fit with what we all – before the coming of the hippo theory – thought a whale ancestor ought to be. An additionally nice thing about the mesonychids is that they had minature hooves on each toe. They were hoofed carnivores, perhaps a bit like wolves but running on true hooves! Could they, then, have given rise to the even-toed ungulates, as well as to the whales? Unfortunately, the idea doesn't fit with the hippo theory specifically. Even though the mesonychids seem to be cousins of today's even-toed ungulates (and there are reasons for believing this over and above their hooves) they are no closer to hippos than they are to all the rest of the cloven-hoofed animals. We keep coming back to the molecular shocker: whales are not just cousins of all the artiodactyls, they are buried within the artiodactyls, closer to hippos than hippos are to cows and pigs.

Gathering all this together, we can sketch a forward chronology as follows. Molecular evidence puts the split between camels (plus llamas) and the rest of the artiodactyls at 65 to 70 million years ago, ap-proximately when the last dinosaurs died. Don't imagine, by the way, that the shared ancestor looked anything like a camel. In those days, all mammals looked more or less like shrews. But about 70 million years ago, the 'shrews' that were going to give rise to camels split from the 'shrews' that were going to give rise to all the rest of the artiodactyls. The split between pigs and the rest (mostly ruminants) took place 65 million years ago. The split between ruminants and hippos took place about 60 million years ago. Then the whale lineage split off from the hippo lineage round about 55 million years ago, which gives time for primitive whales such as the semi-aquatic *Pakicetus* to have evolved by 50 million years ago. Toothed whales and baleen whales parted com-pany much later, around 34 million years ago, around the time when the earliest baleen whale fossils are found.

Perhaps I was exaggerating a little when I implied that a traditional zoologist like me should be positively upset at the discovery of the hippo–whale connection. But let me try to explain why I was genuinely disconcerted when I first read about it a few years ago. It wasn't just that it was different from what I had learned as a student.

That wouldn't have worried me at all, in fact I would have found it positively exhilarating. What worried me, and still does to some extent, was that it seemed to undermine all generalisations that one might wish to make about groupings of animals. The life of a molecular taxonomist is too short to allow a pairwise comparison of every species with every other species. Instead what one does is take two or three whale species, say, and assume that they are representative of whales as a group. It is tantamount to the assumption that the whales are a clade, sharing a common ancestor which is not shared by the other animals with which one is making the comparison. It is assumed not to matter, in other words, which whale you take to stand for all. Similarly, lacking the time to test every species of rodent, say, or artiodactyl, we might take blood* from a rat, and from a cow. It doesn't matter which artiodactyl you take to compare with the representative whale because, yet again, we assume that the artiodactyls are a good clade, so it doesn't make any difference whether we take a cow, a pig, a camel or a hippo.

But now we are told that it does matter. Camel blood and hippo blood really will give a different comparison with whale blood because hippos are closer cousins to whales than they are to camels. See where this lands us. If we can't trust the artiodactyls to hang together as a group, represented by any one of their number, how can we be sure that any group will hang together? Can we even assume that hippos hang together, such that it doesn't matter whether we choose a pygmy hippo or a common hippo for comparison with whales? What if whales are closer to pygmy hippos than to common hippos? Actually we probably can rule that out, because fossil evidence suggests that the two hippo genera split apart about as recently as our split from chimpanzees, and that really does leave too little time to evolve all the different kinds of whales and dolphins.

It is more problematical whether all the whales hang together. On the face of it, the toothed whales and the baleen whales might well represent two entirely separate returns to the sea from the land. Indeed, that very possibility has often been advocated. The molecular taxonomists who demonstrated the hippo connection very wisely did take DNA from both a toothed whale and a baleen whale. They found that the two whales are indeed much closer cousins to each

---

* Actually, blood is not great for genetic comparisons in mammals because, unusually among vertebrates, their red blood corpuscles don't contain DNA.

other than they are to a hippo. But again, how do we know that 'the toothed whales' hang together as a group? And the same for 'the baleen whales'? Maybe all the baleen whales are related to a hippo except the minke whale, which is related to a hamster. No, I don't believe that, and I really do think the baleen whales are a united clade, sharing a common ancestor which is not shared by anything that is not a baleen whale. But can you see how the hippo–whale discovery shakes the confidence?

We could regain our confidence if we could think of a good reason why whales might be special in this respect. If whales are glorified artiodactyls, they are artiodactyls that suddenly took off, evolutionarily speaking, leaving the rest of the artiodactyls behind. Their closest cousins, the hippos, remained relatively static, as normal, respectable artiodactyls. Something happened in the history of the whales that made them shift into evolutionary overdrive. They evolved so much faster than all the rest of the artiodactyls that their origin within that group was obscured, until molecular taxonomists came along and uncovered it. So, what is special about the history of the whales?

When you write it down like that, the solution leaps off the page. Leaving the land and becoming wholly aquatic was a bit like going into outer space. When we go into space we are weightless (not, by the way, because we are a long way from the Earth's gravity, as many people think, but because we are in free fall like a parachutist before he pulls the ripcord). A whale floats. Unlike a seal or a turtle, which still comes on land to breed, a whale never stops floating. It never has to contend with gravity. A hippo spends time in the water, but it still needs stout, treetrunk-like legs and strong leg muscles for the land. A whale doesn't need legs at all, and indeed it doesn't have any. Think of a whale as what a hippo would like to be if only it could be freed from the tyranny of gravity. And of course there are so many other odd things about living the whole time in the sea that it comes to seem far less surprising that whale evolution should have spurted as it did, leaving hippos behind, stranded on land and stranded in the middle of the artiodactyls. This suggests that I was unduly alarmist a few paragraphs back.

Much the same thing happened in the other direction, 300 million years earlier, when our fish ancestors emerged from the water onto the land. If whales are glorified hippos, we are glorified lungfish. The emergence of legless whales from within the middle of the artiodactyls, leaving the rest of the artiodactyls 'behind', should not seem more surprising than the emergence of four-legged land animals from one particular group of fish, leaving those fish 'behind'. That, at any rate, is how I rationalise the hippo–whale connection, and recover my lost zoological composure.

EPILOGUE TO THE HIPPO'S TALE

Zoological composure be blowed. My attention was drawn to the following while the first edition of this book was being prepared. In 1866, the great German zoologist Ernst Haeckel drew up the schematic evolutionary tree of mammals (see below). I had often seen the full tree reproduced in histories of zoology, but I had never before noticed the position of the whales and hippos in Haeckel's scheme. Whales are 'Cetacea', as today, and Haeckel presciently placed them close to the artiodactyls. But the real stunner is where he put the hippos. He called them by the unflattering name 'Obesa' and he

No new thing under the sun. Detail from Ernst Haeckel's evolutionary tree of mammals, published in 1866 [170], showing hippos closely related to whales.

classified them not in the artiodactyls but as a tiny twig on the branch leading to Cetacea.* Haeckel classified hippos as the sister group to the whales: hippos, in his vision, were more closely related to whales than they were to pigs, and all three were more closely related to each other than to cows.

> . . . there is no new thing under the sun. Is there anything whereof it may be said, See, this is new? It hath been already of old time, which was before us.
>
> ECCLESIASTES 1: 9–10

THE SEAL'S TALE

Most wild animal populations have approximately equal numbers of males and females. There's a good Darwinian reason for this, which was clearly seen by the great statistician and evolutionary geneticist R. A. Fisher. Imagine a population in which the numbers were unequal. Now, individuals of the rarer sex will on average have a reproductive advantage over individuals of the commoner sex. This is not because they are in demand and have an easier time finding a mate (although that might be true too). Fisher's reason is a deeper one, with a subtle economic slant. Suppose there are twice as many males as females in the population. Now, since every child born has exactly one father and one mother, the average female must, all other things being equal, have twice as many children as the average male. And vice versa if the population sex ratio is reversed. It is simply a question of allocating the available posterity among the available parents. So, any general tendency for parents to favour sons rather than daughters, or daughters rather than sons, will immediately be counteracted by natural selection for the opposite tendency. The only evolutionarily stable sex ratio is 50/50.

But it isn't quite that simple. Fisher spotted an economic subtlety in the logic. What if it costs twice as much to rear a son, say, as to rear a daughter, presumably because males are twice as big? Well, now, the reasoning changes. The choice that faces a parent is no longer, 'Shall I have a son or a daughter?' It is now, 'Shall I have a son or – for the same price – two daughters?' The balanced sex ratio in the population is now twice as many females as males. Parents who

---

* Haeckel didn't get everything right, however. He put the sirenians (dugongs and manatees) in with whales.

favour sons on the grounds that males are rare, will see their advantage precisely undermined by the extra cost of making males. Fisher divined that the true sex ratio equalised by natural selection is not the ratio of numbers of males to numbers of females. It is the ratio of economic spending on rearing sons to economic spending on rearing daughters. And what does economic spending mean? Food? Time? Risk? Yes, in practice all these are likely to be important, and for Fisher the agent doing the spending was always parents. But economists use a more general expression of cost, which they call opportunity cost. The true cost to a parent of making a child is measured in lost opportunities to make other children. This opportunity cost was named Parental Expenditure by Fisher. Under the name Parental Investment, Robert L. Trivers, a brilliant intellectual successor to Fisher, used the same idea to elucidate sexual selection. Trivers was also the first to understand clearly the fascinating phenomenon of parent–offspring conflict, in a theory that has been carried further in startling directions by the equally brilliant David Haig.

As ever, and at the risk of boring those of my readers not handicapped by a little learning in philosophy, I once again must stress that the purposeful language I have used is not to be treated literally. Parents do not sit down and discuss whether to have a son or a daughter. Natural selection favours, or disfavours, genetic tendencies to invest food or other resources in such a way as to lead eventually to equal or unequal parental expenditure on sons and daughters, over the whole of a breeding population. In practice this will often amount to equal numbers of males and females in the population.

But what about those cases where a minority of males holds the majority of females in harems? Does this violate Fisher's expectations? Or those cases where males parade in front of females in a 'lek', and the females look them over and choose their favourite? Most females have the same favourite, so the end result is the same as for a harem: polygyny – disproportionate access to a majority of females by a privileged minority of males. That minority of males ends up fathering most of the next generation, with the rest of the males hanging about as bachelors. Does polygyny violate Fisher's expectations? Surprisingly, no. Fisher still expects equal investment in sons and daughters, and he is right. Males may have a lower expectation of reproducing at all, but if they do reproduce they reproduce in spades. Females are unlikely to have no children but they are also unlikely to have very many. Even under conditions of extreme polygyny it evens out and Fisher's principle holds.

Some of the most extreme examples of polygyny are to be found among the seals. Seals haul themselves out onto beaches to breed, often in huge 'rookeries', heaving with intense sexual and aggressive activity. In a famous study of elephant seals by the California zoologist Burney LeBoeuf, 4 per cent of the males accounted for 88 per cent of all copulations seen. No wonder the rest of the males are dissatisfied, and no wonder elephant-seal fights are among the fiercest in the animal kingdom.

Elephant seals are named for their trunks (short, by elephant standards, and used for social purposes only), but it could equally be for their size. Southern elephant seals can weigh 3.7 tonnes, more than some cow elephants. Only the bull seals reach this weight, however, and that is one of the central points of the tale. Cow elephant seals are typically less than a quarter the weight of bulls, by whom they, and the pups, are regularly flattened as the bulls charge about fighting each other.*

Why are males so much bigger than females? Because large size helps them to win harems. Most young seals, of whichever sex, are born to a giant father who won a harem, rather than a smaller male who failed to win a harem. Most young seals, of whichever sex, are born to a relatively small mother whose size was optimised to the business of giving birth and rearing babies, rather than the business of winning fights.

The separate optimisation of male and female characteristics comes about through selection of genes. People are sometimes surprised to learn that the genes concerned are present in both sexes. Natural selection has favoured so-called sex-limited genes. Sex-limited genes are present in both sexes but turned on in only one sex. For example, genes that tell the developing seal: 'If you are male grow very big and fight' are favoured at the same time as genes that say, 'If you are female, grow small and don't fight.' Both classes of genes are passed on to sons and to daughters, but each is expressed in one sex and not the other.

If we look at mammals overall, we notice a generalisation. Sexual dimorphism – meaning a big difference between males and females – tends to be most marked in polygynous species, especially those with a harem-style society. As we've seen, there are good theoretical

---

* Don't be surprised that bull seals flatten pups of their own species. Any pup squashed is no more likely to be the bull's own child than the child of any rival bull. Therefore there is no Darwinian selection against squashing.

Relationship between sexual dimorphism and harem size. Each point represents a species of seal or sea lion. Adapted from Alexander *et al.* [5].

reasons why this should be so, and we've also seen that the seals and sea lions go farthest out along this particular limb.

The graph above comes from a study made by the eminent zoologist Richard D. Alexander of the University of Michigan and his colleagues. Each blob in the graph represents one species of seal or sea lion, and you can see that there is a strong relationship between sexual dimorphism and harem size. In extreme cases, for example southern elephant seals and northern fur seals which are the two blobs at the top of the graph, males may be more than six times the weight of females. And, sure enough, in these species the success-ful males – a minority, to put it mildly – have huge harems. Two extreme species can't be used to draw general conclusions. But a statistical analysis of the known data for seals and sea lions confirms that the trend we think we see is real (the odds against its being a chance effect are more than 5,000 to 1). There is weaker evidence in the same direction from ungulates and from monkeys and apes.

To repeat the evolutionary rationale for this, males have a great deal to gain, and also a great deal to lose, from fighting other males. Most individuals born, of either sex, are descended from a long line of male ancestors who succeeded in winning harems, and a long line of female ancestors who were members of them. Therefore most individuals, whether male or female, and whether they themselves go on to be winners or losers, inherit genetic equipment for helping male bodies to win harems and female bodies to join them. Size is at a premium, and the successful males can be very large indeed (see plate 11). Females, by contrast, have little to gain from fighting other females, and they are only as large as they need to be to survive and be good mothers. Individuals of both sexes inherit genes that make females avoid fighting and concentrate on child-rearing. Individuals of both sexes inherit genes that make males fight against other males, even at the expense of time that could have been spent helping to rear children. If only males could agree to settle their disputes by the toss of a coin, they would presumably shrink over evolutionary time to the size of the females or smaller, with great economic savings all round, and they could give their time to looking after children. Their surplus mass, which in extreme cases must cost a great deal of food to build up and maintain, is the price they pay to be competitive with other males.

Of course, not all species are like seals. Many are monogamous and the sexes are much more alike. Species in which the sexes are the same size tend, with some exceptions such as horses, not to have harems. Species in which males are markedly bigger than females tend to have harems, or to practise some other form of polygyny. Most species are either polygynous or monogamous, presumably depending on their different economic circumstances. Polyandry (females mated to more than one male) is rare. Among our close relatives, gorillas have a harem-based polygynous breeding system and gibbons are faithfully monogamous. We could have guessed this from their sexual dimorphism, and lack of it respectively. A large male gorilla weighs twice as much as a typical female, while gibbon males and females are approximately equal in size. Chimpanzees are more indiscriminately promiscuous.

Can the Seal's Tale tell us something about our own natural breeding system, before civilisation and custom obliterated the traces? Our sexual dimorphism is moderate but undeniable. Lots of women are taller than lots of men, but the tallest men are taller than the tallest women. Lots of women can run faster, lift heavier weights, throw javelins further, play better tennis, than lots of men. But for humans, unlike for racehorses, the underlying sexual dimorphism precludes sex-blind open competition at the top level in almost any sport you care to name. In most physical sports, every single one of the world's top hundred men would beat every single one of the world's top hundred women.

Even so, by the standards of seals and many other animals, we are only slightly dimorphic. Less so than gorillas, but more than gibbons. Perhaps our slight dimorphism means our female ancestors lived sometimes monogamously, sometimes in small harems. Modern societies vary so much that you can find examples to support almost any preconception. The *Ethnographic Atlas* of G. P. Murdock, published in 1967, is a brave compilation. It lists particulars of 849 human societies, surveyed all over the world. From it we might hope to count numbers of societies that permit harems versus numbers that enforce monogamy. The problem with counting societies is that it is seldom obvious where to draw lines, or what to count as independent. This makes it hard to do proper statistics. Nevertheless, the atlas does its best. Of those 849 societies, 137 (about 16 per cent) are monogamous, four (less than 1 per cent) are polyandrous, and a massive 83 per cent (708) are polygynous (males can have more than one wife). The 708 polygynous societies are divided about equally into those where polygyny is permitted by the rules of the society but rare in practice, and those where it is the norm. To be brutally precise, of course, 'norm' refers to harem membership for females and harem aspiration for males. By definition, given equal numbers of men and women, the majority of men miss out. The harems of some Chinese Emperors and Ottoman Sultans broke the most extravagant records of the elephant seals and fur seals. Yet our physical dimorphism is small when compared with the seals, and also probably – although this evidence is disputed – when compared with the australopithecines. Would this mean that australopithecine chiefs had harems even larger than Chinese Emperors?

No. We mustn't apply the theory in a naive way. The correlation between sexual dimorphism and harem size is only a loose one. And physical size is only one indicator of competitive strength. For elephant seals, male size is presumably important, because they win their harems by physically fighting other males, biting them or overpowering them by sheer weight of blubber. Size is probably not negligible in hominids. But any sort of differential power, which enables some males to control a disproportionate number of females, can take the place of physical size. In many societies, political clout plays this role. Being a friend of the chief – or, better, being the chief – empowers an individual: enables him to intimidate rivals in a way that is equivalent to the physical intimidation of a large bull seal over a smaller one. Or there may be massive inequalities in economic wealth. You don't fight for wives, you buy them. Or you pay soldiers to fight for them on your behalf. The Sultan or the Emperor may be a physical wimp, yet he may still secure a harem larger than any bull seal. The point I am moving towards is that even if australopithecines were much more dimorphic in size than us, our evolution from them may not, after all, have been a move away from polygyny itself. It may just have been a shift in the weapons used for male competition: from sheer size and brute force to economic power and political intimidation. Or, of course, we also may have shifted towards more genuine sexual equality.

For those of us with a distaste for sexual inequality, it is a consoling hope that cultural polygyny, as distinct from brute-force polygyny, might be rather easy to get rid of. On the face of it, this seems to have happened in those societies, such as (non-Mormon) Christian societies, which became officially monogamous. I say 'on the face of it' and 'officially', because there is also some evidence that apparently monogamous societies are not quite what they seem. Laura Betzig is a historian with a Darwinian turn of mind, and she has uncovered intriguing evidence that overtly monogamous societies like ancient Rome and medieval Europe were really polygynous under the surface. A rich nobleman, or Lord of the Manor, may have had only one legal wife but he had a *de facto* harem of female slaves, or housemaids and tenants' wives and daughters. Betzig cites other evidence that the same was true of priests, even those who were notionally celibate.

These historical and anthropological facts have been seen by some scientists as suggesting, together with our moderate sexual dimorphism, that we evolved under a polygynous breeding regime. But sexual dimorphism is not the only clue we can get from biology. Another interesting signal from the past is testis size.

Our closest relatives, the chimpanzees and bonobos, have extremely large testes. They are not polygynous like gorillas, nor are they monogamous like gibbons. Female chimpanzees in oestrus normally copulate with more than one male. This promiscuous mating pattern is not polyandry, which means the stable bonding of one female with more than one male. It does not predict any simple pattern of sexual dimorphism. But it did suggest to the British biologist Roger Short an explanation for the large testes: chimpanzee genes have been passed down the generations via spermatozoa that had to battle it out in competition with rival sperms from several males inside the same female. In such a world, sheer numbers of

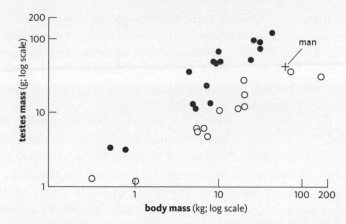

Relationship between testis mass and body mass. Each point represents a species of primate. Adapted from Harvey and Pagel [185].

spermatozoa matter, and this demands big testes. Male gorillas, on the other hand, have small testes but powerful shoulders and huge resonating chests. Gorilla genes do their competing via male fights and chest-thumping threats to win females, which pre-empts subsequent sperm competition inside females. Chimpanzees compete via sperm proxies inside vaginas. This is why gorillas have pronounced sexual dimorphism and small testes, while chimpanzees have large testes and weak sexual dimorphism.

My colleague Paul Harvey, with various collaborators including Roger Short, tested the idea using comparative evidence from monkeys and apes. They took 20 genera of primates and weighed their testes. Well, actually, they went into the library and gathered published information on testis masses. Large animals obviously tend to have larger testes than small animals, so they had to correct for that. Their method was the one explained in the Handyman's Tale for brains. They placed each monkey or ape genus as a point on a graph (above) of testis mass against body mass and, for the same reasons as we saw in the Handyman's Tale, they took logarithms of both. The points fell around a straight line, from marmosets at the bottom to gorillas at the top. As with the brains, the interesting question was which species have relatively big testes for their size, and which smaller. Of all the points scattered about the line, which ones fall above the line and which below?

The results are suggestive. The filled black symbols all represent animals that are like chimpanzees in that females mate with more than one male, and where there is consequently likely to be sperm competition. The chimpanzee itself is the black blob at the top. The open circles are all from animals whose breeding system does not involve much sperm competition, either because they are harem breeders like gorillas (the open circle to the far right) or they are faithfully monogamous like gibbons.

The separation between the open circles and the filled blobs is satisfying.* We seem to have support for the sperm competition hypothesis. And now, of course, we want to know where we fall on the graph. How big are our testes? Our position on the graph (see small vertical cross) is close to the orang utan. We seem to cluster with the open circles rather than with the black blobs. We are not like chimpanzees, and probably have not had to contend with much sperm competition in our evolutionary history. But this graph says nothing about whether the breeding system of our evolutionary past was like a gorilla's (harem) or like a gibbon's (faithful monogamy). That sends us back to the evidence of sexual dimorphism and anthropology, both of which suggest mild polygyny: a small tendency in the direction of harems.

If there is indeed evidence that our recent evolutionary ancestors were weakly polygynous, I hope it needs no saying that this should not be used to justify a moral or political stance, one way or the other. 'You can't get an ought from an is' has been said so often it is in danger of becoming tedious. It is none the less true for that. Let's hasten on to our next rendezvous.

---

* In plots of this kind, it is important to include only data that are independent of each other, otherwise you can unfairly inflate the result. Harvey and colleagues sought to avoid this danger by counting genera instead of species. It is a step in the right direction, but the ideal solution is that urged by Mark Ridley in *The Explanation of Organic Diversity,* and fully endorsed by Harvey: look at the family tree itself and count neither species nor genera but independent evolutions of the characteristics of interest.

XENARTHRANS AND AFROTHERES

The past 15 years have seen a revolution in our understanding of mammal evolution. We now know – as we did not last century – that the living placental mammals divide naturally into four great pilgrim bands, 'superorders'. This pleasing rationalisation is marred only by their ungainly names. We met Laurasiatheria at the last rendezvous, where they merged with the superorder comprising the primates, rodents and their assorted kin (the Euarchontoglires). Here at Rendezvous 13 we meet the two remaining superorders, the Xenarthra and the Afrotheria, associated respectively with South America and with Africa. The geographical link is no coincidence. As we shall see in the Sloth's Tale and its epilogue, around the time of these rendezvous the world's land masses were in the process of splitting apart to form the continents we know today. But the precise time that afrotheres and xenarthrans diverged, even their exact relationships, are the subject of continuing disagreement. In the first edition of this book we had them join the pilgrimage successively. Recent evidence moves us, albeit with little confidence, to have them share their rendezvous with us. We will return to this controversy once we have met the animals themselves.

The Xenarthra take their name from their 'strange joints': a peculiar extra set of contacts between their lower vertebrae which reinforce the backbone and help in digging. Today, they comprise only 31 described species: 21 armadillos, six sloths and four anteaters; but until recently – very recently in some cases – they were much

Giant armadillos with comical tweed caps. This glyptodont species grew to over three metres long.

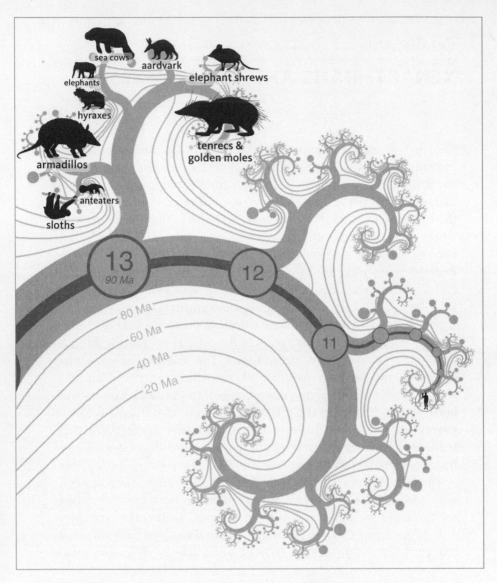

Xenarthrans and afrotheres join. Of the four major placental mammal superorders identified by molecular taxonomists, the earliest divergences involve the South American xenarthrans (approximately 30 species of sloths, anteaters and armadillos) and the afrotheres (80 or so species). The afrotheres fall into two main subgroups: aardvarks, elephant shrews, golden moles and tenrecs on one hand, and hyraxes, elephants and sirenians on the other. The branching order within this second subgroup is unsure.

The order of joining of these two superorders is much debated, as described on page 257. We show one well-supported arrangement, but different genes show different relationships, and many show that we are more closely related to xenarthrans than afrotheres, as we had in the first edition of this book. The conflict between genes almost certainly reflects a rapid set of speciation events.

more diverse. Most famous among the extinct forms are the glyptodonts: car-sized armadillos with what look comically like tweed caps on their heads. They had heavy armoured shells and clubbed tails, sometimes with fearsome spikes, which they wielded against each other and presumably against any large predators foolish enough to attack. If you came across one you might be forgiven for thinking you had met a dinosaur, for they look superficially like the anklyosaurs of the Cretaceous. And we humans probably did meet them in the Americas – quite likely hunting them to extinction and perhaps using their enormous shells for temporary shelter. There were also giant ground sloths the size of elephants, but with the typical long curved claws that we see in modern sloths. One fanciful suggestion is that they used them for carnivorous scavenging, but they were almost certainly pure herbivores, perhaps using their claws to pull branches down from trees. A site from Uruguay shows evidence of humans butchering them, which may be why ground sloths disappeared at the same time as the giant armadillos, a mere 10,000 years ago. Smaller versions colonised Caribbean islands – sloths are surprisingly good swimmers – avoiding human extirpation until perhaps even the time of the conquistadores.

Talking of swimming, the extinct xenarthrans also include, rather astonishingly, 2-metre-long amphibious sloths. A series of fossil species show a beautiful gradation over time from semiaquatic forms to a fully marine, deeper-water specialist, a sloth equivalent of the dugongs and manatees that will be introduced when we meet the afrotheres. Like dugongs, amphibious sloths were strictly herbivorous, probably using their claws to anchor themselves to the bottom where they could graze on marine vegetation. Disappointingly, unlike the hippo's relatives, they never evolved into gargantuan whale-size forms, instead going extinct about 2 million years ago, well before the advent of humans.

As for the anteaters, few fossils exist, but we must count ourselves fortunate that their curvaceous forms still grace the trees and soil of South America. Eating ants and termites requires no teeth, and New World anteaters lack them, as do the Old World pangolins – which have independently hit upon on the same diet (the aardvark, another convergently ant-consuming mammal, proves an enchanting exception to this toothless rule, of which more later). It's not just teeth. Over evolutionary time the entire skull has been modified. In the case of *Myrmecophaga*, the large ground-dwelling anteater with the feather-boa tail, the jaws have become little more than a long, curved tube, a kind of straw for hoovering up ants and termites chivvied out of their nests by means of a long sticky tongue. And let me tell you something else amazing about them. Most mammals, like us, secrete hydrochloric acid into our stomachs to aid digestion, but South American anteaters don't. Instead, they rely upon the formic acid from the ants that they eat. This is typical of the opportunism of natural selection.

The xenarthrans have long been recognised as a natural group. Not so the afrotheres, whose common roots have only come to light as a result of DNA analysis. Starting from a shrew-like ancestor, they have evolved into seven radically different subgroups. Famously first, at least in dictionaries, is the aardvark. The other afrotheres are the elephants, the hyraxes, the golden moles of southern Africa, Madagascan tenrecs, dugongs and manatees (also known as sea cows or sea elephants), and the charming elephant shrews.

I never saw an elephant shrew until I revisited the beautiful country of Malawi which, as Nyasaland, had been my childhood home. My wife and I spent some time in the Mvuu Game Reserve, just south of the great Rift Valley lake which gave the country its name and on whose sandy beaches I had spent my first bucket-and-spade holidays long ago. In the game reserve, we benefited from our African guide's encyclopaedic knowledge of the animals, his unbelievably sharp eye for spotting them, and his engaging turns of phrase in calling them to our attention. Elephant shrews always elicited from him the same joke, which seemed to improve with each repeating: 'One of the small five.'

Elephant shrews, which are named for their long trunk-like noses, are larger than European 'true' shrews, and they run higher, on longer legs – a little suggestive of miniature antelopes. The smaller of the 15 species jump. Elephant shrews used to be more numerous and diverse, and included some plant-eating species as well as the insectivorous ones that survive today. Elephant shrews have the prudent habit of devoting time and attention to making runways for themselves to use later when escaping predators. This sounds like foresight and so, in a way, it is. But it should not be taken to imply deliberate intention (although, as always, that cannot be ruled out). Animals often behave as if they know what is good for them in the future, but we must be careful not to forget that 'as if'. Natural selection is a beguiling counterfeiter of deliberate purpose.

In spite of their dear little trunks, it never occurred to anyone that elephant shrews might be particularly close to elephants. It was always assumed that they were just African versions of Eurasian shrews. Recent molecular evidence, however, astonishes us with the information that elephant shrews are closer cousins to elephants than they are to shrews, although the resemblance of the 'trunk' is almost certainly coincidental. To distance them from true shrews, some people now prefer their alternative name, sengi.

From the small five to the big, we come next to the elephants themselves. Today the elephants are reduced to two genera: *Elephas* the Indian elephant, and *Loxodonta* the African elephant, but many species of elephantids including mastodons and mammoths once roamed through almost every continent except Australia. There are even tantalising hints that they may have made it to Australia too. Fragments of elephant fossils have been reported there, but perhaps they were flotsam from Africa or Asia. Mastodons and mammoths were in America until around 12,000 years ago when they were exterminated, probably by the Clovis people. Mammoths died out so recently in Siberia that they are occasionally found frozen in the permafrost and have even, poets sing, been made into soup:

THE FROZEN MAMMOTH
This Creature, though rare, is still found to the East
Of the Northern Siberian Zone.
It is known to the whole of that primitive group
That the carcass will furnish an excellent soup,
Though the cooking it offers one drawback at least
(Of a serious nature I own):
If the skin be but punctured before it is boiled
Your confection is wholly and utterly spoiled.
And hence (on account of the size of the beast)
The dainty is nearly unknown.

HILAIRE BELLOC

As with all of the Afrotheria, Africa is the ancient home of elephants, mastodons and mammoths, the root of their evolution and the site of most of their diversification.

Officially, the elephant order is called Proboscidea, after their long proboscis or trunk, which is an enlarged nose. The many purposes for which the trunk is used include drinking, which may have been its primitive use. Drinking, when you are a very tall animal like an elephant or a giraffe, is a problem. Food, for elephants and giraffes, mostly grows on trees which may be partly why they are so tall in the first place. But water finds its own level, which tends to be uncomfortably low. Kneeling down to the water is a possibility. Camels do it. But it is hard work getting up again, even more so for elephants or giraffes. Both solve the problem by sucking the water up through a long siphon. Giraffes stick their head on the end of the siphon – the neck. Giraffe heads therefore have to be rather small. Elephants keep their head – which can therefore be bigger and brainier – at the base of the siphon. Their siphon is, of course, the trunk, and it comes in handy for lots of other things as well. I have elsewhere quoted Oria Douglas-Hamilton on the elephant trunk. Much of her life has been devoted, with her husband Iain, to the study and preservation of wild elephants. It is an angry passage, prompted by the horrifying sight of a mass 'cull' of elephants in Zimbabwe.

> I looked at one of the discarded trunks and wondered how many millions of years it must have taken to create such a miracle of evolution. Equipped with fifty thousand muscles and controlled by a brain to match such complexity, it can wrench and push with tonnes of force. Yet, at the same time, it is capable of performing the most delicate operations such as plucking a small seed-pod to pop in the mouth. This versatile organ is a siphon capable of holding four litres of water to be drunk or sprayed over the body, as an extended finger and as a trumpet or loud speaker. The trunk has social functions, too; caresses, sexual advances, reassurances, greetings and mutually inter-twining hugs . . . And yet there it lay, amputated like so many elephant trunks I had seen all over Africa.

Proboscideans also run to tusks, which are greatly enlarged incisor teeth. Modern elephants have tusks only in the upper jaw, but some extinct proboscideans had tusks in the lower jaw as well, or instead. *Deinotherium* had large down-curving tusks in the lower jaw and no tusks at all in the upper jaw. *Amebelodon*, a North American member of the large group of early proboscideans called gomphotheres, had elephant-like tusks in the upper jaw and flat, spade-like tusks in the lower jaw. Perhaps they were indeed used as spades, to dig up tubers. This speculation doesn't, by the way, conflict with the one about the evolution of the trunk as a siphon to obviate the need to kneel for a drink. The lower jaw, with its two flat spades on the end, was so long that a standing gomphothere could easily have used it to dig in the ground.

In *The Water Babies*, Charles Kingsley wrote that the elephant 'is first cousin to the little hairy coney of Scripture . . .' The primary meaning of coney in the English dictionary is rabbit, and two of the word's four occurrences in the Bible are explanations of why the rabbit is not kosher: 'And the coney, because he cheweth the cud, but divideth not the hoof; he is unclean unto you' (Leviticus 11:5 and the very similar passage in Deuteronomy 14:7). But Kingsley can't have meant rabbit, because he goes on to say that the elephant is thirteenth or fourteenth cousin to the rabbit. The other two biblical references refer to an animal that lives among rocks: Psalm 104 ('The high hills are a refuge for the wild goats; and the rocks for the conies') and Proverbs 30:26 ('The conies are but a feeble folk, yet make they their houses in the rocks'). Here, it is generally agreed, coney means hyrax, dassie or rock badger, and Kingsley, that admirably Darwinian clergyman, was right.

Well, he was right at least until those tiresome modern taxonomists burst in. Textbooks say that the elephants' closest living cousins were the hyraxes, which agrees with Kingsley. And indeed both are afrotheres. But recent analysis shows that we must also include dugongs and manatees in the mix, perhaps even as the closest living relatives of the elephants, with hyraxes a close sister to both. Dugongs and manatees are purely marine mammals who never come ashore even to breed, and it looks as though we were misled in the same way as over the hippos and the whales. Purely marine mammals are freed from the constraints of terrestrial gravity and can evolve rapidly in their own special direction. Hyraxes and elephants, left behind on the land, have remained more similar to each other, just as hippos and pigs did. With hindsight, dugongs' and manatees' slightly trunk-like nose and small eyes in a wrinkled face give them a faintly elephantine appearance, but it is probably an accident.

Dugongs and manatees belong to the order Sirenia. The name comes from their supposed resemblance to the sirens of myth, although this is not, it has to be said, very convincing. Their slow, sleepy-lagoon style of swimming could perhaps have been thought mermaid-like, and they suckle their young with a pair of breasts under the flippers. But one can't help feeling that the sailors who first spotted the likeness must have been at sea for a very long time. Sirenians are, with whales, the only mammals that never come on land at any time.

One species, the Amazon manatee, lives in freshwater; the other two manatees are found in the sea as well. Dugongs are exclusively marine, and all four species are vulnerable to extinction, which inspired my wife Lalla to design a T-shirt: Dugoing Dugong Dugone. Until recently we could have included a fifth species, the enormous Steller's sea cow, which lived in the Bering Straits and weighed over 5 tonnes. Tragically, it was hunted to extinction a mere 27 years after its discovery by Bering's ill-fated crew in 1741, an ominous portent of sirenian-human relations.

As with whales and dolphins, the front limbs of sirenians have become flippers and they have no hind limbs at all. Sirenians are also known as sea cows, but they are not closely related to cows and they don't ruminate. Their vegetarian diet requires an immensely long gut and a low energy budget. The high-speed aquabatics of a carnivorous dolphin contrast dramatically with the lazy drifting of the vegetarian dugong: guided missile to dirigible balloon.

There are small afrotheres, too. Golden moles and tenrecs seem to be related to each other, and most modern authorities place them within the afrotheres. Golden moles live in southern Africa where they do the same job as moles do in Eurasia, and do it beautifully, swimming through the sand as if it were water. Tenrecs live mostly in Madagascar. There are some semi-aquatic 'otter shrews' that are actually tenrecs in western Africa. As we saw in the Aye-Aye's Tale, Madagascan tenrecs include shrew-like forms, hedgehog-like forms and an aquatic species which probably returned to the water independently of the African ones.

Finally we come to the solitary aardvark, named from the Afrikaans: 'earth pig'. It does indeed look rather like a small, long-eared pig whose snout has become elongated in that way so common to dedicated eaters of ants. It is also a superb digger, repeatedly digging burrow systems that are large enough for a small human to enter, and which are also used for shelter by other creatures of the veldt. The aardvark is the sole remaining species in the order Tubulidentata, which means 'tubular teeth'. As previously intimated, although it eats ants and termites it nevertheless retains a few molars. The delightful suggestion is that aardvarks have kept them specifically to consume the subterranean fruit of the 'aardvark cucumber', which in turn is entirely dependent on the aardvark to unearth, disperse, then liberally manure its seeds. Of course, this sort of tight mutualism can lead to problems down the line. If the aardvark were to go extinct, chances are the aardvark cucumber would too. Natural selection grants no possibility of foresight.

These then are the afrotheres, in all their glorious shapes and sizes. Africa has since become the home of lots of other mammals such as rhinos and hippos, antelopes and zebras, and the carnivores that prey on them, but those are laurasiatheres, who came into Africa later, from the great northern continent of Laurasia. The afrotheres represent the ancient inhabitants of Africa, just as the xenarthrans do for South America.

We have yet to discuss the actual relationships between these two groups and ourselves. As social media sites would put it, 'it's complicated'. There are three potential arrangements. The afrotheres could join our pilgrimage first; the xenarthrans could join first; or both could be united into a single band already, closer to each other than either is to the rest of the placental mammals. Different studies disagree as to which is correct. One Japanese research group attempted to resolve the issue once and for all by focusing on mostly unique DNA insertions and deletions. Unexpectedly, they found simultaneous and equal support for all three arrangements! An escape from this apparent impasse is explicitly to adopt the gene's eye view of history. As we saw in the Epilogue to the Gibbon's Tale, different parts of the genome (and by extension, different components of a creature's outward appearance) truly can have different affinities. If the shrew-like concestors of ourselves, afrotheres, and xenarthrans went their separate ways within only a few million years, we positively expect some of our genes to be closer to afrotheres, others closer to xenarthrans, and yet others to be equally distant from either. Whatever phylogenetic tree we draw at this point is likely to be a simplification, an illustration of an imperfect genetic consensus rather than a comprehensive picture of reality. For simplicity, we have grouped the afrotheres and xenarthrans together, perhaps the most common arrangement, and one which is suggested by a recent, rather sophisticated genomic analysis. Nevertheless, there is no hiding the fact that different genes almost certainly take different routes at this point. It's even possible that the majority of the genome flowed down different routes to those indicated here.

Whatever the relationships, to the untrained eye the concestors involved would have looked similarly shrew-like. We have produced a reconstruction (plate 12), and in 2013 another was put forward by a team from the American Museum of Natural History (plate 13). As for when they lived, estimates vary from 65 to 120 million years ago, but the most recent molecular clock dates fall somewhere in the middle, perhaps 90 to 100 million years ago, around the time of our 30-million-greats-grandparent. This is during the time that the continents were rifting apart, when Africa and South America were becoming isolated from other land masses and from each other. The deep divisions among mammals are commonly attributed to the slothful drift of the continents, which is the cue for a tale.

PROLOGUE TO THE SLOTH'S TALE

The theory of plate tectonics, as it is now called, is one of the success stories of modern science. When my father was a biology undergraduate at Oxford in the 1930s, what was then called the theory of continental drift was widely, though not universally, ridiculed. It was associated with the German meteorologist Alfred Wegener (1880–1930) but others had suggested something similar before him. Several people had noticed the snug fit of the east coast of South America and the west coast of Africa, but it was generally written off as coincidence. There were some even more striking coincidences in distributions of animals and plants, which had to be explained by postulating land bridges between the continents. But scientists mostly thought that the map had altered through the sea level fluctuating up and down, rather than the continents themselves drifting sideways. Wegener's idea that the continents themselves had drifted was far more revolutionary – and controversial.

Even when I was an undergraduate, in the 1960s instead of the 1930s, it was not an open-and-shut case. Charles Elton, the veteran Oxford ecologist, lectured us on the subject. At the end of his lecture he took a vote (I am sorry to say, because democracy is no way to establish a truth) and I think we were fairly evenly split. That all changed shortly after I graduated. Wegener turns out to have been much closer to the truth than most of his contemporaries who ridiculed him. The main thing he got wrong was that he thought the existing continental masses floated in the semi-liquid mantle, and ploughed their way through it like rafts through the sea. The modern theory of plate tectonics sees the whole surface of the Earth – the seafloor as well as the visible continents – as a set of *plates*. Continents are thick, less dense parts of plates that bulge up above the sea forming mountains, and down into the mantle. Plate boundaries are, as often as not, under the sea. Indeed, we shall understand the theory best if we forget all about the sea: pretend it isn't there. We'll bring it back to flood the low ground later.

Plates do not plough through a sea, either of water or of molten rock. Instead, the entire surface of the Earth is armoured, covered by plates, sliding over the surface, sometimes diving under another plate in the process known as subduction. As a plate moves, it does not leave a gap behind it as Wegener imagined. Instead, the 'gap' is continuously filled by new material welling up from the deep layers of Earth's mantle and contributing to the substance of the plate, in the process called seafloor spreading. In some ways, plate seems too rigid an image: a better metaphor is a conveyor belt, or a roll-top desk. I'll describe it using the neatest, clearest example – the Mid-Atlantic Ridge.

The Mid-Atlantic Ridge is an underwater canyon 16,000 kilometres long, which snakes its way in a huge S-bend down the middle of the North and South Atlantic. The ridge is a zone of volcanic upwelling. Molten rock pushes up from the deeps of the mantle. It then fountains out sideways to east and west, like two desk roll-tops. The east-going roll-top pushes Africa away from the middle of the Atlantic. The west-going roll-top pushes South America in the other direction. That is why these two continents are moving away from one another, at a rate of about one centimetre per year which is, as somebody has imaginatively pointed out, about the rate at which fingernails grow, although rates of movement, from plate to plate, are quite variable. There are similar zones of volcanic upwelling in the floor of the Pacific Ocean, of the Indian Ocean, and in various other places (though they are sometimes called rises rather than ridges). These spreading ocean ridges are the driving engines of plate movement.

The language of 'pushing', however, is grossly misleading, if it suggests that the upwelling from the seafloor pushes the plate from behind. Indeed, how could an object as massive as a continental plate be moved by shoving from behind? It isn't. Rather, the crust and top part of the mantle are moved by the circulating currents in the molten rock beneath. A plate is not so much pushed from behind as dragged by the current in the fluid on which it floats, tugging on the underside of the whole expanse of the plate.

The evidence for plate tectonics is elegantly compelling and the theory is now proved beyond reasonable doubt. If you measure the age of the rocks on either side of a ridge such as the Mid-Atlantic Ridge, you notice a truly remarkable thing. The rocks that are closest to the ridge are the youngest. The further you travel, sideways from the ridge, the older are the rocks. The result is that if you plot 'isochrons' (that is, contour lines of equal age) they run parallel to the ridge itself, snaking with it down the North Atlantic and then the South Atlantic. This is true on both sides of the ridge. The isochrons on one side of the ridge are almost perfectly mirrored on the other side (see plate 14).

Imagine that we set off to cross the bottom of the Atlantic in a submersible tractor, due east at the tenth parallel, from the Brazilian port of Maceio towards the cape of Barra do Cuanza in Angola, just missing Ascension Island on the way. As we go, we sample the rocks beneath our caterpillar tracks (tyres couldn't stand the pressure). For reasons that follow from the theory of volcanic seafloor spreading, we shall be interested only in the igneous basalt (solidified lava) lying at the base of whatever sedimentary rocks may have been deposited above it. It is these igneous rocks which, according to the theory, constitute the roll-top or conveyor belt, as South America moves westward and Africa east. We shall drill down through the sediments – which may be quite thick in places, having been laid down over millions of years – and take samples from the hard volcanic rocks beneath.

For the first 50 kilometres of our eastward journey, we are on the continental shelf. This doesn't count as sea bottom at all, for our purposes. We haven't left the continent of South America, it is just that there is some shallow water above our heads. In any case, for the purpose of explaining plate tectonics, we are ignoring water. But now we descend rapidly to the sea bottom proper, take our first sample from the true seafloor, and analyse radiometrically the date of the basalt under the sediments. Here on the western edge of the Atlantic, it turns out to be of Lower Cretaceous age, some 140 million years old. We continue our eastward journey, taking samples at regular intervals from the volcanic rocks at the base of the sediments, and we find a remarkable fact: they become steadily younger. Five hundred kilometres from our starting point, we are well into the Upper Cretaceous, younger than 100 million years. About 730 kilometres into our journey, although we shall see no discrete border because we are looking only at volcanic rock, we cross the 66-million-year boundary between the Cretaceous and the Palaeogene Period, the geological instant when, on land, the dinosaurs suddenly disappeared. The sequence of decreasing age continues. As we drive due east, the volcanic rocks under the sea get steadily younger and younger. Sixteen hundred kilometres from our starting point we are in the Pliocene, looking at young rocks, contemporary with woolly mammoths in Europe and Lucy in Africa.

When we reach the Mid-Atlantic Ridge itself, about 1,620 kilometres from South America and slightly further (at this latitude) from Africa, we notice that the rocks of our sample are now so young, they are of our own time. They have only just erupted out of the depths of the sea bottom. Indeed, if we are very lucky we may see an eruption in the particular part of the Mid-Atlantic Ridge that we are crossing. But we would have to be lucky because, notwithstanding the image of a continuously moving roll-top conveyor belt, it isn't literally continuous. How could it be, given that the roll-top moves on average at one centimetre per year? When there is an eruption, the rocks are shifted more than one centimetre. But correspondingly, eruptions occur less often than once per year in any one place along the ridge.

Having crossed the Mid-Atlantic Ridge, we continue our eastward journey in the direction of Africa, again drawing up samples of volcanic rock from beneath the sediments. And what we now notice is that the ages of the rocks are a mirror image of what we measured before. The rocks are now getting progressively older as we move away from the central ridge, and this continues all the way to Africa and the eastern margin of the Atlantic. Our last sample, just short of the African continental shelf, shows rocks of Lower Cretaceous age, just like their mirror images on the western side, hard by South America. Indeed, the whole sequence is reflected about the Mid-Atlantic Ridge, and the mirroring is even more precise than you could know from radiometric dating alone. What follows is extremely elegant.

In the Redwood's Tale, we shall meet the ingenious dating technique known as dendrochronology. Tree rings result from the fact that trees have an annual growing season, and not all years are equally favourable, so a signature pattern of thick and thin rings develops. Such fingerprint signatures, when they occasionally arise in nature, are a natural gift to science, to be seized eagerly whenever we encounter them. It is a particularly fortunate fact that something like tree rings, although on a larger timescale, is imprinted into volcanic

Magnetic stripes on either side of an oceanic ridge. Dark stripes represent normal polarity; white, reverse polarity. Geologists group these into magnectic intervals dominated by normal or reverse polarity. The symmetry of the stripes was first identified as evidence for seafloor spreading by Fred Vine and Drummond Matthews in a classic 1963 paper in *Nature* [430]. The crust and rigid top layer of the mantle, together known as the lithosphere, is pushed apart by convection currents in magma in the semi-rigid layer of the mantle below (the asthenosphere). The distinctive pattern of stripes allows us to identify the age of rocks on the sea bed back to about 150 million years ago. Older seafloor has been destroyed by subduction.

lava as it cools and solidifies. It works like this. While lava is still liquid, molecules within it behave like tiny compass needles, and become aligned to the magnetism of the Earth. When the lava solidifies as rock, the compass needles are petrified in their current position. Igneous rock therefore acts as a weak magnet, whose polarity is a frozen record of the Earth's magnetic field at the moment of solidification. This polarity, which is easy to measure, tells us the direction of the magnetic North Pole at the moment when the rock solidified.

Now comes the fortunate fact. The polarity of the Earth's magnetic field reverses at irregular but, by geological standards, quite frequent intervals, on a timescale of tens, or hundreds, of thousands of years. You can immediately see the exciting consequence of this. As the two conveyor belts stream out west and east from the Mid-Atlantic Ridge, their measured magnetic polarity will exhibit stripes, reflecting the flips of the Earth's field, frozen at the moment of solidification of the rock. The pattern of stripes on the west side will mirror the precise pattern of stripes on the east side, because both sets of rocks shared the same magnetic field when they spewed together, as liquid, out of the mid-ocean ridge. It is possible to match up exactly which stripe on the east side of the ridge belongs with which stripe on the west side, and the two stripes can be dated (they have the same date as each other, of course, because they were liquid at the same time when they gushed together out of the ridge). The same pattern of stripes will be found on either side of the spreading zones on all the other ocean floors, although the distances between the mirrored stripes will vary because not all conveyor belts move at the same speed. You could not ask for more compelling evidence.

There are complications. The pattern of parallel stripes does not snake down the seafloor in a simple unbroken way. It is subject to numerous fractures – 'faults'. I chose the tenth parallel south of the equator deliberately for our caterpillar tractor journey, because it happens not to be complicated by any fault lines. At another latitude, our sequence of gradually changing age would have been interrupted by occasional hiccups as we crossed a fault line. But the general picture of parallel isochrons is entirely clear from the geological map of the whole Atlantic floor.

The evidence for the seafloor spreading theory of plate tectonics is, then, very solid, and the dating of the various tectonic events such as the parting of particular continents, is, by geological standards, accurate. The plate tectonics revolution has been one of the swiftest, yet at the same time most decisive, in the whole history of science.

THE SLOTH'S TALE

The Sloth's Tale is a story of the global movement of continents, oceans and animals: a story of biogeography. To Alfred Russel Wallace and Charles Darwin, co-discoverers of natural selection, the geographical aspect of natural history gave away the fact of evolution. If species were created independently, why should a creator choose to put 50 species of lemur on Madagascar, *but nowhere else*? Why should the Galapagos host a set of finches so different from species on other oceanic islands, but strikingly similar both to each other and to *birds on the nearest mainland*? Why indeed did creation on islands involve flying species like birds and bats, but rarely frogs and land mammals?

And of course there was the geographical distribution of the xenarthrans: anteaters, armadillos and sloths. They occur only in South America, the destination for the twentysomething Darwin and, 16 years later, the twentysomething Wallace. For Darwin, it was his first and last adventure; for Wallace almost so (his brother died of yellow fever, he nearly died of malaria, and his ship caught fire and sank on the way home). Both remarked upon the oddities of the South American fauna, and the mammals in particular. Darwin dug up a giant sloth skull, Wallace tasted sloth stew. And as for how it inspired them, the very first lines of Darwin's masterwork speak for themselves:

> When on board H.M.S. 'Beagle,' as naturalist, I was much struck with certain facts in the distribution of the inhabitants of South America, and in the geological relations of the present to the past inhabitants of that continent. These facts seemed to me to throw some light on the origin of species – that mystery of mysteries, as it has been called by one of our greatest philosophers.

Despite his disastrous initial trip, Wallace continued travelling. It was in the islands of South East Asia that he independently hit upon the idea of natural selection. And it was there that he finally grasped the power of combining geographical distributions with biological taxonomy.

> It is certainly a wonderful and unexpected fact that an accurate knowledge of the distribution of birds and insects should enable us to map out lands and continents which disappeared beneath the ocean long before the earliest traditions of the human race.
> ALFRED RUSSEL WALLACE
> *The Malay Archipelago*

Wallace didn't know the half of it. When I was a small child in Africa my father regaled my little sister and me with bedtime stories, as we lay under our mosquito nets and marvelled at his luminous wristwatch, about a 'Broncosaurus' who lived faaaaaaaaaaaar away in a place called Gonwonky-land. I forgot all about this until much later when I learned about the great southern continent of Gondwanaland, or Gondwana*.

Gondwana accounted for over half the globe's land area. It came together deep in the Precambrian, and lasted for 400 million years,

---

* 'Gondwanaland' was coined before it was known that continents moved, and originally described a continent consisting of Africa and South America in their present positions but with the south Atlantic drained. I will follow the current scientific consensus and use the shorter name Gondwana, which avoids a potential tautology, since *vana* in Sanskrit means land (actually forest). But the original coining has the virtue of distinguishing the giant continent from the region of central Madhya Pradesh where the Gonds live, which is still called Gondwana and which gave its name to the Gondwana geological series.

160 Ma   120 Ma   80 Ma

The breakup of Godwana. Snapshots based on data from [383], as made available in the free GPlates software package (www.gplates.org).

existing both on its own and sometimes as part of a larger Pangaea. Our concern is its demise. This was rather sudden and explosive, by the standards of geological time. One hundred and sixty million years ago, the north-western half of Gondwana consisted of the vast area that was to become South America and Africa. It extended right up to the equator and had almost separated from the northern supercontinent of Laurasia. Forming the south-eastern portion of Gondwana, and not quite grazing the South Pole, was the combined area of what we now know as Antarctica, India and Madagascar, and Australasia.

Powered by convection currents deep in the Earth's mantle, the forces of plate tectonics were splitting Gondwana in two. By 150 million years ago, the north-western and south-eastern halves were partially separated by a 2,000-mile inlet that had appeared down the east coast of what is now Africa. As it widened, the two halves of Gondwana swung apart, so that by 130 million years ago they were only joined at a tiny point, where South America touched Antarctica. But instead of simply continuing to separate, each part started fragmenting too. The north-western part of Gondwana cracked from top to bottom, to one side South America, the other Africa, so that by 120 million years ago there existed an immensely long and narrow, angled channel between the two, which was to widen into the Atlantic Ocean of today. Meanwhile, on the south-eastern part, India (with Madagascar attached) split off and started heading north. The final fragmentation kicked off 90 million years ago, when the remaining south-eastern land mass started splitting into Antarctica and Australia, while India divested itself of Madagascar, beginning a spectacularly fast migration north, eventually to crunch into the south coast of Asia and raise the Himalayas.

Tectonic activity also had a profound effect on the climate, especially for the southern hub of Antarctica. Although long dark winters caused seasonal snowfall near the South Pole, fossils attest that Cretaceous Antarctica was mostly covered with temperate woodland, and a fine place for animals to live – not just warm-blooded mammals but crocodiles and dinosaurs too. This was in part due to globally high temperatures. Volcanoes and ocean ridge eruptions meant carbon dioxide levels many times higher than today, resulting in a major 'greenhouse effect' of the sort that currently threatens to return. And Antarctica was particularly warm because of its Gondwanan connections. Land stretching from the pole to the equator directed warm ocean currents from the tropics to far southern latitudes, in a more dramatic version of the way the Gulf Stream fosters palm trees in western Scotland today. From about 50 million years ago, as Antarctica's land connections gradually disappeared, water began to circulate around the entire continent, increasing its thermal isolation and setting it down the path to icy separation.

Gondwana and the forests of Antarctica still echo in the distribution of many plants and animals today. One example is the Gondwanan distribution of *Nothofagus*, the southern beech tree. It no longer grows in Antarctica of course, but is still found in South America, Australia and New Zealand, carried there by continental drift. The flightless birds known as ratites used to be touted as another classic case, but at Rendezvous 16 we will find that the elephant bird mounts a refreshing challenge on that front. Perhaps the best-known Gondwanan group is the marsupials, whose formal introduction must wait until the next rendezvous. Although classically associated with Australia, they seem to have arrived there via the Americas, where some species still live today. The standard explanation is that the Australian forms came via Antarctica, before the continents rifted apart.

There are two contrasting theories used to explain the distribution of related species. The first emphasises the importance of long-distance dispersal, such as the trans-oceanic colonisation of South America by monkeys and rodents. In contrast 'vicariance biogeography' emphasises the role of geological processes, such as shifting land masses, in separating biological populations. The choice between these alternatives can arouse surprisingly bitter dispute among biogeographers.

At first glance Rendezvous 13 – an almost simultaneous split between xenarthrans, afrotheres and the northern hemisphere placental mammals – seems to fit rather neatly with the vicariance theory. The last links between South America, Africa and Laurasia were severed at almost exactly the same time, when a land bridge from Brazil to West Africa, and another land bridge through Gibraltar, were inundated by a combination of land movement and rising sea levels. Perhaps the ancestors of xenarthrans scurried westwards out of Africa, and our own ancestors northwards, just before these routes disappeared, leaving the ancestors of afrotheres stranded on the isolated island of Africa? It is a popular idea, with one major drawback. The last time you could actually walk between the continents has now been pushed back to about 120 million years ago, substantially earlier than our estimate of 90 million years for Rendezvous 13.

If the times of the rendezvous and the land bridge disappearances are even approximately correct, then the ancestors of sloths must have crossed to South America by some form of dispersal over the burgeoning Atlantic, a trip of 100 miles or more. Here we must reiterate a point we made previously. Over millions of years, crossing narrow straits of sea becomes a virtual certainty. It is only when those straits widen into oceans that the probability of dispersal dwindles into insignificance. In other words, on geological timescales, isolation is a probabilistic continuum, rather than an all-or-nothing occurrence. And even very distant dispersal always remains an outside possibility, as evidenced by the monkeys and rodents of the New World, which colonised South America from Africa much later and consequently across a much wider expanse of ocean. Our conclusion is that it is the combination of dispersal and vicariance which is important. The unique South American fauna was probably a result of dispersal, yes, but dispersal gradually restricted by the slow separation of continents.

South America did indeed become substantially more isolated by its continued westward drift. We have talked about the spectacular xenarthrans that resulted from its long period of purdah – the elephantine ground sloths and the heavily armoured glyptodonts. But there were *three* extraordinary groups of mammals that radiated in South America following the demise of the dinosaurs, and, as it turns out, the other two also lend support to a mix of dispersal and vicariance.

We have already mentioned that the marsupials trace back to South America. There they evolved various dog-, bear- and cat-like carnivorous forms. Unlike xenarthrans, they reached South America not from Africa but probably from North America, which is where earlier marsupial fossils are found. The fossils pin this down to about 65 million years ago, when North and South America were not part of the same land mass (although there may have been a temporary chain of islands between them). So the marsupials probably dispersed into South America over water too.

The third group of these South American 'old timers' – to borrow from the book *Splendid Isolation* by the great American zoologist G. G. Simpson – are all now extinct. More's the pity, because they were amazing creatures. They can be loosely called 'ungulates': as we saw at Rendezvous 12, this is not a taxonomically precise word. These South American 'notoungulates' did the same herbivorous job as horses, rhinos and camels, but they evolved independently of them. The litopterns split early into horse-like and camel-like forms, which probably (from the position of the nose bones) had a trunk like an elephant. Another group, the pyrotheres, also seem to have had trunks, and may have been quite elephant-like in other respects. They were certainly very large. The notoungulates included huge rhinoceros-like toxodons – some of whose fossil bones were first found by Darwin – and smaller rabbit-like and rodent-like forms.

Evolution pushed the South American ungulates in such a bewildering variety of directions that until recently it was hard to deduce their exact affinities. But in 2015 Frido Welker and a host of international collaborators (including Sam Turvey, a previous researcher on this book) managed to analyse molecular sequences from museum specimens of *Macrauchenia*, a camel-like pyrothere, and *Toxodon*, the species that Darwin collected. Given that these fossil bones were neither collected from nor stored in a cold environment, you might express incredulity at being able to sequence their DNA. Rightly so. Instead of DNA, the researchers extracted collagen proteins which, as we have previously mentioned, may even survive in dinosaur fossils. Like the DNA sequences from which they have been 'translated', the sequence of amino acids that constitutes a collagen molecule can be used to construct evolutionary trees. The ancient collagen sequences indicate that these two species of South American ungulate were most closely related to the odd-toed Old World ungulates such as horses and rhinos. Their ancestors must have arrived after the diversification among the laurasiatheres. That pins it down to within the last 70 million years, hinting at another overseas dispersal.

The history of South America ended somewhat differently from Madagascar and Australia, those other gigantic rafts cut adrift by the breakup of Gondwana. Like India before it, South America's island isolation came to an end naturally, before human travel brought all zoological isolation more or less to an end. It was a recent thing. The creation of the Isthmus of Panama, around 3 million years ago, led to the Great American Interchange. No longer was dispersal a rare event. Instead, the separate faunas of North and South America were able to flow freely along the narrow corridor formed by the Isthmus, to each other's continents. This enriched the two faunas, but then some extinctions occurred on both sides, presumably at least partly as a result of competition.

Because of the Great American Interchange, there are now tapirs (odd-toed ungulates) and peccaries (even-toed ungulates) in South America – animals that entered from North America even though tapirs are now extinct there and peccaries greatly reduced. Because of the Interchange, there are now jaguars in South America. Before, there were no cats, or any members of the order Carnivora. Instead there were carnivorous marsupials, some (such as *Thylacosmilus*) bearing a fearsome resemblance to the sabretooths (true cats), which were their contemporaries in North America. Since the Interchange there have been ground sloths in North America, as well as armadillos (including the enormous glyptodonts). Coming the other way, llamas, alpacas, guanacos and vicuñas, all members of the camel family, are now confined to South America, but camels originally evolved in North America. They spread into Asia and then Arabia and Africa quite recently, presumably via Alaska, where they gave rise to the Bactrian camels of the Mongolian steppe, and the dromedaries of the hot deserts. The horse family, too, did most of their evolving in North America but then went extinct there, which makes poignant the baffled surprise with which the Native Americans responded to the horses, reintroduced from Eurasia under the infamous conquistadores.

The Great American Interchange happened so recently in geological time that the traces of the story were evident to Wallace, the first to piece the picture together. Our deeper understanding of the evolutionary roots of the South American old timers would surely have pleased him immensely. But perhaps he would have one thing to add. It is not just long-distance dispersal and the movement of continents which can explain species distributions. On a much faster timescale, the rise and fall of sea levels can also explain the links between the flora and fauna of adjacent regions. We will see this in the next rendezvous, when we meet the marsupials and the Australian flora and fauna that they represent. These species spread into Indonesia over land, at times of low sea levels during recent ice ages. Unlike the other examples in this tale, and for reasons unknown, their dispersal was stopped by narrow but deep ocean channels (perhaps there was not enough time to cross successfully, or the ecological niches on the other side were all occupied). Whatever the reason, a clear zoological and botanical line divides east from west Indonesia. This clue to evolutionary history was first noticed by Wallace and rightly bears his name, the Wallace Line. And the zone of transition – the islands between Borneo and New Guinea – has become known to biologists as Wallacea. His geographical namesakes testify that the factors driving the biogeography of sloths and their wider kin extend to a vast range of other species, in areas all over the world.

MARSUPIALS

Here we are, 160 million years ago, in the late Jurassic when Concestor 14, in round figures our 100-million-greats-grandparent, lived in the shadow of the dinosaurs. As we saw in the Sloth's Tale, the northern landmass of Laurasia had almost detached from the great southern continent of Gondwana, which itself was starting to tear down the middle, with Africa and south America to one side, and Antarctica, India and Australia to the other. The climate was warmer than today, although snow and ice may have blanched the poles during the winter months. Only a few flowering plants grew in the temperate forests of coniferous trees and the plains of ferns that covered the northern and southern parts of the globe, and there were correspondingly few of the pollinating insects that we know today. It is in such a world that the entire massed pilgrims of the placental mammals – horses and cats, sloths and whales, bats and armadillos, camels and hyenas, rhinoceroses and dugongs, mice and men – all now represented by a small insectivore, greet the other great group of mammals, the marsupials.

Marsupium means pouch in Latin. Anatomists use it as a technical term for any pouch, such as the human scrotum. But the most famous pouches in the animal kingdom are those in which kangaroos and other marsupials keep their young. Marsupials are born as tiny embryos equipped only to crawl – crawl for their tiny lives through the forest of their mother's fur, into the pouch where they clamp their mouths to a teat.

The other main group of mammals are called placentals because they nourish their embryos with various versions of a placenta: a large organ through which miles of capillary blood vessels belonging to the baby are brought into close contact with miles of capillary blood vessels belonging to the mother. This excellent exchange system (for it serves to remove wastes from the foetus as well as to feed it) enables the baby to be born very late in its career. It enjoys the protection of its mother's body until, in the case of hoofed herbivores for example, it is capable of keeping up with the herd on its own legs, and even running away from predators. Marsupial

bandicoots & bilbies

tasmanian devil, numbat etc.

marsupial moles

monito del monte

koala, kangaroos, wallabies etc.

shrew opossums

opossums

13

12

14
160 Ma

50 Ma

100 Ma

150 Ma

Marsupials join. Three major lines of living mammals are recognised, based upon their method of reproduction. These are the egg-laying mammals (monotremes), the pouched mammals (marsupials) and the placental mammals (including ourselves). Morphology and most DNA studies agree in grouping the marsupials and placentals together, making Rendezvous 14 the divergence of the 340 or so species of marsupial from the 5,000 or so placental mammals.

It is generally accepted that marsupials fall into the seven orders shown here, with the two South American 'opossum' groups at the base (although studies disagree which diverged first). The other interrelationships are not firmly established, especially the branching of the three orders at the top of the diagram here. The South American monito del monte also jumps around on the phylogenetic tree, but recent studies place it as shown here, unifying the remaining four Australian marsupial orders in a single exclusive group.

embryos have placentas too, but these are short-lived and often rudimentary. They swap the womb early on for the pouch – effectively an external womb. And the large teat, to which the baby becomes attached as a semipermanent appendage, works a bit like an umbilical cord. Later, the joey detaches itself from the nipple and sucks only occasionally like a placental infant. It emerges from the pouch as if in a second birth, and uses it less and less frequently as a temporary refuge. Kangaroo pouches open forwards but many marsupial pouches open backwards.

The marsupials, as we have seen, are one of the two great groups into which the surviving mammals are divided. We normally associate them with Australia which, from a faunistic point of view, can conveniently be deemed to embrace New Guinea and its nearby islands, which (as we touched upon in the conclusion to the Sloth's Tale) have been recently connected during times of low sea level. It is unfortunate that no widely recognised word exists to unite these two land masses. 'Meganesia' and 'Sahul' are not memorable or evocative enough. Australasia won't do because it includes New Zealand which, zoologically, has little in common with Australia and New Guinea. I shall coin Australinea for my purpose. An Australinean animal hails from mainland Australia, Tasmania or New Guinea, but not New Zealand. From a zoological point of view, though not a human one, New Guinea is like a tropical wing of Australia, and the mammal faunas of both are dominated by marsupials. Strange though it may seem, marsupials are relative newcomers to Australinea. As we saw in the Sloth's Tale, they have a long and older history of association with South America, where they still occur, mostly in the shape of some dozens of species of opossums.

Even older marsupial fossils have been found in North America, but the oldest so far discovered is *Sinodelphys*, a 125-million-year-old-fossil from China, beautifully preserved even down to its coat of fur. Even more astonishing because of its great age is the recently discovered *Juramaia*, a 160-million-year-old-Chinese fossil currently assigned to the placental mammals, which pushes the marsupial/placental divergence deep into the Jurassic. It seems likely that most marsupial evolution actually took place in the northern hemisphere, perhaps in Asia. Today's Australinean marsupials are the result of a circuitous world tour via North America then South America then Antarctica. They eventually went extinct in the northern hemisphere, but survived in two of the main relics of Gondwana, namely South America and Australinea. And it is Australinea which is the main stage for modern marsupial diversity.

It happens that Australinea, for much of its history since it split off from Gondwana, has had no placental mammals. It is not unlikely that all Australia's marsupials stem from a single introduction of an opossum-like founder species from South America, via Antarctica. We don't know exactly when, but it can't have been much later than 55 million years ago, which is approximately when Australia (more especially Tasmania) pulled far enough away from Antarctica to be inaccessible to island-hopping mammals. It could have been much earlier, depending upon how inhospitable Antarctica was to mammals. American opossums are no more closely related to the animals that Australians call possums than they are to any other Australian marsupials. Other American marsupials, mostly fossils, seem to be more distantly related. Most of the deep branches in the marsupial family tree, in other words, are American, which is one reason why we think the marsupials migrated to Australinea from America, rather than the other way around. But the Australinean branch of the family diversified mightily after their homeland became isolated. The isolation came to an end around 15 million years ago when Australinea (specifically New Guinea) reached close enough to Asia to allow the arrival of bats and (presumably island-hopping) rodents. Then, much more recently, dingos arrived (in trading canoes, we must suppose) and finally a whole host of other animals, such as rabbits, camels and horses, introduced by European immigrants. Most ridiculously of all, these included foxes in order that they should be hunted – an eloquent comment on the claim that this pursuit can be justified as pest control.

Together with the monotremes who join us next, the evolving Australian marsupials were carried, on the great raft that Australia became, out into South Pacific isolation. There, for the next 40 million years the marsupials (and monotremes) had Australia to themselves. If there were other mammals at the beginning* they died out early. Dinosaur shoes were waiting to be filled, in Australia as well as in the rest of the world. From our point of view the exciting thing about Australia is that it was isolated for a long time, and it had

---

* A couple of teeth which seem to belong to condylarths (a group of extinct placental mammals) have been found, but nothing younger than 55 million years.

a very small founder population of marsupial mammals, conceivably even a single species.

And the results? They were dazzling. Of approximately 340 surviving species of marsupial in the world, about three-quarters are Australinean (the rest are all American, mostly opossums plus a few other species such as the enigmatic *Dromiciops*, the monito del monte). The 240 Australinean species (give or take a few depending on whether we are lumpers or splitters)* have branched to fill the whole range of 'trades' formerly occupied by the dinosaurs, and independently occupied by other mammals in the rest of the world. The Marsupial Mole's Tale goes through some of these trades, one by one.

THE MARSUPIAL MOLE'S TALE

There is a living to be made underground, a living which moles (family Talpidae) have made familiar to us in Eurasia and North America. Moles are dedicated burrowing machines, their hands modified into spades, their eyes, which would be useless underground, all but completely degenerate. In Africa, the mole niche is filled by golden moles (family Chrysochloridae). These are superficially very similar to Eurasian moles and for years they have been placed in the same order: Insectivora. In Australia, as we would expect, the niche is filled by a marsupial, *Notoryctes*, the marsupial mole.†

Marsupial moles look like true (talpid) moles and golden moles, they feed on worms and insect larvae like true moles and golden moles, and they burrow like true moles and even more like golden moles. True moles leave an empty tunnel behind them as they dig looking for prey. Golden moles, at least the ones that live in deserts, 'swim' through sand, which collapses behind them, and marsupial moles do the same. Evolution has fashioned the 'spades' of talpid moles out of all five fingers of the hand. Marsupial moles and golden moles use two (or in some golden moles three) claws. The tail

---

* These more-or-less self-evident terms have become technical terms for taxonomists who habitually lump animals (or plants) into a few large groups, or who habitually split them into lots of small groups. Splitters proliferate names, in extreme cases where fossils are concerned, elevating almost every specimen they discover to species status.

† *Necrolestes*, a South American mammal of the Miocene Epoch, and thought to be even more distantly related to us than the marsupials, also appears to have been a 'mole'. Its name, rather inappropriately, translates as 'grave robber'.

is short in talpid moles and marsupial moles, and completely invisible in golden moles. All three are blind and have no visible ears. Marsupial moles (it's what marsupial MEANS) have a pouch, in which the prematurely born (by placental standards) young are housed.

The similarities of these three 'moles' are convergent: independently evolved for their digging habit, from different beginnings, from non-digging ancestors. And it is a three-way convergence. Although golden moles and Eurasian moles are more closely related to each other than either is to marsupial moles, their common ancestor was surely not a specialised burrower. All three resemble each other because they all dig. Incidentally, we are so used to the idea that mammals stepped into dinosaur shoes, it is surprising to reflect that no true dinosaurian 'mole' has so far been found. The nearest seems to have been the herbivorous dinosaur *Oryctodromeus* (or 'digging runner') which dug short burrows rather like rabbits do today.

Australinea is home not just to marsupial moles but to a dramatic cast list of marsupials, each of which plays more or less the same role as a placental mammal on another continent. There are marsupial 'mice' (better called marsupial shrews because they eat insects), marsupial 'cats', 'dogs', 'flying squirrels', and a gallery of counterparts to animals familiar in the rest of the world. In some cases the resemblance is very striking. Flying squirrels such as *Glaucomys volans* of the American forests look and behave very like such denizens of the Australian eucalypt forests as the sugar glider (*Petaurus breviceps*) or the mahogany glider (*Petaurus gracilis*). These are also called 'flying phalangers', although actually they are not members of the family Phalangeridae (cuscuses and brushtail possums). The American flying squirrels are true squirrels, related to our familiar tree squirrels. In Africa, interestingly, the flying squirrel trade is plied by the so-called scaly-tailed squirrels or Anomaluridae which, though also rodents, are not true squirrels. The marsupials of Australia, too, have produced three lineages of gliders, which evolved the habit independently. Returning to placental gliders, we have already met, at Rendezvous 9, the mysterious 'flying lemurs' or colugos which differ from the flying squirrels and the marsupial gliders in that their tail is included in the gliding membrane, as well as all four limbs.

Thylacinus, the Tasmanian wolf, is one of the most famous examples of convergent evolution. Thylacines are sometimes called

Tasmanian tigers because of their striped backs, but it is an unfortunate name. They are much more like wolves or dogs. They were once
common all over Australia and New Guinea, and they survived in
Tasmania until living memory. There was a bounty on their scalps
until 1909, the last authenticated specimen sighted in the wild was
shot in 1930, and the last captive thylacine died in Hobart Zoo in
1936. Most museums have a stuffed specimen. They are easy to tell
from a true dog because of the stripes on the back but the skeleton is
harder to distinguish. Zoology students of my generation at Oxford
had to identify 100 zoological specimens as part of the final exam.
Word soon got around that, if ever a 'dog' skull was given, it was safe
to identify it as *Thylacinus* on the grounds that anything as obvious
as a dog skull had to be a catch. Then one year the examiners, to their
credit, double bluffed and put in a real dog skull. In case you are
interested, the easiest way to tell the difference is by the two
prominent holes in the palate bone, which are characteristic of
marsupials generally. Dingos, of course, are not marsupials but real
dogs, probably introduced by aboriginal man. It may have partly
been competition from dingos that drove the thylacines extinct on
mainland Australia. Dingos never reached Tasmania, which may be
why thylacines survived there until European settlers drove them
extinct. But fossils show that there were other species of thylacine in
Australia that went extinct too early for humans or dingos to bear the
blame.

The 'natural experiment' of Australinean 'alternative mammals' is
often demonstrated in a series of pictures, each matching an Australinean marsupial with its more familiar placental counterpart (see
plate 15). But not all ecological counterparts resemble each other.
There doesn't seem to be any placental equivalent of the honey
possum. It is easier to see why there is no marsupial equivalent of
whales: quite apart from the difficulty of managing a pouch underwater, 'whales' would not be subject to the isolation that allowed the
Australian marsupials to evolve separately. Similar reasoning explains why there are not marsupial bats. And although kangaroos
could be described as the Australinean equivalent of antelopes they
look very different, because so much of their body is built around
their unusual hind-legged hopping gait with massive counter-
balancing tail. Yet the range of 68 species of Australinean kangaroos
and wallabies matches the range of 72 species of antelopes and
gazelles in diet and way of life. The overlap is not perfect. Some

kangaroos will take insects if they get a chance, and fossils tell of a large carnivorous kangaroo which must have been terrifying. There are placental mammals outside Australia which leap kangaroo-fashion, but these are mostly small rodents, like the jumping jerboas. The springhare of Africa is also a rodent, not a true hare, and is the only placental mammal that one might actually mistake for a kangaroo (or rather a small wallaby). Indeed my colleague Dr Stephen Cobb, when teaching zoology at the University of Nairobi, was amused to find himself being excitedly contradicted by his pupils when he told them that kangaroos are confined to Australia and New Guinea.

The lessons of the Marsupial Mole's Tale on the importance of convergence in evolution – real convergence in the forward direction, not the backwards coalescence of this book's central metaphor – will be taken up in the final chapter, The Host's Return.

MONOTREMES

Rendezvous 15 takes place approximately 180 million years ago in the half-monsoonal, half-arid world of the Lower Jurassic. The southern continent of Gondwana was still nestled closely against the great northern continent of Laurasia – the first time on our backwards journey that we find all major land masses collected into a more or less contiguous 'Pangaea'. In forward time, the split of Pangaea would have momentous consequences for the descendants of Concestor 15, perhaps our 120-million-greats-grandparent. Our rendezvous is a rather one-sided affair. The new pilgrims that join all the rest of the mammals here represent only three genera: *Ornithorhynchus anatinus*, the duckbilled platypus which lives in Eastern Australia and Tasmania; *Tachyglossus aculeatus*, the short-beaked echidna which lives all over Australia and New Guinea; and *Zaglossus*, the long-beaked echidna, which is confined to the highlands of New Guinea.[*] Collectively the three genera are known as monotremes.

Several tales have developed the theme of island continents as nurseries of major animal groups: Africa for the afrotheres, Laurasia for the laurasiatheres, South America for the xenarthrans, Madagascar for the lemurs, Australia for most of the surviving marsupials. But it is looking increasingly as though there was a much earlier continental separation among the mammals. According to one supported theory, long before the demise of the dinosaurs, the mammals were split into two major groups called the australosphenidans and the boreosphenidans. Australo, once again, doesn't mean Australian, it means southern. And boreo means northern, as in the northern aurora borealis. The australosphenidans were those early mammals that evolved in the great southern continent of Gondwana. And the boreosphenidans evolved in the northern continent of Laurasia, in a sort of earlier incarnation long before the evolution of the laurasiatheres we know today. The monotremes are the only surviving representatives of the australosphenidans. All the

---

[*] Three species of *Zaglossus* have been distinguished, one of them called, I am delighted to say, *Z. attenboroughi*.

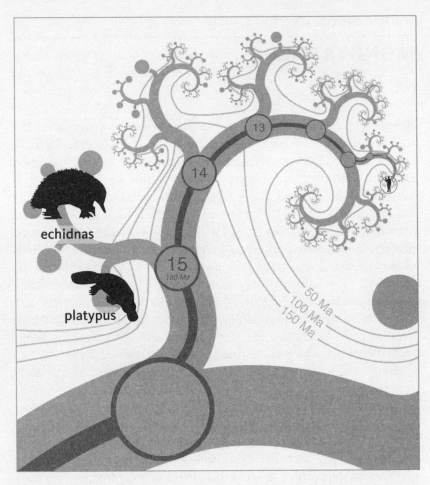

echidnas

platypus

13

14

15
180 Ma

50 Ma
100 Ma
150 Ma

Monotremes join. Living mammals, numbering around 5,500 species, all have fur and suckle their young. Those we have met so far – the placental and marsupial mammals – are thought to share a common northern hemisphere origin in the Jurassic Period. The five monotreme species are the sole survivors of a once diverse lineage of southern-hemisphere mammals which retained the habit of laying eggs.

rest of the mammals, the therians, including the marsupials that we now associate with Australia, are descended from the northern boreosphenidans. Those therians who later became associated with the south, and with the breakup of Gondwana – for instance the afrotheres of Africa and the marsupials of South America and Australia – were boreosphenidans who had migrated south into Gondwana long after their northern origins.

Let's now turn to the monotremes themselves. The echidnas live on dry land and eat ants, termites and sometimes other invertebrates. The platypus lives mostly in water where it feeds on small invertebrates in the mud. Its 'bill' really does look like that of a duck. The echidnas' bill is more tubular. Somewhat surprisingly, by the way, molecular clock evidence suggests that the concestor of echidnas with platypuses lived more recently than the fossil platypus *Obdurodon*, which lived and looked essentially like a modern platypus except that it had teeth inside its duckbill. If we have correctly read the molecular evidence, this would mean that echidnas are modified platypuses who left the water within the last 60 million years or so, lost the webbing between their toes, narrowed the duckbill to make an anteater's probing tube, and developed protective spines.

One respect in which the monotremes resemble reptiles and birds has given them their name. Monotreme means single hole in Greek. As with reptiles and birds, the anus, the urinary tract and the reproductive tract empty into a single shared opening, the cloaca. Even more reptilian is that eggs, not babies, emerge from that cloaca. And not microscopic eggs like all other mammals, but two-centimetre eggs with a tough white leathery shell, containing nutriment to feed the baby until it is ready to hatch, which it eventually does like a reptile or bird with the aid of an egg-tooth on the end of its 'bill'.

Monotremes have some other typically reptilian features too, such as the interclavicle bone near the shoulder, which reptiles, but no therian mammals, possess. On the other hand the monotreme skeleton also has a number of standard mammal traits. Their lower jaw consists of a single bone, the dentary. Reptile lower jaws have three additional bones, around the hinge with the main skull. During the evolution of the mammals, these three bones migrated away from the lower jaw into the middle ear, where, renamed the hammer, the anvil and the stirrup, they transmit sound from the eardrum to the inner ear in a cunning way that physicists call impedance-matching. Monotremes are firmly with the mammals on this point. Their inner

Could your ancestor have looked like this? Drawing of *Henkelotherium*, a eupantothere, by Elke Gröning. (The leaf form shown is that of modern ginkgos; the leaves of Jurassic ginkgos would have been more finely divided.)

ear itself, however, is more reptilian or bird-like, in that the cochlea, the tube in the inner ear that detects sounds of different pitch, is more nearly straight than the snail-shaped coil which all other mammals have, and which gives the organ its name.

Monotremes are again with the mammals in secreting milk for their young: that most proverbially mammalian of substances. But again, they slightly spoil the effect by lacking discrete nipples. Instead, the milk oozes out from pores over a wide area of skin on the ventral surface, where it is licked up by the baby clinging to the hairs on the mother's belly. Our ancestors probably did the same. Monotreme limbs sprawl sideways a little more than those of a typical mammal. You can see this in the weird rolling gait of echidnas: not quite lizard-like, but not entirely mammal-like either. It adds to the impression that the monotremes are sort of inter-mediate between reptiles and mammals.

What did Concestor 15 look like? There is of course no reason to think it was like either an echidna or a platypus. It was, after all, *our* ancestor, as well as theirs, and we've all had a very long time to evolve since. Fossils of the right vintage in the Jurassic Period belong to various types of small shrew-like or rodent-like animals such as

Morganucodon and the large group known as multituberculates. The charming picture on the opposite page is of another of these early mammals, a eupantothere, up a ginkgo tree.

THE DUCKBILL'S TALE

An early Latin name of the platypus was *Ornithorhynchus paradoxus*. It seemed so weird when first discovered that a specimen sent to a museum was thought to be a hoax: bits of mammal and bits of bird stitched together. Others have wondered whether God was having a bad day when he created the platypus. Finding some spare parts left over on the workshop floor, he decided to unite rather than waste them. More insidiously (because they aren't joking) some zoologists write the monotremes off as 'primitive', as though sitting around being primitive was a full-time way of life. To question this is a purpose of the Duckbill's Tale.

Since Concestor 15, platypuses have had exactly the same time to evolve as the rest of the mammals. There is no reason why either group should be more primitive than the other (primitive, remember, precisely means 'resembling the ancestor'). Monotremes might be more primitive than us in some respects, such as laying eggs. But there is no reason at all why primitiveness in one respect should dictate primitiveness in another. There is no substance called Essence of Antiquity that pervades the blood and soaks into the bones. A primitive bone is a bone that has not changed much for a long time. There is no rule that says the neighbouring bone has to be primitive too, not even a faint presumption in that direction – at least unless a further case is made. There's no better illustration than the eponymous duck bill itself. It has evolved far, even if other parts of the platypus have not.

The platypus bill seems comical, its resemblance to that of a duck made the more incongruous by its relatively large size, and also because a duck's bill has a certain intrinsic laughableness, perhaps borrowed from Donald. But humour does an injustice to this wondrous apparatus. If you want to think in terms of an incongruous graft, forget all about ducks. A more telling comparison is the extra nose grafted onto a Nimrod reconnaissance aircraft. The American equivalent is AWACS, more familiar but less appropriate for my comparison in that the AWACS 'graft' is on top of the fuselage rather than at the front like a bill.

The point is that the platypus bill is not just a pair of jaws for dabbling and feeding, as in a duck. It is that too, though it is rubbery rather than horny like a duck's bill. But far more interestingly, the platypus bill is a reconnaissance device, an AWACS organ. Platypuses hunt crustaceans, insect larvae and other small creatures in the mud at the bottom of streams. Eyes aren't much use in mud, and the platypus keeps them tight shut while hunting. Not only that, it closes its nostrils and its ears as well. See no prey, hear no prey, smell no prey: yet it finds prey with great efficiency, catching half its own weight in a day.

If you were a sceptical investigator of somebody claiming a 'sixth sense', what would you do? You'd blindfold him, stop his ears and his nostrils, and then set him some task of sensory perception. Platypuses go out of their way to do the experiment for you. They switch off three senses which are important to us (and perhaps to them on land), as if to concentrate all their attention on some other sense. And the clue is given by one further feature of their hunting behaviour. They swing the bill in movements called saccades, side to side, as they swim. It looks like a radar dish scanning . . .

One of the first scientific descriptions of the platypus, Sir Everard Home's publication in the *Philosophical Transactions of the Royal Society* for 1802, was farsighted. He noticed that the branch of the trigeminal nerve that innervates the face is

> uncommonly large. We should be led by this circumstance to believe that the sensibility of the different parts of the bill is very great, and therefore that it answers the purpose of a hand, and is capable of nice discrimination in its feeling.

Sir Everard didn't know the half of it. It's the reference to a hand that tells. The great Canadian neurologist Wilder Penfield published a famous picture of a human brain, together with a diagram showing the proportions given over to different parts of the body. The map of a part of the brain given over to controlling muscles in different parts of the body, on one side, is shown opposite. Penfield made a similar map of parts of the brain concerned with the sense of touch in different parts of the body. The striking thing about both maps is the huge prominence given to the hand. The face, too, is prominent, especially the parts controlling jaw movements, in chewing and speaking. But it is the hand that you really notice when you see a Penfield 'homunculus'. The image reproduced in plate 16 is another

Penfield brain map.
Adapted from Penfield
and Rasmussen [326].

way of representing the same thing. This grotesque has his body
distorted in proportion to the amount of brain given over to different parts. Again it shows that the human brain is hand-heavy.

Where is all this leading? My account of the Duckbill's Tale is
indebted to the distinguished Australian neurobiologist Jack Pettigrew and his colleagues, including Paul Manger, and one of the
fascinating things they did was to prepare a 'platypunculus', the
platypus equivalent of a Penfield homunculus, reproduced below.
The first thing to say is that it is far more accurate than the Penfield
homunculus, which was based on very scanty data. The platypunculus is a very thorough piece of work. You can see three little platypus
maps on the upper part of the brain: separate representations, in
different parts of the brain, of sensory information from the body

**The platypus brain is
bill-heavy.** 'Platypunctus',
from Pettigrew *et al.* [330].

surface. What matters to the animal is that there is an orderly spatial mapping between each part of the body and the corresponding part of the brain.

Notice that the hands and feet, shaded black on the three maps, are approximately in proportion to the body itself, unlike the case of the Penfield homunculus with its vast hands. What is not in proportion in the platypunculus is the bill. The bill's maps are the huge areas reaching down from the maps of the rest of the body. Where the human brain is hand-heavy, the platypus brain is bill-heavy. Sir Everard Home's guess is looking good. But, as we shall see, in one respect the bill is even better than a hand: it can reach out and 'feel' things that it is not touching. It can feel at a distance. It does it by electricity.

When any animal, such as a freshwater shrimp which is a typical platypus prey, uses its muscles, weak electric fields are inevitably generated. With sufficiently sensitive apparatus these can be detected, especially in water. Given dedicated computer power to handle data from a large array of such sensors, the source of the electric fields can be calculated. Platypuses don't, of course, calculate as a mathematician or a computer would. But at some level in their brain the equivalent of a calculation is done, and the result is that they catch their prey.

Platypuses have about 40,000 electrical sensors distributed in longitudinal stripes over both surfaces of the bill. As the platypunculus shows, a large proportion of the brain is given over to processing the data from these 40,000 sensors. But the plot thickens. In addition to the 40,000 electrical sensors, there are about 60,000 mechanical sensors called push rods, scattered over the surface of the bill. Pettigrew and his co-workers have found nerve cells in the brain that receive inputs from mechanical sensors. And they have found other brain cells that respond to both electrical and mechanical sensors (so far they have found no brain cells that respond to electrical sensors only). Both kinds of cell occupy their correct position on the spatial map of the bill, and they are layered in a way that is reminiscent of the human visual brain, where layering assists binocular vision. Just as our layered brain combines information from the two eyes to construct a stereo percept, the Pettigrew group suggests that the platypus might be combining the information from electrical and mechanical sensors in some similarly useful way. How might this be done?

They propose the analogy of thunder and lightning. The flash of lightning and the crack of thunder happen at the same moment.

We see the lightning instantaneously, but the thunder takes longer to reach us, travelling at the relatively slow speed of sound (and incidentally the bang becomes a rumble because of echoes). By timing the lag between lightning and thunder, we can calculate how far away the storm is. Perhaps the electrical discharges from the prey's muscles are the platypus's lightning, while the thunder is the waves of disturbance in the water caused by the prey animal's movements. Is the platypus brain set up to compute the time lag between the two, and hence calculate how far away the prey is? It seems likely.

As for pinpointing the prey's direction, this must be done by comparing the inputs from different receptors all over the map, presumably aided by the scanning side-to-side movements of the bill, just as a man-made radar uses the rotation of the dish. With such a huge array of sensors projecting to mapped arrays of brain cells, the platypus very likely forms a detailed three-dimensional image of any electrical disturbances in its vicinity.

Pettigrew and his colleagues prepared the contour map below of lines of equal electrical sensitivity around the bill of the platypus. When you think of a platypus, forget duck, think Nimrod, think AWACS; think huge hand feeling its way, by remote pins and needles; think lightning flashing and thunder rumbling, through the watery mud of Australia.

The platypus is not the only animal to use this kind of electrical sense. Various fish do it, including paddlefish such as *Polyodon spathula* (see plate 17). Technically 'bony' fish, paddlefish have secondarily, with their relatives the sturgeons, evolved a cartilaginous

Remote pins and needles. The electric sensory world of the platypus. From Manger and Pettigrew [264].

skeleton like a shark. Unlike sharks, however, paddlefish live in freshwater, often turbid rivers where again eyes are not much use. The 'paddle' is shaped pretty much like the upper jaw of a platypus's bill, though it is not a jaw at all but an extension of the cranium. It can be extremely long, often as much as one-third of the body length. It reminds me of a Nimrod aircraft even more than the platypus does.

The paddle is obviously doing something important in the life of the fish, and it has in fact been clearly demonstrated that it is doing the same job as the platypus bill – detecting electric fields from prey animals. As with the platypus, the electrical sensors are set into pores deployed in longitudinal lines. The two systems are independently evolved, however. Platypus electrical pores are modified mucus glands. Paddlefish electric pores are so similar to the pores used by sharks for electrical sensing, called ampullae of Lorenzini, that they have been given the same name. But where the platypus arranges its sensory pores in a dozen or so narrow stripes along the length of the bill, the paddlefish has two broad stripes, on either side of the midline of the paddle. Like the platypus, the paddlefish has an enormous number of sensory pores – actually even more than the platypus. Both the paddlefish and the platypus are far more sensitive to electricity than any one of their sensors by itself. They must be doing some sort of sophisticated signal summation from different sensors.

There is evidence that the electrical sense is more important for juvenile paddlefish than for adults. Adults who have accidentally lost their paddle have been found alive and apparently healthy, but no juveniles have been found to survive up any creek without a paddle. This may be because juvenile paddlefish, like adult platypuses, target and catch individual prey. Adult paddlefish feed more like planktivorous baleen whales, sieving their way through the mud, catching prey *en masse*. They grow big on this diet, too – not as big as whales, but as long and as heavy as a man, larger than most animals that swim in freshwater. Presumably if you are sieving plankton as an adult, you have less need of an accurate prey-locator than if you are darting after individual prey as a juvenile.

Platypus and paddlefish, then, have independently hit upon the same ingenious trick. Has any other animal discovered it? Whilst doing his D.Phil. work on fossils in China, my sometime research assistant Sam Turvey encountered an extremely unusual trilobite called *Reedocalymene*. Otherwise a 'bog-standard' trilobite (similar to the Dudley Bug, *Calymene*, which features on the coat of arms of the town of Dudley), *Reedocalymene* has one unique and remarkable feature: a huge flattened rostrum, like that of a paddlefish, sticking out a whole body length in front. It can't have been for streamlining, since this trilobite, unlike many others, was obviously unfitted for swimming above the sea bed. A defensive purpose is also unlikely for various reasons. Like a paddlefish, sturgeon or platypus bill, the trilobite's rostrum is studded with what look like sensory receptors, probably used for detecting prey. Turvey knows of no modern arthropods with an electrical sense (interesting in itself, given the versatility of the arthropods), but he would put money on *Reedocalymene* being yet another 'paddlefish' or 'platypus'.

Other fish, though lacking the Nimrod-like 'antenna' of the platypus and the paddlefish, have an even more sophisticated electrical sense. Not content with picking up electrical signals inadvertently given off by prey, these fish generate their own electric fields. They navigate and detect prey by reading the distortions in these self-generated fields. Along with various cartilaginous rays, two groups of bony fish, the gymnotid family of South America and the mormyrid family of Africa, have independently developed this to a high art.

How do these fish make their own electricity? The same way the shrimps and insect larvae and other prey of the platypus inadvertently do it: with their muscles. But whereas the shrimps can't help making a little electricity because that is what muscles just do, the electric fish gang their blocks of muscle together just like batteries in series.* A gymnotid or mormyrid electric fish has a battery of muscle blocks arranged in series along its tail, each generating a low voltage and adding up to a higher voltage. The electric eel (not a true eel but another South American freshwater gymnotid) takes it to an extreme. It has a very long tail into which it can pack a much larger battery of electrical cells than a fish of normal length. It stuns its prey with electric shocks which may exceed 600 volts and can be fatal to people. Other freshwater fish, such as the African electric catfish *Malapterurus* and the marine electric ray *Torpedo* also generate enough volts to kill, or at least knock out, their prey.

---

* Of course, the word battery in its original electrical sense means a battery of cells in series, as opposed to a single cell. If your transistor radio takes six 'batteries', a pedant would insist that it takes one battery of six cells.

These high-voltage fish seem to have pushed, to a literally stunning extreme, a capacity which was originally a kind of radar used by the fish to find its way around and detect prey. Weakly electric fish such as the South American *Gymnotus* and the unrelated African *Gymnarchus* have an electrical organ like the electric eel's but much shorter – their battery consists of fewer modified muscle plates in series – and a weakly electric fish typically generates less than one volt. The fish holds itself like a rigid stick in the water, for a very good reason as we shall see, and electric current flows along curved lines that would have delighted Michael Faraday. All along the sides of the body are pores containing electrical sensors – tiny voltmeters. Obstacles or prey items distort the field in various ways, which are detected by these little voltmeters. By comparing the readings of the different voltmeters and correlating them with the fluctuations of the field itself (sinusoidal in some species, pulsed in others) the fish can calculate the location of obstacles and prey. They also use their electric organs and sensors to communicate with one another.

A South American electric fish such as *Gymnotus* is remarkably similar to *Gymnarchus*, its African opposite number, but there is one revealing difference. Both have a single long fin running the length of the midline, and both use it for the same purpose. They can't throw the body into the normal sinuous waves of a swimming fish because it would distort their electrical sense. Both are obliged to keep the body rigid, so they swim by means of the longitudinal fin, which waves sinuously just like a normal fish should. It means they swim slowly, but presumably it is worth it to get the benefits of a good clear signal. The beautiful fact is that *Gymnarchus* has its longitudinal fin on its back, while *Gymnotus* and the other South American electric fish, including the electric 'eel', keep their longitudinal fin on their belly. It is for such cases that 'the exception that proves the rule' was coined.

Returning to the platypus, the sting in the tale is actually in the hind claws of the male platypus. True venomous stings, with hypodermic injection, are found in various invertebrate phyla, and in fish and reptiles among vertebrates – but never in birds or mammals other than the platypus (unless you count the toxic saliva of solenodons and some shrews that makes their bites slightly venomous). Among mammals, the male platypus is in a class of its own, and it may be in a class of its own among venomous animals too. The fact that the sting is found only in males suggests, rather surprisingly, that it is aimed not at predators (as in bees) nor at prey (as in snakes) but at rivals. It is not dangerous but is extremely painful, and is unresponsive to morphine. It looks as though platypus venom works directly on pain receptors themselves. If scientists could understand how this is done, there is a hope that it might give a clue to how to resist the pain caused by cancer.

This tale began by chiding those zoologists who call the platypus 'primitive' as though that were any kind of explanation for the way it is. At best it is a description. Primitive means 'resembling the ancestor' and there are many respects in which this is a fair description of a platypus. The bill and the sting are interesting exceptions. But the more important moral of the tale is that even an animal that is genuinely primitive in all respects is primitive for a reason. The ancestral characteristics are good for its way of life, so there is no reason to change. As my old tutor Professor Arthur Cain liked to say, an animal is the way it is because it needs to be.

WHAT THE STAR-NOSED MOLE SAID TO THE DUCKBILLED PLATYPUS

The star-nosed mole, who had joined the pilgrimage along with the other laurasiatheres at Rendezvous 12, listened to the Duckbill's Tale with close attention, and with growing recognition in what was left of his vestigial, pin-prick eyes. 'Yes!' he squeaked, too high for some of the larger pilgrims to hear, and he clapped his spades with excitement. 'That's just the way it is for me . . . well, sort of.'

No, it won't do, I wanted to follow Chaucer in having at least one section devoted to what one pilgrim said to another, but I'll limit it to the heading and first paragraph, and now revert to my practice of telling the tale itself in my own words. Bruce Fogle (*101 Questions Your Dog Would Ask Its Vet*) or Olivia Judson (*Dr Tatiana's Sex Advice to All Creation*) might get away with it, but not me.

The star-nosed mole, *Condylura cristata*, is a North American mole which, in addition to burrowing and hunting for worms like other moles, is a good swimmer too, hunting for underwater prey – it often tunnels deep into river banks. It is also more at home above ground than other moles, where it still prefers damp, soggy places. It has large spade hands like other moles.

Pushes the envelope of touch beyond our dreams. An in-your-face view of a star-nosed mole, *Condylura cristata.*

What sets it apart is the remarkable nose that gives it its name. Surrounding the two forwards-pointing nostrils, there is an extraordinary ring of fleshy tentacles, like a baby sea anemone with 22 arms. The tentacles are not used to grasp things. Nor are they an aid to smelling, which is the next hypothesis that might occur to us. Nor, despite the beginning of this section, are they an electrical radar like that of the platypus. Their true nature has been beautifully worked out by Kenneth Catania and Jon Kaas of Vanderbilt University, Tennessee. The star is a touch-sensitive organ, like a super-sensitive human hand, but lacking the grasping function of the hand and emphasising its sensitivity instead. But it isn't just any ordinary touch-sensitive organ. The star-nosed mole pushes the envelope of touch beyond our dreams. The skin of its nose is more sensitive than any other area of skin anywhere among the mammals, not excluding the human hand.

There are 11 tentacles arcing round each nostril, labelled 1 to 11 in order. Tentacle 11, which lies close to the midline and just below the level of the nostril, is special, as we shall see in a moment. Although they are not used for grasping, the tentacles are moved, independently or in particular groupings. The surface of each tentacle is carpeted with a regular array of little round bumps called Eimer's Organs, each one a unit of touch sensitivity, and each one wired up by between seven nerve fibres (for tentacle 11) and four nerve fibres (most of the other tentacles).

The density of Eimer's organs is the same for all tentacles. Tentacle 11, being smaller, has fewer of them, but it has more nerves supplying each one. Catania and Kaas were able to map the tentacles to the brain. They found (at least) two independent maps of the nose star in the cerebral cortex. In each of these two brain areas, the parts of the brain corresponding to each tentacle are laid out in order. And tentacle 11 again is special. It is more sensitive than the rest. Once an object has been first detected by any of the tentacles, the animal then moves the star so that tentacle 11 can examine it carefully. Only then is the decision taken whether to eat it or not. Catania and Kaas refer to tentacle 11 as the 'fovea' of the star.* More generally, they say:

> Although the nose of the star-nosed mole acts as a tactile sensory surface, there are anatomical and behavioral similarities between the mole's sensory system and the visual system of other mammals.

If the star is not an electrical sensor, whence the empathy with the platypus with which I opened this section? Catania and Kaas constructed a schematic model of the relative amount of brain tissue given over to different parts of the body surface. Overleaf is the resulting molunculus, by analogy with Penfield's homunculus and Pettigrew's platypunculus. And just look at it!

You can see where the star-nosed mole's priorities lie. You can get a feel for its world. And *feel* is the right word. This animal lives in a tactile world, dominated by the tentacles of the nose, with a subsidiary interest in the large spade hands and the whiskers.

What is it like to be a star-nosed mole? I am tempted to propose the star-nosed counterpart to an idea I once offered for bats. Bats live in a world of sound, but what they do with their ears is pretty much the same as what, say, insect-hawking birds like swallows do with their eyes. In both cases the brain needs to construct a mental model of a three-dimensional world, to be navigated at high speed, with obstacles to be avoided and small moving targets to catch. The model of the world needs to be the same, whether it is constructed and

---

* The fovea is the small area in the middle of the human retina where cone cells are concentrated so that acuity, and colour vision, are both maximal. We read with our fovea, recognise each other's faces, and do everything that needs fine visual discrimination.

You can see where its priorities lie.
'Molunculus' brain map of the star-nosed mole. Note that some parts, such as tentacle 11, are hidden behind the parts we can see. From Catania and Kaas [60].

updated with the aid of light rays or sound echoes. My conjecture was that a bat probably 'sees' the world (using echoes) in pretty much the same way as a swallow, or a person, sees the world using light.

I even went so far as to speculate that bats hear in colour. The hues that we perceive have no necessary link with the particular wavelengths of light that they represent. The sensation that I call red (and nobody knows if my red is the same as yours) is an arbitrary label for light of long wavelengths. It could equally well have been used for short wavelengths (blue), and the sensation that I call blue used for long wavelengths. Those hue sensations are available in the brain for tying to whatever, in the outside world, is most convenient. In bat brains those vivid qualia would be wasted on light. They are more likely to be used as labels tied to particular qualities of echo, perhaps textures of surfaces on obstacles or prey.

My conjecture now is that a star-nosed mole 'sees' with its nose. And my wild speculation is that it uses those same qualia that we call colour, as labels for tactile sensations. Similarly, I want to guess that duckbilled platypuses 'see' with the bill, and use the qualia we call colour as internal labels for electrical sensations. Could this be why platypuses close their eyes tight shut when they are hunting electrically with the bill? Could it be because the eyes and the bill are competing, in the brain, for internal qualia labels, and to use both senses at once would lead to confusion?

MAMMAL-LIKE REPTILES

The monotremes having joined us, the entire company of mammal pilgrims now walks back 140 million unbroken years, the longest gap yet between any two milestones, to Rendezvous 16 where we are to meet an even larger band of pilgrims than our own, the sauropsids: reptiles and birds. That pretty much means all vertebrates that lay large eggs with a waterproof shell on land. I have to say 'pretty much', partly because monotremes, who have already joined us, also lay that kind of egg. Even turtles that are otherwise wholly marine haul themselves up the beach to lay their eggs. Plesiosaurs may have done the same. Ichthyosaurs, however, were so specialised for swimming that, like the dolphins who later resembled them, they presumably couldn't come on shore at all. They independently discovered how to give birth to live young – as we know from mothers fossilised in the act.

I said that our pilgrims walked through 140 million years without milestones, but of course 'without milestones' is true only within the conventions of this book: we are recognising as milestones only rendezvous with living pilgrims. Our ancestral lineage indulged in fertile evolutionary branching during that time, as we know from the rich fossil record of 'mammal-like reptiles', but none of the branches among these counts as a 'rendezvous' because, as it turned out, none of them survived. There are therefore no modern representatives to set off as pilgrims from the present. When we met a similar problem with the hominids, we decided to give certain fossils honorary status as 'shadow pilgrims'. Since we are pilgrims seeking our ancestors, pilgrims who actually want to know what our 150-million-greats-grandparent looked like, we cannot ignore the mammal-like reptiles and jump straight to Concestor 16. Concestor 16, as we shall see, looked like a lizard (see plate 19). The gap from Concestor 15, which looked like a shrew, is too great to leave unbridged. We have to examine the mammal-like reptiles as shadow pilgrims, as though they were living pilgrims joining our march – although they shall not actually tell tales. But first, some background information on the timespan involved, because it is very long.

The intervening years without rendezvous milestones span half the Jurassic, the whole of the Triassic, the whole of the Permian and the final third of the Carboniferous. As the pilgrimage moves from the Jurassic back into the hotter and drier world of the Triassic – one of the hottest periods in the planet's history, when all the land masses were joined together, forming Pangaea – we pass the late Triassic mass extinction, when three-quarters of all species went extinct. But this is nothing compared to the next transition, from the Triassic Period back into the Permian. At the Permo–Triassic boundary, a staggering 90 per cent of all species perished without descendants, including all the trilobites and several other major groups of animals. The trilobites, to be fair, had already been declining over a long period. But the end-Permian mass extinction was the most devastating of all time. Some believe that, like the Cretaceous one, this was caused by a massive bolide collision, although the prevailing theory blames enormous episodes of volcanism. Even the insects took a severe knock, the only one in their history. At sea, bottom-dwelling communities were almost wiped out. On land, the Noah among the mammal-like reptiles was *Lystrosaurus.* Immediately after the catastrophe, the squat, short-tailed *Lystrosaurus* became extremely abundant over the whole world, rapidly occupying vacant niches.

The natural association with apocalyptic carnage needs to be tempered. Extinction is the eventual fate of nearly all species. Perhaps 99 per cent of all species that have ever existed have gone extinct. Nevertheless, the rate of extinctions per million years is not fixed and only occasionally rises above 75 per cent, the threshold arbitrarily recognised for a 'mass' extinction. Mass extinctions are spikes in the rate of extinction, rising above the background rate.

The diagram on the next page shows rates of extinction per million years.* Something happened at the time of those spikes. Something bad. Perhaps a single catastrophic event, such as the collision with a massive celestial rock that killed the dinosaurs 65 million years ago in the Cretaceous–Palaeogene extinction. Or, in other cases among the five spikes, the agony may have been drawn out. What Richard Leakey and Roger Lewin have called the Sixth Extinction is the one

---

* The absolute figures are lower than 75 per cent, because they refer to genera not species. Absolute figures for species are higher than for genera, because each genus contains lots of species so it's harder to extinguish a genus than a species.

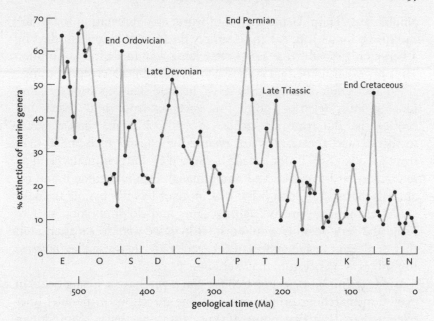

Percent extinction of marine genera throughout the Phanerozoic Eon. Adapted from Sepkoski [382].

now being perpetrated by *Homo sapiens* – or *Homo insipiens* as my old German teacher William Cartwright preferred to say.*

Before we get to the mammal-like reptiles, we face a somewhat tiresome point of terminology. Terms like reptile and mammal can refer to 'clades' or 'grades' – the two are not exclusive. A clade is a set of animals consisting of an ancestor and all its descendants. The 'birds' constitute a good clade. 'Reptile', as traditionally understood, is not a good clade because it excludes birds. Biologists consequently refer to the reptiles as 'paraphyletic'. Some reptiles (e.g. crocodiles) are closer cousins of some non-reptiles (birds) than they are of other reptiles (turtles). To the extent that reptiles all have something in common, they are members of a *grade*, not a clade. A grade is a set of animals that have reached a similar stage in a recognisably progressive evolutionary trend.

Yet another informal grade name, favoured by American zoologists, is 'herp'. Herpetology is the study of reptiles (except birds) and

---

* A remarkable, bushy-browed, slow-spoken man who called a spade a spade and was seldom seen without one, Mr Cartwright discovered environmental activism long before its vogue, and filled his lessons with ecology, to the detriment of our German but the advantage of our humane education.

amphibians. 'Herp' is a rare kind of word: an abbreviation for which there is no long form. A herp is simply the kind of animal studied by a herpetologist, and that is a pretty lame way to define an animal. The only other name that comes close is the biblical 'creeping thing'.

Another grade name is fish. 'Fish' include sharks, various extinct fossil groups, teleosts (bony fish such as trout and pike) and coelacanths. But trout are closer cousins to humans than they are to sharks (and coelacanths are even closer cousins to humans than trout are). So 'fish' is not a clade because it excludes humans (and all mammals, birds, reptiles and amphibians). Fish is a grade name for animals that sort of look fishy. It is more or less impossible to make grade terminology precise. Ichthyosaurs and dolphins look sort of fishy, and very possibly would taste fishy if we were to eat them, but they don't count as members of the fish 'grade', because they *reverted* to fishiness via ancestors that were non-fishy.

Grade terminology works well for you if you have a strong belief in evolution marching progressively in one direction, in parallel lines from a shared starting point. If, say, you think that a whole lot of related lineages were all independently evolving in parallel from amphibianhood through reptilehood towards mammalhood, you could speak of passing through the reptile grade on the way to the mammal grade. Something like that parallel march may have happened. It was the view that I was brought up with, by my own respected teacher of vertebrate palaeontology Harold Pusey. I have a lot of time for it, but it is not something to be taken for granted in general, nor necessarily enshrined in terminology.

If we swing to the other extreme and adopt strict cladistic terminology, the word reptile can be rescued only if it is deemed to include birds. This is the course favoured by the authoritative 'Tree of Life' project founded by the Maddison brothers.* There's a lot to be said for following them, and also for the very different tactic of replacing 'mammal-like reptile' by 'reptile-like mammal'. But the word reptile has become so ingrained in its traditional sense that I fear it would confuse to change it now. Also, there are times when strict cladistic purism can give ludicrous results. Here's a *reductio ad absurdum*. Concestor 16 must have had an immediate descendant on the mammal side and an immediate descendant on the lizard/crocodile/

---

* This excellent resource is continually updated at *http://tolweb.org* but is sadly now without its previous, delightful disclaimer: 'The Tree is under construction. Please have patience: the real Tree took over 3,000,000,000 years to grow.'

1. Something very special began to happen . . .
This painting of a bull is from the Lascaux
Caves in the Dordogne, France. Discovered
in 1940, the paintings are over 16,000 years
old. They show a deep understanding of animal
forms and movement, and a fine artistic sense.
The purpose of the paintings is unknown
(see page 48).

2. Did they hold hands?
These 3.6-million-year-old hominid footprints at Laetoli, Tanzania, were discovered
by Mary Leakey in 1978. They were fossilised
in volcanic ash. The trail extends for some
70 metres and was probably made by
Australopithecus afarensis (see page 90).

3. Lucky to have fossils at all

A fossil worm, *Palaeoscolex sinensis*, from the Chengjiang fossil beds, showing fine details of soft body parts. The Chengjiang fossils date back to the Lower Cambrian, about 525 million years ago (see page 91).

4. Hope of life

Skull of *Sahelanthropus tchadensis*, or 'Toumai', discovered in the Sahel region of Chad by Michael Brunet and colleagues in 2001 (see page 116).

5. Concestor 3

Imaginative reconstruction of Concestor 3, a large quadrupedal ape which probably spent much of its time up in the trees. Like all great apes, it would have displayed considerable intelligence (see page 145).

7. Would a Martian feel at home in Madagascar?

Avenue of baobabs, Morondava, Madagascar. This species of baobab, *Adansonia grandidieri*, is one of six unique to Madagascar (see page 200).

6. Concestor 8

A cat-sized primate active at night and probably during the day too, Concestor 8 would have searched for food in the end of branches of trees, using its forward-pointing eyes and grasping hands and feet (see page 195).

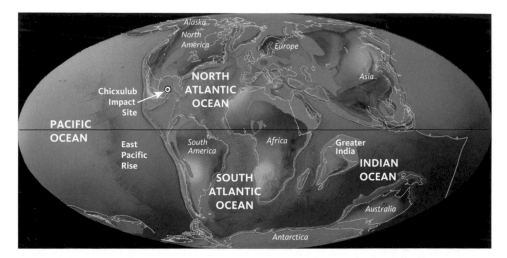

8. The Earth at the time of the K/T extinction, showing the position of the Chicxulub bolide impact [378]. By the end of the Cretaceous, Laurasia and Gondwana had broken up into continental shapes broadly familiar to us today, though Europe was still a large island and India was still making its way towards Asia. The climate was warm and mild, even in the polar regions, as it was throughout the Mesozoic Era, partly as a result of the pattern of warm ocean currents (see pages 195–204).

9. Swimming in its extended phenotype

European beaver, *Castor fiber* (see page 227).

10. A whale of a surprise

Hippopotamus amphibius in its element. Two species of hippo survive today in Africa (the other is the pygmy hippo, *Hexaprotodon liberiensis*), but fossil remains indicate that perhaps three species of hippo lived in Madagascar right up into the Holocene (see page 232).

11. Size is at a premium

Male and female southern elephant seals, *Mirounga leonina* (see page 244).

12. *(left)* &
13. *(below)* Concestor 13

The reconstruction from the first edition of this book *(left)* shows a nocturnal shrew-like creature, climbing along low branches to catch insects. It had the features of a typical placental mammal. Note that angiosperms, the last of the major plant groups to evolve, are present in the background. Below is a more recent reconstruction from 2013 (see pages 257–8).

14. These two maps represent the ages of rocks on the seafloor, based on their remnant magnetism. The map at the top shows the Earth at 68 million years ago. The map below shows the Earth today. The false-colour bands show seafloor rocks from the Cretaceous period pushed back as new seafloor formed, widening the Atlantic. Similar movement can be seen in the Pacific and Indian Oceans (see page 260).

15. Natural experiment

Composite of marsupials (left photo of each pair) and their placental 'equivalents' (see pages 276–7). *From top:* thyalcine (*Thyalcinus cynocephalus*)/grey wolf (*Canis lupus*); striped possum (*Dactylopsila trivirgata*)/skunk (*Conepatus humboldtii*); marsupial mole (*Notoryctes typhlops*)/golden mole (*Eremitalpa granti*); sugar glider (*Petaurus breviceps*)/flying squirrel (*Glaucomys sabrinus*).

16. The human brain is hand-heavy

A Penfield homunculus showing parts of the human body blown up in proportion to the area of brain cortex dedicated to their sensory perception (see page 284).

17. The same ingenious trick?

The paddlefish, *Polyodon spathula* (see page 287).

18. Feeling the fingerprints of past time

A lava lizard, *Microlophus albemarlensis*, basking on rocks formed in the seconds following a volcanic eruption (see page 306).

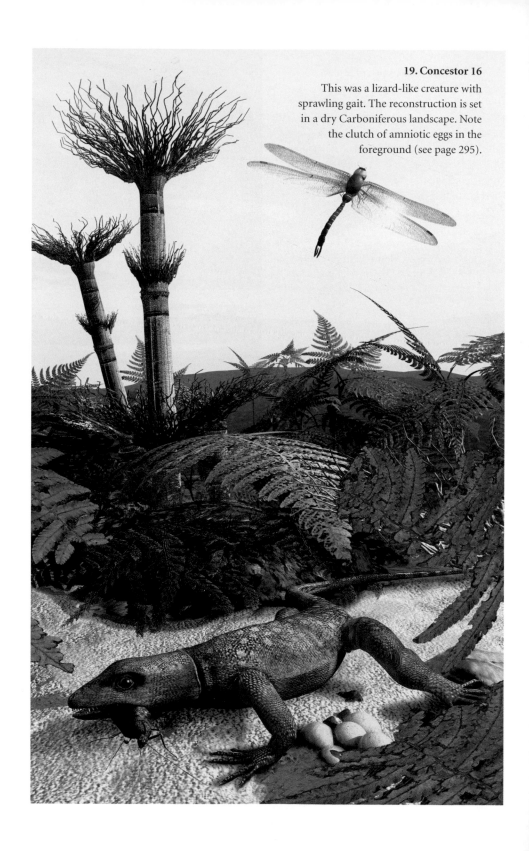

19. Concestor 16

This was a lizard-like creature with sprawling gait. The reconstruction is set in a dry Carboniferous landscape. Note the clutch of amniotic eggs in the foreground (see page 295).

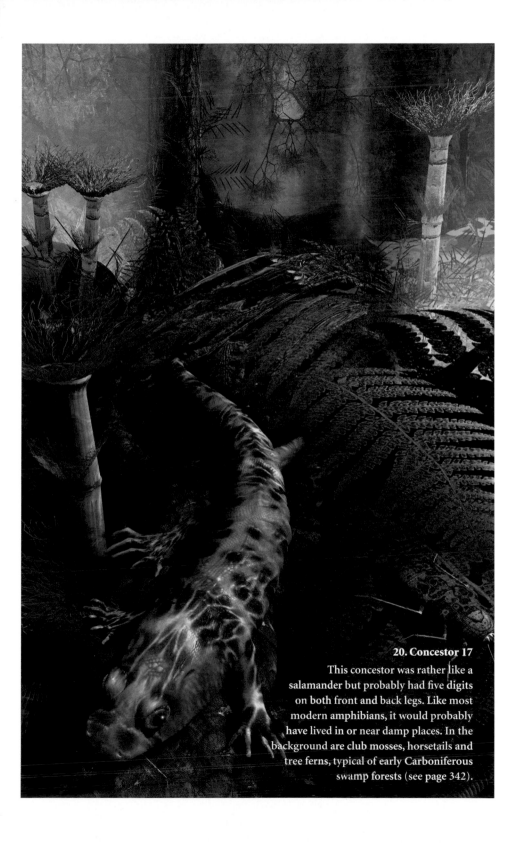

20. Concestor 17

This concestor was rather like a salamander but probably had five digits on both front and back legs. Like most modern amphibians, it would probably have lived in or near damp places. In the background are club mosses, horsetails and tree ferns, typical of early Carboniferous swamp forests (see page 342).

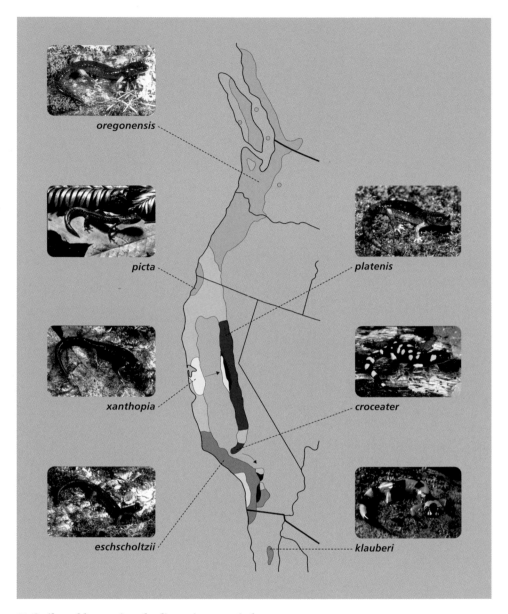

oregonensis

picta

platenis

xanthopia

croceater

eschscholtzii

klauberi

21. Strikes a blow against the discontinuous mind

Ensatina populations around the Central Valley, California. Stippled areas indicate zones of transition. Map adapted from Stebbins (2003) (see page 344).

22. Concestor 18

The land vertebrates evolved from lobe-finned fish, such as the one reconstructed here. The name comes from the prominent lobes on all fins, except the dorsal fin and the heterocercal (asymmetrical) tail (see page 364).

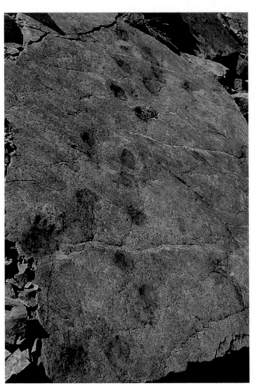

23. Nearly 200 million years older than they 'should' be.

The 390-million-year-old Załchemie trackway, the earliest four-legged footprints found in the fossil record (see page 366). These tracks are thought to have been made underwater, in a shallow lagoon. The track-maker has yet to be found, but it clearly had toes, as seen in the leftmost footprint.

24. 'I would not have been more surprised if I had seen a dinosaur walking down the street'

Coelacanth, *Latimeria chalumnae*, photographed off the Comoros Islands in the Indian Ocean (see page 373).

25. Animal shapes are malleable like plasticine

The variety of teleost forms. *Clockwise from top left:* Snipe eel (*Avocettina infans*); black swallower (*Chiasmodon niger*) after meal; plaice (*Pleuronectes platessa*); sunfish (*Mola mola*); leafy sea dragon (*Phycodurus equus*); gulper eel (*Eurypharynx pelecanoides*) (see pages 382–4).

26. A particularly neat experiment

The reddish *Pundamilia nyereirei* (*left*) and bluish *Pundamilia pundamilia* (*below*), interbreed successfully when their colours are dulled under monochromatic light (see pages 389–90).

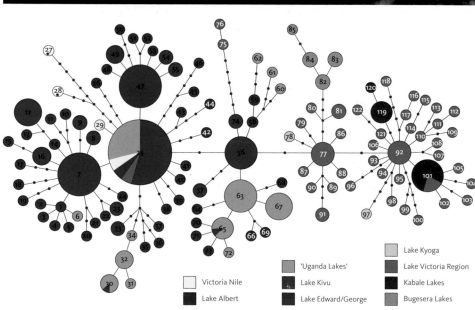

27. Each gene is represented by one circle

Unrooted haplotype network of cichlid DNA, from Verheyen et al. [295] (See page 393).

28. Designed by an artist with a drugged imagination?

Great hammerhead shark, *Sphyrna mokarran* and freshwater sawfish, *Pristis microdon*, (see page 400).

29. Ghostly appearance

Elephant fish, *Callorhynchus milii*, showing the characteristic large head and flapping pectoral fins of chimaeras (see page 402).

30. Like an associate professor getting tenure?
Adult blue sea squirt, *Rhopalaea crassa* (see page 415).

31. A 'Martian', whose very strangeness helps us to see ourselves more clearly, by showing us what we are not

A red cushion sea star, *Oreaster reticulatus*, a beautiful example of five-fold symmetry (see page 428).

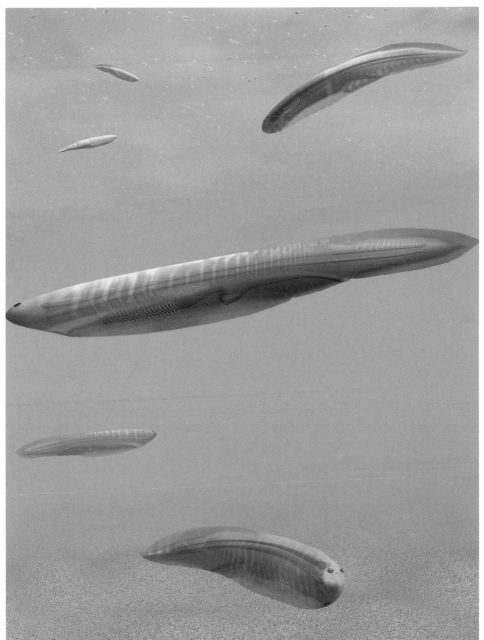

32. Concestor 23

This concestor is thought to have had a notochord (a stiff cartilaginous rod) extending the length of its body from below its rudimentary brain. Like the modern lancelet, it would have had thick myomeres (V-shaped muscle blocks), and would have filtered food through its gills (see page 423).

33. Concestor 26

This worm-like concestor is shown living on the sea bed. Its body is formed from serially repeating parts, with a head at one end and a flow-through gut. The head would have had a mouth opening that might have been surrounded by appendages that aided feeding. It is also likely to have had eyes (see page 439).

34. Carried home by workers in broad, rustling rivers of green

Leaf cutter ants, *Atta* sp., carrying leaf fragments back to the nest. Note the small (minim) worker riding on the leaf (see page 450).

35. Wouldn't a Martian split them three against one?

Left to right: Condoleeza Rice, Colin Powell, George W. Bush, Donald Rumsfeld (see page 455).

36. Cells 'think' they are in the wrong segment

Homeotic mutant fruit fly (see page 474).

37. An evolutionary scandal

Light micrograph of a bdelloid rotifer, *Philodina gregaria*, found in Antarctica (see page 481).

38. Velvet worm

The modern onychophoran, *Peripatopsis moseleyi* (see pages 494–5).

39. What are we to make of this?

Dickinsonia costata, part of the Ediacaran fauna (see page 506).

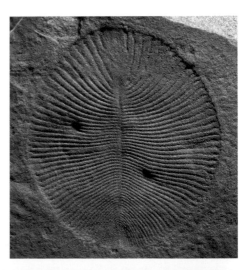

40. Sail by the wind

The blue button or blue sea star *Porpita* has a central gas-filled float surrounded by tentacles. Like its close relative *Velella*, *Porpita* is now thought to be a highly modified polyp rather than a colony (see page 527).

41. Jelly armada

Mastigias jellyfish amassed at the water surface, Palau, western Pacific (see page 529).

42. There aren't many animals who could claim to have redrawn the map of the world
Heron Island, on the Great Barrier Reef (see page 530).

43. Trust, at the barber's shop of the sea
Cleaner fish, *Labroides dimidiatus*, working on a rosy goatfish, *Parupeneus rubescens*, Red Sea (see page 534).

44. Beautifying the oceans of the world

A ctenophore, *Dryodora glandformis*, whose long sticky tentacles trail leftwards out of shot. The iridescent colours result from light diffracting through combs of propulsive 'hairs' along its body (see page 537).

45. Too good for a goddess

Venus's girdle, *Cestum veneris* (see page 539).

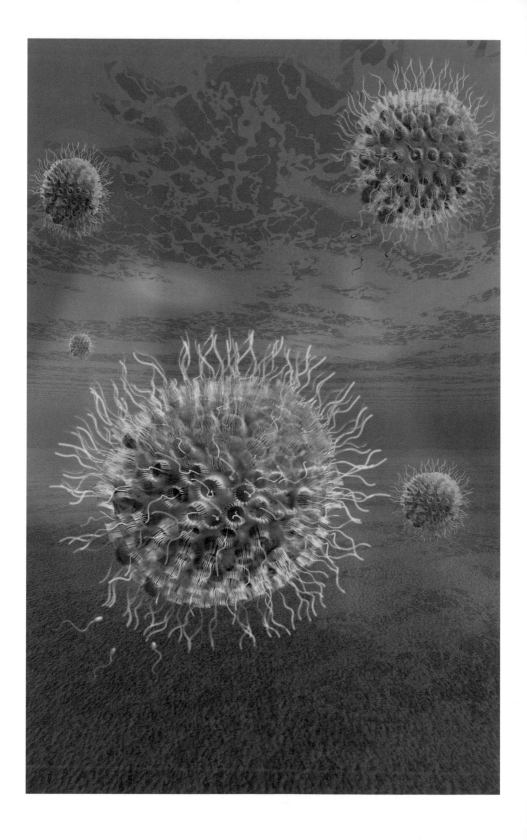

46. Concestor 31

This is thought to have been a ball of outward-pointing choanocytes (see the Sponge's Tale), collecting bacteria in their collars by waving their hair-like cilia. These multicellular animals reproduced sexually, and this reconstruction shows free-swimming sperm and eggs embedded in the colony (see page 548).

47. Venus's flower basket

Detail of spicule skeleton of the glass sponge, *Euplectella aspergillum* (see page 548).

48. Orgy of mushrooms

Stinkhorn mushroom, *Phallus impudicus*, a basidiomycete (see page 565).

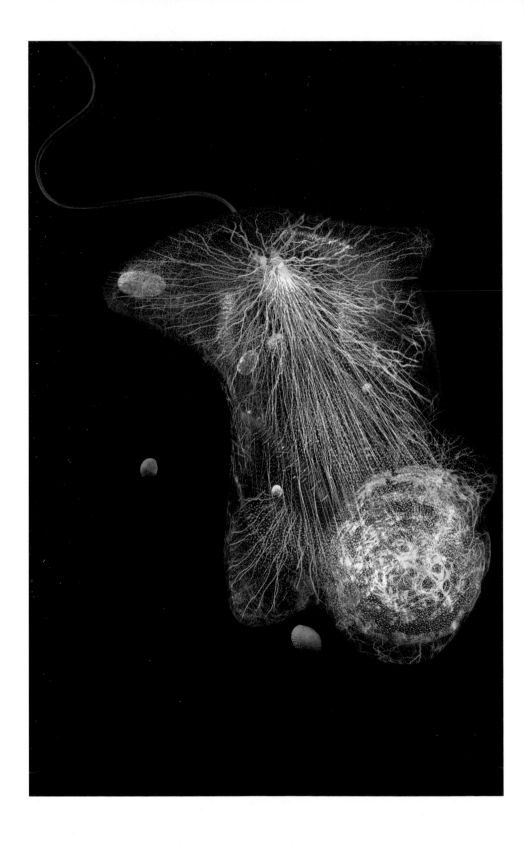

49. Concestor 38

The single-celled eukaryote shows a nucleus at bottom right, surrounded by sheets of endoplasmic reticulum. The cell structure is maintained by a cytoskeleton (the network of white threads). The concestor would probably have moved both by using its whip-like flagellum and by extending parts of its body (see page 616).

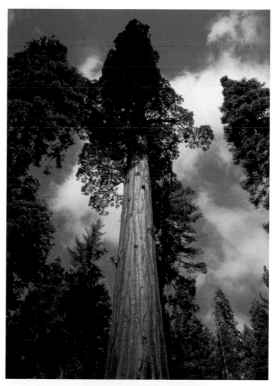

50. If you've seen one . . .

Sequoiadendron giganteum, Sequoia National Park, California (see page 589).

51. Free to tie itself in knots

Computer graphic of transfer RNA, paired with itself to make a miniature double helix (see page 659).

52. Almost indecently eager to evolve eyes

Some examples of eyes. *Clockwise from top left: Nautilus pompilius* (pinhole eye); fossil trilobite (*Phacops*, compound eye made of calcite lenses – some of these can be seen in position in the top part of the eye); simulian blackfly (*Simulium damnosum*, compound eye); stoplight parrotfish (*Sparisoma viride*, fish eye); great horned owl (*Bubo virginianus*, corneal eye) (see page 673).

dinosaur/bird, or 'sauropsid', side. These two must have been all but identical to each other. In fact there must have been a time when they could hybridise with each other. Yet the strict cladist would insist on calling one of them a sauropsid and the other one a mammal. Fortunately we don't often reach such a *reductio* in practice, but such hypothetical cases are good to quote when cladistic purists start getting above themselves.

We are so used to the idea of mammals as successors to the dinosaurs, that we may find it surprising that the mammal-like reptiles flourished before the rise of the dinosaurs. They filled the same range of niches as the dinosaurs were later to fill, and as the mammals themselves were to fill even later still. Actually they filled those niches not once but several times in succession, separated by large-scale extinctions. In the absence of milestones supplied by rendezvous with living pilgrims, I shall recognise three shadowy milestones to bridge the gap between the shrew-like Concestor 15 (which unites us to the monotremes) and the lizard-like Concestor 16 (which unites us to birds and dinosaurs). My colleague Tom Kemp is one of our leading authorities on the mammal-like reptiles, and his picture of some of them and their relationships is overleaf.

Your 150-million-greats-grandmother might have been something a bit like a *Thrinaxodon*, which lived in the Middle Triassic and whose fossils have been found in Africa and Antarctica, then joined to each other within Gondwana. It is too much to hope that it was *Thrinaxodon* itself, or any other particular fossil that we happen to have found. *Thrinaxodon*, like any fossil, should be thought of as a cousin of our ancestor, not the ancestor itself. It was a member of a group of mammal-like reptiles called the cynodonts. The cynodonts were so mammal-like, it is tempting to call them mammals. But who cares what we call them? They are almost perfect intermediates. Given that evolution has happened, it would be weird if there were not intermediates like the cynodonts.

The cynodonts were among several groups that radiated from an earlier group of mammal-like reptiles called the therapsids. Your 160-million-greats-grandfather was probably a therapsid, living in the Permian Period, but it is hard to pick out a particular fossil to represent it. The therapsids dominated the land trades before the dinosaurs arrived in the Triassic Period, and even in the Triassic itself they gave the dinosaurs a run for their money. They included some huge animals: herbivores three metres long, with large and probably

Before the dinosaurs. The phylogenetic relationships of the mammal-like reptiles. adapted from Tom Kemp [218].

Did it eat your 160-million-greats-grandfather? Gorgonopsid skull. Drawing by Tom Kemp.

ferocious carnivores to prey on them, including the gorgonopsids (see above), whose fearsome canine teeth make one think of the sabretoothed cats and marsupials of later times. But our therapsid ancestor was probably a smaller and more insignificant creature. It seems to be a rule that large or specialised animals, such as this fanged gorgonopsid or the tusk-bearing herbivorous dicynodonts (see opposite), don't have a long-term evolutionary future but belong to the 99 per cent of species destined for extinction. The Noah species, the 1 per cent from which we later animals are all descended – whether we ourselves are large and spectacular in our own time or not – tend to be smaller and more retiring.

The early therapsids were a bit less mammal-like than their successors, the cynodonts, but more mammal-like than their predecessors, the pelycosaurs, who constituted the early radiation of mammal-like reptiles. Before the therapsids, your 165-million-greats-grandmother was almost certainly a pelycosaur although, once again, it would be foolhardy to attempt to single out a particular fossil for that honour. The pelycosaurs were the earliest wave of mammal-like reptiles. They flourished in the Carboniferous Period, when the great coalfields were being laid down. The best-known pelycosaur is *Dimetrodon*, the one with the great sail on its back, which grown-ups often mistake for a dinosaur. Nobody knows how *Dimetrodon* used its sail. It may have been a solar panel to help the animal warm up to a temperature where it could use its muscles, and/or perhaps it was a radiator to cool down in the shade, when things got too hot. Or it could have been a sexual advertisement, a bony equivalent of a peacock's fan. The pelycosaurs mostly went extinct during the Permian – all except for the Noah-pelycosaurs who sprouted the second wave of mammal-like reptiles, the therapsids. The therapsids then spent the early part of the Triassic Period 'reinventing many of the lost body forms of the Late Permian'.*

The pelycosaurs were considerably less mammal-like than the therapsids, which in turn were less mammal-like than the cynodonts. For example, the pelycosaurs sprawled on their bellies like lizards, with legs splayed out sideways. Like modern echidnas, they probably had a fish-like wiggle to their gait. The therapsids, and then the cynodonts and finally the mammals, raised their bellies progressively higher off the ground – their legs became more vertical and their gait less reminiscent of a fish on land. Other 'mammalisation' trends – perhaps recognised as progressive only with the hindsight of the mammals that we are – include the following. The lower jaw became reduced to a single bone, the dentary, as its other bones were commandeered by the ear (as discussed at Rendezvous 15). At some point, though fossils can't always pin it down, our ancestors developed hair and a thermostat, milk and advanced parental care, and complex teeth specialised for different purposes.

I have dealt with the evolution of our mammal-like reptile anchors – 'shadow pilgrims' – as three successive waves: pelycosaurs, therapsids and cynodonts. The mammals themselves are the fourth wave, but their evolutionary invasion into the familiar range of ecotypes was postponed 150 million years. First, the dinosaurs had to have their go, which lasted twice as long as all three waves of mammal-like reptiles put together.

On our backwards march, the earliest of our three groups of 'shadow pilgrims' have brought us to a rather lizard-like pelycosaur 'Noah', our 165-million-greats-grandparent, who lived in the Triassic Period, about 300 million years ago. We have almost penetrated back to Rendezvous 16.

---

* This happy turn of phrase is from *Mass Extinctions and Their Aftermath* by A. Hallam and P. B. Wignall.

SAUROPSIDS

Concestor 16, our approximately 170-million-greats-grandparent would have looked superficially like some sort of lizard (see plate 19). She lived some 320 million years ago in the second half of the Carboniferous, a time of vast swamps of giant club moss trees in the tropics (the origin of most coal) and an extensive ice cap at the South Pole. This rendezvous point is where a huge throng of new pilgrims joins us: the sauropsids. The sauropsids are by far the largest contingent of newcomers we have yet had to deal with along our Pilgrims' Way. For most of the years since Concestor 16 lived, sauropsids, in the form of dinosaurs, dominated the planet. Even today, with the dinosaurs gone, there still are almost four times as many sauropsid species as mammals. At Rendezvous 16, approximately 5,500 mammal pilgrims greet 20,000 sauropsids: snakes, lizards, turtles, crocodiles and, above them all, birds. They are the main group of land vertebrate pilgrims. The only reason I am regarding them as joining us, rather than we joining them, is that we arbitrarily chose to see the journey through human eyes.

Seen through sauropsid eyes, these pilgrims belong to two near equal-sized bands: the lizard-like reptiles or lepidosaurs, and the dinosaur-like reptiles or archosaurs.

The lizard-like reptiles number just under 10,000 living species, including the iguanas, Komodo dragons, snakes, wall lizards, skinks, geckos and tuataras. Swimming alongside them are two or three extinct marine groups: the mosasaurs, the plesiosaurs and (possibly) the ichthyosaurs. These three groups independently took to the oceans, evolving in each case into fearsome predators whose largest members outweigh even the biggest carnivorous dinosaurs.

The dinosaurs themselves belong to the archosaurs, the other major branch of the sauropsids. Today, this branch is dominated by the 10,000 species of birds, although it also includes the crocodiles and (controversially) the turtles. The extinct forms include the dinosaurs of course, but also other major groups such as the flying pterodactyls. The word 'extinct' deserves qualification because, as any schoolchild will soon tell you, dinosaurs never entirely disappeared. In fact, they

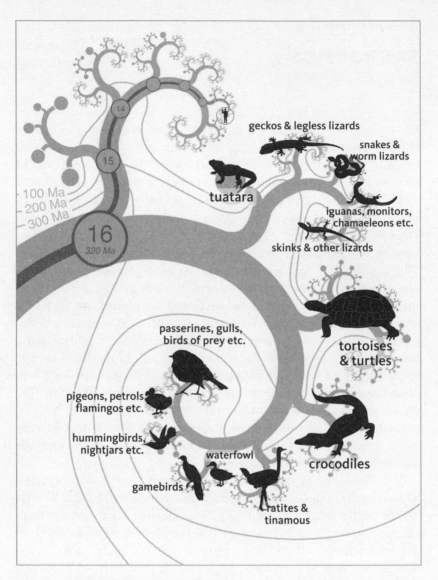

geckos & legless lizards

snakes & worm lizards

tuatara

iguanas, monitors, chamaeleons etc.

skinks & other lizards

14

15

100 Ma
200 Ma
300 Ma

16
320 Ma

tortoises & turtles

passerines, gulls, birds of prey etc.

pigeons, petrols, flamingos etc.

hummingbirds, nightjars etc.

waterfowl

crocodiles

gamebirds

ratites & tinamous

Reptiles (including birds) join. A breakthrough in the evolution of terrestrial vertebrates was the *amnion*, a waterproof yet breathable egg membrane. Two early-diverging lineages of these 'amniotes' survive today: the synapsids (represented by the mammals), and the sauropsids (20,000 living species of 'reptiles' and birds) who join us here. The phylogeny shown here is reasonably secure, save for the position of the turtles (no longer the deepest branch on the sauropsid tree, as was previously thought) and the earliest branching lizards.

became the most successful of all terrestrial (or should that be 'aerial') vertebrates. Birds are an offshoot of one particular order of dinosaurs, the saurischians. The saurischian dinosaurs, such as *Tyrannosaurus* and the freshly reinstated *Brontosaurus*, are closer to birds than they are to the other main group of dinosaurs, the unfortunately named ornithischians* such as *Iguanodon, Triceratops*, and the duckbilled hadrosaurs.

The relationship of birds to saurischian dinosaurs is made secure by recent spectacular finds of feathered dinosaurs in China. Tyrannosaurs (a few species of which certainly sprouted feathers of some sort) are closer cousins to birds than they are even to other saurischians such as the large plant-eating sauropods *Diplodocus* and *Brachiosaurus*.

These, then, are the sauropsid pilgrims, the lizards and snakes, turtles, crocodiles and birds, together with the huge concourse of shadow pilgrims – the pterosaurs in the air, the ichthyosaurs, plesiosaurs and mosasaurs in water, and most famously of all the dinosaurs on land. Focused as this book is on pilgrims from the present, it is not appropriate to expatiate on the dinosaurs, who dominated the planet for so long, and who would dominate it yet, but for the cruel – no, indifferent – bolide that laid them low. It seems added cruelty to treat them now so indifferently.† They do survive after a fashion – the special and beautiful fashion of birds – and we shall do them homage by rounding off this rendezvous with four tales of birds. But first, *in memoriam*, Shelley's well-known 'Ode to a Dinosaur':

> *I met a traveller in an antique land*
> *Who said: 'Two vast and trunkless legs of stone*
> *Stand in the desert . . . Near them, on the sand,*
> *Half-sunk, a shattered visage lies, whose frown,*
> *And wrinkled lip, and sneer of cold command,*
> *Tell that its sculptor well those passions read*
> *Which yet survive, stamped on these lifeless things,*
> *The hand that mocked them and the heart that fed:*
> *And on the pedestal these words appear:*

* Unfortunate because 'ornithischian' means 'bird-hipped', but the resemblance is superficial and confusing.

† Other books have done them proud, for example David Norman's *Dinosaur!* and Robert Bakker's *The Dinosaur Heresies*, not forgetting Robert Mash's delightfully affectionate and witty *How to Keep Dinosaurs*.

"*My name is Ozymandias, king of kings:*
Look on my works, ye Mighty, and despair!"
Nothing beside remains. Round the decay
Of that colossal wreck, boundless and bare
The lone and level sands stretch far away.'

THE LAVA LIZARD'S TALE*

A guide at the Natural History Museum stated confidently that a particular dinosaur was 70,000,008 years old. When asked how he could be so precise he replied, 'Well it was 70 million when I started this job, and that was eight years ago.' The exact date of the great Santiago eruption, on the Galapagos archipelago, is not recorded, but it definitely happened on one particular day in one particular year around 1900. I shall call it SV day (Santiago volcano day). I need to seem as precise as the museum guide, although the exact date doesn't matter. Perhaps it was 19 January 1897, 100 plus eight years before my visit to the island in January 2005.

SV day was one day in the late nineteenth century, a day on which, elsewhere in the world, somebody's grandfather was born at some particular hour. Somebody else died. A moustachioed young man in a straw boater met his true love for the first time and was never the same again. Like every day that has ever been, it was a unique day. Every second of it. It also was the date of the great Santiago volcano, the one that made the lava fields that I walked that January in the company of lava lizards, *Microlophus albemarlensis*, although I knew it only when they moved and betrayed their camouflage (see plate 18).

Lava lizards are pretty much the only things that do move over these barren fields of black, clinker-ringing rock. And as they do so their splayed hands are feeling – though they do not know it – the fingerprints of past time. Fingerprints? Past time? Wait, that is the theme of the Lava Lizard's Tale.

Santiago was one of the four Galapagos islands on which Charles Darwin landed in 1835, and it was the only one where he spent any time, camping for a week while Captain FitzRoy took the *Beagle* to fetch fresh supplies. Darwin knew it as James, for he and his shipmates

* This tale was one of three written for the *Guardian* newspaper when I was on a boat called The Beagle in Galapagos waters in January 2005.

used the English names of all the islands: the evocative Chatham, Hood, Albemarle, Indefatigable, Barrington, Charles and James. He and his small camping party had trouble finding a clear spot to pitch their tent, so thickly did the land iguanas carpet the ground. Today there are no land iguanas left on Santiago. Feral dogs, pigs and rats did for them, although there are still plenty of land iguanas on other islands of this iconic archipelago, while the closely related marine iguanas abound on all the major islands including Santiago.

The black lava fields of Santiago are an unforgettable – almost indescribable – spectacle. Black as a female marine iguana (of course the simile really should go the other way), the rock is called rope lava, and you can soon see why. It is drawn out and plaited in twisted ropes and pleats, folded and gathered like a black silk dress, coiled and whorled in giant fingerprints. Fingerprints, yes, and that brings me to the point of this tale. As the lizard scuttles over the black lava of Santiago it is treading the fingerprints of history, rolled out by the sequence of particular events that transpired, minute by minute, on one particular day late in Darwin's century, marking the minutes of that day, the day of the Santiago volcano.

There cannot be many other ways to see, laid out before you, a complete history, second by second, of one particular day, more than a century ago. Fossils do the same thing but over a much longer timescale. The molecules of a fossil are not the original molecules of the animal that died. Even fossil tracks, like those Mary Leakey found at Laetoli, don't really do it. It is true that Laetoli shows you the exact places where two individual *Australopithecus afarensis* (those diminutive hominids carrying chimpanzee brains around on human legs), perhaps a mated couple, placed their feet during a particular walk together. There is a sense in which these footprints are frozen history, but the rock that you see today is not as it was then. That family walked in fresh volcanic ash which later, over thousands of years, solidified and compacted to make rock. The lava ropes and pleats of Santiago, those giants' fingerprints, are still composed of the very same molecules that were frozen into precisely those positions, only a century ago. And the timescale over which the distinct ropes and pleats were laid down is a timescale of seconds.

As we shall see in the Redwood's Tale, tree rings do it on a timescale of years. Where the whorls of lava fingerprinting are laid down second by second, and fossils are laid down by the millions of years, each tree ring marks exactly one year. Thick rings or thin label good growth years or poor and, because every sequence of half a dozen years or so has its own characteristic pattern of good and poor years, the patterns can be recognised, again and again in different trees, as labels of particular clusters of years. Old trees and young trees show the same fingerprints so, by counting rings and daisy-chaining the patterns from increasingly ancient wooden relics, archaeologists can compile a catalogue of fingerprints outspanning the longest-lived tree.

As the Redwood's Tale also explains, something similar can be done with sediment patterns laid down on the sea bottom and revealed in cores of mud taken up in deep sampling tubes. And, over the longer timespan of hundreds of millions of years, the named strata of the geological series are, in their own way, fingerprints of time. What is so remarkable about the lava fields of Santiago is that these fingerprints were set out on the timescale that we humans deal with every second of our lives, the timescale of musical notes, the timescale of an artist's brush, the timescale of everyday actions and the stream of human thought.

This is a real thought for a surreal landscape. And the Galapagos islands are replete with images that could have come straight from a surrealist's canvas. A tiny desert island off Santa Fe (Barrington to Darwin) looks fit for Man Friday except that instead of palm trees there are giant cactuses. As if the Arizona desert had been transplanted into an azure sea; no surrealist could have done it better. And what are sea lions doing in the Arizona desert, to say nothing of shocking-pink flamingos, equatorial penguins, or flightless cormorants earnestly hanging their impotent, stubby wings out to dry? As for the large flounder that I saw when snorkelling off North Seymour Island, it was pure Salvador Dalí. I would certainly not have spotted it, changing colour to match the corals over which it slid like an oval carpet, if Valentina, our wonderful Ecuadorian guide, had not gracefully dived to point it out to me. It was only later that my wife compared the flounder to the flowing, bending watch of a Dalí painting. And wasn't that very painting, the one with the bent watches, called *The Persistence of Memory*? Not a bad title for the lava fields of Santiago, scuttling ground of the Galapagos lava lizards.

Reality, if you go to the right place, and see it in the right way, can be stranger than a surrealist's imagination. No wonder Darwin drew his inspiration from these enchanted islands.

PROLOGUE TO THE GALAPAGOS
FINCH'S TALE

. . . The lone and level sands stretch far away.

The sweep of geological time is normally so far beyond the ken of poets and archaeologists it is daunting. But geological time is large not only in comparison to the familiar timescales of human life and human history. It is large on the timescale of evolution itself. This would surprise those, from Darwin's own critics on, who have complained of insufficient time for natural selection to wreak the changes the theory requires of it. We now realise that the problem is, if anything, opposite. There has been too much time! If we measure evolutionary rates over a short time, and then extrapolate, say, to a million years, the potential amount of evolutionary change turns out to be hugely greater than the actual amount. It is as though evolution must have been marking time for much of the period. Or, if not marking time, wandering around this way and that, with meandering fluctuations drowning out, in the short term, whatever trends there might be in the long.

Evidence of various kinds, and theoretical calculations, all point towards this conclusion. Darwinian selection, if we impose it artificially as hard as we are able, can drive evolutionary change at a rate far faster than we ever see in nature. To see this, we cash in on the lucky fact that our forebears, whether they fully understood what they were doing or not, have for centuries been selectively breeding domestic animals and plants (see the Farmer's Tale). In all cases these spectacular evolutionary changes have been achieved in no more than a few centuries or, at most, millennia: far faster than even the fastest evolutionary changes that we can measure in the fossil record. No wonder Charles Darwin made much of domestication in his books.

We can do the same thing under more controlled experimental conditions. The most direct test of a hypothesis about nature is an *experiment*, in which we deliberately and artificially mimic the crucial element of nature in the hypothesis. If you have a hypothesis that, say, plants grow better in soil that contains nitrates, you don't just analyse soils to see if there are nitrates there. You experimentally *add* nitrate to some soils but not others. So it is with Darwinian selection. The hypothesis about nature is that non-random survival over generations leads to a systematic shift in average form. The experimental test is to engineer just such non-random survival, in an attempt to steer evolution in some desired direction. That is what artificial selection is. The neatest experiments simultaneously select two lines in opposite directions from the same starting point: say one line making larger animals and the other making smaller ones. Obviously if you want to get decent results before dying of old age you must choose a creature with a faster life cycle than your own.

Fruit flies and mice measure their generations in weeks and months, not decades as we do. In one experiment, *Drosophila* fruit flies were split into two 'lines'. One line was bred, over several generations, for a positive tendency to approach light. In each generation, the most strongly light-seeking individuals were allowed to breed. The other line was bred systematically in the opposite direction, over the same number of generations, for a tendency to shun light. In a mere 20 generations, dramatic evolutionary change was achieved, in both directions. Would the divergence go on for ever at the same rate? No, if only because the available genetic variation would eventually run out and we'd have to wait for new mutations. But before this happens, a great deal of change can be achieved.

Maize has a longer generation time than *Drosophila*. But in 1896 the Illinois State Agricultural Laboratory started breeding for oil content in maize seeds. A 'high line' was selected for increased oil content, and a low line simultaneously selected for decreased oil (see opposite). Fortunately this experiment has been continued far longer than the research career of any normal scientist, and it is possible to see, over 90 or so generations, an approximately linear increase in oil content in the high line. The low line has decreased its oil content less rapidly, but that is presumably because it is hitting the floor of the graph: you can't have less oil than zero.

This experiment, like the *Drosophila* one and like many others of the same type, brings home the potential power of selection to drive evolutionary change very fast indeed. Translate 90 generations of maize, or 20 generations of *Drosophila*, even 20 elephant generations, into real time, and you have something that is still negligible on the geological scale. One million years, which is too short to notice in most parts of the fossil record, is 20,000 times as long as it takes to triple the oil content of maize seeds. Of course this doesn't mean a million years of selection could multiply the oil content by 60,000. Quite apart from running out of genetic variation, there's a limit to how much oil a maize seed can pack in. But these experiments serve

The power of selection. Impact of selection for high and low oil content in maize seeds over 90 generations. Adapted from Dudley and Lambert [117].

to warn against looking at apparent trends spread over millions of fossil years, and naively interpreting them as responses to steadily sustained selection pressures.

Darwinian selection pressures are out there, for sure. And they are immensely important, as we shall see throughout this book. But selection pressures are not sustained and uniform over the sort of timescales that can normally be resolved by fossils, especially in older parts of the fossil record. The lesson of the maize and the fruit flies is that Darwinian selection could meander hither and yon, back and forth, ten thousand times, all within the shortest time we can measure in the record of the rocks. My bet is that this happens.

Yet there are major trends over longer timescales, and we have to be aware of them too. To repeat an analogy I have used before, think of a cork, bobbing about off the Atlantic coast of America. The Gulf Stream imposes an overall eastward drift in the average position of the cork, which will eventually be washed up on some European shore. But if you measure its direction of movement during any one minute, buffeted by waves and eddies and whirlpools, it will seem to move west as often as east. You won't notice any eastward bias unless you sample its position over much longer periods. Yet the eastward bias is real, it is there, and it too deserves an explanation.

The waves and eddies of natural evolution are usually too slow for us to see in our little lifetimes, or at least within the short compass of a typical research grant. There are a few notable exceptions. The school of E. B. Ford, the eccentric and fastidious scholar from whom my generation of Oxford zoologists learned our genetics, devoted decades of research to tracking the year-by-year fortunes of particular genes in wild populations of butterflies, moths and snails. Their results in some cases seem to have straightforward Darwinian explanations. In other cases the noise of buffeting waves drowns out the signal of whatever Gulf Streams may have been tugging the undertow, and the results are enigmatic. The point I am now making is that such enigma is to be expected by any mortal Darwinian – even a Darwinian with a research career as long as Ford's. One of the main messages Ford himself drew from his life's work was that the selection pressures actually to be found in nature, even if they don't always pull in the same direction, are orders of magnitude stronger than anything dreamed of by the most optimistic founders of the neo-Darwinian revival. And this again underlines the point: why doesn't evolution go much faster than it does?

THE GALAPAGOS FINCH'S TALE

The Galapagos archipelago is volcanic, and no more than 10 million years old. During that brief existence, a spectacular quantity of diversity has evolved – most famously among the 14 species of finches widely, though wrongly, believed to have been Darwin's principal inspiration.* The Galapagos finches are among the most

---

* Stephen Gould discusses Darwin's alleged epiphany in 'Darwin at sea – and the virtues of port', one of the essays collected in *The Flamingo's Smile*. It is worth noting that isolated islands are renowned for generating diversity. Hawaii is an even more remote volcanic archipelago, whose Robinson Crusoe bird was a honey-creeper. Her descendants rapidly evolved to 'do a Galapagos' – even evolving a 'woodpecker'. Similarly, about 400 original insect immigrant species spawned all 10,000 endemic Hawaiian species, including a unique carnivorous caterpillar and a semi-marine cricket. Apart from a bat and a seal, there are no native Hawaiian mammals. Alas, to quote E. O. Wilson's beautiful book, *The Diversity of Life*, 'Most of the honeycreepers are gone now. They retreated and vanished under pressure from overhunting, deforestation, rats, carnivorous ants, and malaria and dropsy carried in by exotic birds introduced to "enrich" the Hawaiian landscape.'

thoroughly studied wild animals in existence. Peter and Rosemary Grant have devoted their professional lives to following the year-by-year fortunes of these small island birds. And in the years between Charles Darwin and Peter Grant (who himself bears a pleasing facial resemblance to Darwin) the great (but clean-shaven) ornithologist David Lack also paid them a perceptive and productive visit.[*]

The Grants and their colleagues and students have been returning yearly to the Galapagos Islands for almost half a century, trapping finches, individually marking them, measuring their beaks and wings, and more recently taking blood samples for DNA analysis to draw family trees and track genes under selection. There has probably never been a more complete study of the individuals and genes of any wild population. The Grants know in minute detail exactly what is happening to the bobbing corks which are the finch populations, as they are tossed this way and that in the sea of evolution by selection pressures that change every year.

In 1977 there was a severe drought, and the food supply plummeted. The total number of individual finches of all species on the small island of Daphne Major dropped from 1,300 in January to less than 300 by December. The population of the dominant species, *Geospiza fortis*, the medium ground finch, dropped from 1,200 to 180. The cactus finch, *G. scandens*, fell from 280 to 110. Figures for other species confirmed that 1977 was a finch *annus horribilis*. But the Grant team didn't just count the numbers of each species dying and living. Being Darwinians, they looked at the *selective* mortality figures within each species. Were individuals with certain characteristics more likely to survive the catastrophe than others? Did the drought selectively change the relative composition of a population?

Yes, it did. Within the *G. fortis* population, the survivors were on average more than 5 per cent larger than those that succumbed. And the average beak after the drought was 11.07 mm long compared with 10.68 mm before. The mean depth of beak had similarly gone up from 9.42 mm to 9.96 mm. These differences may seem tiny but, within the sceptical conventions of statistical science, they were too consistent to be due to chance. But why would a drought year favour such changes? The team already had evidence that larger birds with

---

[*] See his 1947 book *Darwin's Finches*. Peter Grant's own classic monograph of 1986, *Ecology and Evolution of Darwin's Finches*, was republished in 1999. The findings have been updated by Peter and Rosemary Grant in their beautiful 2014 book *40 Years of Evolution: Darwin's Finches on Daphne Major Island*.

larger beaks are more efficient than average birds at dealing with the big, tough, spiky seeds such as those of the weed *Tribulus*, which were just about the only seeds to be found during the worst of the drought. A different species, the large ground finch *G. magnirostris*, is the professional when it comes to handling *Tribulus* seeds. But Darwinian survival of the fittest is all about the relative survival of individuals within a species, not the relative survival of one species compared to another. And within the population of medium ground finches, the largest individuals with the largest beaks survived best. The average *G. fortis* individual became a tiny bit more like *G. magnirostris*. The Grant team had observed a small episode of natural selection in action, during a single year.

They witnessed another episode after the drought ended, which pushed the finch populations in the same evolutionary direction, but for a different reason. As with many species of bird, *G. fortis* males are larger than females, and they have larger beaks, which presumably equipped them to survive the drought better. Before the drought there were about 600 males and 600 females. Of the 180 individuals who survived, 150 were male. The rains, when they finally returned in January 1978, unleashed boom conditions which were ideal for breeding. But now there were five males for every female. Understandably, there was fierce competition among the males for the scarce females. And the males who won these sexual competitions, the new winners among the already larger-than-normal surviving males, again tended to be the largest males with the largest beaks. Once again, natural selection was driving the population to evolve larger body size and larger beaks, but for a different reason. As to why females prefer large males, the Seal's Tale has primed us to see significance in the fact that male *Geospiza* – the more competitive sex – are larger than females anyway.

If large size is such an advantage, why weren't the birds just larger in the first place? Because in other years, non-drought years, natural selection favours smaller individuals with smaller beaks. The Grants actually witnessed this in the years following 1982–83 when there happened to be an El Niño flood. After the flood, the balance of seeds changed. The large tough seeds of plants such as *Tribulus* became rare in comparison with the smaller, softer seeds of plants like *Cacabus*. Now smaller finches with smaller beaks came into their own. It wasn't that large birds couldn't eat small, soft seeds. But they needed more of them to maintain their larger bodies. So smaller birds now had a slight edge. And, within the population of medium ground finches, the tables were turned. The evolutionary trend of the drought years was reversed.

The differences in beak size between the successful and the unsuccessful birds in the drought year seem awfully small, don't they? Jonathan Weiner in his excellent book on the Grant's work, *The Beak of the Finch*, quotes a telling anecdote about this, from Peter Grant:

> Once, just as I was beginning a lecture, a biologist in the audience interrupted me: 'How much difference do you claim to see,' he asked me, 'between the beak of a finch that survives and the beak of a finch that dies?'
>
> 'One half of a millimetre, on average,' I told him.
>
> 'I don't believe it!' the man said. 'I don't believe a half of a millimetre really matters so much.'
>
> 'Well, that's the fact,' I said. 'Watch my data and then ask questions.' And he asked no questions.

Peter Grant calculated that it would take only 23 bouts of 1977-style drought on Daphne Major to turn *Geospiza fortis* into *G. magnirostris*. It wouldn't literally be *magnirostris*, of course. But it is a vivid way to visualise the origin of species, and how rapidly it can happen.

Of course, these visible changes shadow the changes going on at the level of DNA. And one of the exciting aspects of living in the genomic era is that interesting and important animals tend to get their genomes sequenced. Dogs and duck-billed platypuses, chimpanzees and coelacanths have all been honoured in the past 10 years. It is no surprise that Darwin's famous finches have now been added to the list.

In 2015, a study by scientists from Uppsala University, in collaboration with the Grants, sequenced the entire genomes of 120 Galapagos finches from various species. They located a number of DNA regions that co-varied with beak shape and homed in on a gene named *ALX1*, which was already known to influence facial structure in humans. How neat it is that they found the same gene influencing beak shape in finches! The *ALX1* gene tree, deduced from Galapagos finch populations, shows branching patterns indicative of natural selection. The gene tree also reveals that influential variants of *ALX1* have been passed between different finch species by hybridisation. These variants originated hundreds of thousands of years ago. But as the Grants' meticulous fieldwork reveals, their frequency in different populations can change incredibly quickly, sweeping back and forth in the population within human lifetimes, responding to the fickle pressures of natural selection.

Today's study of finches on the Galapagos islands sets a beautiful example of how it is possible to marry natural history, ecology, evolution and genetics, to illuminate the workings of nature. Darwin little knew, when he met them and failed to label them properly, what powerful allies 'his' finches would eventually turn out to be.

THE PEACOCK'S TALE

The peacock's 'tail' is not its true morphological tail (the true tail of a bird is the diminutive 'parson's nose'), but a 'fan' made of long back feathers. The Peacock's Tale is exemplary for this book because, in true Chaucerian style, it carries a message or moral from one pilgrim, which helps other pilgrims to understand themselves. In particular, when I was discussing two of the major transitions in human evolution, I looked forward to when the peacock would join our pilgrimage and give us the benefit of his (and I mean his in this case, not her) tale. It is, of course, a tale of sexual selection. Those two hominid transitions were our shift from four legs to two, and the subsequent enlargement of our brain. Let's add a third, perhaps less important but very characteristically human feature: our loss of body hair. Why did we become the Naked Ape?

There were lots of ape species in Africa in the late Miocene. Why did one of them suddenly and rapidly start evolving in a very different direction from the rest – indeed from the rest of the mammals? What picked out this one species and sent it hurtling at high speed in new and strange evolutionary directions: first to become bipedal, then to become brainy, and at some point to lose most of its body hair?

Rapid, apparently arbitrary spurts of evolution in quirky directions say one thing to me: sexual selection. This is where we have to start listening to the peacock. Why does the peacock have a train that dwarfs the rest of its body, quivering and shimmering in the sun with glorious eye-spot motifs of royal purple and green? Because generations of peahens have chosen peacocks who flaunted ancestral equivalents of these extravagant advertisements. Why does the male 12-wired bird of paradise have red eyes and a black ruff with an iridescent green fringe, while Wilson's bird of paradise catches the eye with a scarlet back, yellow neck and blue head? Not because something in their respective diets or habitats predisposes these two species to their different colour schemes. No, these differences, and those that so conspicuously mark out all the other species of bird of paradise, are arbitrary, whimsical, unimportant to anybody – except female birds of paradise. Sexual selection does this kind of thing. Sexual selection produces quirky, whimsical evolution that runs away in apparently arbitrary directions, feeding on itself to produce wild flights of evolutionary fancy.

On the other hand, sexual selection also tends to magnify differences between the sexes – sexual dimorphism (see the Seal's Tale). Any theory that attributes human brains, bipedality or nakedness to sexual selection has got to face up to a major difficulty. There is no evidence that one sex is brainier than the other, nor that one sex is more bipedal than the other. It is true that one sex tends to be more naked than the other, and Darwin made use of this in his own sexual selection theory of the loss of human hair. He supposed that ancestral males chose females rather than the other way around as is normal in the animal kingdom, and that they preferred hairless females. When one sex evolves ahead of the other (in this case the female sex towards hairlessness) the other sex can be thought of as 'dragged in its wake'. It's the kind of explanation we more or less have to offer for that old chestnut, male nipples. It is not implausible to invoke it for the evolution of partial nakedness in man, dragged in the wake of more total nakedness in woman. The 'dragged in its wake' theory works less well for bipedality and braininess. The mind boggles – recoils even – when trying to imagine a bipedal member of one sex walking out with a quadrupedal member of the other. Nevertheless, the 'dragged in its wake' theory has a role to play.

There are circumstances in which sexual selection can favour monomorphism. My own suspicion, and Geoffrey Miller's in *The Mating Mind*, is that human mate choice, unlike perhaps that of peafowl, goes both ways. Moreover, our criteria for choice may be different when we look for a long-term partner than when we seek a one-night stand.

For the moment, we return to the simpler world of peacocks and peahens where females do the choosing and males strut around and aspire to be chosen. One version of the idea assumes that choice of mate (in this case choice by peahens) is arbitrary and whimsical compared with, for example, choice of food or choice of habitat. But you could reasonably ask why this should be so. According to at least one influential theory of sexual selection, that of the great geneticist and statistician R. A. Fisher, there is a very good reason. I have expounded the theory in detail in another book (*The Blind Watchmaker*, Chapter 8) and will not do so again here. The essential point is that male appearance and female taste evolve together in a kind of explosive chain reaction. Innovations in the consensus of female taste within a species, and corresponding changes in male appearance, are amplified in a runaway process which drives both of them in lockstep, further and further in one direction. There is no overweening reason for this one direction to be chosen: it just happens to be the direction in which the evolutionary trend started. The ancestors of peahens happened to take a step in the direction of preferring a larger fan. That was enough for the explosive engine of sexual selection. It kicked in and, within a very short time by evolutionary standards, peacocks were sprouting larger and more iridescent fans, and females couldn't get enough of them.

Every species of bird of paradise, many other birds, and fish and frogs, beetles and lizards, zoomed off in their own evolutionary directions, all bright colours or weird shapes – but different bright colours, different weird shapes. What matters for our purpose is that sexual selection, according to a sound mathematical theory, is apt to drive evolution to take off in arbitrary directions and push things to non-utilitarian excess. The suggestion arose in the chapters on human evolution that this is just what the sudden inflation of the brain looks like. So does the sudden loss of body hair, and even the sudden take-off into bipedality.

Darwin's *Descent of Man* is largely devoted to sexual selection. His lengthy review of sexual selection in non-human animals prefaces his advocacy of sexual selection as the dominant force in the recent evolution of our species. His treatment of human nakedness begins by dismissing – more glibly than his modern followers find comfortable – the possibility that we lost our hair for utilitarian reasons. His faith in sexual selection is reinforced by the observation that in all races, however hairy or however hairless, the women tend to be less hairy than the men. Darwin believed that ancestral men found hairy women unattractive. Generations of men chose the most naked women as mates.* Nakedness in men was dragged along

---

* Of course, like my colleague Desmond Morris, I am using 'naked' to mean hairless rather than unclothed.

in the evolutionary wake of nakedness in women, but never quite caught up, which is why men remain hairier than women.

For Darwin, the preferences that drove sexual selection were taken for granted – given. Men just prefer smooth women, and that's that. Alfred Russel Wallace, the co-discoverer of natural selection, hated the arbitrariness of Darwinian sexual selection. He wanted females to choose males not by whim but on merit. He wanted the bright feathers of peacocks and birds of paradise to be tokens of underlying fitness. For Darwin, peahens choose peacocks simply because, in their eyes, they are pretty. Fisher's later mathematics put that Darwinian theory on a sounder mathematical footing. For Walla-ceans,* peahens choose peacocks not because they are pretty but because their bright feathers are a token of their underlying health and fitness.

In post-Wallace language, a Wallacean female is, in effect, reading a male's genes by their external manifestations from which she judges their quality. And it is a startling consequence of some sophisticated neo-Wallacean theorising that males are expected to go out of their way to make it easy for females to read their quality, even if their quality is poor. This piece of theory – progression of theories, rather – which we owe to A. Zahavi, W. D. Hamilton and A. Grafen, would take us too far afield, interesting as it is. My best attempt at expounding it is in the endnotes to the second edition of *The Selfish Gene*.

This brings us to the first of our three questions about human evolution. Why did we lose our hair? Mark Pagel and Walter Bodmer have made the intriguing suggestion that hairlessness evolved to reduce ectoparasites such as lice and, in keeping with the theme of this tale, as a sexually selected advertisement of freedom from para-sites. Pagel and Bodmer followed Darwin's invocation of sexual selection, but in the neo-Wallacean version of W. D. Hamilton.

Darwin did not try to explain female preference, but was content to postulate it to explain male appearance. Wallaceans seek evolu-tionary explanations for sexual preferences themselves. Hamilton's favoured explanation is all about advertising health. When indivi-duals choose their mates, they are looking for health, freedom from parasites, or signs that the mate is likely to be good at evading or combating parasites. And individuals seeking to be chosen advertise

---

* Helena Cronin's term, in her wonderful book *The Ant and the Peacock*.

their health: make it easy for the choosers to read their health, whether it is good or bad. Patches of bare skin in turkeys and monkeys are conspicuous screens on which the health of their possessors is displayed. You can actually see the colour of the blood through the skin.

Humans don't just have bare skin on their rumps like monkeys. They have bare skin all over, except on the top of the head, under the arms and in the pubic region. When we get ectoparasites such as lice, they are often confined to these very regions. The crab louse, *Phthirus pubis*, is mainly found in the pubic region, but also infests the armpits, beard and even eyebrows. Disquietingly, it seems to have jumped to us from gorillas a few million years ago. The head louse, *Pediculus humanus capitus*, infests only the hairs of the head. The body louse, *P. h. humanus*, is a subspecies in the same species as the head louse which, interestingly, is believed to have evolved from it only after we began to wear clothes. Some workers in Germany have looked at the DNA of head lice and body lice to see when they diverged, with a view to dating the invention of clothes. They put it at 72,000 years, plus or minus 42,000, a date which further analysis has revised upwards to about 170,000 years ago.

Lice need hair, and Pagel and Bodmer's first suggestion is that the benefit of losing our body hair was that it reduced the real estate available to lice. Two questions arise. Why, if losing hair is such a good idea, have other mammals who also suffer from ectoparasites kept theirs? Those, such as elephants and rhinos, that could afford to lose their hair because they are large enough to keep warm without it, have indeed lost it. Pagel and Bodmer suggest that it was the invention of fire and clothes that enabled us to dispense with our hair.* This immediately leads to the second question. Why have we retained hair on our heads, under our arms and in the pubic region? There must have been some overriding advantages. It is entirely plausible that hair on the top of the head protects against sunstroke, which can be very dangerous in Africa where we evolved. As for armpit and pubic hair, it probably helps disseminate the powerful pheromones (airborne scent signals) that our ancestors certainly used in their sex lives, and which we still use more than many of us realise.

---

* If, as discussed earlier in the book, we became hairless over a million years ago, this implicates the use of fire (over a million years ago) rather than the invention of clothing (perhaps low hundreds of thousands of years ago).

So, the straightforward portion of the Pagel/Bodmer theory is that ectoparasites such as lice are dangerous (lice carry typhus and other serious diseases), and ectoparasites prefer hair to bare skin. Getting rid of hair is a good way to make life difficult for these unpleasant and dangerous parasites. It is also much easier for us to see and pick off ectoparasites like ticks if we have no hair. Primates spend a substantial amount of time doing this, to themselves and to each other. It has become, indeed, a major social activity and, as a by-product, a vehicle for bonding.

But I find the most interesting angle on the Pagel/Bodmer theory is one that they treat rather briefly in their paper: sexual selection, which is why it belongs in the Peacock's Tale. Nakedness is not only bad news for lice and ticks. It is good news for choosers trying to discover whether a would-be sexual partner *has* lice or ticks. The Hamilton/Zahavi/Grafen theory predicts that sexual selection will enhance whatever it takes to help choosers tell whether would-be mates have parasites. Hairlessness is a beautiful example. On closing the Pagel/Bodmer paper I thought of T. H. Huxley's famous words: how extremely stupid not to have thought of that.

But nakedness is a small matter. As promised, let's turn now to bipedality and brains. Can the peacock help us to understand those two larger events in human evolution – the rise onto our hind legs, and the inflation of our brain? Bipedality came first, and I shall discuss it first. In Ardi's Tale, I mentioned various theories of bipedality, including the recent squat-feeding theory of Jonathan Kingdon, which I find very convincing. I said that I was postponing my own suggestion to the Peacock's Tale.

Sexual selection, and its power to drive evolution in non-utilitarian arbitrary directions, is the first ingredient in my theory of the evolution of bipedality. The second is a tendency to imitate. The English language even has a verb, to ape, meaning to copy, although I am not sure how apt it is. Among all the apes, humans are the champion copyists, but chimpanzees do it too, and there is no reason to think australopithecines did not. The third ingredient is the widespread habit among apes generally of rising temporarily onto the hind legs, including during sexual and aggressive displays. Gorillas do it to drum on their chests with their fists. Male chimpanzees also thump their chests, and they have a remarkable display called the rain dance which involves leaping about on the hind legs. A captive chimpanzee called Oliver habitually and for preference walks on his hind legs. I have seen a film of him walking, and his stance is surprisingly erect – not a shambling totter, almost a military gait. So un-chimp-like is Oliver's walk that he has been the subject of bizarre speculations. Until DNA tests showed him to be a chimpanzee, *Pan troglodytes*, people have thought he might be a chimp/human hybrid, a chimp/bonobo hybrid, even a relict australopithecine. Unfortunately Oliver's biography is hard to piece together, and nobody seems to know whether he was taught to walk as a trick for a circus or fair-ground sideshow, or whether it is an odd idiosyncrasy: he might even be a genetic mutant. Oliver aside, orang utans are slightly better on their hind legs than chimpanzees; and wild gibbons actually run across clearings bipedally, in a style which is not very different from the way they run along branches in trees – when they are not brachiating under them.

Putting all these ingredients together, my suggestion for the origin of human bipedality is this. Our ancestors, like other apes, walked on all fours when not up in trees, but reared up on their hind legs from time to time, perhaps in something like a rain dance, or to pick fruits off low branches, or to move from one squat-feeding position to another, or to wade across rivers, or to show off their penises, or for any combination of reasons, just as modern apes and monkeys do. Then – this is the crucial additional suggestion I am adding – something unusual happened in one of those ape species, the one from which we are descended. A *fashion* for walking bipedally arose, and it arose as suddenly and capriciously as fashions do. It was a gimmick. An analogy might be found in the legend (probably false, alas) that the Spanish lisp sprang from the fashionable imitation of an admired courtier, or, in another version of the legend, a king of the Habsburg dynasty, or an infanta, who had a speech defect.

It'll be easiest if I tell the story in a sex-biased way, with females choosing males, but remember that it could have been the other way around. In my vision, an admired or dominant ape, a Miocene Oliver perhaps, gained sexual attractiveness and social status through his unusual virtuosity in maintaining the bipedal posture, perhaps in some ancient equivalent of a rain dance. Others imitated his gimmicky habit and it became 'cool', 'it', 'the thing to do' in a local area, just as local bands of chimpanzees have habits of nut-cracking or termite-fishing which spread by fashionable imitation. In my teenage years, a more than usually inane popular song had the refrain,

> *Everybody's talking*
> *'Bout a new way of walking!*

And, while this particular line was probably chosen in the service of a lazy rhyme, it is undoubtedly true that styles of walking have a kind of contagiousness and are imitated because they are admired. The boarding school that I attended, Oundle in central England, had a ritual whereby the senior boys paraded into the chapel after the rest of us were in our places. Their mutually imitated style of walking, a mixture of swagger and lumbering roll (which I now, as a student of animal behaviour and a colleague of Desmond Morris, recognise as a dominance display) was so characteristic and idiosyncratic that my father, who saw it once a term on Parents' Day, gave it a name, 'the Oundle Roll'. That socially observant writer Tom Wolfe has named a particular loose-limbed gait of American dudes, fashionable in a certain social sector, the Pimp Roll.

Returning to our imagined sequence of events among human ancestors, females in the local area of the fashion preferred to mate with males who adopted the new way of walking. They preferred them for the same reason as individuals wanted to join the fashion: because it was admired in their social group. And now the next step in the argument is crucial. Those who were especially good at the fashionable new walk would be most likely to attract mates and sire children. But this would be of evolutionary significance only if there was a genetic component to the variation in ability to do 'the walk'. And this is entirely plausible. We are talking, remember, about a quantitative shift in the amount of time spent doing an existing activity. It is unusual for a quantitative shift in an existing variable *not* to have a genetic component.

The next step in the argument follows standard sexual selection theory. Those choosers whose taste conforms to the majority taste will tend to have children who inherit, from their mothers' choice of mate, skill in walking according to the bipedal fashion. They will also have daughters who inherit their mothers' taste in males. This dual selection – on males for possessing some quality and on females for admiring the self-same quality – is the ingredient for explosive, runaway selection, according to the Fisher theory. The key point is that the precise direction of the runaway evolution is arbitrary and unpredictable. It could have been opposite. Indeed, in another local population perhaps it was in the opposite direction. An explosive evolutionary excursion, in an arbitrary and unpredictable direction, is just the kind of thing we need, if we are to explain why one group of apes (who became our ancestors) suddenly evolved in the direction of bipedality while another group of apes (the ancestors of the chimpanzees) did not. An additional virtue of the theory is that this evolutionary spurt would have been exceptionally fast: just what we need in order to explain the otherwise puzzling closeness in time of Concestor 1 and the supposedly bipedal Toumai and *Orrorin*.

Let's turn now to the other great advance in human evolution, the enlargement of the brain. The Handyman's Tale discussed various theories, and again we left sexual selection till last, postponing it until the Peacock's Tale. In *The Mating Mind*, Geoffrey Miller argues that some very high percentage of human genes, perhaps up to 50 per cent, express themselves in the brain. Yet again, for the sake of clarity, it's convenient to tell the story from one point of view only – females choosing males – but it could go the other way: or both ways simultaneously. A female who seeks a penetrating and thorough reading of the quality of a male's genes would do well to concentrate on his brain. She can't literally look at the brain, so she looks at its works. And, following the theory that males should make it easy by advertising their quality, males will not hide their mental light under a bony bushel but bring it out into the open. They will dance, sing, sweet-talk, tell jokes, compose music or poetry, play it or recite it, paint cave walls or Sistine chapel ceilings. Yes, yes, I know Michelangelo might not, as it happens, have been interested in impressing females. It is still entirely plausible that his brain was 'designed' by natural selection for impressing females, just as – whatever his personal preferences – his penis was designed for impregnating them. The human mind, on this view, is a mental peacock's tail. And the brain expanded under the same kind of sexual selection as drove the enlargement of the peacock's tail. Miller himself favours the Wallacean rather than the Fisherian version of sexual selection, but the consequence is essentially the same. The brain gets bigger, and it does so swiftly and explosively.

The psychologist Susan Blackmore, in her audacious book *The Meme Machine*, has a more radical sexual selection theory of the human mind. She makes use of what have been called 'memes', units of cultural inheritance. Memes are not genes, and they have nothing to do with DNA except by analogy. Whereas genes are transmitted via fertilised eggs (or via viruses), memes are transmitted via imitation. If I teach you how to make an origami model of a Chinese junk, a meme passes from my brain to yours. You may then teach two other people the same skill, each of whom teaches two more, and so on. The meme is spreading exponentially, like a virus. Assuming we have all done our teaching work properly, later 'generations' of the meme will not be consistently different from earlier ones. All will produce the same origami 'phenotype'.* Some junks may be more perfect than others, as some paper-folders take more trouble, say. But quality will not deteriorate gradually and progressively over the 'generations'. The meme is passed on, whole and intact like a gene, even if its detailed phenotypic expression varies. This particular example of a meme is a good analogue for a gene, specifically a gene in a virus. A manner of speaking, or a skill in carpentry, might be more dubious candidates for memes because – I am guessing – progressively later 'generations' in a lineage of imitation will probably become progressively more different from the original generation.

Blackmore, like the philosopher Daniel Dennett, believes that memes played a decisive role in the process that made us human. In Dennett's words:

> The haven all memes depend on reaching is the human mind, but a human mind is itself an artifact created when memes restructure a human brain in order to make it a better habitat for memes. The avenues for entry and departure are modified to suit local conditions, and strengthened by various artificial devices that enhance fidelity and prolixity of replication: native Chinese minds differ dramatically from native French minds, and literate minds differ from illiterate minds.†

It would be Dennett's view that the main difference between anatomically modern brains before the cultural Great Leap Forward and after it is that the latter are swarming with memes. Blackmore goes further. She invokes memes to explain the evolution of the large human brain. It can't be only memes, of course, because we are

---

* As we saw in the Beaver's Tale, phenotype normally means the external appearance by which a gene manifests itself, for example eye colour. Obviously I am here using it in an analogous sense: the visible phenotype of a meme otherwise buried in the brain, as opposed to the phenotype of a gene buried in a chromosome. This is also a good analogy for the 'self-normalising' that I mentioned in the General Prologue, under 'Renewed Relics'. See also my foreword to Blackmore's book.

† Dennett makes constructive use of the theory of memes in various places, including *Consciousness Explained* (from which the quotation is taken), *Darwin's Dangerous Idea*, *Freedom Evolves*, *Breaking the Spell* and *Intuition Pumps*.

talking about major anatomical change here. Memes may manifest themselves in the circumcised penis phenotype (which sometimes passes, in quasi-genetic fashion, from father to son), and they might even manifest themselves in body shape (think of a transmitted fashion for slimming, or elongating the neck with rings). But a doubling in brain size is another matter. This has got to come about through changes in the gene pool. So what role does Blackmore see for memes in the evolutionary expansion of the human brain? This is, again, where sexual selection comes in.

People are most apt to copy their memes from admired models. This is a fact that advertisers bet money on: they pay footballers, film stars and supermodels to recommend products – people who have no expertise to judge them. Attractive, admired, talented or otherwise celebrated people are potent meme donors. The same people also tend to be sexually attractive and therefore, at least in the sort of polygamous society in which our ancestors probably lived, potent gene donors. In every generation, the same attractive individuals contribute more than their fair share of both genes and memes to the next generation. Now Blackmore assumes that part of what makes people attractive is their meme-generating minds: creative, artistic, loquacious, eloquent minds. And genes help to make the kind of brains that are good at generating attractive memes. So, quasi-Darwinian selection of memes in the meme pool goes hand in hand with genuinely Darwinian sexual selection of genes in the gene pool. It is yet another recipe for runaway evolution.

What, on this view, is the exact role of memes in the evolutionary swelling of the human brain? I think the most helpful way to look at it is this. There are genetic variations in brains which would remain unnoticed without memes to bring them out into the open. For example, the evidence is good that there is a genetic component to variation in musical ability. The musical talent of members of the Bach family probably owed much to their genes. In a world full of musical memes, genetic differences in musical ability shine through and are potentially available for sexual selection. In a world before musical memes entered human brains, genetic differences in musical ability would still have been there, but would not have manifested themselves, at least not in the same way. They would have been unavailable for sexual, or natural, selection. Memetic selection cannot change brain size by itself, but it can bring into the open genetic variation that would otherwise have remained under cover.

This could be seen as a version of the Baldwin Effect, which we met in the Hippo's Tale.

The Peacock's Tale has used Darwin's beautiful theory of sexual selection to pick up a number of questions about human evolution. Why are we naked? Why do we walk on two legs? And why do we have big brains? I do not want to go out on a limb for sexual selection as the universal answer to all outstanding questions about human evolution. In the particular case of bipedalism, I am at least as persuaded by Jonathan Kingdon's 'squat feeding' theory. But I applaud the current vogue for giving sexual selection another serious look, after its long neglect since Darwin first proposed it. And it does provide a ready answer to the supplementary question that so often lurks behind the main questions: why, if bipedalism (or braininess or nakedness) was such a good idea for us, do we not see it in other apes? Sexual selection is good at that, because it predicts sudden evolutionary spurts in arbitrary directions. On the other hand, the lack of sexual dimorphism in braininess and in bipedality demands some special pleading. Let's leave the matter there. It needs more thought.

THE DODO'S TALE

Land animals, for obvious reasons, have a hard time reaching remote oceanic islands such as the Galapagos archipelago, or Mauritius. If, through the much invoked freak accident of rafting inadvertently on a detached mangrove, they do happen to find themselves on an island like Mauritius, an easy life is likely to open up. This is precisely because it is hard to get to the island in the first place, so the competition and the predation are usually not so fierce as on the mainland left behind. As we have seen, this is probably how monkeys and rodents arrived in South America.

If I say it is 'hard' to colonise an island, I must hasten to forestall the usual misunderstanding. A drowning individual may try desperately to reach land, but no species ever *tries* to colonise an island. A species is not the kind of entity that tries to do anything. Individuals of a species may happen, by luck, to find themselves in a position to colonise an island previously uninhabited by their kind. The individuals concerned can then be expected to take advantage of the vacuum, and the consequence may be that their species, with hindsight, is said to have colonised the island. The descendants of the species may subsequently change their ways, over evolutionary time, to accommodate the unfamiliar island conditions.

And now here is the point of the Dodo's Tale. It is hard for land animals to reach an island, but it is a lot easier if they have wings. Like the ancestors of the Galapagos finches . . . or the ancestors of the dodo, whoever they were. Flying animals are in a special situation. They don't need the proverbial mangrove raft. Their wings carry them, perhaps as a freak accident, blown on a gale, to a distant island. Having arrived on wings, they often find that they no longer need them. Especially because islands often lack predators. This is why island animals, as Darwin noted on the Galapagos, are often remarkably tame. And this is what makes them easy meat for sailors. The most famous example is the dodo, *Raphus cucullatus*, cruelly renamed *Didus ineptus* by Linnaeus, the father of taxonomy.

The very name dodo comes from the Portuguese for stupid. Stupid is unfair. When Portuguese sailors arrived on Mauritius in 1507, the abundant dodos were completely tame, and approached the sailors in a manner which cannot have been far from 'trusting'. Why would they not trust, for their ancestors had not encountered a predator for thousands of years? Alas for trust. The unfortunate dodos were clubbed to death by Portuguese, and later Dutch, sailors – even though they were deemed 'unpalatable'. Presumably it was 'sport'. Extinction took less than two centuries. As so often, it came about through a combination of killing and more indirect effects. Humans introduced dogs, pigs, rats, and religious refugees. The first three ate dodo eggs, and the last planted sugar cane and destroyed habitats.

Conservation is a very modern idea. I doubt that extinction, and what it means, entered anybody's head in the seventeenth century. I can hardly bear to tell the story of the Oxford Dodo, the last dodo stuffed in England. Its owner and taxidermist, John Tradescant, was induced to bequeath his large collection of curios and treasures to the infamous (some say) Elias Ashmole, which is why the Ashmolean Museum in Oxford is not called the Tradescantian as (some say) it should be. Ashmole's curators (some say, probably falsely) later decided to burn, as rubbish, all of Tradescant's dodo except the beak and one foot. These are now in my sometime place of work, the University Museum of Natural History, where they memorably inspired Lewis Carroll. Also Hilaire Belloc:

The Dodo used to walk around
And take the sun and air.
The sun yet warms his native ground –
The Dodo is not there!

The voice, which used to squawk and squeak
Is now for ever dumb –
Yet you may see his bones and beak
All in the Mu-se-um.

The white dodo, *Raphus solitarius*, was alleged to have met the same fate on the neighbouring island of Réunion.* And Rodriguez, the third island of the Mascarene archipelago, housed, and lost for the same reason, a slightly more distant relative, the Rodriguez solitaire, *Pezophaps solitaria*.

The ancestors of the dodos had wings. Their forebears were flying pigeons who arrived on the Mascarene Islands under their own muscle power, perhaps aided by a freak wind. Once there, they had no need to fly any more – nothing to flee – and so lost their wings. Like Galapagos and Hawaii, these islands are recent volcanic creations, none of them more than 7 million years old. Molecular evidence suggests that the dodo and solitaire probably arrived on the Mascarene Islands from the East, not from Africa or Madagascar as we might otherwise have supposed. Perhaps the solitaire did the bulk of its evolutionary divergence before it finally arrived on Rodriguez, retaining enough wing power to get there from Mauritius.

Why bother to lose the wings? They took a long time to evolve, why not hang on to them in case one day they might come in useful again? Alas (for the dodo) that is not the way evolution thinks. Evolution doesn't think at all, and certainly not ahead. If it did, the

---

* However, the astoundingly knowledgeable Sam Turvey, now at the Zoological Society of London, informs me that the white dodo almost certainly never existed: 'White dodos are figured in a few seventeenth-century paintings, and contemporary travellers made reference to large, white birds on Réunion, but the accounts are vague and possibly confused, and no raphid skeletal material is known from the island. Although the species has been given the scientific name *Raphus solitarius*, and the eccentric Japanese naturalist Masauji Hachisuka defended the occurrence of two dodo species on Réunion (which he named *Victoriornis imperialis* and *Ornithaptera solitaria*), it is more likely that the early accounts either refer to an extinct Réunion ibis (*Threskiornis solitarius*), for which skeletal material is known and which was apparently similar to the living white sacred ibis, or to immature specimens of the grey brown dodo of Mauritius. Alternatively, they may merely be the product of artistic licence.'

dodos would have kept their wings, and the Portuguese and Dutch sailors would not have had sitting targets for their cruel vandalism.

The late Douglas Adams was moved by the sad case of the dodo. In one of the episodes of *Doctor Who* that he wrote in the 1970s, the aged Professor Chronotis's college room in Cambridge serves as a time machine, but he uses it for one purpose only, his secret vice: he obsessively and repeatedly visits seventeenth-century Mauritius in order to *weep for the dodo*. Because of a strike at the BBC, this episode of *Doctor Who* was never broadcast, and Douglas Adams later recycled the haunting dodo *motif* in his novel *Dirk Gently's Holistic Detective Agency*. Call me sentimental, but I must pause for a moment – for Douglas, and for Professor Chronotis and what he wept for.

Evolution, or its driving engine natural selection, has no foresight. In every generation within every species, the individuals best equipped to survive and reproduce contribute more than their fair share of genes to the next generation. The consequence, blind as it is, is the nearest approach to foresight that nature admits. Wings might be useful a million years hence when sailors arrive with clubs. But wings will not help a bird contribute offspring and genes to the next generation, in the immediate here and now. On the contrary wings, and especially the massive breast muscles needed to power them, are an expensive luxury. Shrink them, and the resources saved can now be spent on something more immediately useful such as eggs: immediately useful for surviving and reproducing the very genes that programmed the shrinkage.

That's the kind of thing natural selection does all the time. It is always tinkering: here shrinking a bit, there expanding a bit, constantly adjusting, putting on and taking off, optimising immediate reproductive success. Survival in future centuries doesn't enter into the calculation, for the good reason that it isn't really a calculation at all. It all happens automatically, as some genes survive in the gene pool and others don't.

The sad end of the Oxford Dodo (Alice's Dodo, Belloc's Dodo) is mitigated by a happier sequel. A group of Oxford scientists in the laboratory of my colleague Alan Cooper obtained permission to take a tiny sample from inside one of the foot bones. They also obtained a thigh bone of a solitaire found in a cave on Rodriguez. These bones yielded enough mitochondrial DNA to allow detailed, letter-by-letter sequence comparisons between the two extinct birds and a wide range of living birds. The results confirm that, as long suspected, dodos were modified pigeons. It is also no surprise that, within the pigeon family, the closest relative of the dodo is the solitaire, and vice versa. What is less expected is that these two extinct flightless giants are nested deep inside the pigeon family tree. In other words, dodos are more closely related to some flying pigeons than those flying pigeons are to other flying pigeons; despite the fact that, to look at them, you'd expect all the flying pigeons to be more closely related to each other, with the dodos out on a limb. Among pigeons, the dodos are most close to the Nicobar, *Caloenus nicobarica*, a beautiful pigeon from South-East Asia. In turn, the group consisting of the Nicobar pigeon and the dodos is most closely related to the Victoria crowned pigeon, a splendid bird from New Guinea, and *Didunculus*, a rare Samoan toothbilled pigeon which looks quite like a dodo and whose name even means 'little dodo'.

The Oxford scientists comment that the nomadic lifestyle of the Nicobar pigeon makes it ideally suited to invade remote islands, and Nicobar-type fossils are known from Pacific islands as far east as the Pitcairns. These crowned and toothbilled pigeons, they go on to point out, are large, ground-dwelling birds who rarely fly. It looks as though this whole subgroup of pigeons habitually colonise islands and then lose their power of flight and become larger and more dodo-like. On the Mascarene Islands, the dodo and the solitaire pushed the trend to extremes. And recent fossil discoveries show that complete flightlessness also evolved thousands of miles away on Fiji, in another close relative, the Viti Levu giant pigeon (*Natunaornis gigoura*).

Something like the Dodo's Tale has been repeated on islands all over the world. Many different families of birds, most of which are dominated numerically by flying species, have evolved flightless forms on islands. Mauritius itself had a large flightless rail, *Apha-napteryx bonasia*, also now extinct, which may on occasion have been confused with the dodo. Rodriguez had a related species, *A. leguati*. Rails seem to lend themselves to Dodo's Tale island-hopping followed by flightlessness. In addition to the Indian Ocean forms, there is a flightless rail in the Tristan da Cunha group in the South Atlantic; and most of the Pacific islands have – or had – their own species of flightless rail. Before man ruined the Hawaiian avifauna, there were more than 12 species of flightless rail in that archipelago. More than a quarter of all the world's 60-odd living species of rail are flightless, and all flightless rails live on islands (if you count large islands like New Guinea and New Zealand). Perhaps as many as 200 species have gone extinct on tropical Pacific islands since human contact.

Again on Mauritius, and also now extinct, was a large parrot *Lophopsittacus mauritianus.* This crested parrot was a poor flyer and may have occupied a niche similar to the still (just) surviving kakapo of New Zealand.* New Zealand is, or was, home to a large number of flightless birds belonging to many different families. One of the more striking was the so-called adzebill – a stout, chunky bird, distantly related to cranes and rails. There were different species of adzebill on the North and South Islands, but neither island had any mammals except (for the obvious reason which underlies the Dodo's Tale) bats, and it is easy to imagine that the adzebills made their living in a rather mammal-like way, filling a gap in the market.

In all these cases, the evolutionary story is almost certainly a version of the Dodo's Tale. Ancestral flying birds are carried by their wings to a remote island where an absence of mammals opens up opportunities for making a living on the ground. Their wings are no longer useful in the way that they were on the mainland, so the birds give up flying, and their wings and costly wing muscles degenerate. Until recently there was thought to be one notable exception, one of the oldest and the most famous of all the groups of flightless birds: the ratites, the ostrich order. They have a tale of their own, the Elephant Bird's Tale.

THE ELEPHANT BIRD'S TALE

From the tales of the *Arabian Nights*, the image that most stirred my childish imagination was the roc encountered by Sinbad the Sailor, who at first thought this monstrous bird was a cloud, come over the sun:

> I had heard aforetime of pilgrims and travellers, how in a certain island dwelleth a huge bird, called the 'roc', which feedeth its young on elephants.

The legend of the roc (rucke or rukh) surfaces in several stories of the *Arabian Nights* – two involving Sinbad and two about Abd-al-Rahman. It is mentioned by Marco Polo as living in Madagascar,

* Also memorably celebrated by Douglas Adams, in *Last Chance to See.*

and envoys from the King of Madagascar were said to have presented the Khan of Cathay with a roc feather. Michael Drayton (1563–1631) invoked the monstrous bird's name to contrast it with the proverbially tiny wren:

> All feathered things yet ever knowne to men,
> From the huge Rucke, unto the little Wren . . .

What is the origin of the roc legend? And if it is pure fantasy, whence the recurrent connection with Madagascar?

Fossils from Madagascar tell us that a gigantic bird, the elephant bird *Aepyornis maximus*, lived there, perhaps until as late as the seventeenth century,[*] although more probably around 1000 AD. The elephant bird finally succumbed, perhaps partly through people stealing its eggs which were up to a metre in circumference[†] and would have provided as much food as 200 chicken eggs. The elephant bird was three metres tall and weighed nearly half a tonne – as much as five ostriches. Unlike the legendary roc (which used its 16-metre wingspan to carry Sinbad aloft as well as elephants) the real elephant bird could not fly, and its wings were (relatively) small like an ostrich's. But, though a cousin, it would be wrong to imagine it as a scaled-up ostrich: it was a more robust, heavy-set bird, a kind of feathered tank with a big head and neck, unlike the ostrich's slender periscope. Given how legends readily grow and inflate, *Aepyornis* is a plausible progenitor of the roc.

The elephant bird was probably vegetarian, unlike the fabulously jumbo-phagous roc, and unlike earlier groups of giant carnivorous birds such as the phorusrhachoid family of the New World. These could grow to the same height as *Aepyornis*, with a fearsomely hooked beak which, as if in justification of their nickname of 'feathered tyrannosaurs', looks capable of swallowing a medium-sized lawyer whole. These monstrous cranes seem at first sight better casting for the role of the terrifying roc than *Aepyornis*, but they went extinct too long ago to have started the legend, and in any case Sinbad (or his real-life Arab counterparts) never visited the Americas.

The elephant bird of Madagascar is the heaviest bird known to have lived, but it was not the longest. Some species of moa could

---

[*] Actually there were several related species, in two genera, *Aepyornis* and *Mullerornis*. But *A. maximus*, as its name suggests, most deserves to be called the elephant bird.

[†] Not diameter – it is not quite as surprising as it sounds.

They've gone and there ain't no moa.
Sir Richard Owen with the skeleton of
Dinornis, the giant moa. Owen, to
whom we owe the term dinosaur, was
the first to describe the moa.

reach a height of 3.5 metres, but only if the neck was raised, as in
Richard Owen's mounting (see photo). In life, it seems, they normally carried the head only a little way above the back. But the moa
cannot have generated the roc legend, for New Zealand, too, was well
beyond Sinbad's ken. About ten moa species existed in New Zealand,
ranging in size from turkey to double-sized ostrich. Moas are
extreme among flightless birds in that they have no trace of wings at
all, not even buried vestiges of wing bones. They thrived in both the
North and South Islands of New Zealand until the recent invasion by
the Maori people, about 1250 AD. They were easy prey, no doubt for
the same reason as the dodo. Except for the (extinct) Haast's eagle,
the largest eagle ever to have lived, they had known no predators for
tens of millions of years, and the Maoris slaughtered them all, eating
the choicer parts and discarding the rest, belying, not for the first
time, the wishful myth of the noble savage living in respectful
harmony with his environment. By the time the Europeans arrived,
only a few centuries after the Maoris, the last moa was gone. Legends
and tall stories of sightings persist to this day, but the hope is forlorn.
In the words of a plaintive song, to be sung in a mournful New
Zealand accent:

No moa, no moa
In old Ao-tea-roa.*
Can't get 'em.
They've et 'em;
They've gone and there ain't no moa!

Elephant birds and moas (but not the carnivorous phorusrhachoids nor various other extinct flightless giants) were ratites, an ancient family of birds, which now includes the rheas of South America, the emus of Australia, the cassowaries of New Guinea and Australia, the kiwis of New Zealand, and the ostrich, now confined to Africa and Arabia but previously common in Asia and even Europe.

I take delight in the power of natural selection, and it would have given me satisfaction to report that the ratites evolved their flightlessness separately in different parts of the world, conforming to the message of the Dodo's Tale. In other words, I would have liked the ratites to have been an artificial assemblage, driven to superficial resemblance by parallel pressures in different places. Alas, this is not so. The true tale of the ratites, which I am attributing to the elephant bird, is very different.

No, wait, as they say on the internet. I put the last paragraph in italics to indicate that, copied straight from the first edition, it is now known to be wrong. Or rather, to say it another way, even a little triumphally, my intuitive wish turns out to be bang on right.

The ratites are still an ancient group, although among them we must also count the flying tinamous of South America, quail-like ground birds who have long been suspected to be close relatives of some sort. As evident from the fractal tree which heads this rendez-vous (page 304), this collection of birds, technically know as paleo-gnaths, are more closely related to each other than they are to the rest of the birds: in that sense they are not an artificial assemblage. But they seem to have evolved flightlessness convergently rather than inheriting it from a flightless ancestor. This about-turn in our understanding is due to the elephant bird itself. Or more precisely, DNA from elephant bird bones, painstakingly extracted in the lab of Alan Cooper, leading authority on ancient DNA whom we met in the Dodo's Tale. The evidence is not quite conclusive but it is strongly suggestive.

* The Maori name for New Zealand.

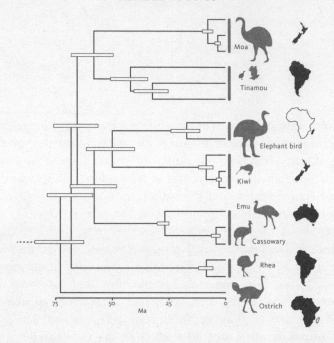

'It would have given me satisfaction to report that ratites evolved their flightlessness separately.' This wish may be true after all. Ancient DNA analysis suggests the evolutionary tree of ratites and tinamous above, which means the fragmentation of Gondwana from 160 to 80 Ma (see page 265) cannot explain their geographical distribution. Hollow bars represent margins of error in the molecular clock dates for each branch in the tree. Adapted from Mitchell *et al.* [291].

The orthodox story, promoted in the first edition of this book, was that a remote ancestor of the ratites evolved its flightlessness, *à la Dodo*, before the break-up of the giant southern continent of Gondwana. Then, when the various fragments of Gondwana went their grinding separate ways, each continent or island took with it its own ratite passengers: Madagascar's ratites became the elephant birds; South America's became the rheas; Australia's became the emus and cassowaries; New Zealand's became the moas and kiwis; and the ostriches ended up in Africa.

If that story of land dispersal had been right, we would expect the Madagascan elephant birds to be most closely related to their Gondwanan neighbours, perhaps ostriches. And since Madagascar broke off from Gondwana early, about 120 million years ago, we would also expect elephant birds to be among the earliest branching ratites. Their newly sequenced DNA confounds both expectations.

The elephant bird's closest relative is . . . the kiwi, from the far side of Gondwana. That means these huge flightless birds perch uncomfortably high in the branches of the ratite tree (see opposite). Worse, fossils hint that kiwis may have evolved from tellingly recent flying ancestors. And to top it all, that other giant flightless bird, the New Zealand moa, turns out to be most closely related to the flying tinamous of distant South America. The most parsimonious suggestion is that the ratite's tale is like an ancient and much larger version of the dodo's. Or as the Cooper group puts it, 'Early ratite evolution appears to have been dominated by flighted dispersal and parallel evolution, with flightlessness evolving a minimum of six times and gigantism a minimum of five.'

Ed Yong, in a lucid article written in 2015, quoted Michael Bunce as saying, 'A number of textbooks need to be rewritten.' And he goes on to quote the first edition of *The Ancestor's Tale*, the passage reproduced in italics above: '. . . "Alas, this is not so." Cheer up, Richard, it is so.' Scientists pay lip service to the idea that they enjoy being proved wrong, because that's how science progresses. It's nice when, once in a while, it turns out to be more than just lip service. I am truly delighted to have been proved wrong – and yet right.

AMPHIBIANS

Three hundred and forty million years ago, in the early Carboniferous Period, only about 20 million years beyond the great milestone of Rendezvous 16, we amniotes (the name that unites mammals with reptiles and birds) meet our amphibian cousins at Rendezvous 17. Pangaea had not yet come together, and northern and southern land masses surrounded a pre-Tethys ocean. A south polar ice cap was beginning to form, there were tropical forests of club mosses around the equator, and the climate was probably something like that of today, although the flora and fauna were of course very different.

Concestor 17, in the vicinity of our 175-million-greats-grandparent, is the ancestor of all surviving tetrapods. Tetrapod means four feet. We who don't walk on four feet are lapsed tetrapods, recently lapsed in our case, much less recently in the case of birds, but we are all called tetrapods. More to the point, Concestor 17 is the grand ancestor of the huge throng of land vertebrates. Despite my earlier strictures on the conceit of hindsight, the emergence of fish onto the land was a major transition in our evolutionary history. We will do it justice at the next rendezvous, in the Lungfish's Tale.

Three main bands of modern amphibian pilgrims have joined forces more recently than their rendezvous with us amniotes. They are the frogs (and toads: the distinction is not a zoologically helpful one), the salamanders (and newts, which are those species that return to the water to breed), and the caecilians (moist, legless burrowers or swimmers, with a superficial resemblance to earthworms or snakes). The frogs have no tail as adults but a vigorously swimming tail as larvae. The salamanders have a long tail in the adult as well as the larval stage, and their body proportions most resemble ancestral amphibians, as judged by fossils. The caecilians have no limbs – not even internal traces of the pectoral and pelvic girdles that supported the limbs of their ancestors. The great length of the caecilian body is achieved by multiplying up the vertebrae in the trunk region (up to 250, compared to 12 in frogs), and their ribs, which provide useful support and protection. The tail, oddly, is very short or even absent: if caecilians had legs, their hind legs would be right at the posterior

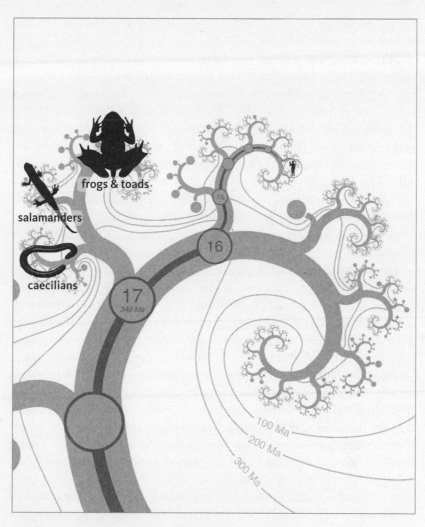

Amphibians join. There are three lineages of living amphibians: the worm-like caecilians, salamanders and frogs. The number of described species ranges from 6,500 to 7,450, around 90 per cent of which are frogs. At odds with several fossil studies, genetic analysis consistently unites these three amphibian orders in a single group, sister to the amniotes, as we show here. Disagreements remain on the order of branching between the three amphibian groups.

tip of the body, which is where some extinct amphibians actually kept them.

Even if they live on land as adults, most amphibians reproduce in water, while amniotes (except in secondarily evolved cases such as whales, dugongs and ichthyosaurs) reproduce on land. Amniotes reproduce either viviparously, giving birth to live young, or with a relatively large, tough-shelled waterproof egg. In both cases the embryo floats in its own 'private pond'. Amphibian embryos are much more likely to float in a real pond, or something equivalent to one. The amphibian pilgrims who join us at Rendezvous 17 may spend part of their time on land, but they are seldom far from water and, at least at some stage in their life cycle, they usually return to it. Those that reproduce on land go to some lengths to contrive watery conditions.

Trees provide relatively safe havens, and frogs have discovered ways to reproduce in them without losing the vital tie to water. Some exploit the small pools of rainwater that form in the rosettes of bromeliad plants. Male African grey tree frogs, *Chiromantis xerampelina*, co-operate to whip up a thick white foam, with their back legs, from a liquid secreted by the females. This foam hardens to a crust on the outside, protecting the moist interior which serves as a nest for the group's eggs. The tadpoles develop inside the wet foam nest, up a tree. When they are ready, in the next rainy season, they wriggle free and drop into puddles of water below the tree, where they develop into frogs. Other species use the foam-nest technique, but they don't co-operate to do so. Instead one male beats up a foam from the secretion of one female.

Some frog species have made interesting transitions in the direction of true viviparity – live birth. The female of the South American marsupial frog (various species of the genus *Gastrotheca*) transfers her fertilised eggs to her back, where they become covered by a layer of skin. There the tadpoles develop and can clearly be seen wriggling under the skin of their mother's back until they eventually burst out. Several other species do something similar, probably independently evolved.

Another South American frog species, named *Rhinoderma darwinii* after its illustrious discoverer, practises a most unusual version of viviparity. The male appears to eat the eggs that he has fertilised. The eggs don't travel down his gut, however. Like many male frogs he has a commodious vocal sac, used as a resonator to amplify the voice, and it is in this moist chamber that the eggs lodge. There they develop, until they are finally vomited out as fully formed froglets, forgoing the freedom to swim as tadpoles.

The key difference between amphibians and amniotes is that amniote skins and eggshells are waterproof. Amphibian skin typically lets water evaporate through it, at the same rate you'd expect from a body of standing water of the same area. As far as the water under the skin is concerned, there might almost as well not be any skin. This is very different from reptiles, birds and mammals, where one of the main roles of the skin is to serve as a barrier to water. There are exceptions among amphibians – most notably among various species of desert frogs in Australia. These exploit the fact that even deserts can have flood times, though brief and well spaced out. During such rare and intermittent times of high rainfall, each frog makes a water-filled cocoon in which it buries itself in a state of torpor, for two years or, by some accounts, even as long as seven years. Some species of frogs can withstand temperatures well below the normal freezing point of water, by manufacturing glycerol as an antifreeze.

Almost no amphibians live in sea water, and it is therefore not surprising that, unlike lizards, they are seldom found on remote islands.* Darwin noted this in more than one of his books, and also the fact that frogs that have been artificially introduced to just such islands thrive there. He presumed that lizard eggs are protected from sea water by their hard shells, whereas sea water promptly kills frogspawn. Frogs are, however, found on all continents except Antarctica, and have probably been there continuously since before the continents broke up. They are a very successful group.

Frogs remind me of birds in one respect. Both have a body plan which is a somewhat bizarre modification of the ancestral one. That is not particularly remarkable, but birds and frogs have taken this bizarre body plan and made it the basis for a whole new range of variation. There are not quite so many frog species as birds, but the more than 6,500 species of frogs, in every part of the world, are impressive enough. Just as the bird body plan is obviously a design

---

* Sam Turvey tells me that the two frog species with the remotest island distribution, the Fijian frogs *Platymantis vitiensis* and *P. vitianus* (closely related and presumably descended from a single colonising ancestor), develop completely in the egg rather than having a free-swimming tadpole. They appear more salt-tolerant than most frogs, with *P. vitianus* sometimes found on beaches. These unusual characteristics, if present in their colonising ancestor as seems likely, would have pre-adapted them for island-hopping.

for flying, even in birds such as ostriches that don't fly, the best way to understand the body plan of an adult frog is as a highly specialised jumping machine. Some species can jump spectacular distances, up to 50 body lengths in the well-named rocket frog of Australia (*Litoria nasuta*). The largest frog in the world, the goliath frog (*Conraua goliath*) of West Africa, which is the size of a small dog, is said to jump three metres. Not all frogs jump, but all are descended from jumping ancestors. They are at least lapsed jumpers, just as ostriches are lapsed fliers. Some tree-dwelling species, such as Wallace's flying frog, *Rhacophorus nigropalmatus*, prolong the jump by spreading their long fingers and toes, whose webbing acts as a parachute. Indeed, they glide a bit like flying squirrels.

Salamanders and newts swim like fish when they are in water. Even on land, their legs are too small and feeble to walk or run in the sense we would recognise, and the salamander uses a sinuous swimming motion like a fish, with the legs just helping it along. Most salamanders today are quite small. The largest reach a respectable 1.5 metres, but this is still far smaller than the giant amphibians of past times, which dominated the land before the rise of the reptiles.

But what did Concestor 17 look like: the ancestor that amphibians share with reptiles and ourselves? Certainly more like an amphibian than an amniote, and more like a salamander than a frog – but probably not quite like either. Our reconstruction in plate 20 is partially based on amphibian fossils of the Lower Carboniferous, such as the half-metre-long *Balanerpeton*. One thing that we can say is that it had legs with five toes. This isn't as obvious a deduction as you might suppose. We tend to think the possession of five digits is deeply stamped in the hands and feet of all tetrapods – the 'penta-dactyl' limb is a classic zoological totem. But the earliest tetrapods, whom we shall meet in the Lungfish's Tale, often had six, seven or even eight toes. Modern amphibians tend to have only four toes on their front legs, and there are hints of a vestigial sixth digit in certain frogs. Nevertheless, it is looking increasingly certain that by the time of Concestor 17, the default number had become five. It is tempting to say the number of digits doesn't matter, is functionally neutral. I doubt that. My tentative guess is that in those early times the different species really did benefit from their respective numbers of toes. They really were more efficient than other numbers would have been, for swimming or walking. Later, the tetrapod limb design hardened at five digits, probably because some internal embryological process came to rely upon that number. In the adult, the number is frequently reduced from the embryonic number – in extreme cases, such as modern horses, to just one, the middle toe.

It is pleasing to speculate on the unintended restrictions imposed upon us by pentadactyly. We would not have invented counting in tens, nor would this book be using hundreds, thousands and millions as standard multiples. Indeed, if we had followed the lead of the frogs or salamanders and defaulted to eight fingers, we'd think naturally in octal arithmetic, binary logic would be easier to understand, and computers might have been invented much earlier.

THE SALAMANDER'S TALE

Names are a menace in evolutionary history. It is no secret that palaeontology is a controversial subject in which there are even some personal enmities. At least eight books called *Bones of Contention* are in print. And if you look at what two palaeontologists are quarrelling about, as often as not it turns out to be a name. Is this fossil *Homo erectus*, or is it an archaic *Homo sapiens*? Is this one an early *Homo habilis* or a late *Australopithecus*? People evidently feel strongly about such questions, but they often turn out to be splitting hairs. Indeed, they resemble theological questions, which I suppose gives a clue to why they arouse such passionate disagreements. The obsession with discrete names is an example of what I call the tyranny of the discontinuous mind. The Salamander's Tale strikes a blow against the discontinuous mind.

The Central Valley runs much of the length of California, bounded by the Coastal Range to the west and by the Sierra Nevada to the east. These long mountain ranges link up at the north and the south ends of the valley, which is therefore surrounded by high ground. Throughout this high ground lives a genus of salamanders called *Ensatina*. The Central Valley itself, about 40 miles wide, is not friendly to salamanders, and they are not found there. They can move all round the rim of the valley but normally not across it, in an elongated ring of more or less continuous population. In practice any one salamander's short legs in its short lifetime don't carry it far from its birthplace. But genes, persisting through a longer timescale, are another matter. Individual salamanders can interbreed with neighbours whose parents may have interbred with neighbours further round the ring, and so on. There is therefore potentially gene flow all around the ring. Potentially. What happens in practice has been elegantly worked out by research at the University of California at Berkeley, initiated by my old colleague Robert Stebbins and continued by David Wake.

In a study area called Camp Wolahi, in the mountains to the south of the valley, there are two clearly distinct species of *Ensatina* which do not interbreed. One is conspicuously marked with yellow and black blotches. The other is a uniform light brown with no blotches. Camp Wolahi is in a zone of overlap, but wider sampling shows that the blotched species is typical of the eastern side of the Central Valley which, here in Southern California, is known as the San Joaquin Valley. The light brown species, on the contrary, is typically found on the western side of the San Joaquin.

Non-interbreeding is the recognised criterion for whether two populations deserve distinct species names. It therefore should be straightforward to use the name *Ensatina eschscholtzii* for the plain western species, and *Ensatina klauberi* for the blotched eastern species – straightforward but for one remarkable circumstance, which is the nub of the tale.

If you go up to the mountains that bound the north end of the Central Valley, which up there is called the Sacramento Valley, you'll find only one species of *Ensatina*. Its appearance is intermediate between the blotched and the plain species: mostly brown, with rather indistinct blotches. It is not a hybrid between the two: that is the wrong way to look at it. To discover the right way, make two expeditions south, sampling the salamander populations as they fork to west and east on either side of the Central Valley. On the east side, they become progressively more blotched until they reach the extreme of *klauberi* in the far south. On the west side, the salamanders become progressively more like the plain *eschscholtzii* that we met in the zone of overlap at Camp Wolahi.

This is why it is hard to treat *Ensatina eschscholtzii* and *Ensatina klauberi* with confidence as separate species. They constitute a 'ring species' (see plate 21). You'll recognise them as separate species if you only sample in the south. Move north, however, and they gradually turn into each other. Zoologists normally follow Stebbins's lead and place them all in the same species, *Ensatina eschscholtzii*, but give them a range of subspecies names. Starting in the far south with *Ensatina eschscholtzii eschscholtzii*, the plain brown form, we move up the west side of the valley through *Ensatina eschscholtzii xanthoptica* and *Ensatina eschscholtzii oregonensis* which, as its name suggests, is also found further north in Oregon and Washington. At the north end of California's Central Valley is *Ensatina eschscholtzii picta*, the semi-blotched form mentioned before. Moving on round the ring and down the east side of the valley, we pass through *Ensatina eschscholtzii platensis* which is a bit more blotched than *picta*, then *Ensatina eschscholtzii croceater* until we reach *Ensatina eschscholtzii klauberi* (which is the very blotched one that we previously called *Ensatina klauberi* when we were considering it to be a separate species).

Stebbins believes that the ancestors of *Ensatina* arrived at the north end of the Central Valley and evolved gradually down the two sides of the valley, diverging as they went. An alternative possibility is that they started in the south as, say, *Ensatina eschscholtzii eschscholtzii*, then evolved their way up the west side of the valley, round the top and down the other side, ending up as *Ensatina eschscholtzii klauberi* at the other end of the ring. Whatever the history, what happens today is that there is hybridization all round the ring, except where the two ends of the line meet, in the far south of California.

As a complication, it seems that the Central Valley is not a total barrier to gene flow. Occasionally, salamanders seem to have made it across, for there are populations of, for example, *xanthoptica*, one of the western subspecies, on the eastern side of the valley, where they hybridise with the eastern subspecies, *platensis*. Yet another complication is that there is a small break near the south end of the ring, where there seem to be no salamanders at all. Presumably they used to be there, but have died out. Or maybe they are still there but have not been found: I am told that the mountains in this area are rugged and hard to search. The ring is complicated, but a ring of continuous gene flow is, nevertheless, the predominant pattern in this genus, as it is with the better-known case of herring gulls and lesser black-backed gulls around the Arctic Circle.

In Britain the herring gull and the lesser black-backed gull are clearly distinct species. Anybody can tell the difference, most easily by the colour of the wing backs. Herring gulls have silver-grey wing backs, lesser black-backs, dark grey, almost black. More to the point, the birds themselves can tell the difference too, for they don't hybridise although they often meet and sometimes even breed alongside one another in mixed colonies. Zoologists therefore feel fully justified in giving them different names, *Larus argentatus* and *Larus fuscus*.

But now here's the interesting observation, and the point of resemblance to the salamanders. If you follow the population of herring gulls westward to North America, then on around the world across Siberia and back to Europe again, you notice a curious fact. The 'herring gulls', as you move round the pole, gradually become less and less like herring gulls and more and more like lesser black-backed gulls until it turns out that our Western European lesser black-backed gulls actually are the other end of a ring-shaped continuum which started with herring gulls. At every stage around the ring, the birds are sufficiently similar to their immediate neighbours in the ring to interbreed with them. Until, that is, the ends of the continuum are reached, and the ring bites itself in the tail. The herring gull and the lesser black-backed gull in Europe never interbreed, although they are linked by a continuous series of interbreeding colleagues all the way round the other side of the world.

Ring species like the salamanders and the gulls are only showing us in the spatial dimension something that must always happen in the time dimension. Suppose we humans, and the chimpanzees, were a ring species. It could have happened: a ring perhaps moving up one side of the Rift Valley, and down the other side, with two completely separate species co-existing at the southern end of the ring, but an unbroken continuum of interbreeding all the way up and back round the other side. If this were true, what would it do to our attitudes to other species? To apparent discontinuities generally?

Many of our legal and ethical principles depend on the separation between *Homo sapiens* and all other species. Of the people who regard abortion as a sin, including the minority who go to the lengths of assassinating doctors and blowing up abortion clinics, many are unthinking meat-eaters, and have no worries about chimpanzees being imprisoned in zoos and sacrificed in laboratories. Would they think again, if we could lay out a living continuum of intermediates between ourselves and chimpanzees, linked in an unbroken chain of interbreeders like the Californian salamanders? Surely they would. Yet it is the merest accident that the intermediates all happen to be dead. It is only because of this accident that we can comfortably and easily imagine a huge gulf between our two species – or between any two species, for that matter.

I have previously recounted the case of the puzzled lawyer who questioned me after a public lecture. He brought the full weight of his legal acumen to bear on the following nice point. If species A

evolves into species B, he reasoned closely, there must come a point
when a child belongs to the new species B but his parents still belong
to the old species A. Members of different species cannot, by definition, interbreed with one another, yet surely a child would not be
so different from its parents as to be incapable of interbreeding
with their kind. Doesn't this, he wound up, wagging his metaphorical finger in the special way that lawyers, at least in courtroom
dramas, have perfected as their own, undermine the whole idea of
evolution?

That is like saying, 'When you heat a kettle of cold water, there is
no particular moment when the water ceases to be cold and becomes
hot, therefore it is impossible to make a cup of tea.' Since I always try
to turn questions in a constructive direction, I told my lawyer about
the herring gulls, and I think he was interested. He had insisted on
placing individuals firmly in this species or that. He didn't allow for
the possibility that an individual might lie half way between two
species, or a tenth of the way from species A to species B. Exactly the
same limitation of thought hamstrings the endless debates about
exactly when in the development of an embryo it becomes human
(and when, by implication, abortion should be regarded as tantamount to murder). It is no use saying to these people that, depending
upon the human characteristic that interests you, a foetus can be
'half human' or 'a hundredth human'. 'Human', to the qualitative,
absolutist mind, is like 'diamond'. There are no halfway houses.
Absolutist minds can be a menace. They cause real misery, human
misery. This is what I call the tyranny of the discontinuous mind, and
it leads me to develop the moral of the Salamander's Tale.

For certain purposes names, and discontinuous categories, are
exactly what we need. Indeed, lawyers need them all the time.
Children are not allowed to drive; adults are. The law needs to
impose a threshold, for example the seventeenth birthday. Revealingly, insurance companies take a very different view of the proper
threshold age.

Some discontinuities are real, by any standards. You are a person
and I am another person and our names are discontinuous labels that
correctly signal our separateness. Carbon monoxide really is distinct
from carbon dioxide. There is no overlap. A molecule consists of a
carbon and one oxygen, or a carbon and two oxygens. None has a
carbon and 1.5 oxygens. One gas is deadly poisonous, the other is
needed by plants to make the organic substances that we all depend

upon. Gold really is distinct from silver. Diamond crystals really are different from graphite crystals. Both are made of carbon, but the carbon atoms naturally arrange themselves in two quite distinct ways. There are no intermediates.

But discontinuities are often far from so clear. My newspaper carried the following item during a recent flu epidemic. Or was it an epidemic? That question was the burden of the article.

> Official statistics show there are 144 people in every 100,000 suffering from flu, said a spokeswoman for the Department of Health. As the usual gauge of an epidemic is 400 in every 100,000, it is not being officially treated as an epidemic by the Government. But the spokeswoman added: 'Professor Donaldson is happy to stick by his version that this is an epidemic. He believes it is many more than 144 per 100,000. It is very confusing and it depends on which definition you choose. Professor Donaldson has looked at his graph and said it is a serious epidemic.'

What we know is that some particular number of people are suffering from flu. Doesn't that, in itself, tell us what we want to know? Yet for the spokeswoman, the important question is whether this counts as an 'epidemic'. Has the proportion of sufferers crossed the rubicon of 400 per 100,000? This is the great decision which Professor Donaldson had to make, as he pored over his graph. You'd think he might have been better employed trying to do something about it, whether or not it counted officially as an epidemic.

As it happens, in the case of epidemics, for once there really is a natural rubicon: a critical mass of infections above which the virus, or bacterium, suddenly 'takes off' and dramatically increases its rate of spreading. This is why public health officials try so hard to vaccinate more than a threshold proportion of the population against, say, whooping cough. The purpose is not just to protect the individuals vaccinated. It is also to deprive the pathogens of the opportunity to reach their own critical mass for 'take-off'. In the case of our flu epidemic, what should really worry the spokeswoman for the Ministry of Health is whether the flu virus has yet crossed its rubicon for take-off, and leapt abruptly into high gear in its spread through the population. This should be decided by some means other than reference to magic numbers like 400 per 100,000. Concern with magic numbers is a mark of the discontinuous mind, or qualitative mind. The funny thing is that, in this case, the discontinuous mind overlooks a genuine discontinuity, the take-off point for an epidemic. Usually there isn't a genuine discontinuity to overlook.

Many Western countries at present are suffering what is described as an epidemic of obesity. I seem to see evidence of this all around me, but I am not impressed by the preferred way of turning it into numbers. A percentage of the population is described as 'clinically obese'. Once again, the discontinuous mind insists on separating people out into the obese on one side of a line, the non-obese on the other. That is not the way real life works. Obesity is continuously distributed. You can measure how obese each individual is, and you can compute group statistics from such measurements. Counts of numbers of people who lie above some arbitrarily defined threshold of obesity are not illuminating, if only because they immediately prompt a demand for the threshold to be specified and maybe redefined.

The same discontinuous mind also lurks behind all those official figures detailing the numbers of people 'below the poverty line'. You can meaningfully express a family's poverty by telling us their income, preferably expressed in real terms of what they can buy. Or you can say 'X is as poor as a church mouse' or 'Y is as rich as Croesus' and everybody will know what you mean. But spuriously precise counts or percentages of people said to fall above or below some arbitrarily defined poverty *line* are pernicious. They are pernicious because the precision implied by the percentage is instantly belied by the meaningless artificiality of the 'line'. Lines are impositions of the discontinuous mind. Even more politically sensitive is the label 'black', as opposed to 'white', in the context of modern society – especially American society. This is the central issue in the Grasshopper's Tale, and I'll leave it for now, except to say that I believe race is yet another of the many cases where we don't need discontinuous categories, and where we should do without them unless an extremely strong case in their favour is made.

Here's another example. Universities in Britain award degrees that are classified into three distinct classes, First, Second and Third Class. Universities in other countries do something equivalent, if under different names, like A, B, C etc. Now, my point is this. Students do not really separate neatly into good, middling and poor. There are not discrete and distinct classes of ability or diligence. Examiners go to some trouble to assess students on a finely continuous numerical scale, awarding marks or points that are designed to be added to other such marks, or otherwise manipulated in mathematically continuous ways. The score on such a continuous numerical scale conveys far more information than classification into one of three categories. Nevertheless, only the discontinuous categories are published.

In a very large sample of students, the distribution of ability and prowess would normally be a bell curve with few doing very well, few doing very badly and many in between. It might not actually be a symmetrical bell like the picture below, but it would certainly be smoothly continuous, and it would become smoother as more and more students are added in.

A few examiners (especially, I hope I'll be forgiven for adding, in non-scientific subjects) seem actually to believe that there really is a discrete entity called the First-Class Mind, or the 'alpha' mind, and a student either definitely has it or definitely hasn't. The task of the examiner is to sort out the Firsts from the Seconds and the Seconds from the Thirds, just as one might sort sheep from goats. The likelihood that in reality there is a smooth continuum, sliding from pure sheepiness through all intermediates to pure goatiness, is a difficult one for some kinds of mind to grasp.

If, against all my expectations, it should turn out that the more students you add in, the more the distribution of exam marks approximates to a discontinuous distribution with three peaks (see lower picture), it would be a fascinating result. The awarding of First, Second and Third Class degrees might then actually be justifiable.

But there is certainly no evidence for this, and it would be very surprising given everything we know about human variation. As things are, it is clearly unfair: there is far more difference between the top of one class

and the bottom of the same class, than there is between the bottom of one class and the top of the next class. It would be fairer to publish the actual marks obtained, or a rank order based upon those marks. But the discontinuous or qualitative mind insists on forcing people into one or other discrete category.

Returning to our topic of evolution, what about sheep and goats themselves? Are there sharp discontinuities between species, or do they merge into each other like first-class and second-class exam performances? If we look only at surviving animals, the answer is normally yes, there are sharp discontinuities. Exceptions like the gulls and the Californian salamanders are rare, but revealing because they translate into the spatial domain the continuity which is normally found only in the temporal domain. People and chimpanzees are certainly linked via a continuous chain of intermediates and a shared ancestor, but the intermediates are extinct: what remains is a discontinuous distribution. The same is true of people and monkeys, and of people and kangaroos, except that the necklace of extinct intermediates is longer, and most links lived longer ago. Because the intermediates are nearly always extinct, we can usually get away with assuming that there is a sharp discontinuity between every species and every other. But in this book we are concerned with evolutionary history, with the dead as well as the living. When we are talking about all the animals that have ever lived, not just those that are living now, evolution tells us there are lines of gradual continuity linking literally every species to every other. When we are talking history, even apparently discontinuous modern species like sheep and dogs are linked, via their common ancestor, in unbroken lines of smooth continuity.

Ernst Mayr, distinguished elder statesman of twentieth-century evolution, has blamed the delusion of discontinuity – under its philosophical name of Essentialism – as the main reason why evolutionary understanding came so late in human history. Plato, whose philosophy can be seen as the inspiration for Essentialism, believed that actual things are imperfect versions of an ideal archetype of their kind. Hanging somewhere in ideal space is an essential, perfect rabbit, which bears the same relation to a real rabbit as a mathematician's perfect circle bears to a circle drawn in the dust. To this day many people are deeply imbued with the idea that sheep are sheep and goats are goats, and no species can ever give rise to another because to do so they'd have to change their 'essence'.

There is no such thing as essence.

No evolutionist thinks that modern species change into other modern species. Cats don't turn into dogs or vice versa. Rather, cats and dogs have evolved from a common ancestor, who lived tens of millions of years ago. If only all the intermediates were still alive, attempting to separate cats from dogs would be a doomed enterprise, as it is with the salamanders and the gulls. Far from being a question of ideal essences, separating cats from dogs turns out to be possible only because of the lucky (from the point of view of the essentialist) fact that the intermediates happen to be dead. Plato might see the irony in the fact that it is actually an imperfection – the sporadic ill-fortune of death – that makes the separation of any one species from another possible. This of course applies to the separation of human beings from our nearest relatives – and, indeed, from our more distant relatives too. In a world of perfect and complete information, fossil information as well as recent, discrete names for animals would become impossible. Instead of discrete names we would need sliding scales, just as the words hot, warm, cool and cold are better replaced by a sliding scale such as Celsius or Fahrenheit.

Evolution is now universally accepted as a fact by thinking people, so one might have hoped that essentialist intuitions in biology would have been finally overcome. Alas, this hasn't happened. Essentialism refuses to lie down. In practice, it is usually not a problem. Everyone agrees that *Homo sapiens* is a different species (and most would say a different genus) from *Pan troglodytes*, the chimpanzee. But everyone also agrees that if you follow human ancestry backward to the shared ancestor and then forward to chimpanzees, the intermediates all along the way will form a gradual continuum in which every generation would have been capable of breeding with its parent or child of the opposite sex.

By the interbreeding criterion every individual is a member of the same species as its parents. This is an unsurprising, not to say platitudinously obvious conclusion, until you realise that it raises an intolerable paradox in the essentialist mind. Most of our ancestors throughout evolutionary history have belonged to different species from us by any criterion, and we certainly couldn't have interbred with them. In the Devonian Period our direct ancestors were fish. Yet, although we couldn't interbreed with them, we are linked by an unbroken chain of ancestral generations, every one of which could have interbred with their immediate predecessors and immediate successors in the chain.

In the light of this, see how empty are most of those passionate arguments about the naming of particular hominid fossils. *Homo* *ergaster* is widely recognised as the predecessor species that gave rise to *Homo sapiens*, so I'll play along with that for what follows. To call *Homo ergaster* a separate species from *Homo sapiens* could have a precise meaning in principle, even if it is impossible to test in practice. It means that if we could go back in our time machine and meet our *Homo ergaster* ancestors, we could not interbreed with them.* But suppose that, instead of zooming directly to the time of *Homo ergaster*, or indeed any other extinct species in our ancestral lineage, we stopped our time machine every thousand years along the way and picked up a young and fertile passenger. We transport this passenger back to the next thousand-year stop and release her (or him: let's take a female and a male at alternate stops). Provided our one-stop time traveller could accommodate to local social and linguistic customs (quite a tall order) there would be no biological barrier to her interbreeding with a member of the opposite sex from 1,000 years earlier. Now we pick up a new passenger, say a male this time, and transport him back another 1,000 years. Once again, he too would be biologically capable of fertilising a female from 1,000 years before his native time. The daisy chain would continue on back to when our ancestors were swimming in the sea. It could go back without a break, to the fishes, and it would still be true that each and every passenger transported 1,000 years before its own time would be able to interbreed with its predecessors. Yet at some point, which might be a million years back but might be longer or shorter, there would come a time when we moderns could not interbreed with an ancestor, even though our latest one-stop passenger could. At this point we could say that we have travelled back to a different species.

The barrier would not come suddenly. There would never be a generation in which it made sense to say of an individual that he is *Homo sapiens* but his parents are *Homo ergaster*. You can think of it as a paradox if you like, but there is no reason to think that any child was ever a member of a different species from its parents, even though the daisy chain of parents and children stretches back from humans to fish and beyond. Actually it isn't paradoxical to anybody but a dyed-in-the-wool essentialist. It is no more paradoxical than the statement that there is never a moment when a growing child ceases to be short and becomes tall. Or a kettle ceases to be cold and

---

* I am not asserting that as a fact. I don't know if it is a fact, although I suspect that it is. It is an implication of our plausibly agreeing to give *Homo ergaster* a different species name.

becomes hot. The legal mind may find it necessary to impose a barrier between childhood and majority – the stroke of midnight on the eighteenth birthday, or whenever it is. But anyone can see that it is a (necessary for some purposes) fiction. If only more people could see that the same applies to when, say, a developing embryo becomes 'human'.

Creationists love 'gaps' in the fossil record. Little do they know, biologists have good reason to love them too. Without gaps in the fossil record, our whole system for naming species would break down. Fossils could not be given names, they'd have to be given numbers, or positions on a graph. Or instead of arguing heatedly over whether a fossil is 'really', say, an early *Homo ergaster* or a late *Homo habilis*, we might call it *habigaster*. There's a lot to be said for this. Nevertheless, perhaps because our brains evolved in a world where most things do fall into discrete categories, and in particular where most of the intermediates between living species are dead, we often feel more comfortable if we can use separate names for things when we talk about them. I am no exception and neither are you, so I shall not bend over backwards to avoid using discontinuous names for species in this book. But the Salamander's Tale explains why this is a human imposition rather than something deeply built into the natural world. Let us use names as if they really reflected a discontinuous reality, but let's privately remember that, at least in the world of evolution, it is no more than a convenient fiction, a pandering to our own limitations.

THE NARROWMOUTH'S TALE

Microhyla (sometimes confused with *Gastrophryne*) is a genus of small frogs, the narrowmouthed frogs. There are several species, including two in North America: the eastern narrowmouth *Microhyla carolinensis*, and the Great Plains narrowmouth *Microhyla olivacea*. These two are so closely related that they occasionally hybridise in nature. The eastern narrowmouth's range extends down the east coast from the Carolinas to Florida, and west until half way across Texas and Oklahoma. The Great Plains narrowmouth extends from Baja California in the west, as far as eastern Texas and eastern Oklahoma, and as far north as northern Missouri. Its range is therefore a western mirror of the eastern narrowmouth's, and it might as well be called the western narrowmouth. The important point is that their ranges meet in the middle: there is an overlap zone running up the eastern half of Texas and into Oklahoma. As I said, hybrids are occasionally found in this overlap zone, but mostly the frogs distinguish just as well as herpetologists do. This is what justifies our calling them two different species.

As with any two species, there must have been a time when they were one. Something separated them: to use the technical term, the single ancestral species 'speciated' and became two. It is a model for what happens at every branch point in evolution. Every speciation begins with some sort of initial separation between two populations of the same species. It isn't always a geographical separation, but, as we shall see in the Cichlid's Tale, an initial separation of some kind makes it possible for the statistical distribution of genes in the two populations to move apart. This usually results in an evolutionary divergence with respect to something visible: shape or colour or behaviour. In the case of these two populations of American frogs, the western species became adapted to life in drier climates than the eastern, but the most conspicuous difference lies in their mating calls. Both are squeaky buzzes, but each buzz of the western species lasts about twice as long (2 seconds) as the eastern species, and its predominant pitch is noticeably higher: 4,000 cycles per second as against 3,000. That is to say, the predominant pitch of the western narrowmouth is about top C, the highest key on a piano, and the eastern predominant pitch is around the F# below that. These sounds are not musical, however. Both calls contain a mixture of frequencies, ranging from far below the predominant to far above. Both are buzzes, but the eastern buzz is lower. The western call, as well as being longer, begins with a distinct peep, rising in pitch before the buzz takes over. The eastern frog goes straight into its shorter buzz.

Why go into so much detail about these calls? Because what I have described is true only in the zone of overlap where the comparison between them is clearest, and that is the whole point of the tale. W. F. Blair tape-recorded frogs from a good spread of sampling locations across the United States, with fascinating results. In areas where the two species of frogs never meet one another – Florida for the eastern species and Arizona for the western – their songs are much more similar to each other in pitch: the predominant pitch of both is around 3,500 cycles per second: top A on the piano. In areas close to the zone of overlap but not quite in it, the two species are more different, but not as different as they are in the zone of overlap itself.

The conclusion is intriguing. Something is pushing the calls of these two species apart in the zone where they overlap. Blair's interpretation, which not everybody accepts, is that hybrids are penalised. Anything that helps potential miscegenators to distinguish the species and avoid mating with the wrong one is favoured by natural selection. Such small differences as there may be are exaggerated in just that part of the country where it matters. The great evolutionary geneticist Theodosius Dobzhansky called this 'reinforcement' of reproductive isolation. Not everybody accepts Dobzhansky's reinforcement theory, but the Narrowmouth's Tale, at least, seems to offer support.

There is another good reason why closely related species might be pushed apart when they overlap. They are likely to be competing for similar resources. In the Galapagos Finch's Tale, we saw how different species of finch have partitioned the available seeds. Species with larger bills take larger seeds. Where they don't overlap, both species can take a broader range of resources – large seeds and small seeds. Where they do overlap, each species is forced, by competition from the other, to become more different from the other. The large-billed species might evolve an even larger bill, the small-billed species an even smaller bill. As usual, by the way, don't be misled by the metaphorical idea of being forced to evolve. What actually happens is that within each species, when the other species is present, individuals who happen to be more different from the competing species thrive.

This phenomenon, where two species differ from each other more when they overlap than when they don't, is called 'character displacement' or 'reverse cline'. It is easy to generalise from biological species to cases where any class of entities differ more when they encounter one another than when they are alone. The human parallels are tempting, but I shall resist. As authors used to say, this is left as an exercise for the reader.

THE AXOLOTL'S TALE

We think of young animals as small versions of the adults they are to become, but this is far from the rule. Probably a majority of animal species manage their life stories very differently. The young make their own living, as specialists in a totally different way of life from their parents. A substantial fraction of plankton consists of swimming larvae whose adulthood – if they survive, which is statistically unlikely – will be very different. In many insects the larval stage is the one that does the bulk of the feeding, building up a body that will eventually metamorphose into an adult whose only roles are dispersal and reproduction. In extreme cases such as mayflies, the adult doesn't feed at all, and – for nature is ever niggardly* – lacks a gut and other costly feeding apparatus.

A caterpillar is a feeding machine which, when it has grown to a good size on plant food, in effect recycles its own body and reconstitutes itself as an adult butterfly that flies, sucks nectar as aviation fuel, and reproduces. Adult bees, too, power their flight muscles on nectar while they gather pollen (a very different kind of food) for the worm-like larvae. Many insect larvae live underwater before hatching into adults, who fly through the air and disperse their genes to other bodies of water. A huge diversity of marine invertebrates have adult stages that live on the sea bottom, sometimes permanently moored to one spot, but very different larval stages that disperse the genes by swimming in the plankton. These include molluscs, echinoderms (sea urchins, starfish, sea cucumbers, brittle stars), sea squirts, worms of many kinds, crabs and lobsters, and barnacles. Parasites typically have a series of distinct larval stages, each with its own characteristic way of life and diet. Often the different life stages are also parasitic, but parasitic on very different hosts. Some parasitic worms have as many as five completely distinct juvenile stages, each of which makes its living in a different way from all the others.

All this means that a single individual must carry within it the full genetic instruction set for each of the larval stages, with their different ways of making a living. A caterpillar's genes 'know' how to make a butterfly, and a butterfly's genes know how to make a caterpillar. Doubtless some of the very same genes are involved, in different ways, in making both these radically different bodies. Other genes lie dormant in the caterpillar and are turned on in the butterfly. Yet others are active in the caterpillar and are turned off and forgotten when it becomes a butterfly. But the whole set of genes is there, in

---

* I use this word advisedly. In 1999 the Mayor of Washington DC accepted the resignation of an official whose description of a budget proposal as niggardly gave offence. Julian Bond, distinguished Chairman of the NAACP, correctly described the Mayor's judgement as niggardly. Inspired by the case, a nasty little student at the University of Wisconsin brought an official complaint against her professor, who had used 'niggardly' in a lecture on Chaucer. Such ignorant witch-hunting is not peculiar to the USA. In 2001, a mob of British vigilantes stoned the house of a consultant paediatrician, mistaking her for a paedophile.

both bodies, and is passed on to the next generation. The lesson is that we shouldn't be too surprised if animals as different from each other as caterpillars and butterflies occasionally evolve directly one into the other. Let me explain what I mean.

Fairy stories are filled with frogs turning into princes, or pumpkins turning into coaches drawn by white horses metamorphosed from white mice. Such fantasies are profoundly unevolutionary. They couldn't happen, not for biological reasons but mathematical ones. Such transitions would have an inherent improbability value to rival, say, a perfect deal at bridge, which means that for practical purposes we can rule them out. But for a caterpillar to turn into a butterfly is not a problem: it happens all the time, the rules having been built up over the ages by natural selection. And although no butterfly has ever been seen to turn into a caterpillar, it should not surprise us in the same way as, say, a frog turning into a prince. Frogs don't contain genes for making princes. But they do contain genes for making tadpoles.

My former Oxford colleague John Gurdon dramatically demonstrated this in 1962 when he transformed an adult frog (well, an adult frog cell!) into a tadpole (it has been suggested that this first-ever experimental cloning of a vertebrate deserves a Nobel Prize*). Similarly, butterflies contain genes for turning into caterpillars. I don't know what embryological hurdles would need to be surmounted in order to persuade a butterfly to metamorphose into a caterpillar. No doubt it would be very difficult. But the possibility is not completely ludicrous in the same way as the frog/prince transformation. If a biologist claimed to have induced a butterfly to turn into a caterpillar, I would study his report with interest. But if he claimed to have persuaded a pumpkin to turn into a glass coach, or a frog into a prince, I'd know he was a fraud without even looking at the evidence. The difference between the two cases is important.

Tadpoles are larvae of frogs or salamanders. Aquatic tadpoles change radically, in the process called 'metamorphosis', into a terrestrial adult frog or salamander. A tadpole may not be quite as different from a frog as a caterpillar is from a butterfly, but there's not a lot in it. A typical tadpole makes its living as a small fish, swimming with its tail, breathing underwater with gills, and eating vegetable matter. A typical frog makes its living on land, hopping

---

* That's what I said in 2004, and in 2012 Sir John Gurdon received one.

rather than swimming, breathing air rather than water, and hunting live animal prey. Yet, different as they seem, we could easily imagine a frog-like adult ancestor evolving into a tadpole-like adult descendant, because all frogs contain the genes for making a tadpole. A frog 'knows' genetically how to be a tadpole, and a tadpole how to be a frog. The same is true of salamanders and they are rather more like their larvae than frogs are like theirs. Salamanders don't lose their tadpole tails, although the tails tend to lose their vertical keel shape and become rounder in cross section. Salamander larvae are often carnivorous like the adults. And, like the adults, they have legs. The most conspicuous difference is that the larvae have long, feathery external gills, but there are lots of less obvious differences too. Actually, to turn a salamander species into a species whose adult stage was a tadpole would be easy – all it would take is for the reproductive organs to mature early, with metamorphosis suppressed. Yet, if it were only the adult stages that fossilised, it would look like a major, and apparently 'improbable' evolutionary transformation.

And so we come to the axolotl, whose tale this is. It is a strange creature, native to a mountain lake in Mexico. It is of the essence of its tale that it is hard to say exactly what an axolotl is. Is it a salamander? Well, sort of. Its name is *Ambystoma mexicanum*, and it is a close relative of the tiger salamander *Ambystoma tigrinum*, which is found in the same area and more widely in North America as well. The tiger salamander, named for obvious reasons, is an ordinary salamander with a cylindrical tail and dry skin, which walks around on land. The axolotl is not at all like an adult sala-mander. It is like a larval salamander. In fact it is a larval salamander except for one thing. It never turns into a proper salamander and never leaves the water, but mates and reproduces while still looking and behaving like a juvenile. I nearly said the axolotl mates and reproduces while still *being* a juvenile, but this might violate the definition of a juvenile.

Definitions apart, there seems little doubt about what happened in the evolution of the modern axolotl. A recent ancestor was just an ordinary land salamander, probably very like the tiger salamander. It had a swimming larva, with external gills and a deep-keeled tail. At the end of larval life it would metamorphose, as expected, into a dry-land salamander. But then a remarkable evolutionary alteration occurred. Probably under the control of hormones, something shift-ed in the embryological calendar such that the sex organs and sexual behaviour matured earlier and earlier (or it may even have been a sudden change). This evolutionary regression continued until sexual maturity was arriving in what was, in other respects, clearly the larval stage. And the adult stage was chopped off the end of the life history. Alternatively, you may prefer to see the change not as an acceleration of sexual maturity relative to the rest of the body ('progenesis'), but as a slowing down of everything else, relative to sexual maturity ('neoteny').*

Whether the means is neoteny or progenesis, the evolutionary consequence is called paedomorphosis. It is not difficult to see its plausibility. Slowing-down or speeding-up of developmental processes, relative to other developmental processes, happens all the time in evolution. It is called heterochrony and it presumably, if you think about it, must underlie many, if not all, evolutionary changes in anatomical shape. When reproductive development varies heterochronically relative to the rest of development, what may evolve is a new species that lacks the old adult stage. This seems to be what happened with the axolotl.

The axolotl is just an extreme among salamanders. Many species seem to be, at least to some extent, paedomorphic. And others do other heterochronically interesting things. The various species of salamander colloquially called 'newts' have an especially revealing life history.† A newt first lives as a gilled larva in water. Then it emerges from the water and lives for two or three years as a kind of salamander on dry land, having lost its gills and the keel on its tail. But, unlike other salamanders, newts don't reproduce on land. Instead, they return to water, regaining some, but not all, of their larval characteristics. Unlike axolotls, newts don't have gills, and their need to come to the surface to breathe air is an important and competitive constraint on their underwater courtship. Unlike the larval gills, they do regain the keel of the larval tail, and in other respects they resemble a larva. But unlike a typical larva, their reproductive organs develop and they court and mate underwater. The dry land phase never reproduces and, in this sense, one might prefer not to call it the 'adult'.

You might ask why newts bother to turn into a dry land form at all, given that they are going to return to water to breed. Why not

---

* Stephen Jay Gould helpfully sorts out the terminology, in his classic *Ontogeny and Phylogeny*.
† A. Fink-Nottle, *in litt.*

just do what axolotls do: start in water and stay in water? The answer seems to be that there is an advantage to breeding in temporary ponds which form in the wet season and are destined to dry up, and you have to be good on dry land in order to reach them. Having reached a pond, how do you then reinvent your aquatic equipment? Heterochrony comes to the rescue: but heterochrony of a peculiar kind, involving going into reverse after the 'dry adult' has served its purpose of dispersing to a new, temporary pond.

Newts serve to emphasise the flexibility of heterochrony. They remind us of the point I made about how genes in one part of the life cycle 'know' how to make other parts. Genes in dry land salamanders know how to make an aquatic form because that is what they once were; and, to prove it, that is precisely what newts do.

Axolotls are, in one respect, more straightforward. They have lopped the dry land phase off the end of the ancestral life cycle. But the genes for making a dry land salamander still lurk in every axolotl. It has long been known, from the classic work of Laufberger and of Julian Huxley mentioned in the Epilogue to Ardi's Tale, that they can be activated by a suitable dose of hormones in the laboratory. Axolotls treated with thyroxine lose their gills and become dry land salamanders, just as their ancestors once did naturally. Perhaps the same feat could be achieved by natural evolution, should selection favour it. One way might be a genetically mediated raising of the natural production of thyroxine (or an increase in sensitivity to the existing thyroxine). Maybe axolotls have undergone paedomorphic and reverse-paedomorphic evolutions repeatedly during their history. Maybe evolving animals in general are continually, though less dramatically than the axolotl, moving one way or another along an axis of paedomorphosis/reverse-paedomorphosis.

Paedomorphosis is one of those ideas of which, once you get the hang of it, you start seeing examples everywhere you look. What does an ostrich remind you of? During the Second World War my father was an officer in the King's African Rifles. His batman Ali, like many Africans of the time, had never seen most of the large wild animals for which their homeland is famous, and his first glimpse of an ostrich sprinting across the savannah elicited a shriek of astonishment: 'Big chicken, BIG CHICKEN!' Ali had it nearly right, but more penetrating would have been 'Big baby chicken!' The wings of an ostrich are silly little stubs, just like the wings of a newly hatched chick. Instead of the stout quills of a flying bird, ostrich feathers are coarse versions of the fluffy down of a baby chick. Paedomorphosis illuminates our understanding of the evolution of flightless birds such as the ostrich and the dodo. Yes, the economy of natural selection favoured downy feathers and stubby wings in a bird that did not need to fly (see the Elephant Bird's Tale and the Dodo's Tale). But the evolutionary route that natural selection employed to achieve its advantageous outcome was paedomorphosis. An ostrich is an overgrown chick.

Pekinese dogs are overgrown puppies. Pekinese adults have the domed forehead and the juvenile gait, even the juvenile appeal, of a puppy. Konrad Lorenz has wickedly suggested that Pekineses and other babyfaced breeds like King Charles spaniels appeal to the maternal instincts of frustrated mothers. The breeders may or may not have known what they were trying to achieve, but they surely didn't know that they were doing it through an artificial version of paedomorphosis.

Walter Garstang, a well-known English zoologist of a century ago, was the first to emphasise the importance of paedomorphosis in evolution. Garstang's case was later taken up by his son-in-law Alister Hardy, who was my professor when I was an undergraduate. Sir Alister delighted in reciting the comic verses which were Garstang's preferred medium for communicating his ideas – a fragment of one of them heads the Lancelet's Tale. They were slightly funny at the time but not, I think, quite funny enough to justify the elaborate zoological glossary which would have to accompany a reprinting here. Garstang's idea of paedomorphosis, however, is today as interesting as ever – which doesn't necessarily mean it is right.

We can think of paedomorphosis as a kind of evolutionary gambit: Garstang's Gambit. It can in theory herald a whole new direction in evolution: can even, Garstang and Hardy believed, permit a dramatic and, by geological standards, sudden breakout from an evolutionary dead end. This seems especially promising if the life cycle sports a distinct larval phase like a tadpole. A larva that is already adapted to a different way of life from the old adult is primed to swerve evolution into a whole new direction by the simple trick of accelerating sexual maturity relative to everything else.

Among the close cousins of the vertebrates are the tunicates or sea squirts. This seems surprising because adult sea squirts are sedentary filter-feeders anchored to rocks or seaweeds. How can these soft bags of water be cousins to vigorously swimming fishes? Well, the adult sea squirt may look like a bag, but the larva looks like a tadpole. It is even called a 'tadpole larva'. You can imagine what Garstang made of this, and we shall revisit the point, and unfortunately cast doubt on Garstang's theory, at Rendezvous 23 when we meet the sea squirts.

Bearing in mind the adult Pekinese as an overgrown puppy, think of the heads of juvenile apes. What do they remind you of? Wouldn't you agree that a juvenile chimpanzee or orang utan is more humanoid than an adult chimpanzee or orang utan? Admittedly it is controversial, but some biologists regard a human as a juvenile ape. An ape that never grew up. An ape axolotl. We have already met the idea in the Epilogue to Ardi's Tale, and I shall not spell it out again here.

LUNGFISH

Rendezvous 18 takes place underwater, in the warm and shallow seas of the lower Devonian Period, about 415 million years ago. This marks a radical change in the environment to which we must acclimatise, for our ancestry will remain entirely marine for the rest of our backwards journey.

The concestor that we meet here, approximately our 185-million-greats-grandparent, was a sarcopterygian, a lobefin fish (see plate 22). In the same way that we accept whales as mammals, descended from a mammalian common ancestor, we land-lubbing tetrapods should strictly classify ourselves as lobe-finned fish, albeit highly modified by our full-bodied commitment to life in air. It is a modification that happened piecemeal, some time between Rendezvous 18 and 17, as our ancestors meandered their way from water to land. The story demands a tale, recounted in a moment by the trickle of pilgrims who join us here, six species who never really left the water. They are our closest fishy relatives, the lungfish.

Lungfish still retain the basic appearance of our shared common ancestor, so it may surprise you to learn that they have subsequently evolved enormous genomes. In fact one species, the marbled lungfish, currently holds the record for the largest genome of any animal: 133 billion letters, compared to our measly 3 billion. This highlights the rather subtle link between a creature's DNA and its outward appearance, an important topic to which we shall return when we meet the only other living lobe-fins, the deep-sea coelacanths.

The six species of lungfish that have survived to the present all live in freshwater and range in length from half a metre to a whopping 2 metres. They are *Neoceratodus forsteri* from Australia, *Lepidosiren paradoxa* from South America, and four species of *Protopterus* from Africa. The Australian lungfish is the most distantly related to the others, and looks really quite excitingly like an ancient sarcopterygian, with fleshy lobed fins. The similarity is weaker in the African and South American species, because their fins have been reduced to long trailing tassels. These fins may appear tassel-like, but captive West African lungfish have recently been observed using these seemingly

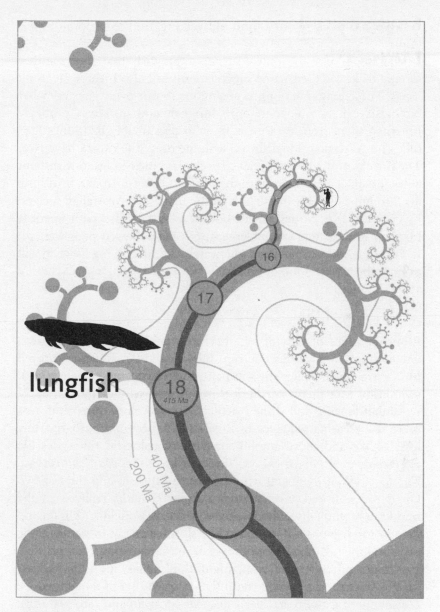

Lungfish join. Humans and the other 'tetrapods' could be said to be lobe-finned fish, whose arms, wings, or legs are modified lobed fins. The two other living lineages of lobefins are the coelacanths and the lungfish. The division of these three lineages at the end of the Silurian is thought to have happened in a very short space of time. This makes the order of branching difficult to sort out, even using genetic data. Nevertheless, genetic and fossil studies are starting to agree that the six lungfish species are the closest living relatives of the tetrapods, as shown here.

vestigial fins to carry out a form of 'walking' on the bottom of their tank.

Walking is not the only feature of lungfish that seems rather tetrapod-like. As their name suggests, lungfish also breathe air using lungs. These lungs are not just analogous to our own. They are what biologists call homologous: both lungfish and ourselves probably inherited them from Concestor 18, who presumably had lungs herself. The Australian lungfish has a single lung, the others have two. The African and South American species use their lungs to withstand a dry season. They burrow into the mud and stay dormant, breathing air through a little breathing hole in the mud. The Australian species, by contrast, lives in permanent bodies of water filled with weed. It takes air into its lung to supplement its gills in oxygen-poor water.

From their vantage point in the water, the air-gulping lungfish will ponder the colonisation of the land.

THE LUNGFISH'S TALE

From the point of view of a lungfish, we land-dwelling vertebrates, we tetrapods, are weirdly aberrant lobe-finned fishes, who ventured into a strange and alien environment, walking on our fins instead of swimming with them. We stem from an otherwise extinct group of Devonian lobefins called the osteolepiforms. Among osteolepiforms there is a pleasing sequence of fossil forms, becoming progressively less fishlike and more amphibian-like as geological time goes by: *Eusthenopteron, Panderichthys, Tiktaalik, Acanthostega, Ichthyostega.* The first two of these you would want to call fish, and the last two you'd probably recognise as amphibians. *Tiktaalik* really is in the middle, a recently discovered and beautiful 'missing link'. Missing no longer, for it plugs a notorious gap, slotting neatly in between the fishy and the salamandery members of the series.

A neat and tidy story, you might think. Alas, life is never that simple, or that elegant. Very recently a splendid set of footprints, the 'Zachełmie trackway', has been discovered in Poland (see plate 23). The spoors are of an undoubted tetrapod – you can distinguish actual toeprints – and unfortunately for our 'neat and tidy' story they are nearly 20 million years older than they 'should' be if they are to conform to it. The tracks were made underwater, on the bottom of what was probably a shallow lagoon, but whoever made them was unequivocally walking. Presumably the neat series of fossils mentioned above are late survivors who have retained their ancestral way of life. In any case it's always too much to hope that we are literally descended from any particular fossil species such as *Tiktaalik*. Rather, the fossil series should be taken as telling us something about the kind of changes that took place in those remote Devonian waters, changes which may even have taken place more than once.

Whoever they were, and whenever they did it, we need to ask why those ancient lobefins developed the changes that permitted the move out of water onto the land. Lungs, for example? And fins that you could walk on rather than, or as well as, swim with? It wasn't that our ancestors were trying to initiate the next big transition in evolution! That's not how evolution works. Ancient creatures such as *Tiktaalik* and *Acanthostega* may not even have had much to do with the land. In these species, limbs and lungs seem to have been adaptations for specific sorts of watery lifestyle, which remained successful for over 20 million years after legs came on the scene. In the same vein, we previously noted that modern lungfish with highly reduced fins have been observed 'walking' on surfaces underwater. Perhaps other modern fish can provide clues too? They include species who seem to be travelling the same evolutionary journey yet again, in our own time – not too surprising, given the great versatility of fish.

A number of modern teleost (not lobefin) fish live in swampy water, poor in oxygen. Their gills cannot extract enough, and they need help from the air. Familiar aquarium fish from the swamps of South East Asia, such as the Siamese fighting fish *Betta splendens*, frequently come to the surface to gulp air, but they still use their gills to extract oxygen. I suppose, since the gills are wet, you could say the gulping is equivalent to locally oxygenating their gill water, as you might bubble air through your aquarium. It goes further than that, however, because the gill chamber is furnished with an auxiliary air space, richly supplied with blood vessels. This cavity is not related to our own lungs, or those of a lobe-finned lungfish from which ours are descended. As we shall see in the Pike's Tale, teleost fish have transformed that original 'lung' into the swim bladder, which they use for keeping their buoyancy neutral. Those modern teleost fish that breathe air through their gill chamber have rediscovered air-breathing by a completely different route. Perhaps the most advanced exponents of the air-breathing gill chamber are the climbing perches *Anabas*. These fish also live in poorly oxygenated water, but their air-breathing abilities allow them to migrate over land into flooded areas during the wet season, and return to deeper waters as their temporary homes dry up. They can survive out of water for days at a time.

Using fins to walk on land is not a difficult recourse for fish to evolve, as a temporary measure. There's even a shark, the epaulette shark (*Hemiscyllium ocellatum*) which does it. Lacking a lung, it survives oxygen depletion by other means, and it uses its fins to walk between tide pools.

Another group of walking (teleost) fish are the mudskippers, for example *Periophthalmus*. Some mudskippers actually spend more time out of water than in it. They eat insects and spiders, which are not normally found in the sea. It is possible that our Devonian ancestors enjoyed similar benefits when they first left the water, for they were preceded onto the land by both insects and spiders. A mudskipper flaps its body across the mudflats, and it can also crawl using its pectoral (arm) fins, whose muscles are so well developed that they can support the fish's weight. Indeed, mudskipper courtship takes place partly on land, and a male may do push-ups, as some male lizards do, to show off his golden chin and throat to females. The fin skeleton, too, has evolved convergently to resemble that of a tetrapod such as a salamander. Mudskippers can jump more than half a metre by bending the body to one side and suddenly straightening it – hence some of their many vernacular names, including 'mud-hopper', 'johnny jumper', 'frogfish' and 'kangaroo fish'. Another common name, 'climbing fish', comes from their habit of climbing mangrove trees looking for prey. They cling to the trees with the pectoral fins, aided by a kind of sucker which is made by bringing the pelvic fins together under the body.

Like the swamp fish already mentioned, mudskippers breathe by taking air into their moist gill chambers. They also take in oxygen through the skin, which has to be kept moist. If a mudskipper is in danger of drying out, it will roll about in a puddle. Their eyes are especially vulnerable to dryness, and they sometimes wipe them with a wet fin. The eyes bulge close together near the top of the head, where, as with frogs and crocodiles, they can be used as periscopes to see above the surface when the fish is underwater. When out on land, a mudskipper will frequently withdraw its bulging eyes into their sockets to moisten them. Before leaving the water on a land sortie, the fish will fill its gill cavities with water. In a popular book on the conquest of the land, the author mentions an account by an eighteenth-century artist living in Indonesia who kept a 'frogfish' alive for three days in his house: 'It followed me everywhere with great familiarity, much like a little dog.' I like the idea that we are descended from some creature which, even if it was different from a modern mudskipper in many other respects, was as adventurous and enterprising as a little dog: the nearest thing, perhaps, to a dog that the Devonian had to offer? A girlfriend of mine from long ago explained why she loved dogs: 'Dogs are such good sports.' The first fish to venture out onto the land must surely have been an archetypal good sport.

Mudskippers may provide an even better analogy than anticipated. New 3D computer reconstructions of the appropriately four-foot-long *Ichthyostega*, one of the most venerable and terrestrial of the Devonian fossils, reveal that it propelled itself somewhat like an enormous mudskipper. It certainly couldn't slither like a salamander: its weak rear legs may not even have been able to touch the ground properly. *Ichthyostega* must have spent a good deal of time in the water, indeed its ears were specialised for underwater use, but when in the shallows or on land, it advanced by skipping or crawling using its strong front legs.

What selection pressures might have impelled its ancestors, and our own, to venture out of water onto the uncharted land? For years, the favoured answer to the question was one that the eminent American palaeontologist Alfred Sherwood Romer derived from the geologist Joseph Barrell. The idea was that these fish, far from struggling to colonise the land, were actually trying to get back to water. In times of drought, fish can easily become stranded in drying pools. Individuals capable of walking and of breathing air have the enormous advantage that they can forsake a doomed, drying pond and set out for a deeper one elsewhere. This admirable theory has become unfashionable but not, I think, for uniformly good reasons. Unfortunately, Romer sought to buttress his theory by quoting the prevailing belief of his day that the Devonian was a time of drought, a belief that has more recently been called into question. But I don't think Romer needed his Devonian desiccated. Even at times of no particular drought, there will always be some ponds shallow enough to be in danger of becoming too shallow for some particular kind of fish. If ponds three feet deep would have been at risk under severe drought conditions, mild drought conditions will render ponds one foot deep at risk. It is sufficient for the Romer hypothesis that there are some ponds that dry up, and therefore some fish that could save their lives by migrating. Even if the world of the late Devonian was positively waterlogged, one could say this simply increases the number of ponds available to dry up, thereby increasing opportunities for saving the life of walking fish and the Romer theory. Nevertheless, it is my duty to record that the theory is now unfashionable.

Surely there are plenty of reasons other than drought for a fish to emerge, temporarily or permanently, onto land. Streams and ponds can become unusable for reasons other than drying up. They can become choked with weeds, in which case, again, a fish that can migrate over land to deeper water might benefit. If, as has been suggested *contra* Romer, we are talking Devonian swamps rather than Devonian droughts, swamps provide plenty of opportunities for a fish to benefit by walking, or slithering or flip-flopping or otherwise travelling through the marshy vegetation, in search of deep water or, indeed, food. This still retains the essential Romer idea that our ancestors left the water, not at first to colonise land, but to return to water.

Since the first edition of this book was published, my colleague the Savilian Professor of Astronomy at Oxford, Steven Balbus, has proposed a provocative idea which, if correct, would vindicate both Romer's theory and my earlier questioning of its unfashionableness. His theory is all about tides. He suggests that our fishy ancestors first started to walk from tide pool to tide pool. And it is important for his theory that our planet has a dramatically fluctuating tidal reach.

The essence of the Romer theory is that our lobefin ancestors ventured onto the land, not following some yearning to graze pastures new and pioneer major evolutionary breakthroughs. They walked on land in order to return to water. Setting drought aside, where else do we see pools separated from other pools by inhospitable land? In the intertidal zone where pools are flooded at high water and separated as the tide recedes.

But that, by itself, won't do for a theory. Balbus's idea is a lot more ingenious. If the tides simply ebbed and flowed to the same high and low water points every day, the inducement to walk from pool to pool would be slight. A fish stranded in a high tide pool would have to wait only a few hours to be connected to the sea again. That's the situation we'd be in if the tides were governed by the gravitational pull of the moon alone, or the sun alone. But they aren't, and that's the core of Balbus's theory. Our tides are governed by a complex interplay of gravitational forces exerted by both the sun and the moon, sometimes adding to each other, sometimes subtracting. The fluctuations come about because of the interaction between our orbit around the sun, the moon's orbit around us, and the Earth's daily rotation about its own axis. The combined effect is that very high tide pools are not refreshed every day but can be separated from the sea, and from other pools, for weeks or even months at a stretch.

But even this version of the theory wouldn't work if the moon were small, as most planetary moons are. Balbus notes, however, that our moon is much bigger than it 'should' be for a planet of Earth's size. This probably stems from the fact, now increasingly securely established, that the moon was formed from an early collision with another planet, which then disintegrated and re-formed with some of Earth's outer layers into a sphere. Our moon is so large that its apparent diameter, as seen by us, is almost the same as the sun's. This is why solar eclipses are so spectacular: the moon almost exactly covers the sun.

The tidal force exerted by a body is proportional to the cube of its apparent diameter, multiplied by its density. This means that, all other things being equal, the sun and the moon should exert the same tidal force on the Earth, because they appear the same size to us. But the moon is more dense than the sun, so it has a stronger influence on the tides, although the two forces are of comparable order of magnitude. The consequence is that we have spring tides (high amplitude) and neap tides (low amplitude) depending on whether the sun and moon line up to combine tidal forces, or are at right angles to the Earth, causing their forces partially to cancel out. Spring tides create and refresh tide pools much further inland than do neap tides. While a fish could benefit from feeding in a high spring tide pool, it would risk being stranded. It might have to wait weeks before the sea washes over the spring tide pool again. Such a fish could use some rudimentary legs to walk seawards where it might find a lower pool.

Balbus points out that the moon is gradually drifting away from Earth. When it was first formed, soon after Earth itself, the moon would have seemed gigantic, filling the sky and eclipsing the sun for much of most days. During those Hadean times its tidal force would have swamped that of the sun. Had there been any oceans at that time, every tide would have been a titanically huge 'spring' tide. By the time the Devonian came along, the moon had receded far enough to have looked only 10 per cent larger than it is today. At that time the 'Balbus Effect' would have been significant: not only would the moon's tidal reach have been larger, there would have been a strong difference between spring and neap tides. Balbus adds that the geographical disposition of the supercontinents, Laurasia to the north and Gondwana to the south, would have fostered extra-large tidal surges in the narrow Tethys Sea funnelled between them.

Balbus's conclusion reinforces mine from the first edition of this book, that Romer's theory was too hastily discarded. Devonian lobefins did indeed develop their eponymous leg-like fins to help them – if not to walk at least to flounder and squirm – from pool to pool. But the pools were isolated not necessarily by drought but perhaps by dramatically varying tides, a tidal variation which would not be normal for a planet of our size but which came to pass by the fortunate happenstance of our abnormally large moon. You could (and he does) express the point anthropically. Our kind of terrestrial life could be found only on a planet that has a large moon whose gravitational pull was comparable to that of its star.

There is a tide in the affairs of Osteolepiforms which, taken at the flood, leads on to fortune . . . ? Well, not really. At least to begin with that would have seemed exaggeratedly prophetic. The stubbily eponymous lobed fins that helped our ancestors move from pool to pool, or helped them to flap and flounder their way across Devonian swamps as others would have it, gradually evolved to become stronger and stronger until they became pre-adapted to a true emergence onto the land.

COELACANTHS

Concestor 19, perhaps our 190-million-greats-grandparent, lived around 420 million years ago, just as plants were colonising the land and coral reefs expanding in the sea. At this rendezvous we meet one of the sparsest, most tenuous bands of pilgrims in this story. We know of only one genus of coelacanth alive today, and its discovery was a huge surprise when it happened. The episode is well described by Keith Thomson in his *Living Fossil: the Story of the Coelacanth.*

The coelacanths were well known in the fossil record, but thought to have gone extinct before the dinosaurs. Then, astoundingly, a living coelacanth turned up in the catch of a South African trawler in 1938. By good fortune Captain Harry Goosen, skipper of the *Nerita*, was friendly with Marjorie Courtenay-Latimer, the enthusiastic young curator of the East London Museum near Port Elizabeth. It was Goosen's habit to put aside interesting finds for her, and on 22 December 1938 he telephoned to tell her he had something. She went down to the quay, and an old Scotsman of the crew showed her a motley collection of discarded fish, which at first didn't seem of any interest. She was about to leave when:

> I saw a blue fin and pushing off the fish, the most beautiful fish I had ever seen was revealed. It was 5 feet long and a pale mauve blue with iridescent silver markings.

She made a sketch of the fish, which she sent to a friend, a chemistry professor named Dr J. L. B. Smith, who she knew to be a fish expert. It knocked him out. 'I would not have been more surprised if I had seen a dinosaur walking down the street.' (See plate 24.) Unfortunately, Smith took his time going to the scene, for reasons that are hard to fathom. By Keith Thomson's account, Smith didn't trust his judgement until he had sent off for a particular reference book, from Dr Barnard, a colleague in Cape Town. Smith hesitantly confessed his secret to Barnard, who was immediately sceptical. It seems that it was weeks before Smith could bring himself to go to East London and actually see the fish. Meanwhile, poor Miss Courtenay-Latimer was coping with its noisome decay. As it was too large to fit in a

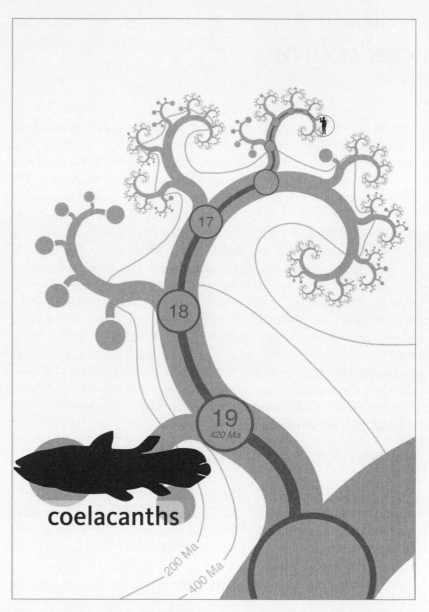

Coelacanths join. Growing consensus places coelacanths (of which there are two living species known) as the earliest diverging of living lobe-fin lineages.

formalin jar, she wrapped it in formalin-soaked cloths. These were inadequate to stave off decay, and eventually she had to have it stuffed. It was in this form that Smith finally saw it:

> Coelacanth, yes, God! Although I had come prepared, that first sight hit me like a white-hot blast and made me feel shaky and queer, my body tingled. I stood as if stricken to stone . . . I forgot everything else and then almost fearfully went close up and touched and stroked, while my wife watched in silence . . . It was only then that speech came back, the exact words I have forgotten, but it was to tell them that it was true, it really was true, it was unquestionably a coelacanth. Not even I could doubt any more.

Smith named it *Latimeria* after Marjorie. It was to be 14 years until he saw another, thanks to reward posters placed on the Comoros Islands near Madagascar. It turned out that a few coelacanths each year were being landed there as bycatch. Nearly all specimens that have been since collected come from this Comoros population, and none are alive: no coelacanth has ever survived more than 24 hours at the surface.

The coelacanth's celebrity status has – regrettably but perhaps understandably – led to acrimony, accusations of fakery, and even the deaths of three divers. The problems spring largely from the fact that coelacanths tend to live 150 metres or more underwater, making them exceptionally difficult to find and study. It was only in 1997 that a second, Indonesian species was discovered by Mark Erdmann off Sulawesi. And only at the turn of the century, four years before Marjorie Latimer died, did divers stumble across a permanent South African population, 500 miles north from her first dazzling discovery, the most beautiful fish she had ever seen.

THE COELACANTH'S TALE

One of the commonest phrases used to describe the coelacanth is 'living fossil'. The expression, due to Darwin, has been somewhat criticised of late,[*] and not just because it derives its redolent charm from a contradiction in terms. As we shall see, it has the potential to be rather misleading, not least in the case of the coelacanth itself.

There are some living things that, while being as alive as you or

[*] Including by Richard Fortey in his book on this very topic: *Survivors*.

me, strongly resemble their ancient ancestors. It seems sensible to have a term for this, and 'living fossil' does the job. It is one of those random, pointless facts that the four most famous living fossils all begin with L: *Limulus*, *Lingula*, lungfish and *Latimeria*. *Limulus*, the so-called 'horseshoe crab' (not a crab at all, but its own thing, superficially resembling a large trilobite) is placed in the same genus as *Limulus darwini* of the Jurassic, 150 million years ago. *Lingula* belongs to the phylum Brachiopoda, sometimes called lamp-shells. The kind of lamp they resemble, if any, is the Aladdin variety with its wick coming out of a kind of teapot spout, but what *Lingula* spectacularly resembles is its own ancestors of 400 million years ago. Its assignment to the very same genus has been revised, but the fossil forms are still remarkably similar to their modern representatives. The various lungfish were introduced at the previous rendezvous, where we noted the resemblance of ancient sarcopterygian fish to the Australian lungfish, *Neoceratodus*. Indeed, when it was first discovered in 1870, it was initially classified in the same genus as fossils from over 200 million years ago, *Ceratodus*: the *Neo* is, appropriately, a new addition.

That leaves *Latimeria*, the coelacanth. Unlike the other three, it has never been placed in the same genus as a 'dead fossil'. Its nearest known relatives, from 70 million years ago, are classified in the genus *Macropoma*. They differed in several ways from living coelacanths, not least because they were two and a half times shorter. A much greater range of forms exists in *Latimeria*'s slightly more distant relatives too. There are trout-shaped freshwater species, eel-like ones, even one coelacanth species that looked a bit like a piranha.

That provides our first glimmer of uncertainty about applying the term 'living fossil' to the coelacanth. Still, it is indisputably true that the bodies of the surviving fishy lobefin species – in particular, lungfish – have evolved at a snail's pace relative to us transformed, terrestrial lobefins. Yet at the same time – here is the main point this whole tale is leading up to – this is not the case for their genomes.

We know from fossils the approximate timings of the ancestral splits between coelacanths, lungfish and ourselves. The first split, at about 420 million years ago, is that between the coelacanths and the rest. Then, perhaps 5 million years later, the lungfish split off, leaving the rest of us, now called tetrapods, to make our own evolutionary way. As evolutionary time goes, these splits occurred at pretty nearly the same time, at least compared to the long time over which the groups have been evolving ever since. What changes in the DNA have occurred in the different lineages over that time?

We can answer that question pretty comprehensively, because the coelacanth has now joined that ever-growing band of species to have had its full genome sequenced. That is not yet the case for any lungfish species because, as previously noted, they have unmanageably big genomes – possibly the largest of any animal. That provides one quick answer to our question. Lungfish demonstrate that it is perfectly possible to undergo enormous changes in DNA without much happening on the outside. There is no obvious link between the overall structure of a genome and the overall structure of the body that it generates. One reason for this disconnect is that much of the genome is composed of essentially useless DNA, in particular various parasitic sequences to which we shall return in the Humped Bladderwort's Tale. These 'transposable elements' probably account for most of the bloat in lungfish genomes. Coelacanth transposable elements have not gone quite so wild. Their genome is about the same size as ours. Nevertheless, as in humans, various genomic parasites still make up a significant fraction of the coelacanth genome, and are still actively jumping around, causing substantial changes in evolutionary time. In this sense, the speed of evolution of the coelacanth genome seems broadly typical of other vertebrates, indeed most other animals, fungi and plants.

Perhaps this is not a surprise. After all, these parasitic sequences usually have little impact on the exterior appearance of their hosts. As long as they don't disrupt biochemical function they can replicate rather freely around the genome. Arguably, to see if the coelacanth's DNA has been evolving slower than expected, we should not be looking at large but essentially invisible changes to the genome. Instead, we should look at the evolution of genes which code for proteins, which we know to have an impact on an organism's form and function. That was the focus of a study by Dan Liang and colleagues from China. The tree diagram overleaf shows their results from 1,290 protein-coding genes, compared between a number of different vertebrate genomes (including partial extracts from an African lungfish genome). The branch lengths are a measure of the numbers of letter-by-letter mutations that have accumulated in these specific genes: they should not be particularly affected by larger-scale changes elsewhere, such as the accumulated genetic debris in lungfish genomes.

If protein-coding genes evolved at a constant rate, regardless of

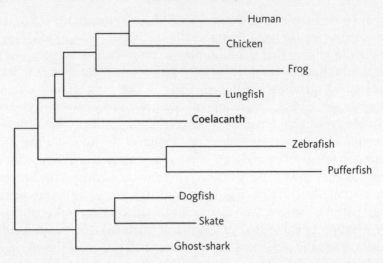

A molecular caucus race. Rate of protein evolution in various vertebrates, including two 'living fossils'. Modified from Liang *et al.* [248]

species, then we would expect all the branches to line up at the right-hand edge. This clearly isn't the case. But neither do the organisms that show the least morphological change have the shortest branches. True, the coelacanth seems to lie on quite a short branch. But that's not as a result of its famed 'living fossil' status, because the lungfish, arguably a better candidate for the term, has seen about as much evolutionary change as a chicken – and a chicken's lifestyle is perhaps the most radically different of any of the animals on the tree. We land-colonising lobefins have experienced a generally faster rate of gene evolution than our aquatic lobe-finned cousins, but even this is not obviously linked to morphological change. And the winners' podium in this molecular caucus race is occupied by a pufferfish, a zebrafish, and lastly a frog, none of which have evolved morphologic-ally as fast as, say, chickens or (vanity cannot help whispering) us.

The diagram illustrates an important fact. The rate of DNA evolu-tion is not always constant, but neither is it obviously correlated with morphological change. The tree above is just one example. Another is provided by the tuatara, a rare lizard-like animal that lives only in New Zealand. This 'living fossil' is generally thought to resemble the common ancestor of the snake and lizard branch of the sauropsid pilgrims who joined us at Rendezvous 16. Yet analysis of tuatara sequences has shown that their rate of mitochondrial DNA evolution is among the fastest known of any land vertebrate.

Some time ago, Lindell Bromham of the University of Sussex and her colleagues attempted a more general study, collecting many studies and comparing morphological change against equivalent trees based on DNA change. And what they found confirmed the message of the Coelacanth's Tale. The overall rate of genetic change is independent of morphological evolution.*

This is not to say that it is constant – that would have been too good to be true. Certain lineages, such as sea squirts, nematode worms and flatworms, seem to have a rather fast overall rate of molecular evolution compared to close relatives. In other instances, such as in the mitochondria of sea anemones, the rate is much slower than related lineages. But the Coelacanth's Tale encourages a hope that, a few decades ago, no zoologist would have dared to entertain. With due caution in choosing genes, and with available methods of correcting for lineages that show variable rates of evolution, we should be able to put a figure, in millions of years, on the time of separation of any species from any other species. This bright hope is the technique known as the 'molecular clock', which we have mentioned a number of times on our pilgrimage, and is responsible for most of the quoted dates on our rendezvous points in this book. The principle of the molecular clock, and the controversies that still bedevil it, will be explained in the Epilogue to the Velvet Worm's Tale. Now, however, it is time to meet the fastest evolving of the animals in our little contest, who join us in a collective band of over 30,000 species. By some counts that makes them most successful of all vertebrates. They are the ray-finned fishes.

---

* An earlier study had obtained a different result. But Bromham and her colleagues convincingly showed that the previous study had failed to allow for non-independence of data – the multiple counting problem that we met in the Seal's Tale.

RAY-FINNED FISH

Rendezvous 20 is a big one, 430 million years ago in the mid Silurian, still with a southern ice cap left over from the cold Ordovician. Concestor 20, which I am estimating to be our 195-million-greats-grandparent, is the one that unites us to the actinopterygian or ray-finned fish, most of whom belong to the large and successful group known as teleosts. The teleost fish are the great success story among modern vertebrates – there are more than 30,000 species of them. They are prominent at many levels of underwater food chains, in both salt- and freshwater. They have managed to invade hot springs at one extreme, and the icy waters of the Arctic seas and high mountain lakes at the other. They thrive in acid streams, stinking marshes and saline lakes.

'Ray' refers to the fact that their fins have a skeleton similar to a Victorian lady's fan. Ray-fins lack the fleshy lobe at the base of each fin – eponym for the lobefin fish like coelacanths and Concestor 18. Unlike our arms and legs, which have relatively few bones, and muscles that can move them relative to one another within the limb, actinopterygian fins are moved mostly by muscles in the main body wall. In this respect, we are more like lobefin fish – as well we should be, for we are lobefins adjusted for life on land. Lobefin fish have muscles in the fleshy fins themselves, just as we have biceps and triceps muscles in our upper arms and Popeye muscles in our lower arms.

The ray-finned fish are mostly teleosts, plus a few odds and ends, including the sturgeons, and the paddlefish whom we met in the Duckbill's Tale. It is right and proper that such a hugely successful group should contribute several tales and I shall relegate most of what I have to say about them to the tales. The teleost pilgrims arrive in a jostling crowd, brilliant in their variety. The magnitude of that variety is the inspiration for the Leafy Sea Dragon's Tale.

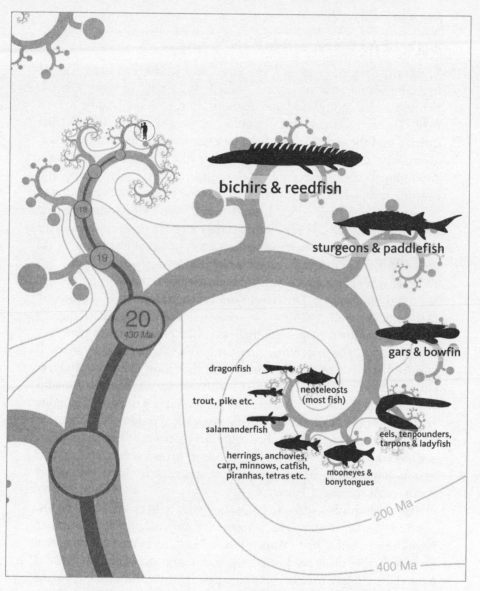

Ray-finned fish join. The ray-finned fish are the closest relatives of we lobefins, and if we count all the tetrapods in with the lobefins, both groups contain roughly the same number of described species – about 30,000. More precisely, as of April 2015 the vast online dataset known as FishBase numbers 33,000 'fish' species, of which over 32,000 are ray-fins. The relationships within the ray-fins are the focus of a project called DeepFin, the results of which are shown here. The deepest branching ray-fins are the bichirs and reed fish (12 species), followed by the sturgeons and paddlefish (27 species), followed by the gars and bowfin (8 species). All the rest are teleosts.

THE LEAFY SEA DRAGON'S TALE

When my daughter was tiny, she loved to ask adults to draw fish for
her. She would rush up to me when I was trying to write a book,
thrust a pencil in my hand and clamour, 'Draw a fish. Daddy, draw a
fish!' The cartoon fish that I would immediately draw to keep her
quiet – and the only kind of fish she ever wanted me to draw – was
always the same: a regulation-issue fish like a herring or a perch,
streamlined side view, pointed at the front end, triangular fin top and
bottom, triangular tail at the back, finally dotted with an eye bracketed
by the curve of a gill cover. I don't think I ran to pectoral or pelvic
fins, which was remiss of me because they all have them. The standard
fish is indeed an extremely common shape, one that obviously works
well over the full range of sizes from minnow to tarpon.

What would Juliet have said if I had possessed the skill to draw for
her a leafy sea dragon, *Phycodurus equus*? 'NO, Daddy. NOT seaweed.
Draw a fish. Draw a FISH.' The message of the Leafy Sea Dragon's
Tale is that animal shapes are malleable like plasticine (see plate 25).
A fish can change in evolutionary time to whatever unfishy shape is
required for its way of life. Those fish that look like the standard-
issue Juliet fish do so only because it suits them. It is a good shape for
swimming through open water. But if survival is a matter of hanging
motionless in beds of gently swaying kelp, the standard fish shape can
be twisted and kneaded, pulled out in fantastically branched projections whose resemblance to the fronds of brown seaweed is so great
that a botanist might be tempted to narrow it down to species
(perhaps of the genus *Fucus*).

The shrimpfish, *Aeoliscus strigatus*, which lives on reefs in the
western Pacific, is again much too cunningly disguised for Juliet to
have been satisfied, had I drawn it as a 'fish'. Its extremely elongated
body is further prolonged by a long snout, and the effect is enhanced
by a dark stripe running right through the eye and straight to the
very untail-like tail. The fish looks like a long shrimp, or a little like a
cut-throat razor – which accounts for its other name of razorfish. It is
covered with a transparent armour which, my late colleague George
Barlow who watched them in the wild told me, even feels like that of
a shrimp. The resemblance to a shrimp is probably, however, no part
of their camouflage. Like many teleosts, shrimpfish swim around in
coordinated groups, and with military synchrony. But unlike any

other teleost you might think of, shrimpfish swim with the body pointing straight down. I don't mean they swim in a vertical direction. They swim in a horizontal direction, but with the body vertical. The whole effect of this synchronised swimming is a resemblance to a stand of weeds, or, even more strikingly, to the tall spines of a giant sea urchin, among which they often take refuge. Swimming head down is a deliberate decision. When alarmed, they are perfectly capable of flipping into more conventional, horizontal mode and they then flee with surprising speed.

Or, what would Juliet have said if I had drawn for her a snipe eel (Nemichthyidae) or a gulper eel (*Eurypharynx pelecanoides*), two deep-sea eels with birds in their names? The snipe eel looks like a joke, ludicrously long and thin, with bird-like jaws that curve away from each other like a megaphone. So dysfunctional do these diverging jaws look, I can't help wondering how many of the fish have been seen alive. Could the megaphone jaws be a distortion in a dried-up museum specimen?

The gulper looks like a nightmare. With jaws ludicrously too large for its body, or so it would seem, it is capable of swallowing whole prey larger than itself – one of several deep-sea fish with this remarkable talent. It is not unusual, of course, for predators to *kill* prey larger than themselves, and then eat them in bits. Lions do it; so do spiders.* But to *swallow* a larger animal than yourself whole is hard to imagine. The gulper eel, and other deep-sea fish – such as the closely related swallower eel, and the unrelated black swallower, which is not an eel – achieve the trick. They do it by a combination of disproportionately oversized jaws and a slack distensible stomach that hangs down only when full, looking rather like some gross external tumour. After the long digestion period, the stomach shrinks again. Why the trick of prodigious swallowing should be peculiar to snakes† and deep-sea fish is not obvious to me. The gulper and the swallower eel lure prey into the vicinity of their mouth with a luminous lure at the tip of the tail.

The teleost body plan seems almost indefinitely malleable over evolutionary time, tolerant of being pulled or squashed into any shape, however distantly removed from the 'standard' fish shape. The

---

* Spiders eat the large prey not so much in bits as in the form of soup. They inject it with digestive juices and then suck it in as if through a straw.
† Snakes do it by disarticulating their skulls. For a snake, eating a meal must be an ordeal comparable to giving birth for a woman.

oceanic sunfish's Latin name, *Mola mola*, means millstone, and it is easy to see why. Seen from the side, it looks like a huge disc, up to an astonishing four metres in diameter and weighing up to two tonnes. The circularity of its outline is broken only by two gigantic fins on top and underneath, each one up to two metres long.

The Hippo's Tale invoked, in explanation of its dramatic difference from its whale cousins, the liberation from gravity that whales must have enjoyed, as soon as they severed all ties with the land. No doubt something similar explains the great variety of shapes that the teleost fish display. But in exploiting that liberation, teleosts have one other advantage over, for example, sharks. Teleosts cope with buoyancy in a very special way, and the pike will tell the tale.

THE PIKE'S TALE

In the sad province of Ulster, where 'the Mountains of Mourne sweep down to the sea', I know a beautiful lake. A party of children were swimming naked there one day, when somebody shouted that they had seen a large pike. Instantly all the boys – but not the girls – fled to dry land. The northern pike, *Esox lucius*, is a formidable predator of small fish. It is beautifully camouflaged, not against predators but to help it steal up on its prey. A stealth predator, and not particularly fast over a distance, it hangs almost motionless in the water, creeping imperceptibly forwards until within striking distance. During the deadly creep, it propels itself with imperceptible movements of the rear-mounted dorsal fin.

This whole hunting technique depends upon the ability to hang in the water at the desired level, like a drifting dirigible, without any effort, in perfect hydrostatic equilibrium. All locomotor work is concentrated on the clandestine business of creeping forwards. If a pike needed to swim in order to maintain its level, as many sharks do, its ambush technique would not work. Effortless maintenance, and adjustment, of hydrostatic equilibrium is what teleost fish are supremely good at, and it may be the single most significant key to their success. How do they do it? By means of the swim bladder: a modified lung filled with gas, which provides sensitive dynamic control of the animal's buoyancy. Except for some bottom-dwellers who have secondarily lost the swim bladder, all teleosts have it – not just pike and not just their prey.

The swim bladder is often explained as working like a Cartesian

Diver, but I think that is not quite correct. You can make a Cartesian Diver quite simply by sealing a bubble of air in a weighted straw so that the whole thing only just floats in a bottle of water. If the pressure is increased (say by squeezing the sides of the sealed bottle), the air bubble is compressed and less water is displaced by the diver as a whole. Therefore, by Archimedes' Principle, the diver sinks. If you squeeze less, the pressure in the bottle decreases, the bubble in the diver expands, and the diver rises. With practice you can squeeze just hard enough that the diver hangs in the water at neutral equilibrium. With finer control you can even change the level at which it does so.

The key point about a Cartesian Diver is that the number of air molecules in the bubble remains fixed, while the volume and the pressure are changed (in inverse proportions, following Boyle's Law). If fish worked like Cartesian Divers, they would use muscle power to squeeze, or relax, the swim bladder, thereby changing the pressure and volume but leaving the number of molecules the same. That would work in theory, but it isn't what happens. Instead of keeping the number of molecules fixed and adjusting the pressure, the fish adjust the number of molecules. To sink, the fish absorbs some molecules of gas from its swim bladder into the blood, thereby reducing the volume. To rise, it does the reverse, releasing molecules of gas into the swim bladder.

In fact, the Cartesian Diver effect is responsible for a major problem with air-filled swim bladders: their delicate hydrostatic equilibrium is unstable. A fish that descends in depth will experience more external pressure, which will compress the swim bladder, just like the air bubble in the Cartesian Diver. That will make the fish less buoyant, so it tends to sink, experiencing even greater pressure, and so on in a positive feedback loop. Similarly, all other things being equal, the swim bladder of a rising fish will expand, making it more buoyant and propelling it to the surface with ever increasing rapidity. While it is an urban myth that this expansion can cause fish to explode, it is certainly true that deep-swimming teleosts brought rapidly to the surface by fishermen can have their stomach forced out of their mouths by a hugely overinflated swim bladder. Perhaps for this reason, some teleosts retain the ability to belch excess air from their swim bladder. Others, especially deep-sea fish, go to the expense of filling their swim bladders with relatively incompressible oils. The deep sea coelacanth, incidentally, seems to have convergently filled its 'lung' with lipids for the same reason, which may explain why fishermen report it as tasting disgustingly oily. A few fish such as tuna, whose lifestyles preclude staying still, have even dispensed with the swim bladder altogether. But for most teleosts it is a cheap yet sophisticated way to maintain equilibrium in their watery surroundings.

In some teleosts, the swim bladder is also used to assist in hearing. The fish's body being mostly water, sound waves propagate through it pretty much as they did through the water before they hit the fish. When they strike the wall of the swim bladder, however, they suddenly reach a different medium, gas. The swim bladder therefore acts as a kind of eardrum. In some species it lies right against the inner ear. In others it is connected to the inner ear by a series of small bones called the Weberian ossicles. These do a similar job to our own hammer, anvil and stirrup, but are completely different bones.

The swim bladder seems to have evolved – been 'co-opted' – from a primitive lung, and some surviving teleosts, such as bowfins, gars and bichirs, still use it for breathing. This perhaps comes as a surprise to us, for whom breathing air seems like a significant 'advance' that went with leaving the water for the land. Even Darwin supposed the lung to be a modified swim bladder. On the contrary, it seems that the primitive breathing lung forked in evolution and went two ways. On the one hand, it carried its old breathing function out onto the land, and we use it still. The other branch of the fork was the new and exciting one: the old lung became modified to form a genuine innovation – the swim bladder.

THE CICHLID'S TALE

Lake Victoria is the third largest lake in the world, but it is also one of the youngest. Geological evidence indicates that it is only about 100,000 years old. It is home to a huge number of endemic cichlid (pronounced 'sick-lid') fish. Endemic means that they are found nowhere else than in Lake Victoria, and presumably evolved there. Depending on whether your ichthyologist is a lumper or a splitter, the number of species of cichlid in Lake Victoria is somewhere between 200 and 500, and a recent authoritative estimate puts it at 450. Of these endemic species, the great majority belong to one tribe, the haplochromines. It looks as though they all evolved, as a single 'species flock', during the last hundred thousand years or so.

As we saw in the Narrowmouth's Tale, the evolutionary splitting of one species into two is called speciation. What surprises us about the young age of Lake Victoria is that it suggests an astonishingly high rate of speciation. There is also evidence that the lake dried up completely about 15,000 years ago, and some people even drew the conclusion that the 450 endemic species must have evolved from a single founder in this astonishingly short time. As we shall see, this is probably an exaggeration. But in any case a little calculation helps to get these short times into perspective. What sort of speciation rate would it take to generate 450 species in 100,000 years? The most prolific pattern of speciation in theory would be a succession of doublings. In this idealised pattern, one ancestral species gives rise to two daughter species, each of those splits into two, then each of those splits into two, and so on. Following this most productive ('exponential') pattern of speciation, an ancestral species could easily generate 450 species in 100,000 years, with what seems like the rather long interval of 10,000 years between speciations within any one lineage. Starting with any one modern cichlid pilgrim and going backwards, there would be only ten rendezvous points in 100,000 years.

Of course, it is highly unlikely that real-life speciation would actually follow the ideal pattern of successive doubling. The opposite extreme would be a pattern in which the founder species successively threw one daughter species after another, with none of the daughter species subsequently speciating. Following this least 'efficient' pattern of speciation, in order to generate 450 species in 100,000 years, the interval between speciation events would need to be a couple of centuries. Even that doesn't sound ridiculously short. The truth surely lies between the two extreme patterns: say one or a few millennia as the average interval between speciation events in any one lineage. When you put it like that, the speciation rate doesn't seem so spectacularly high after all, especially in the light of the sorts of evolutionary rates that we saw in the Galapagos Finch's Tale. Nevertheless, as a sustained feat of speciation, it is very fast and prolific by the standards evolutionists have come to expect, and the cichlid fish of Lake Victoria have become legendary among biologists for this reason.[*]

---

[*] Lake Victoria has been the victim of a man-made catastrophe. In 1954, the British colonial administration, hoping to improve fisheries, introduced the Nile perch (*Lates niloticus*) to the lake. This decision was opposed by biologists, who predicted that the perch would disrupt the lake's unique ecosystems. Their prediction came disastrously true. The cichlids had never evolved to cope with a big predator like the Nile perch. Probably 50 cichlid species have gone for good, and another 130 are critically endangered. In a mere half-century, completely avoidable ignorance has devastated local economies around the lake and irreversibly wiped out a priceless scientific resource.

Lakes Tanganyika and Malawi are only slightly smaller than Victoria – smaller in area, that is. But where Victoria is a wide, shallow basin, Tanganyika and Malawi are Rift Valley lakes: long, narrow and very deep. They are not so young as Victoria. Lake Malawi, which I have already nostalgically mentioned as the site of my first 'seaside' holidays, is between 1 and 2 million years old. Lake Tanganyika is the oldest, at 12–14 million. Despite these differences, all three lakes share the remarkable feature that inspires this tale. All are teeming with hundreds of endemic cichlid fish, unique to the particular lake. Victoria cichlids are a completely different set of species from Tanganyika cichlids, and Malawi cichlids are a completely different set from either. Yet, each of the three flocks of hundreds of species has produced, by convergent evolution in its own lake, an extremely similar range of types. It looks as though a single founding haplochromine species (or very few) entered each infant lake, perhaps through a river. From such small beginnings, successive evolutionary subdivisions – 'speciation events' – generated hundreds of species of cichlids, whose range of types closely paralleled those in each of the other great lakes. This sort of rapid diversification into many different types is called 'adaptive radiation'. Darwin's finches are another famous example of an adaptive radiation, but African cichlids are particularly special because it has happened in triplicate.[*]

Much of the variation within each lake is concerned with diet. Each of the three lakes has its specialists in plankton feeding, its specialists in grazing algae off rocks, its predators on other fish, its scavengers, its food robbers, its fish-egg eaters. There are even parallels to the cleaner fish habit, which is better known from tropical coral reef fish (see the Polypifer's Tale). Cichlid fish have a complicated system of double jaws. In addition to the 'ordinary' outer jaws that we can see, there is a second set of 'pharyngeal jaws' buried deep in the throat. It is likely that this innovation primed the cichlids for their dietary versatility and hence their ability to diversify repeatedly in the great African lakes.

Despite their greater age, Lakes Tanganyika and Malawi don't have a noticeably larger number of species than Victoria. It is as though each lake achieves a sort of closure, at an equilibrium number of

---

[*] The whole topic is treated in detail in Dolph Schluter's book, *The Ecology of Adaptive Radiation*.

species, which doesn't go on getting larger as time goes by. Indeed, it may even get smaller. Lake Tanganyika, the oldest of the three lakes, has the fewest species. Lake Malawi, of intermediate age, has the most. It seems likely that all three lakes followed the Victoria pattern of extremely rapid speciation from very small beginnings, generating several hundred new endemic species within the first few hundred thousand years.

The Narrowmouth's Tale touched upon the favoured theory of how speciation happens, the geographical isolation theory. It is not the only theory, and more than one may be right in different cases. 'Sympatric speciation', the separation of populations into separate species in the same geographical area, can happen under some conditions, especially in insects where it may even be the norm. There is some evidence for sympatric speciation of cichlid fish in small African crater lakes. But the geographical isolation model of speciation is still the dominant one, and it will prevail through the rest of this tale.

According to the geographical isolation theory, speciation begins with the accidental geographical division of a single ancestral species into separate populations. No longer able to interbreed, the two populations drift apart, or are pushed by natural selection in different evolutionary directions. Then, if they subsequently meet after this divergence, they either can't interbreed or don't want to. They often recognise their own species by some particular feature, and studiously avoid similar species who lack it. Natural selection penalises mating with the wrong species, especially where the species are close enough for it to be a temptation, and close enough for hybrid offspring to survive, to consume costly parental resources, and then turn out to be sterile, like mules. Many zoologists have interpreted courtship displays as aimed mainly against miscegenation. This may be an exaggeration, and there are other important selection pressures bearing upon courtship. But it is still probably correct to interpret some courtship displays, and some bright colours and other conspicuous advertisements, as 'reproductive isolation mechanisms' evolved through selection against hybridisation.

As it happens, a particularly neat experiment was done on cichlid fish by Ole Seehausen, now at the University of Bern, and his colleague Jacques van Alphen at the University of Leiden. They took two related species of Lake Victoria cichlids, *Pundamilia pundamilia* and *P. nyererei* (named after one of Africa's great leaders, Julius Nyerere of Tanzania). The two species are very similar, except that

P. nyererei has a reddish colour, whereas *P. pundamilia* is bluish (see plate 26). Under normal conditions, females in choice tests prefer to mate with males of their own species. But now, Seehausen and van Alphen did their critical test. They gave females the same choice, but in artificial monochromatic light. This does dramatic things to perceived colour, as I remember vividly from schooldays in Salisbury, a city whose streets happened to be lit by sodium lights. Our bright red caps, and the bright red buses, all looked dirty brown. This is what happened to both the red and the blue *Pundamilia* males in Seehausen and van Alphen's experiment. Red or blue in white light, they all went dirty brown. And the result? The females no longer distinguished between them, and mated indiscriminately. Offspring of these matings were fully fertile, indicating that female choice is the only thing that stands between these species and hybridisation. The Grasshopper's Tale gives a similar example. If the two species were a bit more different, their offspring would probably be infertile, like mules. Later still in the process of divergence, isolated populations reach the point where they couldn't hybridise even if they wanted to.

Whatever the basis of the separation, failure to hybridise defines a pair of populations as belonging to different species. Each of the two species is now free to evolve separately, free from contamination by the genes of the other, even though the original geographical barrier to such contamination is no more. Without the initial intervention of geographical barrriers (or some equivalent), species could never become specialised to particular diets, habitats or behaviour patterns. Notice that 'intervention' does not necessarily mean it is geography itself that made the active change – as when a valley floods or a volcano erupts. The same effect is achieved if geographical barriers existed all along, wide enough to impede gene flow, but not so formidable that they are never crossed by occasional founder populations. In the Dodo's Tale we met the idea of sporadic individuals having the luck to cross to a remote island, where they then breed in isolation from their parent population.

Islands like Mauritius or the Galapagos are the classic providers of geographical separation, but islands don't have to mean land surrounded by water. When we are talking about speciation, 'island' comes to mean any kind of isolated breeding area, defined from the animal's point of view. Not for nothing is Jonathan Kingdon's beautiful book on African ecology called *Island Africa*. To a fish, a lake is an island. How, then, could hundreds of new fish species diverge from a single ancestor, if they all live in the same lake?

One answer is that, from the fish's point of view, there are lots of little 'islands' within a large lake. All three of the great East African lakes have isolated reefs. 'Reef' here doesn't mean coral reef, of course, but 'a narrow ridge or chain of rocks, shingle, or sand, lying at or near the surface of the water' (*Oxford English Dictionary*). These lake reefs are covered with algae, and many kinds of cichlids crop them. To such a cichlid, a reef might well constitute an 'island', separated by deep water from the next reef, at a distance large enough to constitute a barrier to gene flow. Even though they are capable of swimming from one to the other, they don't want to. There is genetic evidence to support this, from a study in Lake Malawi that sampled one species of cichlid, *Labeotropheus fuelleborni*. Individuals from opposite ends of a large reef shared the same distribution of genes: there was abundant gene flow along the length of the reef. But when the investigators sampled the same species from other reefs, separated by deep water, they found significant differences in visible coloration and in genes. A gap of two kilometres was enough to cause a measurable genetic separation; and the larger the physical gap, the more the genetic gap. Further evidence comes from a 'natural experiment' in Lake Tanganyika. A violent storm in the early 1970s created a new reef, 14 kilometres from its nearest neighbour. This should have been prime habitat for reef-dwelling cichlids, but when the reef was examined several years later none had arrived. Evidently, from the fish's point of view, there are indeed 'islands' within these large lakes.

In order for speciation to happen, there must be populations that are sufficiently isolated for gene flow between them to be rare; but not so isolated that no founding individuals arrive there at all. The recipe for speciation is 'Genes flow but not much'. That is a section heading from George Barlow's *The Cichlid Fishes*, the book that has been my main inspiration while writing this tale. The section describes yet another genetic study in Lake Malawi of four species of cichlid, inhabiting four neighbouring reefs, roughly one to two kilometres apart. All four species, known as *mbuna* in the local dialect, were present on all four reefs. Within each of the four species, there were genetic differences between the four reefs. A sophisticated analysis of the distributions of genes showed that there was indeed a trickle of gene flow between reefs, but a very slight one – a perfect recipe for speciation.

Here's another way in which speciation might have happened, and one that seems especially plausible for Lake Victoria. Radiocarbon dating of mud suggests that Lake Victoria dried up about 15,000 years ago. *Homo sapiens*, not long predating the dawn farmers of Mesopotamia, could walk dry-footed from Kisumu in Kenya straight across to Bukoba in Tanzania – a journey that today is a 300-kilometre voyage by the MV *Victoria*, a decent-sized ship popularly known as the 'Queen of Africa'. That was an extremely recent drying up, but who knows how many times the Victoria basin has been drained and flooded, flooded and drained in the millennia before that? On the timescale of thousands of years, the lake level may rise and fall like a yo-yo.

Now, hold that thought in mind, together with the theory of speciation by geographical isolation. When the Victoria basin dries up from time to time, what will be left? It could be a desert if the drying up were complete. But a partial drying up would leave a scattering of little lakes and pools, representing the deeper depressions in the basin. Any fish trapped in these little lakes would have the perfect opportunity to evolve away from their colleagues in other little lakes and become separate species. Then, when the basin flooded again and the large lake reconstituted, the newly distinct species would all swim out and join the larger Victoria fauna. When the yo-yo went down the next time, it would be a different set of species that accidentally found itself separated in each of the smaller refugia. Again, what a wonderful recipe for speciation.

Evidence from mitochondrial DNA supports this theory of rising and falling lake levels for the older Lake Tanganyika. Although a deep rift lake, not a shallow basin like Victoria, there is evidence that Lake Tanganyika's level used to be much lower, and it was at that time separated into three medium-sized lakes. The genetic evidence suggests an early segregation of cichlids into three groupings, presumably one for each of the old lakes, followed by further speciations after the formation of the present large lake.

In the case of Lake Victoria, Erik Verheyen, Walter Salzburger, Jos Snoeks and Axel Meyer have done a very thorough genetic study of the mitochondria of haplochromine cichlid fish, not only in the main lake but in the neighbouring rivers, and the satellite lakes Kivu, Edward, George, Albert and others. They showed that Victoria and its smaller neighbours share a monophyletic 'species flock' that began to diverge about 100,000 years ago. This sophisticated piece of research used the methods of parsimony, maximum likelihood and Bayesian analysis that we met in the Gibbon's Tale. Verheyen and colleagues looked at the distribution in all the lakes and neighbouring rivers of 122 'haplotypes' from the mitochondrial DNA of these fish. A haplotype, as we saw in Eve's Tale, is a length of DNA that lasts long enough to be recognised repeatedly in lots of individuals, who might well belong to lots of different species. For simplicity I shall use the word 'gene' as an approximate synonym for haplotype (although purist geneticists would not). The scientists were, temporarily, ignoring the question of species. They were, in effect, imagining genes swimming around in lakes and rivers, and counting the frequency with which they did so.

It is easy to misunderstand the beautiful diagram (see plate 27), with which Verheyen and his colleagues summarised their work. It is tempting to think that the circles represent species clustered around parent species, as in a family tree. Or that they represent small lakes clustered around larger lakes, as in a stylised route map of an (amphibious!) airline's network of destinations. Neither of these is even close to what the diagram represents. The circles are neither species nor geographical hubs. Each one is a haplotype: a 'gene', a particular length of DNA that an individual fish might or might not possess.

Each gene, then, is represented by one circle. The area of the circle conveys the number of individuals, *regardless of species*, added up over all lakes and rivers surveyed, who possessed that particular gene. The small circles indicate a gene that was found in only a single individual. Gene 25, to judge from the area of its circle (the largest one), was found in 34 individuals. The number of circles, or blobs, on the line joining two circles represents the minimal number of mutational changes you need to go from one to the other. You will recognise from the Gibbon's Tale that this is a form of parsimony analysis, but slightly easier than parsimony analysis of distantly related genes, because the intermediates are still around. The small black blobs represent intermediate genes that have not been found in real fish, but can be inferred as probably existing in the course of evolution. It is an unrooted tree that doesn't commit itself to the direction of evolution.

Geography enters into the diagram only in the colour coding. Each circle is a pie chart showing the number of times the gene concerned was found in each of the lakes or rivers surveyed (see the colour key at bottom right of the diagram). Of the numerous genes, those labelled 12, 47, 7 and 56 were found only in Lake Kivu (all red circles). Genes 77 and 92 were found only in Lake Victoria (all blue). Gene 25, the most abundant of all, turned up mostly in Lake Kivu but also in significant numbers in the 'Uganda lakes' (a cluster of small lakes close to each other and to the west of Lake Victoria). The pie chart shows that gene 25 was also found in the Victoria Nile river, in Lake Victoria itself, and in Lake Edward/George (these two small and neighbouring lakes are united for purposes of the count). Once again, bear in mind that the diagram contains no information at all about species. The blue slice of pie in gene 25's circle indicates that two individuals from Lake Victoria contained this gene. We are given no indication at all on whether those two individuals were of the same species as each other, or the same species as any of the Lake Kivu individuals bearing that gene. That is not what this diagram is about. It is a diagram to delight any enthusiast for the selfish gene.

The results were powerfully revealing. Little Lake Kivu emerges as the fountainhead of the entire species flock. Genetic signals show that Lake Victoria was 'seeded' with haplochromine cichlids on two separate occasions from Lake Kivu. The great drying of 15,000 years ago by no means extinguished the species flock, and very probably enhanced it in the way we were just imagining, through the Victoria basin becoming a 'Finland' of lakelets. As for the origin of the older population of cichlids in Lake Kivu itself (it now has 26 species, including 15 endemic haplochromines), the genetic oracle says they came from Tanzanian rivers.

This work is only just beginning. The imagination at first quails, and then is uplifted, by the contemplation of what will be achieved when such methods are routinely applied not just to cichlid fish in African lakes, but to any animals, in any 'archipelago' of habitats.

THE BLIND CAVE FISH'S TALE

Animals of various kinds have found their way into dark caves, where living conditions are obviously very different from outside. Repeatedly, and in many different animal groups including flatworms, insects, crayfish, salamanders and fish, cave dwellers have independently evolved many of the same changes. Some can be thought of as constructive changes – for instance, delayed reproduction, fewer but larger eggs, and increased longevity. Apparently in compensation for their useless eyes, cave animals typically have enhanced senses of taste and smell, long feelers and, in the case of fish, improvements to the lateral line system (a pressure-related sense organ beyond our empathy but deeply meaningful to fish). Other changes are referred to as regressive. Cave dwellers tend to lose their eyes and their skin pigment, becoming blind and white.

The Mexican tetra *Astyanax mexicanus* (also known as *A. fasciatus*) is particularly remarkable because different populations *within* the one species of fish have independently followed streams into caves and very rapidly evolved a common pattern of cave-related regressive changes, which can be directly contrasted with fellow species members still living outside. These 'Mexican blind cave fish' are found only in Mexican caves – mostly limestone caves in a single valley. Once understandably thought to belong to their own separate species, they are now classified as a race of the same species, *Astyanax mexicanus*, which is common in surface waters from Mexico to Texas. The blind race has been found in 29 separate caves and, to repeat, it looks strongly as though at least some of these cave populations evolved their regressive eyes and white coloration independently of each other: surface-dwelling tetras have on many occasions taken up residence in caves, and independently lost their eyes and their colour on each occasion.

Intriguingly, it appears that some populations have been in their caves longer than others, and this shows itself as a gradient in the extent to which they have pushed in the typical cave-specific direction. The extreme is found in the Pachon cave, believed to hold the oldest cave population. At the 'young' end of the gradient is the Micos cave, whose population is relatively unchanged from the normal surface-dwelling form of the species. None of the populations can have been in their caves very long because this is a South American species which could not have crossed into Mexico before the formation of the Isthmus of Panama 3 million years ago – the Great American Interchange. My guess is that the cave populations of tetras are far younger than that.

It is easy to see why dwellers in darkness might never have evolved eyes in the first place; less easy to see why, given that their recent ancestors certainly had normal, functioning eyes, the cave fish should 'bother' to get rid of them. If there is a possibility, however slight, of a cave fish finding itself washed out of its cave into the light of day, wouldn't there be some benefit in keeping the eyes 'just in case'? That isn't how evolution works, but it can be rephrased in respectable terms. Building eyes – indeed, building anything – is not free of cost. Individual fish that divert resources into some other part of the animal's economy would have an advantage over rival fish that retain full-sized eyes.* If a cave-dweller has insufficient probability of needing eyes to offset the economic costs of making them, eyes will disappear. Where natural selection is concerned, even very slight advantages are significant. Other biologists leave economics out of their reckoning. For them, it is sufficient to invoke an accumulation of random changes in eye development, which are not penalised by natural selection because they make no difference. There are many more ways of being blind than of being sighted, so random changes, for purely statistical reasons, tend towards blindness.

And this leads us to the main point of the Blind Cave Fish's Tale. It is a tale of Dollo's Law, which states that evolution is not reversed. Is Dollo's Law disproved by the cave fish's apparent reversal of an evolutionary trend, shrinking again the eyes that grew, so painstakingly, over past evolutionary time? Is there, in any case, some general theoretical reason to expect evolution to be irreversible? The answer to both questions is no. But Dollo's Law has to be correctly understood, and that is the purpose of this tale.

Except in the very short term, evolution cannot be precisely and exactly reversed, but the emphasis is on 'precisely and exactly'. It is very improbable that any particular evolutionary pathway, specified in advance, will be followed. There are too many possible pathways. An exact reversal of evolution is just a special case of a particular evolutionary pathway, specified in advance. With such a large number of possible paths that evolution might follow, the odds are heavily against any one particular path, and that includes an exact reversal of the forward one just travelled. But there is no law against evolutionary reversal as such.

Dolphins are descended from land-dwelling mammals. They returned to the sea and resemble, in many superficial respects, large, fast-swimming fish. But evolution has not reversed itself. Dolphins resemble fish in certain respects, but most of their internal features clearly label them as mammals. If evolution had truly reversed itself, they would simply be fish. Maybe some 'fish' really are dolphins – the

---

* Eyes can be an even more costly extravagance if they become infected or irritated, which is probably why burrowing moles have reduced them as much as possible.

reversion to fish being so perfect and far-reaching that we haven't noticed? Want a bet? That is the sense in which you can bet heavily on Dollo's Law. Especially if you look at evolutionary change at the molecular level.

This interpretation of Dollo's Law could be called the thermodynamic interpretation. It is reminiscent of the Second Law of Thermodynamics, which states that entropy (or disorder or 'mixed-upness') increases in a closed system. A popular analogy (or it may be more than an analogy) for the Second Law is a library. Without a librarian energetically reshelving books in their correct places, a library tends to become disordered. The books become mixed up. People leave them on the table, or put them on the wrong shelf. As time goes by, the library's equivalent of entropy inevitably increases. That's why all libraries need a librarian, constantly working to restore the books to order.

The great misunderstanding of the Second Law is to assume that there is a driving urge towards some particular goal state of disorder. It isn't like that at all. It is just that there are far more ways of being disordered than of being ordered. If the books are shuffled at random by sloppy borrowers, the library will automatically move away from the state (or the small minority of states) that anybody would recognise as ordered. There is no systematic drive towards a state of high entropy. Rather, the library meanders in some random direction away from the initial state of high order and, no matter where it wanders in the space of all possible libraries, the vast majority of possible pathways will constitute an increase in disorder. Similarly, of all the evolutionary pathways that a lineage could follow, only one out of a vast number of possible pathways will be an exact reversal of the path by which it has come into being. Dollo's Law turns out to be no more profound than the 'law' that if you toss a coin 50 times, you won't get all heads – nor all tails, nor strict alternation, *nor any other particular, prespecified sequence.* The same 'thermodynamic' law would also state that any *particular* evolutionary pathway in a 'forward' direction (whatever that might mean!) will not be precisely followed twice.

In this thermodynamic sense, Dollo's Law is true but unremarkable. It doesn't deserve the title of law at all, any more than there is a 'law' against tossing a coin 100 times and getting heads every time. One could imagine a 'real law' interpretation of Dollo's Law which stated that evolution could not return to anything that was vaguely like an ancestral state, as a dolphin is vaguely like a fish. This interpretation would indeed be remarkable and interesting but it is (ask any dolphin) false. And I cannot imagine any sensible theoretical rationale that would expect it to be true.

THE FLOUNDER'S TALE

An endearing quality of Chaucer is the naive perfectionism of his General Prologue, where he introduces his pilgrims. It wasn't enough to have a Doctour of Physik on the pilgrimage – he had to be the finest doctor in the land:

> *In all this world ne was ther noon hym lik,*
> *To speke of physik and of surgerye.*

The 'verray, parfit gentil knyght' was, it seemed, unmatched in Christendom for bravery, loyalty and even temper. As for his squire and son, he was 'A lovere and a lusty bacheler . . . wonderly delyvere, and of greet strengthe'. To top it all, he was 'as fressh as is the month of May'. Even the knight's yeoman knew all there was to know of woodcraft. The reader comes to take it for granted that, if a profession is mentioned, its practitioner will automatically turn out to be unrivalled in all England.

Perfectionism is a vice of evolutionists. We are so used to the wonders of Darwinian adaptation, it is tempting to believe there could be nothing better. Actually, it is a temptation that I can almost recommend. A surprisingly strong case can be built for evolutionary perfection, but it must be done with circumspection and sophisticated attention.* Here I shall give just one example of a historical constraint, the so-called 'jet engine effect': imagine how imperfect a jet engine would be if, instead of being designed on a clean drawing board, it had to be changed one step at a time, screw by screw and rivet by rivet, from a propeller engine.

The flat fish known as skates, which we will meet at the next rendezvous, might have been designed on a drawing board to be flat, resting on the belly, with wide 'wings' reaching symmetrically out to both sides. Teleost flatfish do it in a different way. They rest on one side, either the left (e.g. plaice) or the right (e.g. turbot and

---

* I have set out the pitfalls in a chapter of *The Extended Phenotype* called 'Constraints on Perfection'.

Picasso would have loved them. The skate (*Raja batis*, top) rests on its belly; the flounder (*Bothus lunatus*) on its right side. The eye on that side has migrated over time to the left (upper) side. Drawing by Lalla Ward

flounder). Whichever the side, the shape of the whole skull is distorted so that the eye on the lower side moves over to the upper side, where it can see. Picasso would have loved them. But, by the standards of any drawing board, they are revealingly imperfect. They have precisely the kind of imperfection you would expect from being evolved rather than designed.

SHARKS AND THEIR KIN

'Out of the murderous innocence of the sea . . .' The context of Yeats's poem was completely other but – I can't help it – the phrase always makes me think of a shark. Murderous, but innocent of deliberate cruelty, just making a living as perhaps the world's most effective killing machine. I know people for whom the great white shark is their worst nightmare. If you are one of them, you may not wish to know that the Miocene shark *Carcharocles megalodon* was three times the size of a great white, with jaws and teeth to scale.

My own recurrent nightmare, having grown up as an exact contemporary of the atomic bomb, is not a shark but a huge, black, futuristic, delta-winged aircraft bristling with high-tech missile launchers, filling the sky with its shade and my heart with foreboding. Almost exactly the shape of a manta ray in fact. The dark shape that roars over the treetops of my dreams, with its twin gun turrets so enigmatically menacing, is a sort of technological cousin to *Manta birostris*. I always found it hard to accept that these seven-metre monsters are harmless filter-feeders, straining plankton through their gills. They are also extremely beautiful.

What of the sawfish, what on earth is that all about? And the hammerhead shark? Hammerheads occasionally attack people, but that is not why they might invade your dreams. It is the bizarre T-shaped head, the eyes set wider than you expect outside science fiction, as though this shark were designed by an artist with a drugged imagination (see plate 28). And the thresher shark, *Alopias*, isn't that another work of art, another candidate for a dream? The upper lobe of the tail is nearly as long as the rest of the body. Threshers use their prodigious tailblades first to herd prey, then to thresh them to death. However, the story that a thresher is capable of decapitating a fisherman with a single swipe of the tail seems to be another of those fishermen's tall tales. Not to overdo the nightmare theme, lantern sharks glide through the darkness of the deep sea, signalling to each other via flashes of light, with military stealth, even extending to cloaking their shadow from below by emitting a dull light from their underside.

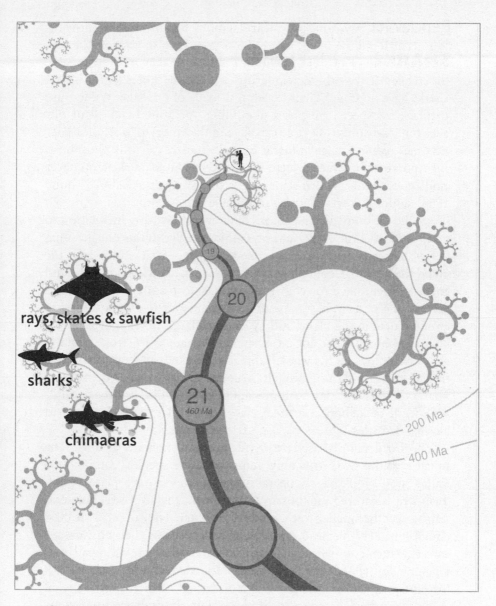

rays, skates & sawfish

sharks

chimaeras

19

20

21
460 Ma

200 Ma

400 Ma

Sharks and their kin join. The cartilaginous fish, who join us here, include the sharks and rays. Fossils leave no doubt about the early split of the jawed vertebrates into bony fish and these cartilaginous fish. Recent, robust data strongly supports this scheme of relationships within the 850 or so species of cartilaginous fish.

The sharks, rays and other cartilaginous fish or chondrichthyans join us at Rendezvous 21, 460 million years ago, in seas off the icy-cold and barren lands of the Middle Ordovician. The most noticeable difference between the new pilgrims and all the others so far is that sharks have no bone. Their skeleton is made of cartilage. We too use cartilage for special purposes like lining our joints, and all of our skeleton starts out as flexible cartilage in the embryo. Most of it later becomes ossified when mineral crystals, mostly calcium phosphate, incorporate themselves. Except for the teeth, the shark skeleton never undergoes this transformation. Nevertheless, their skeleton is quite rigid enough to sever your leg in a single bite.

Unlike the bony fish, no sharks have ever shown any inclination to come on land, and sharks and rays almost all live in the sea, although a few genera venture up estuaries and rivers. Sharks also lack the swim bladder that contributes to the success of the bony fish, and many of them have to swim continuously to maintain their desired level in the water. They assist their buoyancy by retaining the waste product urea in the blood and by having a large, oil-rich liver.

If you should be so incautiously affectionate as to stroke a shark, you would find that its whole skin feels like sandpaper, at least if you stroke it 'against the grain'. It is covered with dermal denticles – sharp, tooth-like scales. Not only are they tooth-like, but the formidable teeth of a shark are themselves evolutionary modifications of dermal denticles.

The cartilaginous fish are divided into two main groups. The smaller of the two, with only about 50 living species, contains the rather weird-looking chimaeras or ghost-sharks (see plate 29). If manta rays feature in nightmares as bombers, the deep-sea chimaeras might play the smaller role of jump-jet fighter. They occupy the class Holocephali (whole head) and can be recognised by their unusual gill covers, which completely encase the separate gills, providing a single opening for all of them. Unusually for sharks, their skin is not covered with dermal denticles but is 'naked', which may be what gives them their ghostly appearance. Their resemblance to a nightmare plane comes from the fact that their tails are not prominent and they swim by 'flying' with their large pectoral fins.

The rest of the cartilaginous fish, of which there are about 800 species, belong in the Elasmobranchii, which itself is now thought to split into two main branches. One branch contains what could be thought of as flattened sharks: the skates, rays and sawfish. The other contains the true sharks, of varying shapes and sizes. However, none of them are tiny. The shark body plan seems to lend itself to large size. Perhaps the smallest species is the dwarf lanternshark *Etmopterus perryi*, which grows up to about 20 centimetres long. The biggest, the whale shark *Rhincodon typus*, can be up to 12 metres long and weigh 12 tonnes. Like the second largest, the basking shark *Cetorhinus maximus*, and like the largest whales, the whale shark is a plankton feeder. *Carcharocles megalodon*, already mentioned as the stuff of nightmares, was not – to use a calculated understatement – a filter-feeder. That Miocene monster had teeth, each one as big as your face. It was a voracious predator, like the majority of sharks today, and sharks have topped the food chains of the sea for hundreds of millions of years with relatively little change.

There have been two major radiations of sharks. The first flourished mightily in the Palaeozoic seas, especially during the Carboniferous Period. This ancient domination of sharks had come to an end by the beginning of the Mesozoic Era (the age of dinosaurs on land). After a lull of about 100 million years, the sharks enjoyed another major resurgence in the Cretaceous, which has continued to this day. But successful as sharks certainly are – and over a spectacularly long time too – teleost fish outnumber them well over thirtyfold when it comes to species numbers.

A word association test that mentioned 'shark' would very probably elicit the response 'jaws', so it is appropriate that Concestor 21, perhaps our 200-million-greats-grandparent, is the grand ancestor of all the vertebrates that have true jaws, the gnathostomes. *Gnathos* in Greek means 'lower jaw', and that is specifically what sharks and all the rest of us share. It was one of the triumphs of classical comparative anatomy to demonstrate that jaws evolved from modified parts of the gill skeleton. The fish we will encounter in the next rendezvous are Agnathans: well endowed with gills but, as their name implies, lacking jaws, as well as lacking other notable features such as paired side fins and mineralised teeth. They are the eel-shaped lampreys and hagfish.

LAMPREYS AND HAGFISH

Rendezvous 22, where we meet the 40 or so species of lamprey and around 70 known species of hagfish, occurs somewhere in the warm seas of the early Cambrian, say 525 million years ago. At a rough guess, Concestor 22 was our 235-million-greats-grandparent. Sans jaws, sans side fins, sans bone, the pilgrims we meet here have clung on as pivotal messengers from the dawn of vertebrates. But there are several important groups of fish who would have joined our pilgrimage at various points between Rendezvous 21 and 22, were it not for the fact they their lineages have now gone extinct. Although our journey focuses on pilgrims that have made it to the present, these extraordinary creatures deserve a little attention.

Perhaps most impressive, and the main predators during the Devonian 'Age of Fish', were the placoderms. Their name means 'plate-skins', in reference to the hard, bony armour plating that covered their head and upper body. Indeed, some were so heavily armoured that even their forelimbs had a tubular, jointed exoskeleton, superficially similar to a crab's leg. If you encountered one in a poor light and an imaginative frame of mind, you could be forgiven for thinking you had stumbled on a strange kind of lobster or crab. As a rather young undergraduate, I used to dream about discovering a living placoderm – it was my equivalent of the scoring-a-century-for-England fantasy. Closely related to us gnathostomes, placoderms were the earliest fish to have hinged jaws, and simplified teeth to fill them. Although most species were only tens of centimetres long, they also included the 6-metre-plus *Dunkleosteus*, with a bite strong enough to snap a great white shark in two. Recent studies show that some placoderms may be more closely related to us than others. Excitingly, this would mean that at one point our ancestors actually were some sort of placoderm. Unfortunately this would turn 'placoderm' into a grade of fish rather than an inclusive natural grouping and, as we discussed with the mammal-like-reptiles, would render the name taxonomically invalid.

It's not just lower jaws that trace back to placoderms. Our legs originate in the paired pelvic fins that are first found in early placoderms. Some male placoderms also had another set of appendages

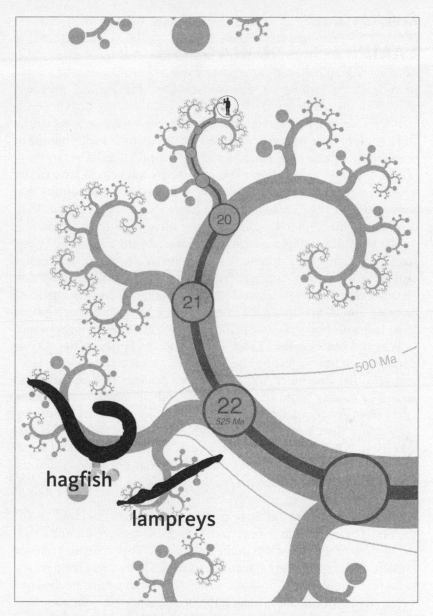

hagfish

lampreys

20

21

22
525 Ma

500 Ma

Jawless fish join. Recent molecular studies of rare genomic changes have convinced researchers that the eel-like lampreys and hagfish group together [188], forming the outgroup to all other vertebrates, as shown here. The number of deep-sea hagfish species is almost certainly an underestimate: difficulty of study means that the phylogenetic relationships within hagfish are rather tentative.

further down the body. These seem to have been used as external genitalia to grasp and impregnate females, in the same manner that sharks use their 'claspers' today. The inference that these ancient fish bore live young has been cemented by the recent discovery of a fossil placoderm containing an embryonic baby: the earliest pregnant mother in the fossil record.

If we owe our legs to placoderms, then we owe our arms to the slightly more distantly related 'ostracoderms', which include some of the earliest fish to have paired front ('pectoral') fins. Also armour-plated, they had neither lower jaws nor teeth and seem to have fed by sucking soft food into their mouths. It is now beyond doubt that some (such as the Osteostraci) were more closely related to us than others, and the term 'ostracoderms' is gradually going out of use.

The bony plates of placoderms and ostracoderms give the lie to any suggestion that calcified bone is an 'advanced' feature of vertebrates that 'took over' from cartilage. Sturgeons and some other 'bony' fish resemble sharks in possessing a skeleton almost entirely made of cartilage, but all are descended from far more bony ancestors – indeed from fish with heavy armour plating. Bone is an ancient innovation.

Why did both the jawed placoderms and the jawless ostracoderms develop such heavily fortified bodies? What was it about those Palaeozoic seas that demanded such formidable protection? The presumed answer is equally formidable predators, and the obvious candidates, apart from other placoderms, are the eurypterids or sea scorpions, some more than two metres in length – the largest arthropods that ever lived. Whether or not any of the eurypterids had venomous stings like modern scorpions (recent evidence suggests not), they still must have been fearsome predators, capable of driving the Devonian fish, both jawless and jawed, to evolve costly armour plating.

Finally, we turn to the only jawless fish to have survived to the present. The eel-shaped lampreys and hagfish are more distantly related to us than the plated fish of the Palaeozoic. They have no bones or armour, making lampreys easy to eat, as King Henry I had good reason to regret (school history books never fail to remind us that he died of a surfeit of them). But it is unclear if their ancestors were quite so soft on the palate. True, all the species today construct their skulls and skeleton from soft, uncalcified cartilage. Even their teeth are made of keratin, like your fingernails. Nevertheless, extracted lamprey cartilage can calcify in a petri dish, lamprey genes produce a type of cartilage which gets calcified in other fish, and to top it all, calcified lamprey-like tissue has been found in fossils. It is quite possible that Concestor 22 had some calcified parts, possibly even true bone, which has since disappeared in the living lampreys and hagfish.

Modern lampreys live part or all of their lives in freshwater. They are unusual in that they spend the first few years of their lives as a simplified larva called an 'ammocoete'. In this guise, toothless, blind and buried in the sediment, they filter-feed on detritus. They then undergo true metamorphosis into adults. Like vertebrate mayflies, in about half the species the adults forgo feeding and devote all their efforts to reproduction. In the remainder, the adults become parasitic on other fish. Instead of jaws they have a circular sucker around the mouth, looking a bit like an octopus sucker but with concentric rings of tiny teeth. The lamprey fastens its sucker to the outside of another fish, the little teeth rasp through the skin, and the lamprey sucks the blood of its victim, like a leech. Lampreys have had serious effects on fisheries, for example in the North American Great Lakes.

Hagfish or 'slime-eels' are deep-sea creatures, and therefore harder to study. They resemble lampreys in their long, eel-like shape, their lack of a lower jaw, their lack of paired limbs and their row of gill port-holes on either side. But hagfish are not parasites. They rummage with their mouth hole around the bottom of the sea for small invertebrates, or they scavenge on dead fish or whales, often wriggling inside to eat from the inside out. They can produce extraordinary amounts of mucus, both for lubrication and defence: deep-sea filming reveals attacking sharks swimming away in disgust, their mouths entirely gummed up by hagfish slime. They use their surprising talent for tying their body in knots both to get a purchase when burrowing into carcasses, and also to escape from their self-inflicted mucus envelope by sliding their body-knot from head to tail, an impressive yet revolting feat of escapology.

Although lampreys and hagfish are the most distant relatives we have yet encountered, their fishiness encourages us to classify them – along with all the pilgrims who have joined us so far – in the vertebrates. Yet, belying this name, their torso is not held together by a vertebral column. Instead, their 'backbone' is a single flexible rod of cartilage called the notochord.* Most vertebrates (including

---

* The word is calculated to confuse, because chord with an h, in modern English, means only something musical, as in 'The Lost Chord', one of my favourite songs. The notochord is a cord, without an h, meaning rope. However, chord is a recognised archaic spelling of cord (rope), and the connection with music may be that *chorda* is the Latin for the string of an instrument.

us) have a notochord as embryos, but it becomes replaced, to a greater or lesser extent, by a set of articulated vertebrae. The notochord itself survives into the adult only in fragments, such as the intervertebral discs whose tendency to slip can cause us so much grief. Lampreys and hagfish have no vertebrae at all in their spine; it could be said to be one long disc. But they do have structures that show affinities to vertebrae. Lampreys have regularly spaced hoops of cartilage that protect the nerves which run above the notochord. It has recently been discovered that in the tail end of hagfish there are also repeated segments of cartilage which lie below their notochord. The implication is that Concestor 22 possessed the rudiments of vertebrae, with the basic features of a skeleton already laid out. As a satisfying – although ultimately semantic – point, this could justify both lampreys and hagfish being granted the name of 'vertebrate'.

Vertebrates were once thought to have arisen long after the Cambrian Period. Maybe it was an aspect of our snobbish desire to arrange the animal kingdom on a ladder of progress. Somehow it seemed right and fitting that there was an age where animal life was limited to invertebrates, setting the scene for the dramatic entrance of the mighty vertebrates. Zoologists of my generation were taught that the earliest known vertebrate was a jawless fish called *Jamoytius* (named, somewhat freely, after J. A. Moy-Thomas) that lived in the middle of the Silurian Period, 100 million years after the Cambrian when most of the invertebrate phyla arose. Obviously vertebrates must have had ancestors living in the Cambrian, but they were assumed to be invertebrate forerunners of the true vertebrates – protochordates. *Pikaia* has been heavily promoted as the oldest fossil protochordate.* It was a delicious surprise, therefore, when apparently true vertebrate fossils started turning up in the Cambrian strata of China, and the Lower Cambrian at that. This has robbed *Pikaia* of some of its mystique. There were true vertebrates, jawless fish, living before *Pikaia*. The vertebrates hark back to the deep Cambrian.

Until 2014 we only knew of three rather similar Cambrian vertebrates, *Haikouichthys*, *Myllokunmingia* and *Zhongjianichthys*. We can now add *Metaspriggina*, a fossil which hails from the younger Canadian strata that originally gave us *Pikaia*. Not surprisingly

---

* This Cambrian fossil, originally classified as an annelid worm, was later recognised as a protochordate, in which role it starred in S. J. Gould's *Wonderful Life*.

given their immense age, these fossils (which we shall meet again in the Velvet Worm's Tale) are not in mint condition, and much is still unknown about these primeval fish. They seem to have had most of the features you'd expect from a relative of the lampreys and hagfish. With gills, a body made of zig-zag muscle blocks, and a notochord, they all vaguely resemble the ammocoete larva of lampreys (although they probably had a functioning pair of eyes). It seems likely that Concestor 22 looked much more like these fossils than like an adult lamprey or hagfish, both of which have acquired various specialisations in their separate half-billion-year evolutionary journey.

Rendezvous 22 is a major milestone. From now on, for the first time, all the vertebrates are united in a single pilgrim band. It is a big event because, traditionally, animals were divided into two major groups, the vertebrates and the invertebrates. As a division of convenience, the distinction has always been useful in practice. From a strictly cladistic point of view, however, the vertebrate/invertebrate distinction is an odd one, nearly as unnatural as the ancient Jewish classification of humanity into themselves and 'gentiles' (literally everybody else). Important though we vertebrates think ourselves, we don't constitute even a whole phylum. We are a subphylum of the phylum Chordata, and the phylum Chordata should be thought of as on a par with, say, the phylum Mollusca (snails, limpets, squids, etc.) or the phylum Echinodermata (starfish, sea urchins, etc.). The phylum Chordata includes creatures that, while not vertebrates, still possess a notochord – for example, the lancelets, whom we shall meet at Rendezvous 24.

Strict cladism notwithstanding, there really is something rather special about vertebrates. My Oxford colleague Professor Peter Holland has commented that the traditional divide between vertebrates and invertebrates deserves revival. He has made to me the powerful point that there is a massive difference in genome complexity between (all) vertebrates and (all) invertebrates: 'It is, at the genetic level, perhaps the biggest change in our metazoan* ancestry.' The Lamprey's Tale explains why this might be so.

---

* Metazoa means many-celled animals, and we shall be meeting the term further on in the pilgrimage.

THE LAMPREY'S TALE

The Lamprey's Tale concerns the creation of genetic novelty, and it takes up a theme we have met before: there is a separate gene's-eye view of ancestry and pedigree that is surprisingly independent of the view we get when we think about family trees in more traditional ways.

Haemoglobin is well known as the vitally important molecule that carries oxygen to our tissues and gives our blood its spectacular colour. Human adult haemoglobin is actually a composite of four protein chains called globins, knotted around each other like an orgy of hagfish to form a beautifully functioning parcel. The sequence of amino acids in the chains shows that the four globins are closely related to each other, but they are not identical. Two of them are called alpha globins (each a chain of 141 amino acids), and two are beta globins (each a chain of 146 amino acids). These globins are encoded by genes on different human chromosomes: alpha on chromosome 11, beta on chromosome 16. These aren't the only globins encoded in our genome. There are other rather different globins which serve different purposes. For example myoglobin, the first protein to be visualised in 3D (earning its discoverers the 1962 Nobel Chemistry Prize), is a muscle component used to store rather than transport oxygen. It is thanks to myoglobin that sperm whales can remain underwater for for over 2 hours without running out of breath. More recent discoveries include cytoglobin, which probably helps our cells deal with low oxygen levels, and other globins whose function is as yet unknown, with appropriately mysterious names such as 'globin X' and 'globin Y'.

Never mind all these names. Here's the fascinating point. The detailed similarities between globin genes reveal that they cannot have arisen independently of each other. They are parallel copies, what a geneticist would call 'paralogous genes'. The alpha and beta sequences, for example, are evidently the result of a copy-and-paste duplication of the sort discussed in the Howler Monkey's Tale. Others, as we shall see, may have been copied in a more radical manner. The various copied and altered globin genes are literally cousins of each other – members of a family. But these distant cousins still co-exist inside you and me. They still sit side by side within every cell of every warthog and every wombat, every owl and every lizard.

On the scale of whole organisms, of course, all vertebrates are cousins of each other too. The tree of vertebrate evolution is the family tree we are all familiar with, its branch-points representing speciation events – the splitting of species into daughter species. In reverse, they are the rendezvous points that punctuate this pilgrimage. But there is another family tree occupying the same timescale, whose branches represent not speciation events but gene duplication events within genomes. And the branching pattern of the globin tree looks very different from the branching pattern of the family tree, if we trace it in the usual, orthodox way, with species branching to form daughter species. There is not just one evolutionary tree in which species divide and give rise to daughter species. Every gene has its own tree, its own chronicle of splits, its own catalogue of close and distant cousins.

The DNA sequences which produce our haemoglobin, the alpha and beta genes, started out as sisters. About half a billion years ago, in an early fish (perhaps of the ostracoderm or placoderm grade that we described previously), an ancestral globin gene accidentally duplicated, both copies remaining in that fish's genome. One copy was destined to give rise to the alphas, on what would eventually become chromosome 11 in our genome, the other to the betas, now on our chromosome 16. There is no point in trying to guess which chromosome either of them sat on in the intermediate ancestors. The locations of recognisable DNA sequences, indeed the number of chromosomes into which the genome is divided, are shuffled and changed with surprisingly gay abandon. Chromosome numbering systems, therefore, do not generalise across animal groups.

Now here's a fascinating point. Given that the duplication which produced alpha and beta versions took place half a billion years ago, it will of course not be just our human genomes that show the split, and possess both alpha genes and beta genes in different parts of our genomes. We should see the same within-individual split if we look at the genomes of any other mammals, at birds, reptiles, amphibians or bony fish – for our common ancestor with all of them lived less than 500 million years ago. Wherever it has been investigated, this expectation has proved correct.

The alpha-beta split represents a single branch point in the globin family tree, but it is by no means the only one. Recent gene duplications include several actually within the alpha and beta lineages: although I have been implying there is a single alpha and a single beta gene, that isn't quite accurate. Both alpha and beta now exist as a small cluster of closely related genes, duplicated in series on the chromosome. In humans, the alpha cluster consists of seven slightly different versions of the gene. Two of these are pseudogenes – disabled versions of the gene with faults in their sequence, never translated into protein. Another gene copy seems to be functional but, oddly, never used – as far as we know. Two are active early in embryonic development. The remaining two are considered the 'alpha' genes proper, used in the production of adult haemoglobin. Surprisingly, both these alpha genes produce identical proteins (although they are switched on and off in different ways). The messiness of the whole set-up is typical of a system that has evolved by accidental duplication, rather than designed by an engineer. Yet these accidental duplicates provide raw material for evolution, as different copies evolve to take on more specialised roles. An appealing example is seen in the other lineage, the beta globin cluster.

In humans, the beta cluster also happens to contain seven globin genes (it is a coincidence that this is the same number of alphas: other vertebrates such as birds or teleost fish have clusters of different sizes, reflecting a different history of duplication and deletion). As with the alphas, not all the genes in the beta globin cluster are functional. Among the ones that do work is the 'standard' beta gene that we find – twice – in adult haemoglobin. The interesting part comes when we look at two other working copies, the result of duplications around 200 million years ago. They are called gamma globins, and duplication has left them free to evolve a role specific to the placental mammals. When you and I were foetuses in our mothers' wombs, we needed oxygen, which we extracted from our mothers' blood via the placenta. This requires a special type of foetal haemoglobin, which binds oxygen more tightly, and hence can grab it from the adult haemoglobin of the mother. Foetal haemoglobin is made by replacing the two adult beta globins with two gamma globins. In this case, gene duplication and subsequent specialisation provided a necessary step on the path that culminated in the placental mammals.

So much for recent duplications of globin genes. How about older ones? Gene duplications leave two traces: the pattern of divergence between genes within our own genomes, and the presence or absence of the various splits in more distantly related species. Lampreys and hagfish are the most distantly related vertebrates of all, the only ones who diverged more than half a billion years ago. And, sure enough, although lampreys and hagfish have a type of oxygen-carrying 'haemoglobin', it is not made up of two alphas and two betas like the rest of the vertebrates. It isn't even that their single haemoglobin gene diverged from ours just before the alpha/beta split, although that would have been neat. Instead, the gene that reddens lamprey blood branches much lower in the globin family tree, prior to the divergence of our haemoglobin from our myoglobin. Not only that, but lampreys have developed a muscle-based oxygen-storing globin, their 'myoglobin', which is more closely related to their 'haemoglobin' than it is to myoglobins from any other vertebrate. They seem to have convergently evolved oxygen management systems from ancestral globin genes, in the same way that both bats and birds have independently evolved wings from ancestral tetrapod limbs.

In pursuit of the history of these genes, we can examine more distant species and more distant globin genes too. All the globins which we have mentioned so far, save 'globin X', only occur in vertebrates, including lampreys and hagfish (globin X is the outgroup, and also occurs in insects, crustaceans and flatworms). More specifically, the vertebrate-only globins, which presumably arose between Concestor 22 and Concestor 23, fall into four major lineages. These are represented by haemoglobin (alpha and beta), myoglobin, cytoglobin (also present in lampreys), and globin Y. None of these branches of the globin tree are seen in other animals, not even the tunicates and lancelets that we will meet at the next two rendezvous points. And the real revelation comes when we look beyond the globins at completely different gene families. In vertebrates, DNA sequences of all sorts show duplication at the base of the vertebrate tree (the classic example is a fourfold duplication of the Hox gene cluster, the subject of the Fruit Fly's Tale). The simplest explanation is that these genes were not simply duplicated independently. Instead, large chunks of the ancestral vertebrate genome were duplicated, probably not just once, but two times. Such duplications, incidentally, are common in plants but rare in animals. On one momentous day, more than half a billion years ago, many, perhaps all, of our ancestor's chromosomes were copied, and on a later day in a descendant of that ancestor those copies themselves were recopied, to produce a genome containing four instances of most genes. The pattern is not always easy to spot today, for many copies have been deleted, shuffled around, or further reduplicated. Nevertheless, phylogenetic analysis of entire genomes is increasingly supporting this '2R' hypothesis – two rounds of whole genome duplication in the ancestors of vertebrates. It is in this sense that Peter Holland identifies the vertebrate genome as being massively more complex than the genomes of their nearest invertebrate relatives. Massive duplication and divergence of DNA may mark the start of the rise of the vertebrates.

Our gene divergences go back much further in time too, predating the vertebrates, predating the animals, even predating the split between animals and plants, back to our time as single-celled bacteria, and to the origin of life itself. Each one of our genes, if you go back far enough, owes its origin to the splitting of some ancient gene. Something like this entire book could be written for each gene. We arbitrarily decided that this should be a human pilgrimage, and we defined our milestones as meeting points with other lineages, which means, in the forward direction, speciation events at which our human ancestors split away from the others. We've already made the point that we could equally have begun our pilgrimage with a modern dugong, or a modern blackbird, and counted a different set of concestors back to Canterbury. But we are now making a more radical point. We could also write a backward pilgrimage for any *gene*.

We could choose to follow the pilgrimage of alpha haemoglobin, or cytochrome-c, or any other named gene. Rendezvous 1 would have been the milestone at which our chosen gene most recently duplicated to make a copy of itself elsewhere in the genome. Rendezvous 2 would have been the previous duplication event, and so on. Each of the rendezvous milestones would have taken place inside some particular animal or plant, just as the Lamprey's Tale has identified a Cambrian fish as the likely receptacle for the split between alpha and beta haemoglobin.

The gene's eye view of evolution keeps forcing itself upon our attention.

SEA SQUIRTS

The sea squirts seem, at first, unlikely recruits to our human-centred pilgrimage. Previous arrivals have not been too dramatically different from those already on the march. Vertebrates all retain a certain fishiness, even if in our case it has been modified by the demands of terrestriality. A sea squirt is something else. It doesn't swim like a fish. It doesn't swim like anything. It doesn't swim. It is far from clear why it deserves the illustrious name of chordate at all. A typical sea squirt is a bag filled with sea water, plus a gut and reproductive organs, anchored to a rock. The bag is topped by two siphons – one for drawing water in, the other for exhaling it. Day and night, water streams in through one siphon and out again through the other. On the way, it passes through the pharyngeal basket, a filtering net that strains out particles of food. Some sea squirts are packed together in colonies, but each member does essentially the same thing. No sea squirt is even faintly reminiscent of a fish, or indeed of any vertebrate (see plate 30).

No adult sea squirt, that is. However, like other sedentary, bottom-dwelling, filter-feeding animals, sea squirts have a dispersal phase, a larva that swims in the plankton. And the sea squirt's larva looks like . . . a tadpole. Or like a diminutive 'ammocoete' lamprey larva. Like an ammocoete, the 'tadpole larva' propels itself by a tail that undulates from side to side. Moreover, this tail extends well past the anus, unlike most invertebrate swimmers whose anus is right at the tip of the body. Most invertebrates have a nerve cord which runs along the belly (see the Ragworm's Tale), but like vertebrates the nerve cord of the sea squirt larva runs along its back. Most telling of all, it has a cartilaginous notochord. The larva, though not the adult sea squirt, has the appearance of at least a rudimentary chordate. When it is ready to metamorphose into an adult, the larva fastens itself onto a rock (or whatever is to be its adult resting place) head first, loses its tail, its notochord and most of its nervous system, and settles down for life.

The significance of this was known to Darwin. He gave the sea squirts the following unpromising introduction, under their scientific name of ascidians:

21

500 Ma

22

23
535 Ma

sea squirts

Sea squirts join. Animals with a stiff cartilaginous 'notochord' are classified together as chordates (in humans, remnants of this rod linger as the discs between our vertebrae). Since 2006, studies of large numbers of genes and even whole genomes shown increasing support for the theory that the sea squirts and their allies (of which there are about 3,000 described species) are even more closely related to us than the fish-like chordates known as lancelets, whom we shall meet at Rendezvous 24. The order of Rendezvous 23 and 24 has thus been reversed compared to the first edition of this book. This grouping of the vertebrates and sea squirts into an exclusive group (sometimes called 'Olfactores') seems to have been met with remarkable acceptance among most taxonomists.

They hardly appear like animals, and consist of a simple, tough, leathery sack, with two small projecting orifices. They belong to the Molluscoidea of Huxley – a lower division of the great kingdom of the Mollusca; but they have recently been placed by some naturalists amongst the Vermes or worms. Their larvae somewhat resemble tadpoles in shape, and have the power of swimming freely about.

I should say that neither Molluscoidea nor Vermes are any longer recognised, and sea squirts are no longer placed close to molluscs or worms. Darwin goes on to mention his own satisfaction in discovering such a larva in the Falkland Islands in 1833, and he continues as follows:

> M. Kovalevsky has lately observed that the larvae of Ascidians are related to the Vertebrata, in their manner of development, in the relative position of the nervous system, and in possessing a structure closely like the *chorda dorsalis* of vertebrate animals . . . We should then be justified in believing that at an extremely remote period a group of animals existed, resembling in many respects the larvae of our present Ascidians, which diverged into two great branches – the one retrograding in development and producing the present class of Ascidians, the other rising to the crown and summit of the animal kingdom by giving birth to the Vertebrata.

But now we have a division of opinion among experts. There are two theories of what happened: the one Darwin voiced, and a later one, which the Axolotl's Tale has already attributed to Walter Garstang. You remember the message of the axolotl, the message of neoteny. Sometimes the juvenile stage in a life-cycle can develop sex organs and reproduce: it becomes sexually mature, while remaining immature in other aspects of its being. We have previously applied the axolotl's message to Pekinese dogs, to ostriches and to ourselves: we humans appear to some scientists to be juvenile apes who have accelerated their reproductive development and chopped off the adult phase of the life cycle.

Garstang applied the same theory to sea squirts at this much older juncture in our evolutionary history. The adult phase of our remote ancestor, he suggested, was a sedentary sea squirt, which evolved the tadpole larva as an adaptation to disperse, in the same way as a dandelion seed has a little parachute to carry the next generation far away from the site of its parent. We vertebrates, Garstang suggested, are descended from sea squirt *larvae* – larvae that never grew up: or rather larvae whose reproductive organs grew up but who never turned into sea squirt adults.

A second Aldous Huxley might project fictional human longevity to the point where some super-Methuselah finally settles down on his head and metamorphoses into a giant sea squirt, fastened permanently to the sofa in front of a television. The plot would gain added satirical punch from the popular myth that a sea squirt larva, when it abandons pelagic activity for sedentary adulthood, 'eats its own brain'. Somebody must once have colourfully expressed the more mundane fact that, like a caterpillar in its chrysalis, the metamorphosing sea squirt larva breaks down its larval tissues and recycles them into the adult body. This includes breaking down the head ganglion, which was useful when it was an active swimmer in the plankton. Mundane or not, a literary metaphor as promising as that was never going to pass unnoticed – a meme as fecund would not go unspread. More than once I have seen a reference to the larval sea squirt which, when the time comes, settles down to a sedentary life and 'eats its brain, like an associate professor getting tenure'.

There is a group of modern animals within the sea squirt subphylum called the Larvaceae, which are reproductively adult but resemble sea squirt larvae. Garstang pounced on them, seeing them as a more recent rerun of his ancient evolutionary script. In his view, the larvaceans had ancestors that were bottom-dwelling, sedentary sea squirts, with a planktonic larval phase. They evolved the capacity to reproduce in the larval stage, and then chopped the old adult stage off the end of their life cycle. This could all have transpired rather recently, giving us a fascinating glimpse of what perhaps happened to our ancestors half a billion years ago.

Garstang's theory is certainly an attractive one, and it was much in favour for many years, especially in Oxford under the influence of Garstang's persuasive son-in-law, Alister Hardy. Alas, it now seems unlikely to be correct. The evidence comes, as often nowadays, from careful study of DNA. As somewhat of a recompense, these studies have uncovered some intriguing and unexpected features of sea squirt evolution.

One suggestive point was made to me by Peter Holland, the current holder of Alister Hardy's Chair. If the larvaceans constitute a recent re-enactment of an ancient Garstang scenario, they should find closer kinship with some modern sea squirts than with others.

Yet DNA reveals that among all the sea squirts, the deepest split may be between the larvaceans, and everything else, as you might expect if the other sea squirts became sedentary later.

The point is reinforced by a recent reordering of rendezvous points. Since 2006, studies of whole genomes have increasingly shown that we are more closely related to sea squirts that to lancelets (whom we now meet at the next rendezvous). This was a bit of a surprise because lancelets retain a fishy appearance throughout their lives, and it was always assumed *they* were our closest non-vertebrate cousins. With sea squirts joining our pilgrimage more recently than the lancelets, Garstang's scenario would need to be applied three times over. The adult phase would need to have been lost in the vertebrates, the larvaceans, and now the lancelets too. Somewhat improbable. It is more reasonable – more parsimonious – to propose that that the sedentary adult phase evolved later, just once, as Darwin suggested.

As for the timing, the paucity of fossils from this point on makes dating so difficult and controversial that my courage fails me. The estimate I have adopted for the date of Concestor 23 is 535 million years ago, which would put it around our 245-million-greats-grand-parent, but I could easily be wrong. We can nevertheless agree with Darwin's tacit assumption that the adult of that remote species looked like a tadpole. One branch of its descendants stayed tadpole-shaped and evolved into fish. The other branch got tenure, settled down on the sea bottom and became a sedentary filter-feeder, retaining its former adult form only in the larval stage.

However dull and sedentary we might find the life of a sea squirt, its DNA is anything but. The Coelacanth's Tale warned us against naively extrapolating from whole genomes to whole bodies. Sea squirt genomes seem almost designed* to hammer home the point. For example, they have some of the fastest recorded rates of molecular evolution in all the animal kingdom. Perhaps, you might think, this is because of the need for an additional life stage. But of the three sea squirt genomes sequenced to date, the fastest evolving is *Oiko-pleura dioica*, a larvacean with no sedentary adult phase. You might have thought that larvaceans would need fewer genes, compared to other sea squirts with a more complex life cycle. Yet *Oikopleura* is estimated to have around 18,000 genes, whereas the other two, *Ciona intestinalis* and *Ciona savignyi*, only have around 16,000 genes each,

---

* Creationists, note the word 'seem'.

having lost many important animal genes (including substantial parts of the normally vital Hox cluster, the focus of the Fruit Fly's Tale). These two sedentary *Ciona* species look similar enough that they are sometimes confused, and they can even be made to hybridise in the lab. Yet their DNA sequences turn out to be as different from each other, on average, as the human genome is from that of a chicken. As a parting surprise, note that while *Oikopleura* has almost the same number of genes as humans, its genome is over 40 times smaller. In part it has achieved this by trimming much of the 'junk' DNA both between and within genes – a strategy to which we shall return in the Humped Bladderwort's Tale. More radically, *Oikopleura* has also deleted excess DNA by clumping multiple genes together into a single 'operon', activated by a single set of switches. Perhaps because of this, the order of genes on its chromosomes has seen wild evolutionary change.

The sea squirts, with their radical changes in lifestyle and their equally radical genomes, push the concept of being a chordate to extremes. They are a natural experiment in anatomy and genetics – a perfect counterpoint to help us understand how genomes evolve and the methods deployed to control the development of our bodies.

LANCELETS

All the pilgrims we have met belong to a single great phylum, the Chordata, which is rounded off by a tidy little animal who wriggles up to join us here. It is the lancelet or amphioxus. *Amphioxus* used to be its Latin name, but the precedence-happy rules of nomenclature imposed *Branchiostoma* on it. Nevertheless, it had become so well known as *Amphioxus* that the name lives on. It is a protochordate, not a vertebrate, but it is clearly related to the vertebrates, and placed with them in the phylum Chordata. There are a few other related genera, but they are very similar to *Branchiostoma*, and I shall not distinguish them but call them all, informally, amphioxus.

I call amphioxus tidy because it elegantly lays out the features that proclaim it to be a chordate. It is a living, swimming (well, mostly buried in sand, actually) textbook diagram. There is the notochord running the length of the body, but not a trace of a vertebral column. There is the nerve tube on the dorsal side of the notochord, but no brain unless you count the small swelling at the front end of the nerve tube (where there is also an eye spot), and no skeletal brain case. There are the gill slits at the sides, and the segmental muscle blocks along the length of the body, but no trace of limbs. There is the tail, stretching back behind the anus, unlike a typical worm, which has the anus at the very end of the body. Amphioxus is also unlike a worm, but like many fish, in being shaped like a vertical blade, rather than cylindrical. It swims like a fish, with side-to-side undulations of the body, using the fish-like muscle blocks. The gill slits are part of the feeding apparatus, not primarily for breathing at all. Water is drawn in through the mouth and passed out through the gill slits, which act as filters to catch food particles. This is very likely how Concestor 24 used its gill slits, which would mean that gills for breathing came later, as an afterthought. If so, it is a pleasing reversal that, when the lower jaw eventually evolved, it was modified from a part of the gill apparatus.

If forced to put a date on Rendezvous 24, I would guess about 540 million years ago, the vintage of our 250-million-greats-grandparent, but dates of 775 million years or more are commonly quoted too. For

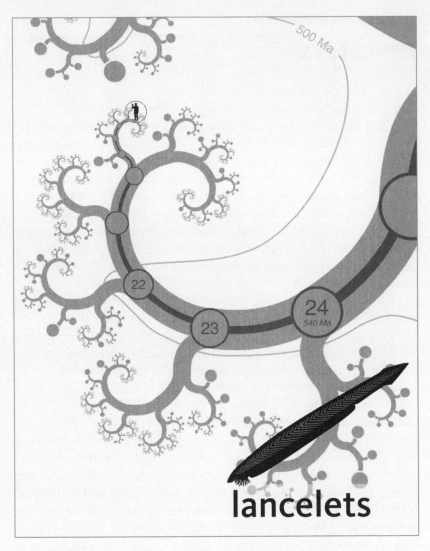

lancelets

Lancelets join. The 32 known species of fish-like animals commonly known as lancelets are known to be the last chordate group to join our pilgrimage. Note that the dates of rendezvous points from now on backwards are much disputed (see the Velvet Worm's Tale and its epilogue).

this reason I shall from now on abandon my attempts to describe the state of the world at the time of the concestor. As for what it looked like, I don't think we shall ever know for certain, but it is not implausible that Concestor 24 really may have been quite like a lancelet (see plate 32). If that is so, it is equivalent to saying that the lancelet is primitive. But that demands an immediate cautionary tale, from the Lancelet itself.

THE LANCELET'S TALE

If just one touch of sunshine more should make his gonads grow
The lancelet's claim to ancestry would get a nasty blow.
WALTER GARSTANG (1868–1949)

We have already met Walter Garstang, the distinguished zoologist who idiosyncratically expressed his theories in verse. I quote the couplet above not to develop Garstang's own theme, which we tackled extensively when we met the sea squirts ('his' gonads in the poem refers not to the lancelet but to a juvenile 'ammocoete' lamprey). I am concerned only with the last line, and especially the phrase 'claim to ancestry'. The lancelet, *Branchiostoma* or amphioxus has enough features in common with true vertebrates to have been long regarded as a surviving relative of some remote ancestor of the vertebrates. Or even – which is the real butt of my criticism – as the ancestor itself.

I am being unfair to Garstang, who knew perfectly well that the lancelet, as a surviving animal, could not be literally ancestral. Nevertheless, such talk really does sometimes mislead. Students of zoology delude themselves into imagining that when they look at some modern animal, which they call 'primitive', they are seeing a remote ancestor. This delusion is betrayed by phrases such as 'lower animal', or 'at the bottom of the evolutionary scale', which are not only snobbish but evolutionarily incoherent. Darwin's advice to himself would serve us all: 'Never use the words higher and lower.'

Lancelets are live creatures, our exact contemporaries. They are modern animals who have had exactly the same time as we have in which to evolve. Another telltale phrase is 'a side branch, off the main line of evolution'. All living animals are side branches. No line of evolution is more 'main' than any other, except with the conceit of hindsight.

Modern animals like lancelets, then, should never be revered as ancestors, nor patronised as 'lower', nor, for that matter, flattered as 'higher'. Slightly more surprisingly – and here we come to the second main point of the Lancelet's Tale – it is probably in general safest to say the same of fossils. It is theoretically conceivable that a particular fossil really is the direct ancestor of some modern animal. But it is statistically unlikely, because the tree of evolution is not a Christmas tree or a Lombardy poplar, but a densely branched thicket or bush. The fossil you are looking at probably isn't your ancestor, but it may help you to understand the *kind* of intermediate stage your real ancestors went through, at least in respect of some particular bit of the body, such as the ear, or the pelvis. A fossil, therefore, has something like the same status as a modern animal. Both can be used to illuminate our guesses about some ancestral stage. Under normal circumstances, neither should be treated as though it really is ancestral. Fossils as well as living creatures are usually best treated as cousins, not ancestors.

Members of the cladistic school of taxonomists can become positively evangelical about this, proclaiming the non-specialness of fossils with the zeal of a puritan or a Spanish inquisitor. Some go right over the top. They take the sensible statement, 'It is unlikely that any particular fossil is an ancestor of any surviving species', and interpret it to mean 'There never were any ancestors!' Obviously this book stops short of such an absurdity. At every single moment in history there must have been at least one human ancestor (contemporary with, or identical to, at least one elephant ancestor, swift ancestor, octopus ancestor, etc.), even if any particular fossil almost certainly isn't it.

The upshot is that, on our backward journey towards the past, the concestors we have been meeting have not, in general, been particular fossils. The best we can normally hope for is to put together a list of attributes that the ancestor probably had. We have no fossil of the common ancestor we share with the chimpanzees, even though that was less than 10 million years ago. But we were able to guess, with misgivings, that the ancestor was most likely to have been, in Darwin's famous words, a hairy quadruped, because we are the only ape that walks on its hind legs and has bare skin. Fossils can help us with our inferences, but mostly in the same kind of indirect way that living animals help us.

The moral of the Lancelet's Tale is that it is vastly harder to find an ancestor than a cousin. If you want to know what your ancestors looked like 100 million years ago, or 500 million years ago, it is no use reaching down to the appropriate depth in the rocks and hoping to come up with a fossil labelled 'Ancestor', as if from some Mesozoic or Palaeozoic bran tub. The most we can normally hope for is a series of fossils that, some with respect to one part, others with respect to another part, represent the *kind* of thing the ancestors probably looked like. Perhaps this fossil tells us something about our ancestors' teeth, while that fossil a few million years later gives us an inkling about our ancestors' arms. Any particular fossil is almost certainly not our ancestor but, with luck, some parts of it may resemble the corresponding parts of the ancestor just as, today, the shoulder-blade of a leopard is a reasonable approximation to the shoulder-blade of a puma.

AMBULACRARIANS

Our pilgrimage is now a milling horde, having amassed all the vertebrates, together with their primitive chordate cousins, amphioxus and the sea squirts. It comes as quite a surprise that the next pilgrims to join us, our closest relatives among the invertebrates, include those strange creatures – we shall soon refer to them as 'Martians' – the starfish, sea urchins, brittle stars and sea cucumbers. These, together with a largely extinct group called the crinoids or sea lilies, comprise the phylum Echinodermata, the spiny-skinned ones. Along with the echinoderms come a few miscellaneous worm-like groups which, in the absence of molecular evidence, had been placed elsewhere in the animal kingdom. The acorn worms and their kind (Enteropneusta and Pterobranchia) had previously been classified with the sea squirts as protochordates. Molecular evidence now links them, not so very far away, with the echinoderms in a super-phylum called Ambulacraria.

There is also a curious wormy genus called *Xenoturbella* which may belong here. Nobody knew where to put little *Xenoturbella* – it seems to lack most of the things that a respectable worm ought to have, like a proper excretory system and a through-flow gut. Zoologists shuffled this obscure little worm from phylum to phylum, and had pretty well given up on it when, in 1997, somebody announced that, despite all appearances, it was a highly degenerate bivalve mollusc, with affinities to cockles. This confident statement came from molecular evidence. *Xenoturbella*'s DNA closely resembled that of a cockle and, as if to clinch it, *Xenoturbella* specimens were found to contain mollusc-type eggs. Terrible warning! In what looks like the classic nightmare of the modern forensic detective – contamination of the suspect's DNA by that of the murder victim – it has now turned out that the reason *Xenoturbella* contained mollusc DNA is that it eats molluscs! When researchers sequenced another one, this time taking care to exclude all parts of its gut, they discovered an even more surprising affinity: with the Ambulacraria. It is still debated, with some researchers now insisting they belong among the acoelomorph worms whom we shall meet at Rendezvous 27. But

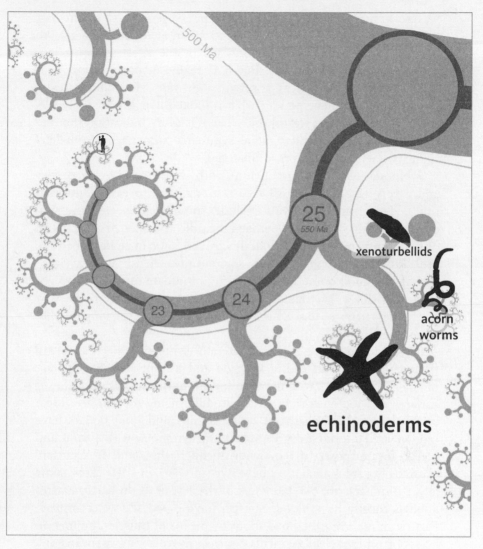

Starfish and their kin join. We chordates belong to the major branch of animals known as the deuterostomes. Recent molecular studies suggest that the other deuterostome phyla form a sister group to the chordates. This group, named the ambulacraria, includes the echinoderms, a phylum with about 7,000 described species, the hemichordates (a phylum of about 120 species of acorn worms and pterobranchs), and may also include *Xenoturbella*, two species of which are shown above (a few more deep-sea species have recently been discovered). Save for the xenoturbellids, which could conceivably join us at the next rendezvous, the ambulacrarian 'superphylum' is now generally accepted.

we are going with the opinion that puts them here, as the last member to join them 'before' we greet them at Rendezvous 25. We are tentatively placing this rendezvous in the latest Precambrian, about 550 million years ago, but it is really anyone's guess. That would make Concestor 25 approximately our 260-million-greats-grandparent. We have no idea what it looked like, but it surely was more worm-like than starfish-like. There is every indication that the echinoderms evolved their radial symmetry secondarily from left-right symmetrical ancestors – 'Bilateria'.

Echinoderms are a large phylum, with about 6,000 living species and a very respectable fossil record going back to early Cambrian times. Those ancient fossils include some weirdly asymmetrical creatures. Indeed, weird is perhaps the adjective that first occurs to one contemplating the echinoderms. A colleague once described the cephalopod molluscs (octopuses, squids and cuttlefish) as 'Martians'. He made a good point, but I think my candidate for the role might be a starfish. A 'Martian', in this sense, is a creature whose very strangeness helps us to see ourselves more clearly by showing us what we are not (see plate 31).

Earth's animals are mainly bilaterally symmetrical: they have a front end and a rear end, a left side and a right side. Starfish are radially symmetrical, with the mouth right in the middle of the lower surface, and the anus right in the middle of the top surface. Most echinoderms are similar, but heart urchins and sand dollars have rediscovered a modest degree of bilateral symmetry with a front and a rear for purposes of burrowing through the sand. If 'Martian' starfish have sides at all, they have five sides (or, in a few cases, some larger number), not two like most of the rest of us on Earth. Earth's animals mostly have blood. Starfish have piped sea water instead. Earth's animals mostly move about by means of muscles, pulling on bones or other skeletal elements. Starfish move about by means of a unique hydraulic system, using pumped sea water. Their actual propulsive organs are hundreds of small 'tube feet' on their under surface, arrayed in avenues along the five axes of symmetry. Each tube foot looks like a thin tentacle with a little round sucker on the end. On its own it is too small to move the animal, but the whole array pulling together can do it, slowly but powerfully. A tube foot is extended by hydraulic pressure, exerted by a little squeezed bulb at its near end. Each individual tube foot has a cycle of activity rather like a tiny leg. Having exerted its pull, it releases its sucker, picks itself up and swings forward to take a new grip with the sucker, and pull again.

Sea urchins get around by the same method. Sea cucumbers, which are shaped like warty sausages, can move this way too, but burrowing ones move the whole body as earthworms do, by alternately squeezing the body so it elongates forward, then pulling the rear up behind. Brittle stars, which have (usually) five slender, waving arms radiating out from a nearly circular central disc, move by rowing with whole arms, rather than dragging themselves along by tube feet. Starfish too have muscles that swing whole arms about. They use them, for example, to engulf prey and pull mussel shells apart.

'Forward' is arbitrary for these 'Martians', and that includes brittle stars and most urchins as well as starfish. Unlike most Earth life forms, who have a definite front end with a head, a starfish can 'lead' with any one of its five arms. The hundreds of tube feet somehow manage to 'agree' to follow the lead arm at any one time, but the lead role can change from arm to arm. The co-ordination is achieved by a nervous system, but it is a different pattern of nervous system from any others we are accustomed to on this planet. Most nervous systems are based upon a long trunk cable running from front to rear, either along the dorsal side (like our spinal cord) or along the ventral side, in which case it is often double, with a ladder of connections between the left and right sides (as in worms and all arthropods). In a typical Earth creature, the main longitudinal trunk cable has side nerves, often paired in segments repeated serially from front to rear. And it usually has ganglia, local swellings which, when sufficiently large, are dignified with the name of brain. The starfish nervous system is utterly different. As we have come to expect by now, it is radially arranged. There is a complete ring going right round the mouth, from which five (or however many arms there are) cables radiate out, one along each arm. As you would expect, the tube feet along each arm are controlled by the trunk nerve running along it.

In addition to the tube feet, some species also have hundreds of so-called pedicellariae (singular pedicellaria), scattered over the lower surface of the five arms. These have tiny pincers, and are used for catching food, or in defence against small parasites.

Alien 'Martians' though they may appear, starfish and their kind are still our relatively close cousins. Less than 4 per cent of all animal species are closer cousins to us than starfish are. By far the greater part of the animal kingdom is yet to join our pilgrimage. And they mostly arrive all together, at Rendezvous 26, in one gigantic influx of pilgrims. The protostomes are about to overwhelm even the multitude of pilgrims who are already on the march.

PROTOSTOMES

In the deeps of geological time, and increasingly deprived of the hard support of fossils, we are now entirely reliant on the technique that we referred to in the General Prologue as triangulation. The upside is that the genetic techniques for triangulating back to the past using living organisms are getting ever more sophisticated. Molecular triangulation confirms a belief long held by comparative anatomists, or more strictly comparative embryologists, that the greater part of the animal kingdom is deeply divided into two great subkingdoms, the Deuterostomia and the Protostomia.

Here's how embryology comes in. Animals typically pass through a watershed event in their early life called gastrulation. The distinguished embryologist and scientific iconoclast Lewis Wolpert said:

It is not birth, marriage or death, but gastrulation, which is truly the most important time in your life.

Gastrulation is something that all animals do early in their life. Typically, before gastrulation, an animal embryo consists of a hollow ball of cells, the blastula, whose wall is one cell thick. During gastrulation the ball indents to form a cup with two layers. The opening of the cup closes in to form a small hole called the blastopore. Almost all animal embryos go through this stage, which presumably means it is a very ancient feature indeed. You might expect that so fundamental an opening would become one of the two deep holes in the body, and you'd be right. But now comes the big divide in the animal kingdom, between the Deuterostomia (every pilgrim who arrived before Rendezvous 26, including us) and the Protostomia (the huge throng who are now joining at Rendezvous 26).

In deuterostome embryology, the eventual fate of the blastopore is to become the anus (or at least the anus develops close to the blastopore). The mouth appears later as a separate perforation at the other end of the gut. The protostomes do it differently: in some, the blastopore becomes the mouth, and the anus appears later; in others, the blastopore is a slit that subsequently zippers up in the middle,

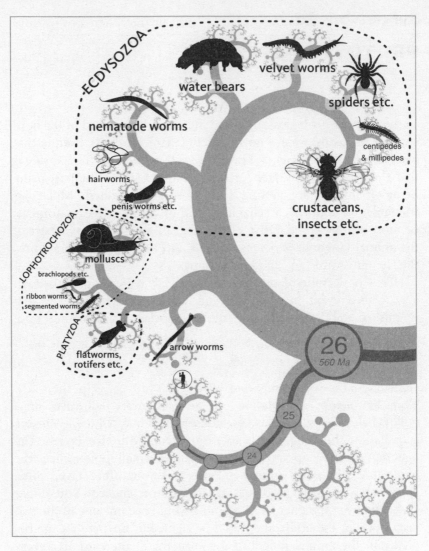

Protostomes join. At this rendezvous the 80,000 or so known deuterostome species join well over a million described protostomes. This protostome phylogeny represents another radical rearrangement brought about by genetics within the past 15 years. We mark three major groups: two of which (Ecdysozoa and Lophotrochozoa) are now universally accepted, the third less so. The remaining group, the predatory arrow worms or chaetognaths, are an ancient phylum with representatives found in the Chengjiang fossil beds. Their position is not yet resolved: we show one possible placement. As well as the dates of rendezvous points since Rendezvous 24, the order of branching within the Lophotrochozoa is also much disputed. For this reason, we have omitted contour lines from here on.

with the mouth at one end and the anus at the other. Protostome means 'mouth first'. Deuterostome means 'mouth second'.

This traditional embryological classification of the animal kingdom has been upheld by modern molecular data. There are indeed two main kinds of animal, the deuterostomes (our lot) and the protostomes (them over there). However, some phyla that used to be included in the deuterostomes have now been moved by molecular revisionists, whom we shall follow, to the protostomes. These are the three so-called lophophorate phyla – the phoronids, brachiopods and bryozoans – now grouped together with the molluscs and annelid worms in the 'Lophotrochozoa' division of the protostomes. For goodness' sake don't bother to remember the 'lophophorates' – they are mentioned here only because zoologists of a certain age might be surprised not to find them among the deuterostomes. There are also some animals that don't belong to either the protostomes or the deuterostomes, but we'll meet them at later rendezvous.

Rendezvous 26 is the biggest of all, more of a gigantic rally of pilgrims than a rendezvous. When does it happen? Such ancient dates are hard to estimate. Our attempt of 560 million years is plus or minus a large margin of error, and probably more plus than minus in this case. The same goes for the estimate that Concestor 26 is our 270-million-greats-grandparent. The protostomes constitute the great bulk of the pilgrimage of animals. Because our own species is of the deuterostome persuasion, we have given them special attention in this book, and we are portraying the protostomes as joining the pilgrimage all together, at one major rendezvous. Not only the protostomes themselves would see it the other way around – a dispassionate observer would too.

The protostomes have a much greater number of animal phyla than the deuterostomes, including the largest phyla of all. They include the molluscs, with twice as many species as the vertebrates. They include the three great worm phyla: flatworms, roundworms and annelid worms, whose species together outnumber the mammal species perhaps thirtyfold. Above all, the protostome pilgrims include the arthropods: insects, crustaceans, spiders, scorpions, centipedes, millipedes and several other smaller groups. The insects alone constitute at least three-quarters of all animal species, and probably more. As Robert May, distinguished former President of the Royal Society, has said, to a first approximation all species are insects.

Before the days of molecular taxonomy, we grouped and divided animals by looking at their anatomy and embryology. Of all the classificatory levels – species, genus, order, class, etc. – phylum had a special, almost mystical status. Animals within one phylum were clearly related to one another. Animals in different phyla were too distinct for any relationships to be taken seriously. The phyla were separated by an all but unbridgeable gulf. Molecular comparison now suggests that the phyla are much more connected than we ever thought they were. In a sense that was always obvious – nobody believed the animal phyla arose separately from primordial slime. They had to be connected to each other, in the same sort of hierarchical patterns as their constituent parts. It was just that the connections were hard to see, lost in deep time.

There were exceptions. The protostome–deuterostome grouping above the phylum level was admitted, based on embryology. And within the protostomes it was widely accepted that the annelid worms (segmented earthworms, leeches and bristle worms) were related to arthropods, both having a segmented body plan. That particular connection now seems to be wrong, as we shall see: nowadays the annelids are partnered with the molluscs. Actually, it was always a bit worrying that marine annelids had a kind of larva that was so similar to the larvae of many marine molluscs that they were given the same name, the 'trochophore' larva. If the annelid–mollusc grouping is right, it means that the segmented body plan was invented twice (by annelids and arthropods), rather than the trochophore larva being invented twice (by annelids and molluscs). The association of annelids with molluscs, and their separation from arthropods, is one of the bigger surprises that molecular genetics has dealt those zoologists brought up on morphologically based taxonomy.

Molecular evidence divides the protostome phyla into two, or perhaps three, main groups: super-phyla, I suppose we could call them. Some authorities have yet to accept this classification, but I shall go along with it while recognising that it could still be wrong. The two super-phyla are called the Ecdysozoa and the Lophotrochozoa. The possible third super-phylum, which is less widely acknowledged, but which I shall accept rather than lumping them in with the Lophotrochozoa as some prefer, is the Platyzoa.

The Ecdysozoa are named after their characteristic habit of moulting, or ecdysis (from a Greek word meaning roughly to get your kit off). That gives an immediate hint that the insects, crustaceans, spiders, millipedes, centipedes, trilobites and other arthropods are ecdysozoans, and this means that the ecdysozoan faction of the protostome pilgrimage is very large indeed, far more than three-quarters of the animal kingdom.

The arthropods dominate both the land (especially insects and spiders) and the sea (crustaceans and, in earlier times, trilobites). With the exception of the eurypterids, those Palaeozoic sea scorpions* which, we conjectured, terrorised the Palaeozoic fishes, arthropods have not achieved the enormous body size of some extreme verte-brates. This is often attributed to limits set by their method of encasing themselves in an armour-plated exoskeleton, with their limbs in hard jointed tubes. It means they can grow only by ecdysis: casting their outer casing aside at regular intervals and hardening a new, larger one. How the eurypterids managed to exempt themselves from this alleged size limitation is not entirely clear to me.

Dispute lingers about how the sub-contingents of arthropods are arranged. Most zoologists now reject the earlier view that the insects belong with the myriapods (centipedes, millipedes and their kind), separated off from the crustaceans. There is a growing consensus that insects are a derived sort of crustacean. The myriapods and spiders have now been pushed off as outgroups, who colonised the land independently. Everyone agrees that spiders and scorpions, together with the terrifying eurypterids, belong together in the group called chelicerates. *Limulus*, the living fossil known, unfortunately, as the horseshoe crab, is also placed in the chelicerates, despite its superficial resemblance to the extinct trilobites, which are separated off in their own group.

Allied to the arthropods within the Ecdysozoa, and sometimes called panarthropods, are two small contingents of pilgrims, the onychophorans and the tardigrades. Onychophorans or velvet worms, such as *Peripatus*, are now classified in the phylum Lobopo-dia, which has an important fossil contingent, as we shall see in the Velvet Worm's Tale. *Peripatus* itself looks a bit like a caterpillar of

---

* There were also giant land scorpions in the Palaeozoic, estimated at a metre long, a fact that I do not greet with equanimity (one of my earliest memories, before I fainted, is of being stung by a modern African scorpion). The largest known trilobite, *Isotelus rex*, reached 72 centimetres in length. Dragonflies with wingspans up to 70 centimetres flourished in the Carboniferous. Today's largest arthropod, the Japanese spider crab *Macrocheira kaempferi* has a 30-centimetre body, and the span between its enormously elongated claw-bearing limbs can be up to four metres.

rather endearing mien, although in this respect it is outdone by the tardigrades. Whenever I see a tardigrade I want to keep it as a pet. Tardigrades are sometimes called water bears, and they have the cuddly appearance of a baby bear. A very baby bear indeed: you can only just see them without a microscope, waving their eight stubby legs with a charming air of infantile ineptitude.

The other major phylum in the superphylum Ecdysozoa is that of the nematode worms. They too are extremely numerous, a fact made memorable long ago by the American zoologist Ralph Buchsbaum:

> If all the matter in the universe except the nematodes were swept away, our world would still be dimly recognisable . . . we should find its mountains, hills, vales, rivers, lakes, and oceans represented by a film of nematodes . . . Trees would still stand in ghostly rows representing our streets and highways. The location of the various plants and animals would still be decipherable, and, had we sufficient knowledge, in many cases even their species could be determined by an examination of their erstwhile nematode parasites.

I was delighted by this image when I first read Buchsbaum's book, but I must confess, returning to reread it now, I find myself sceptical. Let's just say that nematode worms are extremely numerous and ubiquitous.

Smaller phyla in the Ecdysozoa include various other kinds of worms, including the priapulid or penis worms. These are quite aptly named, although the champion in this vein is the fungus whose Latin name is *Phallus* (wait for Rendezvous 35). It is superficially surprising that the priapulids are now classified so far from the annelid worms.

The lophotrochozoan pilgrims may be outnumbered by the Ecdysozoa, but even they decisively outnumber our own deuterostome pilgrims. The two big lophotrochozoan phyla are the molluscs and the annelids. The annelid worms are not long confusible with the nematode worms, for the annelids are segmented – like the arthropods, as we have seen. This means that their body is arranged as a series of segments fore and aft, like the trucks of a train. Many body parts, for example nerve ganglia and blood vessels running around the gut, are repeated in every segment along the length of the body. The same is true of arthropods, most obviously millipedes and centipedes because their segments are all pretty much the same as each other. In a lobster or, even more, a crab, many of the segments are different from each other, but you can still clearly see that the body is segmented in the fore-and-aft direction. Their ancestors surely had more uniform segments like a woodlouse or a millipede.* Annelid worms are like millipedes or woodlice in this respect, although the worms are more closely related to the non-segmented molluscs. The most familiar annelid worms are common or garden (for once the phrase is strictly apt) earthworms. I am privileged to have seen giant earthworms (*Megascolides australis*), in Australia, said to be capable of growing to four metres long.

The Lophotrochozoa include other worm-like phyla, for instance the nemertine worms, not to be confused with the nematodes. The similarity of name is unfortunate and unhelpful, compounded by further confusion with two other worm phyla, the Nematomorpha and the Nemertodermatida. Nema (*nematos*) in Greek means 'thread', while Nemertes was the name of a sea nymph. Unfortunate coincidence, that. On a school marine biology field trip to the Scottish coast with our inspiring zoology teacher Mr I. F. Thomas, we found a bootlace worm, *Lineus longissimus*, a species of nemertine legendarily capable of growing to 50 metres. Our specimen was at least 10 metres long.

There are various other more-or-less worm-like phyla, but the biggest and most important phylum of the Lophotrochozoa is the Mollusca: the snails, oysters, ammonites, octopuses and their kind. The mollusc contingent of the pilgrimage mostly creeps at snail's pace, but squids are among the fastest swimmers in the sea, using a form of jet propulsion. They, and their cousins the octupuses, are the most spectacularly proficient colour-changers in the animal kingdom, streets better than the proverbial chameleons, not least because they change in quick time. The ammonites were relatives of the squids who lived in coiled shells that served them as flotation organs, as with the still surviving *Nautilus*. Ammonites once thronged the seas but went finally extinct at the same time as the dinosaurs. I hope they changed colour too.

Another major group of molluscs is the bivalves: oysters, mussels, clams and scallops, with two shells or valves. Bivalves have a single extremely powerful muscle, the adductor, whose function is to close the valves and lock in the closed position against predators. Don't

---

* There are some wonderful millipedes – pill millipedes – that look and behave just like woodlice. This is one of my favourite examples of convergent evolution.

Ceci n'est pas une coquille. Fossil brachiopod (*Doleorthis*) from the Silurian.

put your foot in a giant clam (*Tridacna*) – you'll never get it back. The bivalves include *Teredo*, the shipworm, which uses its valves as cutting tools to bore through driftwood, wooden ships and the pilings of piers and quays. You will probably have seen their holes, of cleanly circular cross-section. Piddocks do something similar through rock.

Superficially like bivalve molluscs are the brachiopods, the lamp shells. They are also part of the great lophotrochozoan contingent of the protostome pilgrimage, but are not closely related to the bivalve molluscs. We have already met one of them, *Lingula*, in the Coelacanth's Tale, as a 'living fossil'. There are now only about 350 species of brachiopod, but in the Palaeozoic Era they rivalled the bivalve molluscs.* The resemblance between them is superficial: the bivalve molluscs' two shells are left and right, where the two brachiopod shells are top and bottom. The status of the brachiopod pilgrims, and two allied 'lophophorate' groups called phoronids and bryozoans, is still disputed. As already mentioned, I am following the dominant contemporary school of thought in placing them in the Lophotrochozoa (to which name, indeed, they have contributed). Some zoologists leave them where they used to be, outside the protostomes altogether and in the deuterostomes, but I suspect theirs is a losing battle.

The third major branch of the protostome super-phylum, the Platyzoa, would be joined by some authorities to the Lophotrochozoa. 'Platy' means 'flat', and the name Platyzoa comes from one of the component phyla, the flatworms or Platyhelminthes. 'Helminth' means 'intestinal worm', and while some flatworms are parasitic (tapeworms and flukes), there is also a large group of freeliving flatworms, the turbellarians, which are often extremely beautiful.

---

* Stephen Gould compared them in a nice essay called 'Ships That Pass in the Night'.

Recently, some of the animals traditionally classified as flatworms, for example the acoels, have been removed by molecular taxonomists out of the protostomes altogether. We shall meet them presently.

Other phyla are provisionally placed in the Platyzoa, but for the moment it is for want of anywhere more certain to put them, and they are mostly not flat. Belonging to the so-called 'minor phyla', they are fascinating in their own right and each deserves a whole chapter in a textbook of invertebrate zoology. Unfortunately, however, we have a pilgrimage to complete and must press on. Of these minor phyla I shall just mention the rotifers, because they have a tale to tell.

Rotifers are so small that they were originally grouped with single-celled protozoan 'animalcules'. They are actually multicellular and quite complicated in miniature. One group of them, the bdelloid rotifers, are remarkable because no male has ever been seen. This is what their tale is about, and we shall come to it soon.

So this vast flood of protostome pilgrims, composite of tributaries far and wide, truly the dominant stream of animal pilgrims, converges on its rendezvous with the deuterostomes, the junior (by comparison) contingent, whose progress we have hitherto followed for the sufficient reason that it is our own. The grand ancestor of both, Concestor 26 from our human point of view, is extremely hard to reconstruct at such a remote distance of time.

It seems likely that Concestor 26, which we have attempted to reconstruct in plate 33, was some kind of worm. But that is only to say a long thing, bilaterally symmetrical, with a left and a right side, a dorsal and a ventral side, and a head and a tail end. Indeed, the name Bilateria is commonly used to refer to the grand grouping of protostomes and deuterostomes (plus the little worms we will meet at the next rendezvous). Why is this pattern, the worm form, so common? The most primitive members of all three protostome subgroups, and the most primitive deuterostomes, are all of the form that we should generally call worm-shaped. So let's have a tale about what it means to be a worm.

I wanted to put the worm's tale into the grey and muddy mouth of the lugworm.* Unfortunately, the lugworm spends most of its

---

* A lugworm, with its grey and muddy mouth
Sang that somewhere to north or west or south
There dwelt a gay, exulting, gentle race . . .
                    W. B. YEATS
                    (1865–1939)

time in a U-shaped burrow, which is just what we don't need for the
tale, as will soon become apparent. We need a more typical worm,
which actively crawls or swims in a forward direction: for whom
front and rear, left and right, and up and down have a clear
meaning. So the lugworm's close cousin, *Nereis* the ragworm, shall
take over the role. An 1884 magazine article for anglers said, 'The
bait used is that damp kind of centipede called a ragworm.' It is not
a centipede, of course, but a polychaete worm. It lives in the sea,
where it normally crawls over the bottom but is capable of swimming if necessary.

THE RAGWORM'S TALE

Any animal that moves, in the sense of covering the ground from A
to B rather than just sitting in one place and waving its arms or
pumping water through itself, is likely to need a specialised front
end. It might as well have a name, so let's call it the head. The head
hits novelty first. It makes sense to take in food at the end that
encounters it first, and to concentrate the sense organs there too –
eyes perhaps, some kind of feelers, organs of taste and smell. Then
the main concentration of nervous tissue – the brain – had best be
near the sense organs, and near the action at the front end, where
the food-catching apparatus is. So we can define the head end as the
leading end, the one with the mouth, the main sense organs and the
brain, if there is one. Another good idea is to void wastes somewhere
near the back end, far from the mouth, to avoid re-imbibing what
has just been passed out. By the way, although all this makes sense if
we think worm, I should remind you that the argument evidently
does not apply to radially symmetrical animals such as starfish. I am
genuinely puzzled why starfish and their kind opt out of this
argument, which is one reason why I referred to them as 'Martians'.

To return to our primeval worm, having dealt with its fore and aft
asymmetry, how about up-down asymmetry? Why is there a dorsal
side and a ventral side? The argument is similar, and this one applies
to starfish just as much as to worms. Gravity being what it is, there
are lots of inevitable differences between up and down. Down is
where the sea bottom is, down is where the friction is, up is where the
sunlight comes from, up is the direction from which things fall on
you. It is unlikely that dangers will threaten equally from below and

above, and in any case those dangers are likely to be qualitatively different. So our primitive worm should have a specialised upper or 'dorsal' side and a specialised 'ventral' or lower side, rather than simply not caring which side faces the sea bottom and which side faces the sky.

Put our front-rear asymmetry together with our dorsal-ventral asymmetry, and we have automatically defined a left side and a right side. But unlike the other two axes, we find no general reason to distinguish the left side from the right side: no reason why they should be anything other than mirror images. Danger is not more likely to threaten from the left than the right, or vice versa. Food is not more likely to be found on the left or the right side, though it may well be more likely above or below. Whatever is the best way for a left side to be, there is no general reason to expect any difference for the right. Limbs or muscles that were not mirrored left-right would have the unfortunate effect of driving the animal round in circles, instead of in direct pursuit of some goal.

Perhaps revealingly, the best exception I can think of is fictional. According to a Scottish legend (probably invented for the amusement of tourists, and said to be believed by many of them), the haggis is a wild animal living in the Highlands. It has short legs on one side and long legs on the other, in accordance with its habit of running only one way round the sides of steep Highland hillsides. The prettiest real-life example I can think of is the wonky-eyed jewel squid of Australian waters, whose left eye is much larger than its right. It swims at a 45-degree angle, with the larger, telescopic left eye looking upwards for food, while the smaller right eye looks below for predators. The wrybill is a New Zealand sandpiper whose bill curves markedly to the right. The bird uses it to flick pebbles to the side and expose prey. Striking 'handedness' is to be seen in fiddler crabs, who have one hugely enlarged claw for fighting or, more to the point, displaying their ability to fight. But perhaps the most intriguing story of asymmetry in the animal kingdom was told me by Sam Turvey. Trilobite fossils often display bite marks, indicating narrow escapes from predators. The fascinating thing is that about 70 per cent of these bite marks are on the right-hand side. Either trilobites had an asymmetrical awareness of predators, like the wonky-eyed jewel squid, or their predators had handedness in their attack strategy.

But those are all exceptions, mentioned for their curiosity value and to make a revealing contrast with the symmetrical world of our primitive worm and its descendants. Our crawling archetype has a left and a right side which are mirror images of each other. Organs tend to arise in pairs, and where there are exceptions, such as the wonky-eyed jewel squid, we notice it and comment.

How about eyes? My guess is that Concestor 26 did have eyes, but it isn't enough just to claim this on the basis that all its modern descendants have them. It isn't enough, because the various kinds of eyes are very diverse: so much so that it has been estimated that 'the eye' has evolved independently more than 40 times in various parts of the animal kingdom.* How do we reconcile this with the statement that Concestor 26 had eyes?

To give intuition a steer, let me say first that what is claimed to have evolved 40 times independently is not light-sensitivity per se, but image-forming optics. The vertebrate camera eye and the crustacean compound eye evolved their optics (working on radically different principles) independently of one another. But both these eyes are descended from one organ in the common ancestor (Concestor 26), which was probably an eye of some kind.

The evidence is genetic, and it is persuasive. In the fruit fly *Drosophila* there is a gene called *eyeless*. Geneticists have the perverse habit of naming genes by what goes wrong when they mutate. The *eyeless* gene normally negates its name by making eyes. When it mutates and fails to have its normal effect on development, the fly has no eyes, hence the name. It is a ludicrously confusing convention. To avoid it, I shall not refer to the *eyeless* gene, but will use the comprehensible abbreviation *ey*. The *ey* gene normally makes eyes, and we know this because when it goes wrong the flies are eyeless. Now the story starts to get interesting. There is a very similar gene in mammals, called *Pax6*, also known as *small eye* in mice and *aniridia* (no iris) in humans (again named for the negative effect of its mutant form).

The DNA sequence of the human *aniridia* gene is more similar to the fruit fly's *ey* gene than it is to other human genes. They must be inherited from the shared ancestor which was, of course, Concestor

---

* I have discussed this at length in *Climbing Mount Improbable*, in a chapter called 'The Fortyfold Path to Enlightenment', and I return to it at the end of this book.

26. Again, I shall call it *ey*. Walter Gehring and his colleagues in Switzerland did an utterly fascinating experiment. They introduced the mouse equivalent of the *ey* gene into fruit fly embryos, with astounding results. When introduced into the part of a fruit fly embryo that was destined to make a leg, it caused the eventual adult fly to grow an extra 'ectopic' eye on its leg. It was a fly eye, by the way: a compound eye, not a mouse eye. I don't think there is any evidence that the fly could see through it, but it had the unmistakable properties of a respectable compound eye. The instruction given by the *ey* gene seems to be 'grow an eye here, of the kind that you would normally grow'. The fact that the gene is not only similar in mice and flies, but induces the development of eyes in both, is very strong evidence that it was present in Concestor 26; and moderately strong evidence that Concestor 26 could see, even if only the presence versus the absence of light. Perhaps, when more genes have been investigated, the same argument can be generalised from eyes to other bits. In fact, in one sense, this has already been done – we'll deal with it in the Fruit Fly's Tale.

The brain, sitting at the front end for the reasons we have argued, needs to make nervous contact with the rest of the body. In a worm-shaped animal, it is sensible that it should do so via a main cable, a principal nerve trunk, running along the length of the body, probably with side branches at intervals along the body to exercise local control and take in local information. In a bilaterally symmetrical animal like a ragworm or a fish, the trunk nerve must run either dorsal or ventral of the digestive tract, and here we strike one of the main differences between us deuterostomes on the one hand and the protostomes, who have joined us in such strength, on the other. In us, the spinal nerve cord runs along the back. In a typical protostome like a ragworm or a centipede, it is on the ventral side of the gut.

If Concestor 26 was indeed some kind of worm, it presumably followed either the dorsal nerve pattern or the ventral nerve pattern. I can't call them deuterostome and protostome patterns because the two separations don't quite coincide. The acorn worms (those rather obscure deuterostomes who arrived with the echinoderms at Rendezvous 25) are hard to interpret, but at least on some views they have a ventral nerve cord like a protostome, although for other reasons they are classified as deuterostomes. Let me instead divide the animal kingdom into the dorsocords and the ventricords. The dorsocords are all deuterostomes. The ventricords are mostly protostomes, plus some early deuterostomes perhaps including the acorn worms. The echinoderms, with their remarkable reversion to radial symmetry, don't fit into this classification at all.

The difference between dorsocords and ventricords extends to other things than just the position of the main nerve running along the body. Dorsocords have a ventral heart, whereas ventricords have a dorsal heart, pumping blood forward along a main dorsal artery. These and other details suggested in 1820 to the great French zoologist Geoffroy St Hilaire that a vertebrate could be thought of as an arthropod, or an earthworm, turned upside down. After Darwin and the acceptance of evolution, zoologists from time to time suggested that the vertebrate body plan had actually evolved through a worm-like ancestor literally turning upside down.

That is the theory that I want to support here, on balance and with some caution. The alternative, which is that a worm-like ancestor gradually rearranged its internal anatomy while staying the same way up, seems to me less plausible because it would have involved a greater amount of internal upheaval. I believe a change in behaviour came first – suddenly by evolutionary standards – and it was followed by a whole lot of consequential evolutionary changes. As so often, there are modern equivalents to make the idea vivid for us today. The brine shrimp is one example, and we hear its tale next.

THE BRINE SHRIMP'S TALE

Brine shrimps, *Artemia*, and the closely related fairy shrimps are crustaceans that swim on their backs, and therefore have their nerve cord (the 'true' zoological ventral side) on the side that now faces the sky. The upside-down catfish, *Synodontis nigriventris*, is a deuterostome that does the same thing the other way round. It is a fish that swims on its back, and therefore has its main trunk nerve on the side facing the river bottom, which is the 'true' zoological dorsal side. I don't know why brine shrimps do it, but the catfish swim upside down because they take food from the water surface, or from the undersides of floating leaves. Presumably, individual fish discovered that this was a good source of food and learned to turn over. My conjecture* is that, as the generations went by, natural selection favoured those individuals who learned to perform the trick best, their genes 'caught up' with the learning, and now they never swim any other way.

The brine shrimp's inversion is a recent re-enactment of something that happened, in my view, more than half a billion years ago. An ancient, long-lost animal, some kind of worm with a ventral nerve cord and a dorsal heart like any protostome, turned over and swam, or crawled, upside down like a brine shrimp. A zoologist who happened to be present at the time would have died rather than relabel the main nerve trunk dorsal just because it now ran along the side of the body facing the sky. 'Obviously', all his zoological training would have told him, it was still a ventral nerve cord, corresponding to all the other organs and features that we expect to see on the ventral surface of a protostome. Equally 'obvious' to this Precambrian zoologist, the heart of our inverted worm was, in the deepest sense, a 'dorsal' heart, even though it now beat under the skin nearest the sea bottom.

Given enough time, however – given enough millions of years of swimming or crawling 'upside down' – natural selection would come to reshape all the organs and structures of the body to fit in with the upside-down habit. Eventually, unlike our modern brine shrimp, which has only recently turned over, the traces of the original dorsal/ventral homologies would become obliterated. Later generations of palaeo-zoologists who encountered the descendants of this early maverick, after some tens of millions of years of upside-down habit, would start to redefine their concepts of dorsal and ventral. This is because so many anatomical details would have changed over evolutionary time.

Other animals that swim on their backs are sea otters (especially when engaging in their remarkable habit of smashing shellfish with stones on the belly), and the aptly named backswimmers (all the time). Backswimmers are a kind of bug,† sometimes known as

---

* Following the already mentioned theoretical idea known as the Baldwin Effect. Superficially, it sounds like Lamarckian evolution and the inheritance of acquired characteristics. Not so. Learning doesn't imprint itself into the genes. Instead, natural selection favours genetic propensities to learn certain things. After generations of such selection, evolved descendants learn so fast that the behaviour has become 'instinctive'.

† Bug means something precise, not just anything small. A bug is an insect of the order Hemiptera.

greater water boatmen, which row themselves underwater with their legs. The related lesser water boatmen do the same kind of thing, but they swim the right way up.

Imagine that the descendants of our modern water boatmen or brine shrimps on the one hand, and the descendants of our modern upside-down catfish on the other, were to maintain their habits of swimming upside down for 100 million years into the future. Isn't it entirely likely that they might each give rise to a whole new sub-kingdom, each body plan so radically reshaped by the upside-down habit that zoologists who didn't know the history would define the brine shrimps' descendants as having a 'dorsal' nerve cord, and the descendants of the catfish as having a 'ventral' nerve cord.

As we saw in the Ragworm's Tale, the world presents important practical differences between up and down, and these would start to imprint themselves, by natural selection, on the sky-pointing side and the floor-pointing side respectively. What had once been the zoologically ventral side would start to look more and more like a zoologically dorsal side, and vice versa. I believe this is exactly what happened somewhere along the line leading to vertebrates, and that is why we now have a dorsal nerve cord and a ventral heart. Modern molecular embryology offers some supporting evidence from the ways in which genes that define the dorso-ventral axis are expressed – genes a bit like the Hox genes that we shall meet in the Fruit Fly's Tale – but the details are beyond our scope here.

The upside-down catfish, recent though its inverted habit undoubtedly is, has already taken one revealing little step in this evolutionary direction. Its Latin name is *Synodontis nigriventris*. *Nigriventris* means 'dark belly', and it introduces a fascinating vignette at the end of the Brine Shrimp's Tale. One of the main differences between up and down in the world is the predominant direction of light. While not necessarily directly overhead, the sun's rays generally come from above rather than below. Hold your fist up and you'll find, even under an overcast sky, that its upper surface is better lit than its lower surface. This fact opens a key way in which we and many other animals can recognise solid, three-dimensional objects. A uniformly coloured curved object, such as a worm or a fish, looks lighter on top, darker below. I am not talking about the hard shadow cast by the body – it is a more subtle effect than that. A gradient of shading, from lighter above to darker below, smoothly betrays the curvature of the body.

Turn the fish upside down. Upside-down catfish (*Synodontis nigriventris*) in characteristic pose.

It works in reverse. The photograph of moon craters is printed upside down. If your eye (well, to be more precise, your brain) works in the same way as mine, you will see the craters as hills. Turn the book upside down, so that the light appears to come from another direction, and the hills will turn into the craters that they truly are.

One of my very first experiments as a graduate student demonstrated that newly hatched baby chicks seem to see the same illusion, straight out of the egg. They peck at photographs of simulated grains, and strongly prefer them if lit as if from above. Turn the photograph

Turn the book upside down. Craters on the far side of the moon.

over and they shun it. This seems to show that baby chicks 'know' that light in their world normally comes from above. But since they have only just hatched out of the egg, how do they know? Have they learned it during their three days of life? It is perfectly possible, but I tested it experimentally and found it not to be so. I raised chicks and tested them in a special cage in which the only light they ever saw came from below. Experience of pecking grain in this upside-down world would, if anything, teach them to prefer upside-down photographs of solid grains. Instead, they behaved exactly like normal chicks raised in the real world with light coming from above. Apparently because of genetic programming, all the chicks prefer to peck at photographs of solid objects lit from above. The solidity illusion (and hence, if I am right, the 'knowledge' of the predominant direction of light in the real world) seems to be genetically programmed in chicks – what we used to call 'innate' – rather than learned as (I'm guessing) it probably is in us.

Whether learned or unlearned, there is no doubt that the surface shading illusion of solidity is a powerful one. It has provoked a subtle form of camouflage called countershading. Look at any typical fish, out of water on a slab, and you'll notice that the belly is much lighter in colour than the back. The back may be dark brown or grey, while the belly is light grey, verging on white in some cases. And what is all this about? There seems no doubt that it is a form of disguise, based on counteracting the shading gradient that normally gives the game away for curved, solid things like fish. In the best of all possible worlds, a countershaded fish, when viewed in a normal light coming from above, will look perfectly flat. The expected gradient from light above to dark below will be exactly countered by the gradient in the fish's colour from light below to dark above.

Taxonomists often name species from dead specimens in museums.* Presumably that explains *nigriventris* as opposed to *invertus* or whatever is the Latin for 'upside down'. If you examine the upside-down catfish on a slab, you find that it is *reverse* countershaded. Its belly, which faces the sky, is darker than its back, which faces the bottom. Reverse countershading is one of those wonderfully elegant exceptions that prove the rule. The first catfish to swim upside down would have been horribly conspicuous. Its skin coloration would have

---

* Although I have to admit that the habits of the upside-down catfish have long been known. It is portrayed in its customary position in Ancient Egyptian wall paintings and engravings.

conspired with the natural shading from overhead light to make it look preternaturally solid. No wonder the change of habit was followed, in evolutionary time, by a reversal of the usual skin-colour gradient. A parallel case is that of the nudibranch mollusc (sea slug) *Glaucus atlanticus*. This beautiful creature floats upside down, feeding on Portuguese men-of-war, and it is 'reverse countershaded', just like the catfish.

Countershading does not just work in the sea. My old maestro Niko Tinbergen, before he left the Netherlands for Oxford, had a student called Leen De Ruiter to whom he proposed research on countershading in caterpillars. Caterpillars of many species play exactly the same trick on their (in this case bird) predators as fish do on theirs. These caterpillars are beautifully countershaded, with the result that they look flat when seen in a normal light. De Ruiter took the twigs on which caterpillars were sitting and turned them upside down. Immediately they became far more conspicuous because they suddenly looked much more solid. And birds took them in greater numbers.

If a De Ruiter were to come along and force the catfish to turn over and swim, like any normal fish, with its zoologically dorsal side uppermost, it would suddenly become far more conspicuously solid.* Reverse countershading in upside-down catfish is a single example of a consequential change that has followed, in evolutionary time, their change of habit. In another hundred million years, just think how comprehensively their whole bodies might change. There is nothing sacrosanct about 'dorsal' and 'ventral'. They can reverse, and I think they did reverse in the early ancestry of today's dorsocords. My bet is that Concestor 26 had its main trunk nerve running along the ventral side of the body like any protostome. We are modified worms swimming on our backs, descended from an early equivalent of a brine shrimp which, for some long-forgotten reason, turned over.

The more general moral of the Brine Shrimp's Tale is this. Major transitions in evolution may have begun as changes in behavioural habit, perhaps even non-genetic learned changes of habit, which only

---

* You might ask how an experimenter could force a catfish to turn over against its natural preference, and I do not know. But, to add just one tiny vignette, I do know how to make a brine shrimp swim, like any normal crustacean, with its zoologically dorsal side uppermost. Just shine an artificial light on them from below, and they will instantly turn over. Evidently brine shrimps use light as their cue to decide which way is up. I don't know if the catfish use the same cue. They could equally well use gravity.

later were followed by genetic evolution. I fancy that a comparable tale could be told for the first bird ancestor to fly, the first fish to come out on the land, and the first whale ancestor to return to the water (as Darwin speculated with his fly-catching bear). A change of habit by an adventurous individual is later followed by a long evolutionary catch-up and clean-up. That is the most far-reaching lesson from the Brine Shrimp's Tale.

THE LEAF CUTTER'S TALE

Just as humanity did at the time of our Agricultural Revolution, ants independently invented the town. A single nest of leaf cutter ants, *Atta*, can exceed the population of Greater London. It is a complicated underground chamber, up to 6 metres deep and 20 metres in circumference, surmounted by a somewhat smaller dome above ground. This huge ant city, divided into hundreds or even thousands of separate chambers connected by networks of tunnels, is sustained ultimately by leaves cut into manageable pieces and carried home by workers in broad, rustling rivers of green (see plate 34). But the leaves are not eaten directly, either by the ants themselves (though they do suck some of the sap) or by the larvae. Instead they are painstakingly mulched as compost for underground fungus gardens. It is the small round knobs or 'gongylidia' of the fungi that the ants eat and, more particularly, that they feed to the larvae. Cropping by the ants normally stops the fungi forming spore-bearing bodies (the equivalent of the mushrooms that we eat). This deprives fungus experts of the cues they normally use to identify species, and it means that the fungi themselves are dependent on the ants for their propagation. They have apparently evolved to flourish only in the domesticated environment of an ants' nest, which makes it a true example of domestication by an agricultural species other than our own. When a young queen ant flies off to found a new colony, she takes a precious cargo with her: a small culture of the fungus with which to sow the first crop in her new nest. This reminds me of a story of penicillin, perhaps the most important fungus of all. When Florey, Chain and their Oxford colleagues were developing penicillin, at the height of the Second World War, a German invasion of Britain was thought to be imminent. Florey and his junior colleague Heatley deliberately infected their own clothes with the mould, as the best way of secretly preserving the culture.

The energy for running the fungus ant colony is ultimately gathered from the sun by the leaves used to make the compost, a total leaf area that is measured in acres in the case of a large *Atta* colony. Fascinatingly the termites, that other hugely successful group of town-making insects, have also independently discovered fungus agriculture. In their case the compost is made of chewed-up wood. As with ants and their fungus, the termite fungus species is found only in termite nests and it seems to have been 'domesticated'. On those occasions where a termite fungus (*Termitomyces*) is permitted to produce a fruiting body, it sprouts from the side of the mound. It is said to be delicious and is on sale as a delicacy in the markets of Bangkok. A West African species, *Termitomyces titanicus*, is in *The Guinness Book of Records* as the largest edible mushroom in the world, with a cap diameter of up to one metre.

Several groups of ants have independently evolved the habit of keeping domestic 'dairy' animals in the form of aphids. Unlike other symbiotic insects that live inside ants' nests and don't benefit the ants, the aphids are pastured out in the open, sucking sap from plants as they normally do. As with mammalian cattle, aphids have a high throughput of food, taking only a small amount of nutriment from each morsel. The residue that emerges from the rear end of an aphid is sugar-water – 'honeydew' – only slightly less nutritious than the plant sap that goes in at the front. Any honeydew not eaten by ants rains down from trees infested with aphids, and is plausibly thought to be the origin of 'manna' in the Book of Exodus. It should not be surprising that ants gather it up, for the same reason as the followers of Moses did. But some ants have gone further and corralled aphids, giving them protection in exchange for being allowed to 'milk' the aphids, tickling their rear ends to make them secrete honeydew which the ant eats directly from the aphid's anus.

At least some aphid species have evolved in response to their domestic existence. They have lost some of the normal aphid defensive responses and, according to one intriguing suggestion, some have modified their rear end to resemble the face of an ant. Ants are in the habit of passing liquid food to one another, mouth to mouth, and the suggestion is that individual aphids that evolved this rear-end face-mimicry facilitated being 'milked' and therefore gained protection by ants from predators.

The Leaf Cutter's Tale is a tale of delayed gratification as the basis of agriculture. Hunter-gatherers eat what they gather and eat what they hunt. Farmers don't eat their seed corn; they bury it in the ground and wait months for a return. They don't eat the compost with which they fertilise the soil and don't drink the water with which they irrigate it. Again, it is all done for a delayed reward. And the leaf cutter ant got there first. Consider her ways and be wise.

THE GRASSHOPPER'S TALE

The Grasshopper's Tale treats the vexed and sensitive topic of race.

There is a pair of European grasshopper species, *Chorthippus brunneus* and *C. biguttulus*, which are so similar that even expert entomologists can't tell them apart, yet they never cross-breed in the wild although they sometimes meet. This defines them to be 'good species'. But experiments have shown that you need only allow a female to hear the mating call of a male of her own species caged nearby and she will happily mate with a male of the wrong species, 'thinking', one is tempted to say, that he is the singer. When this happens, healthy and fertile hybrids are produced. It doesn't normally happen in the wild because a female doesn't normally find herself near, but unable to reach, a male of her own species at the same time as a male of the wrong species is singing.

Comparable experiments have been done on crickets, using temperature as an experimental variable. Different species of cricket chirp at different frequencies, but the chirp frequency is also temperature-dependent. If you know your crickets, you can use them as a reasonably accurate thermometer. Fortunately, not only the male's chirping frequency but also the female's perception of it is temperature-dependent: the two vary in lockstep, which normally precludes miscegenation. A female in an experiment, offered a choice of males singing at two different temperatures, chooses the one at her own temperature. The male singing at a different temperature is treated as if he belongs to the wrong species. If you heat up a female, her preference shifts to a 'hotter' song, even if that causes her to prefer a cool male of the wrong species. Once again, this normally doesn't happen in nature. If a female can hear a male at all, he can't be far away, and so is likely to be at approximately the same temperature as she is.

Grasshopper song is temperature-dependent in the same kind of way. Using grasshoppers of the same genus, *Chorthippus*, with which we began (though different species of the genus), German scientists did some technically ingenious experiments. They managed to attach tiny thermometers (thermocouples) and tiny electric heaters to the insects. So miniaturised were these, the experimenters could heat the head of a grasshopper without heating its thorax, or heat the thorax without heating the head. Then they tested females' preferences for songs produced by males stridulating at various temperatures.* They found that what matters for the female song preference is the temperature of her head. But it is the temperature of the male's thorax that determines his stridulation rate. Fortunately of course, in nature, where there are no experimenters with tiny electric heaters, the head and thorax will normally be at the same temperature, as will the male and the female. So the system works, and hybridisation doesn't happen.

It is quite common to find pairs of related species that never interbreed under natural conditions but that can do so if humans interfere. The case of *Chorthippus brunneus* and *C. biguttulus* is just one example. The Cichlid's Tale told of a comparable case in fish, where monochromatic light abolished the discrimination between a reddish and a bluish species. And it happens in zoos. Biologists normally classify animals that mate under artificial conditions but refuse to mate in the wild as separate species, as has happened with the grasshoppers. But unlike, say, lions and tigers, which can hybridise in zoos to make (sterile) 'ligers' and 'tigrons', those grass-hoppers look identical. Apparently the only difference is in their songs. And it is this, and only this, that stops them cross-breeding and therefore leads us to recognise them as separate species. Human beings are the other way round. It requires an almost superhuman feat of political zeal to overlook the conspicuous differences between our own local populations or races. Yet we happily interbreed across races and are unequivocally and uncontroversially defined as mem-bers of the same species. The Grasshopper's Tale is about races and

---

* Stridulation is how grasshoppers, and crickets, make sound. Grasshoppers scrape their legs against their wing covers. Crickets scrape the two wing covers against each other. They sound similar, but grasshoppers are generally more buzzy, crickets more musical. Of one nocturnal tree cricket it has been said that if moonlight could be heard, that is how it would sound. Cicadas are quite different. As if buckling a tin lid, they buckle part of the thorax wall, repeatedly and fast, so it sounds like a continuous buzz, usually extremely loud and sometimes patterned in very complex ways, characteristic of the species.

species, about the difficulties of defining both, and what all this has to say about human races.

'Race' is not a clearly defined word. 'Species', as we have seen, is different. There really is an agreed way to decide whether two animals belong in the same species: can they interbreed? Obviously they can't if they are of the same sex, or are too young or too old, or one of them happens to be sterile. But those are pedantries, easy to get around. In the case of fossils, too, which obviously can't breed, we apply the interbreeding criterion in our imaginations. Do we think it *likely* that, if these two animals were not fossils but alive, fertile and of opposite sex, they would be able to interbreed?

The interbreeding criterion gives the species a unique status in the hierarchy of taxonomic levels. Above the species level, a genus is just a group of species that are pretty similar to each other. No objective criterion exists to decide *how* similar they have to be, and the same is true of all the higher levels: family, order, class, phylum and the various 'sub-' or 'super-' names that intervene between them. Below the species level, 'race' and 'subspecies' are used more or less interchangeably and, again, no objective criterion exists that would enable us to decide whether two people should be considered part of the same race or not, nor to decide how many races there are. And of course there is the added complication, absent above the species level, that races interbreed, so there are lots of people of mixed race.

Presumably species, on their way to becoming sufficiently separate to be incapable of interbreeding, usually pass through an intermediate stage of being separate races. Separate races might be regarded as species in the making, except that there is no necessary expectation that the making will continue to its end – to speciation.

The interbreeding criterion works pretty well, and it delivers an unequivocal verdict on humans and their supposed races. All living human races interbreed with one another. We are all members of the same species, and no reputable biologist would say any different. But let me call your attention to an interesting, perhaps even slightly disturbing fact. While we happily interbreed with each other, producing a continuous spectrum of inter-races, we are strangely reluctant to give up our divisive racial language. Wouldn't you expect that if all intermediates are on constant display, the urge to classify people as one or the other of two extremes would wither away, smothered by the absurdity of the attempt, which is continually manifested everywhere we look? Unfortunately, this is not what happens, and perhaps that very fact is revealing.

People who are universally agreed by all Americans to be 'black' may draw less than one-eighth of their ancestry from Africa, and often have a light skin colour well within the normal range for people universally agreed to be 'white'. In the picture of four American politicians in plate 35, two are described in all newspapers as black, the other two as white. Wouldn't a Martian, unschooled in our conventions but able to see skin shades, be more likely to split them three against one? Surely yes. But in our culture, almost everybody will immediately 'see' Mr Powell as 'black', even in this particular photograph which happens to show him with possibly lighter skin than either Bush or Rumsfeld.

It is an interesting exercise to take a photograph, such as the one in plate 33, of Colin Powell standing next to some representative 'white' men (they must be next to each other so the lighting conditions are the same). From each face, cut a small uniform rectangle, say from the forehead, and place the patches side by side. You will find that there is very little difference between Powell and the 'white' men with whom he is standing. He may be lighter or darker, depending upon the particular cases. But now 'zoom out' and look again at the original photograph. Immediately, Powell will look 'black'. What cues are we picking up on?

To ram home the point, do the same 'forehead patch' exercise with Powell standing next to a genuinely black man such as Daniel arap

Next to a genuinely black man. Colin Powell with Daniel arap Moi.

Moi, the recent President of Kenya (previous page). This time, the forehead patches will look dramatically different. But then, when we 'zoom out' and look at the whole faces, we again 'see' Mr Powell as 'black'. The news story that accompanied this picture of Powell, visiting Moi in May 2001, implied that the same conventions are understood in Africa:

> As the first African-American secretary of state, Powell has been accorded near-messianic treatment in Africa. And perhaps because he is black, there was a profound resonance to Powell's harsh criticism . . .

Why do people so readily swallow the apparent contradiction – and there are numerous similar examples – between the verbal statement, 'he is black', and the picture it accompanies. What is going on here? Various things. First, we are curiously eager to embrace racial classification, even when talking about individuals whose mixed parentage seems to make a nonsense of it, and even where (as here) it is irrelevant to anything that matters.

Second, we tend not to describe people as of mixed race. Instead, we plump for one race or the other. Some American citizens are of pure African descent and some are of pure European descent (leaving aside the fact that, in the longer term, we are all of African descent). Maybe it is convenient for some purposes to call them black and white respectively, and I am not proposing any principled objection to these names. But many people – probably more than most of us realise – have both black and white ancestors. If we are going to use colour terminology, many of us are presumably somewhere in between. Yet society insists on calling us one or the other. It is an example of the 'tyranny of the discontinuous mind', which was the subject of the Salamander's Tale. Americans are regularly asked to fill in forms in which they have to tick one of five boxes: Caucasian (whatever that might mean – it certainly doesn't mean from the Caucasus), African-American, Hispanic (whatever that might mean – it certainly does not mean, as the word seems to suggest, Spanish), Native American or Other. There are no boxes labelled half and half. But the very idea of ticking boxes is incompatible with the truth, which is that many, if not most, people are a complicated mixture of the offered categories and others. My inclination is irritably to refuse to tick any boxes, or to add my own box labelled 'human'. Especially when the rubric uses the mealymouthed euphemism 'Ethnicity'.

Third, in the particular case of 'African-Americans', there is something culturally equivalent to genetic dominance in our use of language. When Mendel crossed wrinkled peas with smooth peas, all the first generation progeny were smooth. Smooth is 'dominant', wrinkled is 'recessive'. The first generation progeny all had one smooth allele and one wrinkled, yet the peas themselves were indistinguishable from peas with no wrinkled genes. When an Englishman marries an African, the progeny are intermediate in colour and in most other characteristics. This is unlike the situation in peas. But we all know how society will call such children: 'black' every time. Blackness is not a true genetic dominant like smoothness in peas. But social *perception* of blackness behaves like a dominant. It is a cultural or memetic dominant. That insightful anthropologist Lionel Tiger has attributed this to a racist 'contamination metaphor' within white culture. And no doubt there is also a strong and understandable will on the part of descendants of slaves to identify with their African roots. I have already remarked on this in Eve's Tale – regarding the television documentary where Jamaican immigrants to Britain were emotionally reunited with alleged 'family' in West Africa.

Fourth, there is high inter-observer agreement in our racial categorisations. A man such as Colin Powell, of mixed race and intermediate physical characteristics, is not described as white by some observers and black by others. A small minority will describe him as mixed. All others will without fail describe Mr Powell as black – and the same goes for anybody who shows the *slightest trace* of African ancestry, even if their percentage of European ancestors is overwhelming. Nobody describes Colin Powell as white, unless they are trying to make a political point by the very fact that the word jars against the audience's expectations.

There is a useful technique called 'inter-observer correlation'. It is a measure that is often used in science to establish that there really is a reliable basis for a judgement, even if nobody can pin down what that basis is. The rationale, in the present case, is this. We may not know how people decide whether somebody is 'black' or 'white' (and I hope I have just demonstrated that it isn't because they are black or white!) but there must be *some* sort of reliable criterion lurking there because any two randomly chosen judges will come to the same decision.

The fact that the inter-observer correlation remains high, even

over a huge spectrum of inter-races, is impressive testimony to
something fairly deep-seated in human psychology. If it holds up
cross-culturally, it will be reminiscent of the anthropologists' finding
about perception of hue. Physicists tell us that the rainbow, from red
through orange, yellow, green and blue to violet is a simple continuum of wavelength. It is biology and/or psychology, not physics,
that singles out particular landmark wavelengths along the physical
spectrum for special treatment and naming. Blue has a name. Green
has a name. Blue-green does not. The interesting finding of anthropologists' experiments (as opposed to some influential anthropological theories, by the way) is that there is substantial agreement over
such namings across different cultures. We seem to have the same
kind of agreement over judgements of race. It may prove to be even
stronger and clearer than for the rainbow.

As I said, zoologists define a species as a group whose members
breed with each other under natural conditions – in the wild. It
doesn't count if they breed only in zoos, or if we have to use artificial
insemination, or if we fool female grasshoppers with caged singing
males, even if the offspring produced are fertile. We might dispute
whether this is the only sensible definition of a species, but it is the
definition that most biologists use.

If we wished to apply this definition to humans, however, there is
a peculiar difficulty: how do we distinguish natural from artificial
conditions for interbreeding? It is not an easy question to answer.
Today, all surviving humans are firmly placed in the same species,
and they do indeed happily interbreed. But the criterion, remember,
is whether they choose to do so under natural conditions. What are
natural conditions for humans? Do they even exist any more? If, in
ancestral times, as sometimes today, two neighbouring tribes had
different religions, different languages, different dietary customs,
different cultural traditions and were continually at war with one
another; if the members of each tribe were brought up to believe that
the other tribe were subhuman 'animals' (as happens even today); if
their religions taught that would-be sexual partners from the other
tribe were taboo, 'shiksas', or unclean, there could well be no interbreeding between them. Yet anatomically, and genetically, they could
be completely the same as each other. And it would take only a
change of religious or other customs to break down the barriers to
interbreeding. How, then, might somebody try to apply the interbreeding criterion to humans? If *Chorthippus brunneus* and *C.*

biguttulus are separated as two distinct species of grasshoppers because they prefer not to interbreed although they physically could, might humans, at least in ancient times of tribal exclusivity, once have been separable in the same kind of way? *Chorthippus brunneus* and *C. biguttulus*, remember, in all detectable respects except their song, are identical, and when they are (easily) persuaded to hybridise their offspring are fully fertile.

Whatever we may think as observers of superficial appearances, the human species today is, to a geneticist, especially uniform. Taking such genetic variation as the human population does possess, we can measure the fraction that is associated with the regional groupings that we call races. And it turns out to be a small percentage of the total: between 6 and 15 per cent depending on how you measure it – much smaller than in many other species where races have been distinguished. Geneticists conclude, therefore, that race is not a very important aspect of a person. There are other ways to say this. If all humans were wiped out except for one local race, the great majority of the genetic variation in the human species would be preserved. This is not intuitively obvious and may be quite surprising to some people. If racial statements were as informative as most Victorians, for example, used to think, you would surely need to preserve a good spread of all the different races in order to preserve most of the variation in the human species. Yet this is not the case.

It certainly would have surprised Victorian biologists who, with few exceptions, saw humanity through race-tinted spectacles. Their attitudes persisted into the twentieth century. Hitler was unusual in gaining the power to turn racialist ideas into government policy. Plenty of others, not just in Germany, had the same thoughts but lacked the power. I have previously quoted H. G. Wells's vision of his New Republic (*Anticipations*, 1902), and I do so again because it is such a salutary reminder of how a leading British intellectual, regarded in his time as progressive and left-leaning, could say such horrifying things, only a century ago, and scarcely be noticed doing so.

And how will the New Republic treat the inferior races? How will it deal with the black? . . . the yellow man? . . . the Jew? . . . those swarms of black, and brown, and dirty-white, and yellow people, who do not come into the new needs of efficiency? Well, the world is a world, and not a charitable institution, and I take it they will have to go . . . And the ethical system of these men of the New Republic, the ethical system which will dominate the world state, will be shaped primarily to favour the procreation of what is fine and efficient and beautiful in humanity – beautiful and strong bodies, clear and powerful minds . . . And the method that nature has followed hitherto in the shaping of the world, whereby weakness was prevented from propagating weakness . . . is death . . . The men of the New Republic . . . will have an ideal that will make the killing worth the while.

I suppose we should take comfort from the change that has come over our attitudes during the intervening century. Perhaps, in a negative sense, Hitler can take some credit for this, since nobody wants to be caught saying anything that he said. But what, I wonder, will our successors of the twenty-second century be quoting, in horror, from us? Something to do with our treatment of other species, perhaps?

But that was an aside. We were dealing with the unusually high level of genetic uniformity in the human species, despite superficial appearances. If you take blood and compare protein molecules, or if you sequence genes themselves, you will find that there tends to be much less difference between any two humans living anywhere in the world than there is between two African chimpanzees. We can explain this human uniformity by guessing that our ancestors, but not the chimpanzees, passed through one or more genetic bottlenecks not very long ago. The various human populations around the world were reduced to a small numbers, many probably going extinct, but some just pulled through. That is why we are so genetically uniform today. Similar evidence, of even greater genetic uniformity, suggests that cheetahs passed through an even narrower bottleneck more recently, around the end of the last Ice Age.

Some people may find the evidence of biochemical genetics unsatisfying because it seems not to square with their everyday experience. Unlike cheetahs, we don't 'look' uniform.* Norwegians, Japanese and Zulus really do look rather dramatically different from one another. With the best will in the world, it is intuitively hard to believe what is in fact the truth: that they are 'really' more alike than three chimpanzees who look, to our eyes, much more similar.

This is, of course, a politically sensitive matter, a point I heard

---

* As an aside, leopards don't either. But black 'panthers', once thought to be a separate species, differ from spotted leopards at a single genetic locus.

being amusingly lampooned by a West African medical researcher at a gathering of about 20 scientists. At the beginning of the conference, the chairman asked each of us around the table to introduce ourselves. The African, who was the only black person there – and he really was black, unlike many 'African-Americans' – happened to be wearing a red tie. He finished his self-introduction by laughingly saying, 'You can easily remember me. I am the one with the red tie.' He was genially mocking the way people bend over backwards to pretend not to notice racial differences. I think there was a *Monty Python* sketch along the same lines. Nevertheless, we can't write off the genetic evidence which suggests that, all appearances to the contrary, we really are an unusually uniform species. What is the resolution to the apparent conflict between appearance and measured reality?

It is genuinely true that, if you measure the total variation in the human species and then partition it into a between-race component and a within-race component, the between-race component is a very small fraction of the total. Most of the variation among humans can be found within races as well as between them. Only a small admixture of extra variation distinguishes races from each other. That is all correct. What is not correct is the inference that race is therefore a meaningless concept. This point has been clearly made by the distinguished Cambridge geneticist A. W. F. Edwards in a paper called 'Human genetic diversity: Lewontin's fallacy'. R. C. Lewontin is an equally distinguished Cambridge (Mass.) geneticist, known for the strength of his political convictions and his weakness for dragging them into science at every possible opportunity. Lewontin's view of race has become near-universal orthodoxy in scientific and social science circles. He wrote, in a famous paper of 1972:

It is clear that our perception of relatively large differences between human races and subgroups, as compared to the variation within these groups, is indeed a biased perception and that, based on randomly chosen genetic differences, human races and populations are remarkably similar to each other, with the largest part by far of human variation being accounted for by the differences between individuals.

This is, of course, exactly the point I accepted above, not surprisingly since what I wrote was largely based on Lewontin. But see how Lewontin goes on:

> Human racial classification is of no social value and is positively
> destructive of social and human relations. Since such racial classifica-
> tion is now seen to be of virtually no genetic or taxonomic significance
> either, no justification can be offered for its continuance.

We can all happily agree that human racial classification is of no
social value and is positively destructive of social and human relations. That is one reason why I object to ticking boxes in forms and
why I object to positive discrimination in job selection. But that
doesn't mean that race is of 'virtually no genetic or taxonomic
significance'. This is Edwards's point, and he reasons as follows.
However small the racial partition of the total variation may be, if
such racial characteristics as there are are highly correlated with other
racial characteristics, they are by definition informative, and therefore of taxonomic significance.

Informative means something quite precise. An informative statement is one that tells you something you didn't know before. The
information content of a statement is measured as reduction in prior
uncertainty. Reduction in prior uncertainty, in turn, is measured as a
change in probabilities. This provides a way to make the information
content of a message mathematically precise, but we don't need to
bother with that.* If I tell you Evelyn is male, you immediately know
a whole lot of things about him. Your prior uncertainty about the
shape of his genitals is reduced (though not obliterated). You now
know facts you didn't know before about his chromosomes, his
hormones and other aspects of his biochemistry, and there is a
quantitative reduction in your prior uncertainty about the depth of
his voice, and the distribution of his facial hair and of his body fat
and musculature. Contrary to Victorian prejudices, your prior
uncertainty about Evelyn's general intelligence or ability to learn,
remains unchanged by the news about his sex. Your prior uncertainty about his ability to lift weights or excel at most sports is
quantitatively reduced, but only quantitatively. Plenty of females can
beat plenty of males at any sport, although the best males can
normally beat the best females. Your ability to bet on Evelyn's
running speed, say, or the power of his tennis serve, has been slightly
raised by my telling you his sex, but it has not reached certainty.

---

* As it happens, Lewontin himself was one of the first biologists to use information
theory, and indeed he did so in his paper on race, but for a different purpose. He used it
as a convenient statistic for measuring diversity.

Now to the question of race. What if I tell you Suzy is Chinese, how much is your prior uncertainty reduced? You now are pretty certain that her hair is straight and black (or was black), that her eyes have an epicanthic fold, and one or two other things about her. If I tell you Colin is 'black' this does not, as we have seen, tell you he is black. Nevertheless, it is clearly not uninformative. The high inter-observer correlation suggests that there is a constellation of characteristics that most people recognise, such that the statement 'Colin is black' really does reduce prior uncertainty about Colin. It works the other way around to some extent. If I tell you Carl is an Olympic sprinting champion, your prior uncertainty about his 'race' is, as a matter of statistical fact, reduced. Indeed, you can have a moderately confident bet that he is 'black'.*

We got into this discussion through wondering whether the concept of race was, or had ever been, an information-rich way to classify people. How might we apply the criterion of inter-observer correlation to judging the question? Well, suppose we took standard full-face photographs of 20 randomly chosen natives of each of the following countries: Japan, Uganda, Iceland, Sri Lanka, Papua New Guinea and Egypt. If we presented all 120 people with all 120 photographs, my guess is that every single one of them would achieve 100 per cent success rates in sorting them into six different categories. What is more, if we told them the names of the six countries involved, all 120 subjects, if they were reasonably well educated, would correctly assign all 120 photographs to the correct countries. I haven't done the experiment but I am confident that you will agree with me on what the result would be. It may seem unscientific of me not to bother to do the experiment. But my confidence that you, being human, will agree without doing the experiment, is the very point I am trying to make.

If the experiment were to be done, I do not think Lewontin would expect any other result than the one I have predicted. Yet an opposite prediction would seem to follow from his statement that racial classification has virtually no taxonomic or genetic significance. If there is no taxonomic or genetic significance, the only other way to get a high inter-observer correlation would be a worldwide similarity in cultural bias, and I do not think Lewontin would want to predict

---

* Sir Roger Bannister got into terrible hot water, for no good reason that I could discern except people's hair-trigger sensitivities on matters of race, when he said something similar a few years ago.

that either. In short, I think Edwards is right and Lewontin, not for the first time, wrong. Lewontin did his sums right, of course: he is a brilliant mathematical geneticist. The proportion of the total variation in the human species that falls into the racial partition of variation is, indeed, low. But because the between-race variation, however low a percentage of the total variation, is *correlated*, it is informative in ways that could surely be demonstrated by measuring the inter-observer concordance of judgement.

I must at this point reiterate my strong objection to being asked to fill in forms in which I have to tick a box labelling my 'race' or 'ethnicity', and voice my strong support for Lewontin's statement that racial classification can be actively destructive of social and human relations – especially when people use racial classification as a way of treating people differently, whether through negative or positive discrimination. To tie a racial label to somebody is informative in the sense that it tells you more than one thing about them. It might reduce your uncertainty about the colour of their hair, the colour of their skin, the straightness of their hair, the shape of their eye, the shape of their nose and how tall they are. But there is no reason to suppose that it tells you anything about how well-qualified they are for a job. And even in the unlikely event that it did reduce your statistical uncertainty about their likely suitability for some particular job, it would *still* be wicked to use racial labels as a basis for discrimination when hiring somebody. Choose on the basis of ability, and if, having done so, you end up with an all-black sprinting team, so be it. You have not practised racial discrimination in arriving at this conclusion.

A great conductor, when auditioning instrumentalists for his orchestra, always had them perform behind a screen. They were told not to speak, and they even had to remove their shoes for fear the sound of high heels would betray the sex of the performer. Even if it were statistically the case that women tend to make better harpists, say, than men, this does *not* mean that you should actively discriminate against men when you choose a harpist. Discriminating against individuals purely on the basis of a group to which they belong is, I am inclined to think, always evil. There is near-universal agreement today that the apartheid laws of South Africa were evil. Positive discrimination in favour of 'minority' students on American campuses can fairly, in my opinion, be attacked on the same grounds as apartheid. Both treat people as representative of groups rather than as individuals in their own right. Positive discrimination is sometimes justified as redressing centuries of injustice. But how can it be just to pay back a single individual today for the wrongs done by long-dead members of a plural group to which he belongs?

Interestingly, this kind of singular/plural confusion shows up in a form of words which is tellingly diagnostic of bigots: 'The Jew . . .' instead of 'Jews . . .'

> Your Fuzzy Wuzzy is an excellent fighter, but he can't tell his left from his right. Now, your Pathan . . .

People are individuals, they are individually different, far more different from other members of their group than their groups are from each other. In this, Lewontin is undoubtedly right.

Inter-observer agreement suggests that racial classification is not totally uninformative, but what does it inform *about*? About no more than the characteristics used by the observers when they agree: things like eye shape and hair curliness – nothing more unless we are given further reasons to believe it. For some reason it seems to be the superficial, external, trivial characteristics that are correlated with race – perhaps especially facial characteristics. But why are human races so different in just these superficially conspicuous characteristics? Or is it just that we, as observers, are predisposed to notice them? Why do other species look comparatively uniform whereas humans show differences that, if we encountered them elsewhere in the animal kingdom, might make us suspect we were dealing with a number of separate species?

The most politically acceptable explanation is that the members of any species have a heightened sensitivity to differences among their own kind. On this view, it is just that we *notice* human differences more readily than differences within other species. Chimpanzees whom we find almost identical look just as different, in chimpanzee eyes, as a Kikuyu is different from a Dutchman in our eyes. Expecting to confirm this kind of theory at the within-race level, the eminent American psychologist H. L. Teuber, an expert on the brain mechanisms of facial recognition, asked a Chinese graduate student to study the question, 'Why do Westerners think Chinese people look more alike than Westerners?' After three years' intensive research, the Chinese student reported his conclusion. 'Chinese people really do look more alike than Westerners!' Teuber told the story with much twinkling and wiggling of eyebrows, a sure sign with him that a joke was on the way, so I don't know what the truth of the matter is. But I have no difficulty in believing it, and I certainly don't think it should upset anyone.

Our (relatively) recent worldwide diaspora out of Africa has taken us to an extraordinarily wide variety of habitats, climates and ways of life. It is plausible that the different conditions have exerted strong selection pressures, particularly on externally visible parts, such as the skin, which bear the brunt of the sun and the cold. It is hard to think of any other species that thrives so well from the tropics to the Arctic, from sea level to the high Andes, from parched deserts to dripping jungles, and through everything in between. Such different conditions would be bound to exert different natural selection pressures, and it would be positively surprising if local populations did not diverge as a result. Hunters in the deep forests of Africa, South America and South-East Asia have all independently become small, almost certainly because height is a handicap in dense vegetation. Peoples of high latitude, who, it has been surmised, need all the sun they can get to make vitamin D, tend to have lighter skins than those who face the opposite problem – the carcinogenic rays of the tropical sun. It is plausible that such regional selection would especially affect superficial characteristics like skin colour, while leaving most of the genome intact and uniform. And indeed recent study of ancient European genomes by David Reich's group in Harvard has been able to pinpoint the origin of some of the genes involved. Among the five locations in the genome which have seen the strongest selection, along with lactose tolerance (see the Farmer's Tale) and eye colour, are two genes involved in skin pigmentation and another associated with vitamin D levels. East European farmers and Scandinavian hunter-gatherers seem to be the source of the light-skinned genetic variants, which then spread through the dark-skinned hunter-gatherers in Western Europe.

In theory, environmental factors could be the full explanation for our superficial and visible variety, covering deep similarity. But it doesn't seem enough to me. At the very least, I think it might be helped along by an additional suggestion, which I offer tentatively. It takes off from our earlier discussion about cultural barriers to inter-breeding. We are indeed a very uniform species if you count the totality of genes, or if you take a truly random sample of genes; but perhaps there are special reasons for a disproportionate amount of variation in those very genes that make it easy for us to *notice* variation, and to distinguish our own kind from others. This would include the genes responsible for externally visible 'labels' like skin colour. Yet again, I want to suggest that this heightened discriminability has evolved by sexual selection, specifically in humans because we are such a culture-bound species. Because our mating decisions are so heavily influenced by cultural tradition, and because our cultures, and sometimes our religions, encourage us to discriminate against outsiders, especially in choosing mates, those superficial differences that helped our ancestors to prefer insiders over outsiders have been enhanced out of all proportion to the real genetic differences between us. No less a thinker than Jared Diamond has supported a similar idea in *The Rise and Fall of the Third Chimpanzee*. And Darwin himself more generally invoked sexual selection in explanation of racial differences.

I want to consider two versions of this theory: a strong and a weak one. The truth could be any combination of the two. The strong theory suggests that skin colour, and other conspicuous genetic badges, evolved actively as discriminators in choosing mates. The weak theory, which can be thought of as leading into the strong version, places cultural differences, such as language and religion, in the same role as geographical separation in the incipient stages of speciation. Once cultural differences have achieved this initial separation, with the consequence that there is no gene flow to hold them together, the groups would subsequently evolve apart genetically, as if geographically separated.

Recall from the Cichlid's Tale that an ancestral population can split into two genetically distinct populations only if given a head start by an initial accidental separation, usually assumed to be geographical. A barrier such as a mountain range reduces gene flow between two populated valleys. So the gene pools in the two valleys are free to drift apart. The separation will normally be abetted by different selection pressures; one valley may be wetter than its neighbour on the other side of the mountains, for instance. But the initial accidental separation, which I have so far assumed to be geographical, is necessary.

Nobody is suggesting that there is anything deliberate about the geographical separation. That isn't what 'necessary' means at all. Necessary just means that, if there didn't happen to be an initial geographical (or equivalent) separation, the various parts of the population would be genetically bound together by sexual mixing between them. Speciation couldn't happen without an initial barrier.

Once the two putative species, initially races, have begun to pull apart, genetically speaking, they may then pull even farther apart – even if the geographical barrier subsequently disappears.

There is controversy here. Some people think the initial separation has to be geographical, while others, especially entomologists, emphasise so-called sympatric speciation. Many herbivorous insects eat only one species of plant. They meet their mates and lay their eggs on the preferred plants. Their larvae then apparently 'imprint' on the plant that they grow up eating, and they choose, when adult, the same species of plant to lay their own eggs.* So if an adult female made a mistake and laid her eggs on the wrong species of plant, her daughter would imprint on that wrong plant and would, when the time came, lay her eggs on plants of the same wrong species. Her larvae then would imprint on the same wrong plant, hang around the wrong plant when adult, mate with others hanging around the wrong plant and eventually lay their eggs on the wrong plant.

In the case of these insects, you can see that, in a single generation, gene flow with the parental type could be abruptly cut off. A new species is theoretically free to come into being without the need for geographical isolation. Or – another way of putting it – the difference between two kinds of food plant is, for these insects, equivalent to a mountain range or a river for other animals. It has been argued that this kind of sympatric speciation is commoner among insects than 'true' geographical speciation, in which case, since the majority of species are insects, it could even be that most speciation events are sympatric. Be that as it may, I am suggesting that human culture provides a special way in which gene flow can find itself blocked, which is somewhat analogous to the insect scenario I have just outlined.

In the insect case, plant preferences are handed down from parent to offspring by the twin circumstances of larvae fixating on their food plant, and adults mating and laying eggs on the same food plants. In effect, lineages establish 'traditions' that travel longitudinally down generations. Human traditions are similar, if more elaborate.

---

* Imprinting is the process, often said to have been discovered by Konrad Lorenz, whereby young animals, for instance goslings, take a kind of mental photograph of an object they see during a critical period early in life, and which they follow while young. Usually it will be a parent, but it could be Konrad Lorenz's boots. Later in life, the 'mental photograph' influences choice of mate; this usually means a member of their own species, but they might try to mate with Lorenz's boots. The gosling story isn't as simple as that, but the analogy to the insect case should be clear.

Examples are languages, religions and social manners or conventions. Children usually adopt the language and the religion of their parents although, just as with the insects and the food plants, there are enough 'mistakes' to make life interesting. Again, as with the insects mating in the vicinity of their preferred food plants, people tend to mate with others speaking the same language and praying to the same gods. So different languages and religions can play the role of food plants, or of mountain ranges in traditional geographical speciation. Different languages, religions and social customs can serve as barriers to gene flow. From here, according to the weak form of our theory, random genetic differences simply accumulate on opposite sides of a language or religion barrier, just as they might on opposite sides of a mountain range. Subsequently, according to the strong version of the theory, the genetic differences that build up are reinforced as people use conspicuous differences in appearance as additional labels of discrimination in mate choice, supplementing the cultural barriers that provided the original separation.*

I am certainly not suggesting that humans should be thought of as more than one species. Very much the contrary. What I am suggesting is that human culture – the fact that we depart so strongly from random mating in directions determined by language, religion and other cultural discriminators, has done very odd things to our genetics in the past. Even though, if you take the totality of genes into account, we are a very uniform species, we are astonishingly variable in super-ficial features which are trivial but conspicuous: discrimination fodder. The discrimination might apply not just to mate choice but to choice of enemies and victims of xenophobic or religious prejudice.

THE FRUIT FLY'S TALE

In 1894 the pioneering geneticist William Bateson published a book called *Materials for the Study of Variation, Treated with Especial Regard to Discontinuity in the Origin of Species*. He compiled a fascinating, almost macabre list of genetic abnormalities, and considered how they might illuminate evolution. He had horses with cloven hooves, antelopes with a single horn in the middle of the

---

* A potential problem, which would need sorting out if the idea were to be pursued, is that the theory of mathematical genetics suggests, for geographical separation and by implication this cultural hypothesis too, that the separation has to be pretty complete for genetic differentiation to be maintained.

head, people with an extra hand, and a beetle with five legs on one side. In his book, Bateson coined the term 'homeosis' for a remarkable type of genetic variation. *Homoio* means 'same' in Greek, and a homeotic mutation (as we would now call it, although 'mutation' had not been coined when Bateson wrote) is one that causes a part of the body to appear in some different part.

Bateson's own examples included a sawfly with a leg growing in the place where an antenna should be. As soon as you hear of this remarkable abnormality, you might suspect, with Bateson, that here must be an important clue to how animals develop normally. You and Bateson would be right, and that is the subject of this tale. That particular homeosis – leg in place of antenna – was later discovered in the fruit fly *Drosophila* and named antennapedia. *Drosophila* ('dew lover') has long been the geneticists' favourite animal. Embryology should never be confused with genetics, but recently *Drosophila* has assumed a starring role in embryology as well as genetics, and this is a tale of embryology.

Embryonic development is controlled by genes, but there are two very different ways in which this might theoretically happen. The Mouse's Tale introduced them as blueprint and recipe. A builder makes a house by placing bricks in positions specified by a blueprint. A cook makes a cake not by placing crumbs and currants in specified positions but by putting ingredients through specified procedures, such as sieving, stirring, beating and heating.* Textbooks of biology are wrong when they describe DNA as a blueprint. Embryos do nothing remotely like following a blueprint. DNA is not a description, in any language, of what the finished body should look like. Maybe on some other planet living things develop by blueprint embryology, but I find it hard to imagine how it would work. It would have to be a very different kind of life. On this planet, embryos follow recipes. Or, to change to another equally un-blueprint-like analogy, which is in some ways more apt than the recipe: embryos construct themselves by following a sequence of origami folding instructions.

The origami analogy fits early embryology better than late. The main organisation of the body is initially laid down by a series of foldings and invaginations of layers of cells. Once the main body plan is safely in place, later stages in development consist largely of

---

* This favourite analogy was first used by my friend Sir Patrick Bateson, a relative of Sir William, as it happens.

growth, as if the embryo were being inflated, in all its parts, like a balloon. It is a very special kind of balloon, however, because different parts of the body inflate at different rates, the rates being carefully controlled. This is the important phenomenon known as allometry. The Fruit Fly's Tale is concerned mostly with the earlier, origami phase of development, not the later, inflationary one.

Cells are not laid like bricks to a blueprint, but it is the behaviour of cells that determines embryonic development. Cells attract, or repel, other cells. They change shape in various ways. They secrete chemicals, which may diffuse outwards and influence other cells, even some distance away. Sometimes they die selectively, carving out shapes by subtraction, as if a sculptor were at work. Like termites co-operating to build a mound, cells 'know' what to do by reference to the neighbouring cells with whom they find themselves in contact, and in response to chemicals in gradients of concentration. All cells in the embryo contain the same genes, so it can't be their genes that distinguish one cell's behaviour from another's. What does distinguish a cell is which of the genes are turned on, which usually is reflected in the gene products – proteins – that it contains.

In the very early embryo, a cell needs to 'know' where it lies along two main dimensions: fore and aft (anterior/posterior) and up-down (dorsal/ventral). What does 'know' mean? It initially means that a cell's behaviour is determined by its position along chemical gradients in each of the two axes. Such gradients necessarily start in the egg itself, and are therefore under the control of the mother's genes, not the egg's own nuclear genes. For example, there is a gene called *bicoid* in the *Drosophila* mother's genotype, which expresses itself in the 'nurse' cells that make her eggs. The protein made by the *bicoid* gene is shipped into the egg, where it is concentrated at one end, whence it fades towards the other end. The resulting concentration gradient (and others like it) labels the anterior/posterior axis. Comparable mechanisms at right-angles label the dorsal/ventral axis.

These labelling concentrations persist in the substance of the cells that are produced as the egg subsequently divides. The first few divisions occur without any addition of new material, and the divisions are incomplete: lots of separate nuclei are made, but they are not completely separated by cellular partitions. This multinucleate 'cell' is called a syncitium. Later, partitions form, and the embryo becomes properly cellular. Through all this, as I say, the original chemical gradients persist. It follows that cell nuclei in different parts of the embryo will be bathed in different concentrations of key substances, corresponding to the original two-dimensional gradients, and this will cause different genes to be turned on in different cells (we are now, of course, talking about the embryo's own genes, no longer the mother's). This is how differentiation of cells begins, and projections of the principle lead to further differentiation at later stages of development. The original gradients set up by maternal genes give way to new and more complex gradients set up by the embryo's own genes. Consequent forkings in the lineages of embryonic cells recursively generate further differentiations.

In arthropods there is a larger-scale partitioning of the body, not into cells but into segments. The segments are arrayed in line, from front of head to tip of abdomen. Insects have six head segments, of which the antennae are on segment 2, followed on other segments by the mandibles and then other mouthparts. The segments of the adult head are compressed into a small space, so their fore-and-aft alignment is not too clear, but it can be seen in the embryo. The three thoracic segments (T1, T2 and T3) are more obviously in a line, each bearing a pair of legs. T2 and T3 normally bear wings, but in *Drosophila* and other flies only T2 has wings.* The second pair of 'wings' is modified into halteres, small club-shaped organs on T3, which vibrate and serve as miniature gyroscopes to guide the fly. Some early fossil insects had three pairs of wings, one pair on each of the three thoracic segments. Behind the thoracic segments are a larger number of abdominal segments (11 in some insects, eight in *Drosophila*, depending on how you reckon the genitals at the rear end). Cells 'know' (in the sense already excused) which segment they are in, and they behave accordingly. Each cell is told which segment it is in through the mediation of special control genes called Hox genes, which turn themselves on inside the cell. The Fruit Fly's Tale is mostly a tale of Hox genes.

It would make things neat and easy to explain if I could now tell you that there is one Hox gene for each segment, with all the cells of a given segment having only its own numbered Hox gene turned on. It would be even tidier if the Hox genes were arrayed along the length of a chromosome, in the same order as the segments they influence. Well, it isn't *quite* as tidy as that, but it very nearly is. The Hox genes

---

* Some other insects, such as cockroaches and beetles, fly with only T3 wings, having modified the T2 wings into hardened protective wing cases called elytra. Crickets and grasshoppers, as we have heard, further modified the elytra as sound-producing organs.

are indeed arranged in the right order along one chromosome, and that is wonderful – gratuitously so, given what we know of how genes work. But there aren't enough Hox genes for the segments – only eight. And there's a more messy complication that I must get out of the way. The segments of the adult don't exactly correspond to the so-called *para*segments of the larva. Don't ask me why (perhaps the Designer was having an off day), but each adult segment is made up of the back half of one larval parasegment plus the front half of the next. Unless otherwise stated, I'll use the word segment to mean larval (para) segment. As for the question of how eight Hox genes in a row take charge of some 17 segments in a row, it is partly done by resorting to the chemical gradient trick again. Each Hox gene is mainly expressed in one segment, but it is also expressed, in decreasing concentration as you go backwards, in more posterior segments. A cell knows which segment it is in by comparing the chemical outputs of more than one upstream Hox gene. It is a bit more complicated than that, but there is no need to go into such detail here.

The eight Hox genes are arrayed in two gene complexes, physically separated along the same chromosome. The two are called the Antennapedia Complex and the Bithorax Complex. These names are doubly unfortunate. A complex of genes is named after a single member of that complex, which is no more important than the others. And, worse, the genes themselves are as usual named after what happens when they go wrong, rather than after their normal function. It would be better to call them something like the Front Hox Complex and the Rear Hox Complex. However, we are stuck with the existing names.

The Bithorax Complex consists of the last three Hox genes, which are named, for historical reasons that I shan't go into, *Ultrabithorax, Abdominal-A* and *Abdominal-B*. They affect the back end of the animal, as follows. *Ultrabithorax* itself is expressed from segment 8 all the way to the posterior end. *Abdominal-A* is expressed from segment 10 to the end, and *Abdominal-B* is expressed from segment 13 to the end. The products of these genes are made in decreasing concentration gradient as we move towards the back end of the animal, from their various starting points. So, by comparing the concentrations of the products of these three Hox genes, a cell in the posterior part of a larva can tell which segment it is in and act accordingly. It is a similar story for the front end of the larva, where the five Hox genes of the Antennapedia Complex are in charge.

A Hox gene, then, is a gene whose mission in life is to know whereabouts in the body it is, and so inform other genes in the same cell. We are now armed to understand homeotic mutations. When things go wrong with a Hox gene, the cells in a segment are mis-informed about which segment they are in, and they fashion the segment they 'think' they are in. So, for instance, we see a leg growing in the segment that would normally grow an antenna. This makes perfect sense. The cells in any segment are perfectly capable of assembling the anatomy of any other segment. Why should they not? The instructions for making any segment lurk in the cells of every segment. It is the Hox genes, under normal conditions, that call forth the 'correct' instructions for making the anatomy appropriate to each segment. As William Bateson rightly suspected, homeotic abnormal-ity opens a revealing window on how the system normally works.

Recall that flies, unusually among insects, normally have only one pair of wings, plus a pair of gyroscopic halteres. The homeotic mutation *Ultrabithorax* misleads cells in the third thoracic segment into 'thinking' they are in the second thoracic segment. They there-fore collaborate to make an extra pair of wings, instead of a pair of halteres (see plate 36). There is a mutant flour beetle (*Tribolium*) in which all 15 segments develop antennae, presumably because the cells all 'think' they are in segment 2.

This brings us to the most wonderful part of the Fruit Fly's Tale. After they had been discovered in *Drosophila*, Hox genes started turning up all over the place: not only in other insects such as beetles, but in almost all other animals that have been looked at, including ourselves. And – this really is almost too good to be true – they very often turn out to be doing the same kind of thing, even down to informing cells which segment they are in and (better still) being arrayed in the same order along chromosomes. Let's now turn to the mammal story, which has been most thoroughly worked out in the laboratory mouse – that *Drosophila* of the mammal world.

Mammals, like insects, have a segmented body plan, or at least a modular, repeated plan that affects the backbone and associated structures. Each vertebra can be thought of as corresponding to one segment, but it isn't just bones that are repeated rhythmically as we move from neck to tail. Blood vessels, nerves, muscle blocks, cartilage discs and ribs, where present, all follow the repetitive, modular plan. As with *Drosophila* the modules, though following the same general plan as each other, are different in detail. And like the insect division into head, thorax and abdomen, vertebrae are grouped into cervical (neck), thoracic (upper back vertebrae with ribs), lumbar (lower back vertebrae without ribs) and caudal (tail). As in *Drosophila*, the cells, whether they are bone cells, muscle cells, cartilage cells or anything else, need to know which segment they are in. And as in *Drosophila*, they know because of Hox genes – Hox genes that recognisably correspond to particular *Drosophila* Hox genes – although, unsurprisingly, given the immensity of time since Concestor 26, they are far from identical. Again as in *Drosophila*, the Hox genes are arranged in the right order on the chromosome. Vertebrate modularity is very different from that of insects, and there is no reason to think that their common ancestor, at Rendezvous 26, was a segmented animal. Nevertheless, the evidence of Hox genes suggests, at the very least, that there is some sort of deep similarity between insect and vertebrate body plans, which was also present in Concestor 26. And, indeed, in other body plans, including those that are not segmented.

Given the two rounds of vertebrate genome duplication that we saw in the Lamprey's Tale, it may not surprise you to find that in vertebrates such as the laboratory mouse there is not just one array of Hox genes on one chromosome; there are four separate arrays. The *a* series is on chromosome 6, the *b* series on chromosome 11, *c* on chromosome 15 and *d* on chromosome 2. Resemblances between them show that they have arisen during evolution by duplication: *a4* matches *b4* matches *c4* matches *d4*. There have also been some deletions, for certain slots are missing from each of the four arrays: *a7* and *b7* match each other, but neither the *c* series nor the *d* series has a representative in 'slot' 7. When two, three or four versions of a Hox gene impinge upon one segment, their effects are combined. And, as with *Drosophila*, all mouse Hox genes exert their strongest effect in the first (most anterior) segment in their domain of influence, with a gradient of decreasing expression downstream in more posterior segments.

It gets better. With minor exceptions, each gene from the *Drosophila* array of eight Hox genes resembles its opposite number in the mouse series more than it resembles the other seven genes in the *Drosophila* series. And they are in the same order along their respective chromosomes. Every one of the eight *Drosophila* genes has at least one representative in the mouse series of 13. The detailed gene-for-gene coincidence between *Drosophila* and mouse can only indicate shared inheritance – from Concestor 26, the grand progenitor of all the protostomes and all the deuterostomes. That means the vast majority of animals are descended from an ancestor that had Hox genes arranged in the same linear order as we see in modern *Drosophila* and modern vertebrates. Think of it! Concestor 26 had Hox genes, and in the same order as ours.

As I've already said, it doesn't follow that Concestor 26's body was partitioned into discrete segments. Indeed, it probably wasn't. But there was surely some kind of fore-and-aft gradient running from head to tail that was mediated by the homologous series of Hox genes arranged in the right order along a chromosome. Concestors being dead and beyond the reach of molecular biologists, it now becomes of great interest to look for Hox genes in their modern descendants. Concestor 24 is the ancestor we share with amphioxus. Given that the more distantly related *Drosophila* has the same fore-and-aft series as mammals, it would be positively worrying if amphioxus didn't have it too. Peter Holland, with his research group, has looked into the matter, and their results are gratifying. Yes, amphioxus's modular body plan is mediated by (14) Hox genes, and yes, they are arranged in the right order along the chromosome. Unlike the mouse, but like *Drosophila*, there is only one series, not four parallel series, presumably because Concestor 24 predates the vertebrate genome duplications we discussed in the Lamprey's Tale.

How about other animals, strategically chosen for what they can tell us about other particular concestors? Hox genes have now been found in every animal that has been looked at except ctenophores, placozoans and sponges (whom we shall meet at Rendezvous 29, 30, and 31 respectively). For example, we find them in sea urchins, *Limulus*, shrimps, molluscs, annelid worms, acorn worms, sea squirts, nematode worms and flatworms. This much we could have guessed, knowing that all these animals are descended from Concestor 26, and we already have good reason to think Concestor 26 had Hox genes, like its descendants *Drosophila* and mouse.

Cnidarians, such as *Hydra* (they aren't due to join us until Rendezvous 28), are radially symmetrical – they don't have an anterior/posterior axis, nor a dorsal/ventral one. They have an oral/aboral (mouth versus opposite-to-mouth) axis. It isn't obvious what, if anything, corresponds to their long axis, so what might we expect their Hox genes to do? It would be tidy if they used them to define the oral/aboral axis, but so far it isn't clear that this is so. Most cnidarians only have two Hox genes anyway, to pit against

Drosophila's eight and amphioxus's 14. It is agreeable that one of these two genes resembles the anterior complex of *Drosophila*, while the other resembles the posterior complex. Concestor 28, the one we share with them, presumably had the same. Then one of the two duplicated several times during evolution to produce the Antennapedia Complex, while the other one duplicated within the same animal lineage to produce the Bithorax Complex. That is exactly the kind of way genes increase in the genome (see the Howler Monkey's Tale). But more research is needed before we shall know what, if anything, the two genes are doing in the planning of the cnidarian body.

Echinoderms are radially symmetrical, like cnidarians, but secondarily so. Concestor 25, which they share with us vertebrates, was bilaterally symmetrical, like a worm. Echinoderms have a variable number of Hox genes – ten in the case of sea urchins. What are these genes doing? Does a relic of the ancestral anterior/posterior axis lurk within the body of a starfish? Or do the Hox genes exert their influence successively along the length of each of the five arms? This might seem to make sense. We know that Hox genes express themselves in the arms, and the legs, of mammals. I don't mean that the arrays of Hox genes from 1 to 13 express themselves in order, from shoulder to fingertips. It is more complicated than that – as well it might be – because the vertebrate limb is not arranged in modules that succeed each other along its length. Instead, there is first one bone (the humerus in the arm, femur in the leg), then two bones (radius and ulna in the arm, tibia and fibula in the leg), then lots of little bones culminating in the fingers and toes. This fan arrangement, inherited from the more obvious fan of our fishy ancestors' fins, doesn't lend itself to straightforward Hox linearity. Even so, Hox genes are involved in the limb development of vertebrates.

By analogy, it would not be surprising if Hox genes were also expressed in starfish or brittle star arms (and even sea urchins can be thought of as starfish who have curled their arms up in a five-pronged arch, meeting at the tips and zipped together down the sides). Moreover, starfish arms, unlike our arms or legs, really are serially modular along their length. The tube feet, with all their associated hydraulic plumbing, are units that repeat in two parallel rows along the length of each arm: just the thing for Hox gene expression! Brittle star arms even look and behave like five worms.

T. H. Huxley referred to 'The great tragedy of science – the slaying of a beautiful hypothesis by an ugly fact.' The true facts about echinoderm Hox genes may not be ugly, but they don't follow the pretty pattern I have just suggested. Something else happens, which has its own rather surprising beauty. Echinoderm larvae are tiny, bilaterally symmetrical swimmers in the plankton. The five-way radially symmetrical bottom-dwelling adult doesn't develop as a transformation of the larva. Instead, it starts as a tiny miniature adult *inside* the larva's body, which grows until eventually the rest of the larva is discarded. Hox genes are expressed in the correct linear order, but not along each arm. Instead the order of expression follows a roughly circular route *around* the baby adult. If we think of the Hox axis as a 'worm', there are not five 'worms', one for each arm. There is a single 'worm' curled around inside the larva. The front end of the 'worm' sprouts arm number 1, the back end of the 'worm' sprouts arm number 5. Homeotic mutations in starfish, then, might be expected to grow too many arms. And, sure enough, mutant starfish with six arms are known, and were recorded in Bateson's book. There are also some species of starfish that have much larger numbers of arms, and they have presumably evolved from homeotic mutant ancestors.

Hox genes have not been found in plants, nor in fungi, nor in the single-celled organisms we used to call protozoa. But now we come to a complication of terminology that must be dealt with before we go any further.

Hox genes are not the only genes that can cause homeotic mutations. They are just one family of genes which nests within (and happens to be named after) a much larger class of control genes known as the 'homeobox' genes. The 'box' refers to a diagnostic sequence of 180 DNA letters that all the members of this class have somewhere in their length, which is called the homeodomain. Even homeobox genes are not the only source of homeotic mutations. Other control genes exist, such as those in the 'MADS box' family. These are found in animals and plants, and are responsible for homeotic mutations in flowers, causing petals to grow in the place of the pollen-bearing stamens. Terminologically, Hox genes are a particular family of homeobox genes, and homeobox genes are themselves only one class (strictly superclass) of homeotic gene.

Among the other families of homeobox genes which have now been discovered are the ParaHox genes, also involved in anterior-posterior development (of the gut and nervous system, as it happens) and also arranged in a short linear array. The ParaHox cluster were first clearly defined in amphioxus, but if anything they are more widespread than the Hox genes. It seems that the ParaHox genes are 'cousins' of the Hox genes, in the sense that they correspond to, and are arranged in the same order as, Hox genes. They certainly arose by duplication from the same ancestral set of genes as the Hox genes. Other homeobox genes are more distantly related to Hox and Para-Hox, but form families of their own. The *Pax* family is found in all animals. A particularly notable member of this family is *Pax6*, which corresponds to the gene known as *ey* in *Drosophila*. I've already mentioned that *Pax6* is responsible for telling cells to make eyes. The same gene makes eyes in animals as different as *Drosophila* and mouse, even though the eyes produced are radically different in the two animals. In a similar way to Hox genes, *Pax6* doesn't tell cells *how* to make an eye. It only tells them that here is the *place* to make an eye.

A rather parallel example is the small family of genes called *tinman*. Again *tinman* genes are present in both *Drosophila* and mice. In *Drosophila*, *tinman* genes are responsible for telling cells to make a heart, and they normally express themselves in just the right place to make a *Drosophila* heart. As we have by now come to expect, *tinman* genes are also involved in telling mouse cells to make a heart in the right place for a mouse's heart.

The whole set of homeobox genes constitutes a very large number, divided into families and subfamilies just as animals themselves are divided into families and subfamilies. It is like the case of haemoglobin, which we examined in the Lamprey's Tale. There we learned that human alpha globin is truly a closer cousin to, say, lizard alpha globin than it is to human beta globin – which is in turn a closer cousin to lizard beta globin. Similarly, human *tinman* is a closer cousin to fruit fly *tinman* than it is to human *Pax6*. It is possible to construct a very full family tree of homeobox genes that exists side by side with the family tree of the animals that contain them. Both family trees are equally valid. Both are true trees of ancestry, formed by splitting events that happened at particular moments in geological history. In the case of the animal family trees, the splitting events are speciations. In the case of the homeobox gene family trees (or the globin genes), the splitting events are gene duplications within genomes.

The tree of animal homeobox genes splits into two great classes, the AntP and the PRD classes. I shall not spell out what these abbreviations stand for, because both are perversely confusing. The PRD class includes the *Pax* genes, and various other subclasses. The AntP class includes the Hox and ParaHox, and again various other subclasses. In addition to these two great classes of animal homeobox genes, there are various more distantly related homeobox genes that are (misleadingly) called 'divergent'. These are found not only in animals but in plants, fungi and 'protozoa' as well.

Only animals have true Hox genes, and they are always used in the same kind of way – to specify information about position in the body, whether or not the body is neatly divided into discrete segments. Although we previously asserted that Hox genes have not been found in sponges or ctenophores, they might be lurking in some of the as yet unsequenced species, or perhaps were present in their ancestors and subsequently lost. That's certainly what happened to the ParaHox genes. They were not found in the first sponge genome to be sequenced, but then a bone-fide ParaHox gene was discovered in the genomes of another set of sponges. The implication is that all sponges had ParaHox genes (and possibly Hox genes too) and many sponge groups have lost them entirely, save for some species which have retained one ParaHox gene. It would not be surprising to find that all animals descend from an ancestor that had some form of ProtoHox cluster. This would encourage my colleagues Jonathan Slack, Peter Holland and Christopher Graham, then all at Oxford, who proposed a new definition of the very word 'animal'. Hitherto, animals were defined as opposed to plants, in a rather unsatisfactorily negative way. Slack, Holland and Graham suggested a positive, specific criterion that has the effect of uniting all animals and excluding all non-animals, such as plants and protozoa. The Hox story shows that animals are not a highly varied, unconnected miscellany of phyla, each with its own fundamental body plan acquired and maintained in lonely isolation. If you forget morphology and look only at the genes, it emerges that all animals are minor variations on a very particular theme. What delight to be a zoologist at such a time.

THE ROTIFER'S TALE

The brilliant theoretical physicist Richard Feynman is rumoured to have said, 'If you think you understand quantum theory, you don't understand quantum theory.' I am tempted by an evolutionist's equivalent: 'If you think you understand sex, you don't understand sex.' The three modern Darwinians from whom I believe we have the most to learn – John Maynard Smith, W. D. Hamilton and George C. Williams – all sadly now dead, devoted substantial parts of their long careers to wrestling with sex. Williams began his 1975 book *Sex and Evolution* with a challenge to himself: 'This book is written from a conviction that the prevalence of sexual reproduction in higher plants and animals is inconsistent with current evolution-ary theory . . . there is a kind of crisis at hand in evolutionary biology . . .' Maynard Smith and Hamilton said similar things. It is to resolve this crisis that all three Darwinian heroes, along with others of the rising generation, laboured. I shall not attempt an account of their efforts, and certainly I have no rival solution to offer myself. Instead, the Rotifer's Tale displays an under-explored *consequence* of sexual reproduction for our view of evolution.

Bdelloidea is a large class of the phylum Rotifera (see page 439 and plate 37). The existence of the bdelloid rotifers is an evolutionary scandal. Not my own *bon mot* – it rings with the unmistakable tones of John Maynard Smith. Many rotifers reproduce without sex. In this respect they resemble aphids, stick insects, various beetles and a few lizards, and are not particularly scandalous. What stuck in Maynard Smith's craw is that the bdelloids *as a whole* reproduce only asexually – every last one of them, evidently descended from a bdelloid common ancestor that must have lived long enough ago to beget 18 genera and 360 species. Remains in amber suggest that this male-spurning matriarch lived at least 40 million years ago, very probably more. The bdelloids are a highly successful group of animals, astonishingly numerous and a dominant part of the freshwater faunas of the world. Not a single male has ever been found.[*]

What is so scandalous about that? Well, suppose we take a family tree of the whole animal kingdom. The tips of the twigs, all round the surface of the tree, represent species. Major boughs represent classes

---

[*] To be scrupulously correct, in nearly 300 years of scientific investigation, there has been a single report of a male bdelloid rotifer made by the Danish zoologist C. Wesenberg-Lund (1866–1955). 'With great hesitation I venture to remark, that twice I saw among the thousands of Philodinidae (*Rotifer vulgaris*) a little creature, unquestionably a male rotifer . . . but both times I failed to get it isolated. It moves round and between the numerous females with extreme rapidity' (understandably, no doubt). Even before the strong evidence of Mark Welch and Meselson (see page 484) zoologists were not inclined to take Wesenberg-Lund's never-repeated observation as sufficient proof of the existence of male bdelloids.

or phyla. There are millions of species, which means that the evolutionary tree is far more intricately branched than any woody tree you will ever see. There are only a few tens of phyla, and not that many more classes. The phylum Rotifera is one branch of the tree and it splits into four sub-branches, of which the class Bdelloidea is one. This class branch subdivides, and subdivides further, eventually to yield 360 twigs each representing one species. The same kind of thing is going on for all the other phyla, each with its own classes and so on. The outer twigs of the tree represent the present; slightly inward from the outermost shell of twigs represents a short distance into the past; and so on to the main trunk which represents, say, a billion years ago.

That understood, we now take paint to our grey, wintry tree and colour the tips of certain twigs to label particular features. We might paint red all twigs that represent flying animals – powered flying as opposed to passive gliding, which is much more common. If we now step back and contemplate the whole tree, we shall notice large areas of red separated by even larger areas of grey representing whole major groups of animals that don't fly. Most of the insect twigs, the bird twigs and the bat twigs are red, and they are neighbours of other red twigs. None of the others are. With a few exceptions like fleas and ostriches, three entire classes are made up of flying animals. Redness is distributed in broad patches of uniform red against uniform grey.

Think what this means for evolution. The three patches of red must have begun long ago with three ancestral animals – an early insect, an early bird and an early bat – discovering how to fly. Flying obviously turned out to be a really good idea once it had been discovered, because it persisted and spread through all the descendant branches as the three species eventually gave rise to three large collections of descendant species, almost all of them persisting with their ancestor's ability to fly: the class Insecta, the class Aves and the order Chiroptera.

But now we do the same thing not for flying but for reproducing asexually, without males. (Never without females, by the way. Unlike eggs, sperms are too small to go it alone. Asexual reproduction in animals means dispensing with males.) On our tree of life, we paint the twigs of all asexually reproducing species blue. Now we notice a completely different pattern. Where flying showed up as great swathes of red, asexual reproduction shows up as tiny sporadic dots of blue. An asexual beetle species shows up as a blue twig completely surrounded by grey. Maybe a group of three species in one genus are blue, but neighbouring genera are grey. You see what this means? Asexual reproduction arises from time to time, but it rapidly goes extinct before it has had time to grow into a stout branch with many blue twigs. Unlike flying, the habit of reproducing asexually doesn't persist long enough to give rise to a whole family, order or class of asexual creatures.

With one scandalous exception! Unlike all the other little blips of blue, the bdelloid rotifers are a hefty patch of uninterrupted blue, enough to make the proverbial sailor's trousers. In evolution, what this seems to mean is that an ancestral bdelloid discovered asexual reproduction, just like the odd beetle we talked about. But whereas asexual beetles and hundreds of other asexual species dotted around the tree go extinct long before they can evolve into a larger grouping such as a family or order, let alone a class, the bdelloids seem to have stuck with asexuality and flourished with it through enough evolutionary time to generate an entire asexual class, now numbering 360 species. For the bdelloid rotifers, but not convincingly for any other kind of animal, asexual reproduction is like flying. It seems to be a good and successful innovation for the bdelloids, while in all other parts of the tree it constitutes a fast track to extinction.

The statement that there are 360 species raises a problem. The biological definition of a species is a group of individuals that interbreed with each other and not with others. The bdelloids, being asexual, don't interbreed with anybody; every one is an isolated female, each of whose descendants goes her own sweet way, in genetic isolation from every other individual. So when we say there are 360 species, we can only mean there are 360 types, which we humans recognise as looking sufficiently different from each other that we would expect, if they bred sexually at all, to see them shunning the other types as sexual partners.

Not everyone has accepted that the bdelloid rotifers really are asexual. There is a large gap in logic between the negative statement that males have never been seen, and a positive conclusion that there aren't any. As Olivia Judson recounts in her winningly sophisticated zoological comedy *Dr Tatiana's Sex Guide to All Creation*, naturalists have been caught like that before. Apparently asexual species often turn out to have concealed males. The males of certain angler fish are tiny dwarfs who ride about as parasites on the bodies of females. Had they been even smaller, we might have overlooked them altogether, as nearly happened in the case of certain scale insects where the males are, in the words of my colleague Laurence Hurst, 'wee things that stick to the females' legs'. Hurst went on to quote his mentor Bill Hamilton's astute remark:

> How often do you see humans having sex? If you were a Martian looking around, you'd be pretty sure we were asexual.

So it would be nice to come up with some more positive evidence that the bdelloid rotifers really are anciently asexual. Geneticists are becoming increasingly ingenious in the art of reading the patterns of gene distribution in modern animals and making inferences about their evolutionary history. In Eve's Tale, we saw that we could reconstruct early human migrations by picking up 'signals' in the genomes of living people. The logic is not deductive. We don't deduce from modern genes that the course of history must have been so-and-so. Instead we say, if the course of history was so-and-so, we should expect to see such-and-such a pattern of gene distribution today. That is what is done for human migrations, and something similar has been done for bdelloid rotifers by David Mark Welch and Matthew Meselson of Harvard University. Mark Welch and Meselson used genetic signals to make inferences not about migration but about asexual reproduction. Again, the form of their logic is not deductive. Instead they reasoned that, if the bdelloids had been purely asexual for many millions of years, we should expect the genes of living bdelloids to show a certain pattern.

What pattern? Mark Welch and Meselson's reasoning was ingenious. First you must understand that the bdelloid rotifers, though asexual, are diploid. That is to say, they are like all sexually reproducing animals in having two copies of each chromosome. The difference is that the rest of us reproduce by making eggs or sperms that have only one copy of each chromosome. The bdelloids produce eggs that have both copies of each chromosome. So a bdelloid's egg cell is like any of her other cells, and a daughter is an identical twin of her mother, give or take the odd mutation. It is these odd mutations that, over the millions of years, have gradually built up in diverging lines, presumably under natural selection, to produce the 360 species that we see today.

An ancestral female, whom I shall call the gynarch, mutated in such a way as to dispense with males and with meiosis, and substitute mitosis as the method of producing eggs.* From then on, throughout the cloned population of females, the fact that the chromosomes were originally paired became irrelevant. Instead of five pairs of chromosomes (or whatever the number was – the equivalent of 23 in us), there were now ten chromosomes – each linked to its erstwhile partner only by a kind of receding memory. Chromosome partners used to meet up and exchange genes every time a rotifer made eggs or sperm. But during the millions of years since the gynarch drove out the males and founded the bdelloid gynodynasty, every chromosome has been drifting apart from its erstwhile partner genetically, as their genes have mutated independently of one another. And this has happened even though they have shared cells in shared bodies all that time. In the good old days of males and sex, it didn't happen. In every generation each chromosome paired off with its opposite number and exchanged genes before making eggs or sperm. This held the pairs of chromosomes in a kind of intermittent embrace, preventing them from drifting apart in their gene content.

You and I have 23 pairs of chromosomes. We have two chromosome 1s, two chromosome 5s, two chromosome 17s and so on. With the exception of the X and Y sex chromosomes, there is no consistent difference between members of a pair. Since they exchange genes every generation, the two chromosome 17s are just chromosome 17s and there is no point in calling them, say, left chromosome 17 and right chromosome 17. But from the moment the rotifer gynarch froze her genome, all that changed. Her left chromosome 5 was passed intact to all her daughters, as was her right chromosome 5, and never the twain met for more than 40 million years. Her 100-greats-grand-daughters still had a left chromosome 5 and a right chromosome 5. Although by then they would have picked up some mutations, all the left chromosomes would be identifiable by their resemblance to each other, inherited from the gynarch's left chrosomome 5.

There are now 360 species of bdelloids, all descended from the gynarch and separated from her by exactly the same length of time as each other. All individuals of all species still have a left and a right copy of each chromosome, inherited with plenty of mutational change down the line, but no gene swapping across from left to right. Each left and right pair within each individual will now be far

* Meiosis is the special form of cell division that halves the number of chromosomes in order to make sex cells. Mitosis is the ordinary form of cell division used for making body cells, which duplicates all the chromosomes of a cell.

more different from one another than you'd expect if there had been any sexual activity at all, at any time in their ancestry since the days of the gynarch. They may even be approaching the time when you can no longer recognise that they were originally paired at all.

But now suppose we compare two modern species of bdelloid rotifers, say *Philodina roseola* and *Macrotrachela quadricornifera*. Both belong to the same subgroup of bdelloids, the Philodinidae, and they certainly have a common ancestor who lived much more recently than the gynarch. Given no sex, there has been exactly the same time for the 'left' and the 'right' chromosomes within every individual of either species to drift apart – the time since the gynarch. Left will be very different from right within every individual. But if you compare, say, left chromosome 5 of *Philodina roseola* with left chromosome 5 of *Macrotrachela quadricornifera*, you should find them pretty similar because they haven't had very long to pick up independent mutations. And right compared with right will also yield few differences. We arrive at the remarkable prediction that a chromosomal comparison across once-paired chromosomes *within* individuals should yield a greater difference than a cross-species comparison – of 'left' with 'left', or of 'right' with 'right'. The longer the time since the gynarch, the greater the difference. Given sex, the prediction would be precisely the opposite, essentially because there is no such thing as 'left' or 'right' identity across species, and plenty of gene-swapping between paired chromosomes within species.

Mark Welch and Meselson used these opposite predictions to test the theory that the bdelloids really have been sexless and male-less for a very long time – with stunning success. They looked at modern bdelloids to see whether it was indeed true that paired chromosomes (or chromosomes that had once been paired) were much more unlike each other than they 'should' be, gene for gene, if sexual recombination had been holding them together. They used other rotifers, non-bdelloids who do have sex, as a control for comparison. And the answer is yes. Bdelloid chromosomes are far more different from their pairs than they 'should' be, by an amount compatible with the theory that they gave up sex not just 40 million years ago, which is the age of the oldest amber in which bdelloids have been found, but about 80 million years ago. Mark Welch and Meselson scrupulously bend over backwards to discuss possible alternative interpretations of their results, but these are far-fetched, and I think they are right to conclude that the bdelloid rotifers really are anciently, continuously, universally and successfully asexual. They really are an evolutionary scandal. For perhaps 80 million years they have flourished by doing something that no other group of animals can get away with, except for very short periods before going extinct.

Why would we normally expect asexual reproduction to lead to extinction? Well, that is a big question because it amounts to the question of what is good about sex itself – the question that better scientists than I have spent book after book failing to answer. I shall just point out that the bdelloid rotifers are a paradox within a paradox. In one way, they are like the soldier in the marching platoon whose mother cried out, 'There goes my boy – he's the only one in step.' Maynard Smith called them an evolutionary scandal, but he was the one mainly responsible for pointing out that sex itself, on the face of it, is the evolutionary scandal. At least a naive view of Darwinian theory would predict that sex should be heavily disfavoured by natural selection, outcompeted twofold by asexual reproduction. In that sense the Bdelloidea, far from being a scandal, appear to be the only soldier in step. Here's why.

The problem is the one Maynard Smith dubbed the twofold cost of sex. Darwinism, in its modern form, expects that individuals will strive to pass on as many of their genes as possible. So isn't it just daft to throw half your genes away with every egg or sperm you make, in order to mix the other half with the genes of somebody else? Wouldn't a mutant female who behaves like a bdelloid rotifer, and passes on 100 per cent of her genes to every offspring instead of 50 per cent, do twice as well?

Maynard Smith added that the reasoning breaks down if the male partner works hard, or contributes economic goods, in such a way that a couple can rear twice as many offspring as an asexual loner. In that case, the twofold cost of sex is cancelled out by a doubling in the number of offspring. In a species such as an emperor penguin, with male and female parent contributing approximately equally towards the labour and other costs of childrearing, the twofold cost of sex is abolished, or at least mitigated. In species where economic and labour contributions are unequal, it is nearly always the father who shirks, devoting his energies instead to duffing up other males. This magnifies the cost of sex, up to the full twofold penalty of the original reasoning. This is why Maynard Smith's alternative name, the twofold cost of males, is preferable. In this light – which Maynard Smith himself was largely responsible for shining – it isn't the bdelloid rotifers who are the evolutionary scandal but everything else. More pertinently, the male sex is an evolutionary scandal. Except that it does exist and, indeed, is almost universal throughout the animal kingdom. What is going on? As Maynard Smith wrote, 'One is left with the feeling that some essential feature of the situation is being overlooked.'

The twofold cost is the starting point for masses of theorising by Maynard Smith, Williams, Hamilton and many younger colleagues. The widespread existence of males who don't earn their keep as fathers must mean that there really are very substantial Darwinian benefits to sexual recombination itself. It is not too difficult to think of what they might be in qualitative terms, and lots of possible benefits, some obvious, some esoteric, have been proposed. The problem is to think of a benefit of sufficient quantitative *magnitude* to counteract the massive twofold cost.

To do justice to all the theories would take a book – it has already taken several, including the seminal works I have previously mentioned by Williams and Maynard Smith, and Graham Bell's beautifully written tour de force *The Masterpiece of Nature.* Yet no definitive verdict has emerged. A nice book aimed at a non-specialist audience is Matt Ridley's *The Red Queen.* Though primarily favouring one of the theories on offer, W. D. Hamilton's theory that sex serves an unceasing arms race against parasites, Ridley does not neglect to explain the problem itself and the other answers to it. As for me, I shall swiftly recommend Ridley's book and the others before going straight to the main purpose of this tale, which is to draw attention to an under-appreciated *consequence* of the evolutionary invention of sex. Sex brought into existence the gene pool, made meaningful the species, and changed the whole ball game of evolution itself.

Think what evolution must look like to a bdelloid rotifer. Think how different the evolutionary history of those 360 species must have been from the normal pattern of evolution. We portray sex as raising diversity and so, in a sense, it does: that is the basis of most theories of how sex overcomes its twofold cost. But, paradoxically, it also has a seemingly opposite effect. Sex can act as a kind of barrier to evolutionary divergence. Indeed, a special case of this was the basis of Mark Welch and Meselson's research. In a population of mice, say, any tendency to strike out in some enterprising new evolutionary direction is held in check by the swamping effect of sexual mixing. The genes of the would-be enterprising diverger are swamped into conformity by the inertial mass of the rest of the gene pool. That is why geographical isolation is so important to speciation. It takes a mountain range or a difficult sea crossing to allow a newly striking-out lineage to evolve its own way without being dragged back to the inertial norm.

Think how different evolution must be for the bdelloid rotifers. Far from being swamped into normalcy by the gene pool, they don't even *have* a gene pool. The very idea of a gene pool has no meaning if there is no sex.* 'Gene pool' is a persuasive metaphor because the genes of a sexual population are being continually mixed and diffused, as if in a liquid. Bring in the time dimension, and the pool becomes a river, flowing through geological time – an image that I developed in *River out of Eden*. It is the binding effect of sex that provides the river with its limiting banks, channelling the species into some kind of evolutionary direction. Without sex, there would be no coherently channelled flow, but a shapeless outward diffusion: less like a river than like a smell, wafting out in all directions from some point of origin.

Natural selection presumably takes place among the bdelloids, but it must be a very different kind of natural selection from the one the rest of the animal kingdom is accustomed to. Where there is sexual mixing of genes, the entity that is carved into shape by natural selection is the gene pool. Good genes tend statistically to help the individual bodies in which they find themselves to survive. Bad genes tend to make them die. In sexually reproducing animals, it is the deaths and reproductions of individual animals that constitute the immediate selective events, but the long-term consequence is a change in the statistical profile of genes in the gene pool. So, it is the gene pool, as I say, that is the object of the Darwinian sculptor's attention.

Moreover, genes are favoured for their capacity to co-operate with other genes in building bodies. That is why bodies are such harmonious engines of survival. The right way to look at this, given sex, is that genes are continually being tried out against different genetic backgrounds. In every generation, a gene is shuffled into a new team of companions, meaning the other genes with which it shares a body on any particular occasion. Genes that are habitually good companions, fitting in well with others and co-operating well with them,

* People sometimes confusingly say gene pool when they mean genome. The genome is the set of genes within one individual. The gene pool is the set of all genes in all the genomes of a sexually breeding population.

tend to be in winning teams – meaning successful individual bodies that pass them on to offspring. Genes that are not good co-operators tend to make the teams in which they find themselves become losing teams – meaning unsuccessful bodies that die before reproducing.

The proximal set of genes with which a gene has to co-operate are the ones with which it shares a body – this body. But in the long term, the set of genes with which it has to co-operate are all the genes of the gene pool, for they are the ones that it repeatedly encounters as it hops from body to body down the generations. This is why I say it is the gene pool of a species that is the entity sculpted into shape by the chisels of natural selection. Proximally, natural selection is the differential survival and reproduction of whole individuals – the individuals that the gene pool throws up as samples of what it can do. Once again, none of this could be said of the bdelloid rotifers. Nothing like the sculpting of the gene pool goes on, for there is no gene pool to sculpt. A bdelloid rotifer has just one big gene.

What I have just called attention to is a consequence of sex, not a theory for the benefit of sex, nor a theory of why sex arose in the first place. But if I ever were to attempt a theory of the benefit of sex; if I were ever to essay a serious assault on the 'essential feature of the situation that is being overlooked', it is hereabouts that I would start. And I would listen again and again to the Rotifer's Tale. These tiny, obscure denizens of puddles and mossy moisture may hold the key to the outstanding paradox of evolution. What's wrong with asexual reproduction, if the bdelloid rotifers have run with it for so long? Or, if it's right for them, why don't the rest of us do it and save the massive twofold cost of sex?

THE BARNACLE'S TALE

When I was at boarding school, it was occasionally necessary to apologise to the housemaster for being late for dinner: 'Sorry I'm late, sir: orchestra practice,' or whatever the excuse might be. On those occasions when there really was no good excuse and we had something to hide, we formed the habit of murmuring, 'Sorry I'm late sir: barnacles.' He always nodded kindly, and I don't know whether he ever wondered what this mysterious out-of-school activity might be. It is possible that we were inspired by the example of Darwin, who devoted years of his life to barnacles so single-mindedly that his children were moved to ask, in innocent puzzlement after being shown round the house of some friends, 'Then where does [your father] do his barnacles?' I'm not sure that we knew the Darwin story then, and I suspect that we invented the excuse because there is something about barnacles that seems too implausible to be a bluff. Barnacles are not what they seem. That applies to other animals too. And it is the theme of the Barnacle's Tale.*

Contrary to all appearances, barnacles are crustaceans. The ordinary acorn barnacles, which encrust the rocks like miniature limpets, helping your shoes not to slip if you have them and hurting your feet if you don't, are completely unlike limpets internally. Inside the shell, they are distorted shrimps lying on their backs, kicking their legs in the air. Their feet bear feathery combs or baskets with which they filter particles of food out of the water. Goose barnacles do the same thing, but instead of sheltering under a conical shell like an acorn barnacle, they sit on the end of a stout stalk. They get their name from yet another misunderstanding of the true nature of barnacles. Their wet filtering 'feathers' give them the appearance of a baby bird in its egg. In the days when people believed in spontaneous generation, a folk belief grew that goose barnacles hatched into geese, specifically *Branta leucopsis*, the barnacle goose.

Most deceptive of all – indeed probably holding the record for animals not looking remotely like the thing that zoologists know them to be – are the parasitic barnacles, such as *Sacculina*. *Sacculina* is not what it seems with a vengeance. Zoologists would never have realised that it is in fact a barnacle, but for its larva. The adult is a soft sac that clings to the underside of a crab and sends long, branching, plant-like roots inside to absorb nourishment from the crab's tissues. The parasite not only doesn't look like a barnacle, it doesn't look like a crustacean of any kind. It has completely lost all trace of the armour plating, and all trace of the bodily segmentation that nearly all other arthropods have. It might as well be a parasitic plant or fungus. Yet, in terms of its evolutionary relationships, it is a crustacean, and not just a crustacean but specifically a barnacle. Barnacles are indeed not what they seem.

---

* The great scientist J. B. S. Haldane offered a completely different Barnacle's Tale, a parable in which philosophical barnacles contemplate their world. Reality, they conclude, is everything they can reach with their filtering arms. They are dimly aware of 'visions', but doubt their physical reality because barnacles on different parts of the rock disagree as to their distance and shape. This clever allegory on the limitations of human thought and the growth of religious superstition is Haldane's tale, not mine, and I shall merely recommend it and pass on. It is in the eponymous essay of *Possible Worlds*.

Fascinatingly, the embryological development of *Sacculina*'s extraordinarily uncrustacean-like body is starting to be understood in terms of the kind of Hox genes that were the subject of the Fruit Fly's Tale. The gene called *Abdominal-A*, which normally supervises the development of a typical crustacean abdomen, is not expressed in *Sacculina*. It looks as though you can turn a swimming, kicking, leggy animal into a shapeless fungoid just by suppressing Hox genes.

By the way, *Sacculina*'s branching root system is not indiscriminate in its invasion of the crab's tissues. It heads first for the crab's reproductive organs, which has the effect of castrating the crab. Is this just an accidental by-product? Probably not. Castration not only sterilises the crab. Like a fat bullock, the castrated crab, instead of concentrating on becoming a lean, mean, reproducing machine, diverts resources towards getting larger: more food for the parasite.[*]

To lead into the final tale of this cluster, here's a little fable set in the future. Half a billion years after vertebrate and arthropod life completely perished in the mother of all comet collisions, intelligent life has eventually re-evolved in remote descendants of octopuses. Octopoid palaeontologists come upon a rich fossil bed dating from the twenty-first century A D. Not a fair cross-section of contemporary life, this bounteous shale nevertheless impresses the palaeontologists with its variety and diversity. Carefully weighing the fossils up with eight-arm balanced judgement, and expertly sucking the details, one octopodan scholar goes so far as to suggest that life, during this pre-catastrophe dawn age, was more extravagantly profligate in its diversity than it ever would be again, throwing up weird and wonderful new body plans in gleeful experimentation. You can see what he means by thinking of your own animal contemporaries and imagining that a small sampling of them fossilises. Think of the herculean task facing our future palaeontologist, and empathise with his difficulties in trying to discern their affinities from imperfect and sporadic fossil traces.

Just to take one example, how on earth would you classify the animal at the top of the opposite page? Evidently a new 'weird wonder', probably deserving to have a previously unnamed phylum coined in its honour? A whole new Bauplan, hitherto unknown to zoology?

---

[*] I went to town on such cases of parasites subtly manipulating the intimate physiology of their hosts in the parasite chapter of *The Extended Phenotype*.

Weird wonder? A whole new Bauplan? Female *Thaumatoxena andreinii*. Drawing by Henry Disney.

Well, no. To return from futuristic fantasy to the present, the weird wonder above is actually a fly, *Thaumatoxena andreinii*. Not only that, it is a fly that belongs to the perfectly respectable family Phoridae. A more typical member of the Phoridae is pictured below, *Megaselia scalaris*.

What happened to *Thaumatoxena*, the 'weird wonder', is that it took up residence in a termite nest. The demands of life in that claustral world are so different that – probably in rather a short time – it lost all resemblance to a fly. The boomerang-shaped front end is what is left of the head. Then comes the thorax, and you can see the remains of the wings tucked in between the thorax and the abdomen, which is the hairy bit at the back.

How a fly ought to be? Phorid fly, *Megaselia scalaris* (Loew). Drawing by Arthur Smith.

The moral is that of the barnacle again. But the parable of the palaeontologist of the future, and his seduction by the rhetoric of weird wonders gleefully carousing in morphospace, was not idly spun. It was intended as a softening-up for the next tale, which is all about the 'Cambrian Explosion'.

THE VELVET WORM'S TALE

If modern zoology admits of anything approaching a full-blown origin myth, it is the Cambrian Explosion. The Cambrian is the first period of the Phanerozoic Eon, the last 540 million years, during which animal and plant life as we know it suddenly became manifest in fossils. Before the Cambrian, fossils were either tiny traces or enigmatic mysteries. From the Cambrian onwards, there has been a clamorous menagerie of multicellular life, more or less plausibly presaging our own. It is the suddenness with which multicellular fossils appear at the base of the Cambrian that prompts the metaphor of explosion.

Creationists love the Cambrian Explosion because it seems, to their carefully impoverished imaginations, to conjure a sort of palaeontological orphanage inhabited by parentless phyla: animals without antecedents, as if they had suddenly materialised overnight from nothing, complete with holes in their socks.* At the other extreme, romantically overheated zoologists love the Cambrian Explosion for its aura of Arcadian Dreamtime, a zoological age of innocence in which life danced to a frenzied and radically different evolutionary tempo: a prelapsarian bacchanalia of leaping improvisation before a fall into the earnest utilitarianism that has prevailed since. In *Unweaving the Rainbow* I quoted the following words of a distinguished biologist who may, by now, have thought better of it:

> Soon after multicelled forms were invented, a grand burst of evolutionary novelty thrust itself outward. One almost gets the sense of multicellular life gleefully trying out all its possible ramifications, in a kind of wild dance of heedless exploration.

If there is one animal, more than any other, that stands for this feverish vision of the Cambrian, it is *Hallucigenia*. Stands? Hallucinations apart, you might suspect that such an unlikely creature never stood in its life. And you would be right. It seems that *Hallucigenia* – and Simon Conway Morris chose its name advisedly – was originally reconstructed upside down. That is why it stands on improbably spiky toothpick stilts. Fossils discovered later, from China, suggest

* Bertrand Russell, of course.

Feverish vision of the Cambrian. *Hallucigenia* reconstructed upside down and back to front.

that the single row of 'tentacles' along the 'back' were legs. A single row of legs – did it balance as if on a tightrope? No, these fossils hinted at a second row, which were subsequently found on re-examination of the original specimen. More recent Chinese discoveries have filled out details of the head, which turns out to be the other end to the ghostly blob in the original reconstruction above (the blob is now thought to be material squeezed out of its backside). Modern reconstructions of *Hallucigenia*, such as the splendid drawing kindly created for us by David Legg (see overleaf), look as though they might just have been at home in the real world. It is no longer classified as a 'weird wonder' of uncertain and probably long-vanished affinities. Instead, together with many other Cambrian fossils, *Hallucigenia* is now tentatively placed in the phylum Lobopodia, which has modern representatives in the form of *Peripatus* and the other 'onychophorans' or 'velvet worms' whom we met at Rendezvous 26.

In the days when annelid worms were thought to be close relatives of arthropods, the Onychophora were often touted as 'intermediate' – 'bridging the gap' between them, although that is not an entirely helpful concept if you think carefully about how evolution works. The annelids are now placed in the Lophotrochozoa, while the Onychophora are ecdysozoans with the arthropods. *Peripatus*, with its ancient affinities, is well placed among modern pilgrims to tell the tale of the Cambrian Explosion.

The modern Onychophora (see plate 38) are widely distributed in

Hallucigenia. Modern reconstruction by Dr David Legg from the Museum of Natural History, Oxford.

the tropics and especially in the southern hemisphere. *Peripatus, Peripatopsis,* and all the modern onychophorans live on land, in leaf litter and humid places, where they hunt snails, worms, insects and other small prey. In the Cambrian, of course, *Hallucigenia* and the remote forebears of *Peripatus* and *Peripatopsis* lived – along with everybody else – in the sea.

Hallucigenia's connection with the modern Onychophora is still controversial, and we must remember what a lot of imagination necessarily intervenes between a blurred and squashed fossil in a rock, and the reconstruction that is eventually drawn, often in daring colour, on the page. Before the discovery of well-preserved relatives elsewhere, it was even proposed that *Hallucigenia* might not be a whole animal at all, but a part of some unknown animal. It would not be the first time such a mistake had been made. Some early artists' reconstructions of Cambrian scenes included a swimming jellyfish-like creature, seemingly inspired by tinned pineapple rings, which turned out to be part of the jaw apparatus of the mysterious predatory animal *Anomalocaris*, pictured opposite. Other Cambrian fossils, for example *Aysheaia*, certainly seem quite like marine versions of *Peripatus*, and this reinforces *Peripatus*'s entitlement to tell this Cambrian tale.

Most fossils, in any era, are the remains of hard parts of animals: vertebrate bones, the carapaces of arthropods or the shells of molluscs or brachiopods. But there are three Cambrian fossil beds – one in Canada, one in Greenland and one in China – where freak conditions, with almost miraculous good fortune for us, preserved soft parts as well. These are the Burgess Shale of British Columbia, Sirius Passet of northern Greenland, and the Chengjiang site of

Anomalocaris saron.

southern China.* The Burgess Shale was first discovered in 1909 and was made famous 80 years later by Stephen Gould in *Wonderful Life*. The Sirius Passet site in northern Greenland was discovered in 1984 but is so far less studied than the other two. In the same year, the Chengjiang fossils were discovered by Hou Xian-guang. Dr Hou is one of those who have collaborated on a beautifully illustrated monograph, *The Cambrian Fossils of Chengjiang, China*, published in 2004 – fortunately for me just before the first edition of this book went to press.

The Chengjiang fossils are now dated between 520 and 525 million years old. That's roughly contemporary with Sirius Passet, and some 10 or 15 million years older than the Burgess Shale, but these outstanding fossil sites have a similar fauna. There are lots of lobopods, many looking more or less like marine versions of *Peripatus*. There are algae, sponges, worms of various kinds, brachiopods looking pretty much like modern ones, and enigmatic animals of uncertain kinship. There are large numbers of arthropods, including crustaceans, trilobites and lots of others that loosely resemble crustaceans or trilobites but may have belonged in their own rather separate groups. The large (over a metre in some cases), apparently predatory *Anomalocaris* and its kind are found in Chengjiang as well as the Burgess Shale. Nobody is quite sure what they were – probably distant relations of the arthropods – but they must have been spectacular. Not all the 'weird wonders' of the Burgess Shale have been found at Chengjiang, for example *Opabinia*, with its famous five eyes.

The Sirius Passet fauna from Greenland includes a beautiful creature called *Halkieria*. It has been thought to be an early mollusc

---

* A fourth site, Orsten ('stink stone') in Sweden, preserves soft bodies in a different way.

Breaks down the mystical reverence for the great phlya. *Halkieria evangelista,* from Sirius Passet, Greenland, dating from the Lower Cambrian. Drawing from Simon Conway Morris.

but Simon Conway Morris, who has described many of the strange creatures of the Cambrian, believes it has affinities with three major phyla: molluscs, brachiopods and annelid worms. This gladdens my heart because it helps to break down the almost mystical reverence with which zoologists regard the great phyla. If we take our evolution seriously, it has to be the case that, as we go back in time and approach their rendezvous points, they will become more and more like each other, more and more closely related. Whether or not *Halkieria* fits the bill, it would be worrying if there were *not* an ancient animal that united annelids, brachiopods and molluscs. Note the shells, one at each end, in the illustrations above.

As we saw at Rendezvous 22, there are fossils that appear to be true vertebrates which predate even the amphioxus-like *Pikaia* of the Burgess Shale. Traditional zoological wisdom never had vertebrates

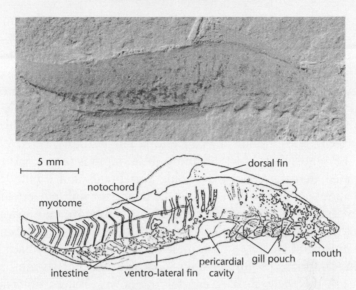

Vertebrates were not supposed to be this old. Fossil *Myllokunmingia fengjiaoa*, Chengjiang. From Shu *et al.* [388].

arising so early. Yet the Chengjiang beds preserve fossils that look pretty much like good jawless fish, such as had previously been thought not to arise until 50 million years later in the middle of the Ordovician. With more than 500 specimens now discovered, *Haikou-ichthys* (alas, not named after the Japanese verse form) is by far the best known. The other two similarly unpronounceable species, *Myllokunmingia* and *Zhongjianichthys*, are known only from single distorted specimens, and it has been argued that one or the other are simply badly formed Haikous. These controversial taxonomic changes are eloquent of how difficult it is to discern the details of very old fossils. Above is a photograph of an individual *Myllokun-mingia* fossil, together with a drawing of it made with a camera lucida. I find myself filled with admiration for the patience that goes into reconstructing ancient animals like these.

The pushing of the vertebrates back into the middle of the Cambrian only strengthens the idea of sudden explosion that is the basis of the myth. It really does appear that most of today's major animal phyla first appear as fossils in a narrow span within the Cambrian. This doesn't mean that there were no representatives of those phyla before the Cambrian. But they have mostly not fossilised. How should we interpret this? We can distinguish various combinations of three main hypotheses, rather like our three hypotheses for the explosion of the mammals after the extinction of the dinosaurs.

1. NO REAL EXPLOSION. On this view there was only an explosion of fossilisability, not of actual evolution. The phyla actually go back a long way before the Cambrian, with concestors spread out through hundreds of millions of years in the Precambrian. This view is supported by some molecular biologists who have used molecular clock techniques to date key concestors. For example, G. A. Wray, J. S. Levinton and L. H. Shapiro, in a famous paper of 1996, estimated that the concestor uniting vertebrates and echinoderms lived about a billion years ago, and the concestor uniting vertebrates and molluscs was 200 million years earlier still, more than twice the age of the so-called Cambrian Explosion. Molecular clock estimates have in general tended to push these deep branchings way back into the Precambrian, far further than most palaeontologists are happy with. On this view, fossils were, for unknown reasons, not readily formed before the Cambrian. Perhaps they lacked easily fossilisable hard parts, such as shells, carapaces and bones. After all, the Burgess Shale and the Chengjiang beds are extremely unusual, among all geological layers, in recording soft parts as fossils. Perhaps Precambrian animals, although long existing in a wide range of complex body plans, were simply too small to fossilise. In favour of this idea, there are some small animal phyla that have left no fossils at all after the Cambrian, until they appear today as live 'orphans'. Indeed, the large and beautiful class Turbellia, free-living flatworms, have left no fossils at all. They might have been born yesterday. Why then should we feel entitled to expect fossils before the Cambrian? The counter-argument is that not all animals fossilise poorly. For example, some sponges have hard internal skeletons (see Rendezvous 31). If they were around billions of years ago, we might expect to detect them. And wormy animals might be expected to leave tracks and burrows, especially if there are no other large animals to disturb the sediment. Yet ancient traces like these only stretch back a few tens of millions of years into the Precambrian.

2. MEDIUM-FUSE EXPLOSION. The concestors uniting the various phyla really did live reasonably close to each other in time, but still spread out over several tens of millions of years before the observed explosion of fossils. From the great distance of the present, Chengjiang at 520 million years seems at first sight rather close to a putative concestor at, say, 560 million. But a full 40 million years separates

them, more than 40 times longer than the species *Homo sapiens* has existed. Even 10 million years is a long time in the light of the extremely rapid evolutionary bursts of the Galapagos Finch's Tale and the Cichlid's Tale. It is all too easy, with hindsight, to think that because we recognise two ancient fossils as belonging to different modern phyla, those two fossils must have been as different from each other as modern representatives of the two phyla are. It is too easy to forget that the modern representatives have had half a billion years in which to diverge. There is no good reason to believe that a Cambrian taxonomist, blessedly free of 500 million years' worth of zoological hindsight, would have placed the two fossils in separate phyla. He might have placed them only in separate orders, notwithstanding the then unknowable fact that their descendants were destined eventually to diverge so far as to warrant separate phylum status.

3. OVERNIGHT EXPLOSION. This third school of thought is, in my opinion, bonkers. Or, to use more parliamentary language, wildly and irresponsibly unrealistic. But I must spend some time on it because it has recently become unaccountably popular, following the rhetoric I attributed to romantically overheated zoologists.

The third school believes that new phyla sprang into existence overnight, in a single macromutational leap. Here are some quotations I have used before, in *Unweaving the Rainbow*, from otherwise reputable scientists.

> It was as if the facility for making evolutionary leaps that produced major functional novelties – the basis of new phyla – had somehow been lost when the Cambrian period came to an end. It was as if the mainspring of evolution had lost some of its power . . . Hence, evolution in Cambrian organisms could take bigger leaps, including phylum-level leaps, while later on it would be more constrained, making only modest jumps, up to the class level.

Or this, from the same distinguished scientist from whom we heard at the beginning of the tale.

> Early on in the branching process, we find a variety of long-jump mutations that differ from the stem and from one another quite dramatically. These species have sufficient morphological differences to be categorized as founders of distinct phyla. These founders also branch, but do so via slightly closer long-jump variants, yielding branches from each founder of a phylum to dissimilar daughter

species, the founders of classes. As the process continues, fitter variants are found in progressively more nearby neighborhoods, so founders of orders, families, and genera emerge in succession.

Those preposterous quotations moved me to retort that it is as though a gardener looked at an old oak tree and remarked, wonder-ingly:

> Isn't it strange that no major new boughs have appeared on this tree for many years? These days, all the new growth appears to be at the twig level!

Here's another quotation, which this time I will attribute because it was published after *Unweaving the Rainbow* and I have therefore not used it before. Andrew Parker's *In the Blink of an Eye* is mainly concerned with advocating his interesting and original theory that the Cambrian Explosion was triggered by animals' sudden discovery of eyes. But before coming to his theory itself, Parker begins by falling hook, line and sinker for the 'wild and irresponsible' version of the Cambrian Explosion myth. He first expresses the myth itself in the most frankly 'explosive' version I have read:

> 544 million years ago there were indeed three animal phyla with their variety of external forms, but at 538 million years ago there were thirty-eight, the same number that exists today.

He goes on to make it clear that he is not talking about extremely rapid gradualistic evolution compressed into a period of 6 million years, which would be an extreme version of our Hypothesis Two, and just barely acceptable. Nor is he saying, as I would, that near the initial divergence of a pair of (what are destined to become) phyla, they would not have been very different – would, indeed, have passed through successive stages of being a pair of species, then genera, and so on until eventually their separation warranted recog-nition at the phylum level. No, Parker gives every appearance of regarding his 38 phyla, at 538 million years, as fully fledged phyla that sprang into existence overnight, at the drop of a macro-mutational hat:

> Thirty-eight animal phyla have evolved on Earth. So only thirty-eight monumental genetic events have taken place, resulting in thirty-eight different internal organisations.

Monumental genetic events are not utterly out of the question. Control genes of the various Hox families that we met in the Fruit Fly's Tale can certainly mutate in dramatic ways. But there's monumental and monumental. A fruit fly with a pair of legs where the antennae should be is about as monumental as it gets, and even then there is a big question mark over survival. There is a powerful general reason for this, which I shall briefly explain.

A mutant animal has a certain probability of being better off as a consequence of its new mutation. 'Better off' means better when compared to the premutated parental type. The parent must have been at least good enough to survive and reproduce, otherwise it wouldn't be a parent. It is easy to see that the smaller the mutation, the more likely it is to be an improvement. 'It is easy to see' was a favourite phrase of the great statistician and biologist R. A. Fisher, and he sometimes used it when it was anything but easy for ordinary mortals to see. In this particular case, however, I think it is genuinely easy to follow Fisher's argument for the case of a simple metric feature – something such as thigh length, which varies in one dimension: some number of millimetres that could grow larger or could grow smaller.

Imagine a set of mutations of increasing magnitude. At one extreme, a mutation of zero magnitude is by definition exactly as good as the parent's copy of the gene which, as we've seen, must have been at least good enough to survive childhood and reproduce. Now imagine a random mutation of small magnitude: the leg, say, gets one millimetre longer or one millimetre shorter. Assuming that the parental gene is not perfect, a mutation that is infinitesimally different from the parental version has a 50 per cent chance of being better and a 50 per cent chance of being worse: it'll be better if it is a step in the right direction, worse if it is a step in the opposite direction, relative to the parental condition. But a very large mutation will probably be worse than the parental version, *even if it is a step in the right direction*, because it will overshoot. To push to the extreme, imagine an otherwise normal man with thighs two metres long.

Fisher's argument was more general than this. When we are talking about macromutational leaps into new phylum territory, we are no longer dealing with simple metric characters like leg length, and we need another version of the argument. The essential point, as I have put it before, is that there are many more ways of being dead than of being alive. Imagine a mathematical landscape of all possible animals. I have to call it mathematical, because it is a landscape in hundreds of dimensions and it includes an almost infinitely large range of conceivable monstrosities, as well as the (relatively) small number of animals that have actually ever lived. What Parker calls a 'monumental genetic event' would be equivalent to a macromutation of huge effect, not just in one dimension as with our thigh example, but in hundreds of dimensions simultaneously. That is the scale of change we are talking about if we imagine, as Parker seems to, an abrupt and immediate change from one phylum to another.

In the multidimensional landscape of all possible animals, living creatures are islands of viability separated from other islands by gigantic oceans of grotesque deformity. Starting from any one island, you can evolve away from it one step at a time, here inching out a leg, there shaving the tip of a horn, or darkening a feather. Evolution is a trajectory through multidimensional space, in which every step of the way has to represent a body capable of surviving and reproducing about as well as the parental type reached by the preceding step of the trajectory. Given enough time, a sufficiently long trajectory leads from a viable starting point to a viable destination so remote that we recognise it as a different phylum, say, molluscs. And a different step-by-step trajectory from the same starting point can lead, through continuously viable intermediates, to another viable destination, which we recognise as yet another phylum, say, annelids. Something like this must have happened for each of the forks leading to each pair of animal phyla from their respective concestors.

The point we are leading up to is this. A random change of sufficient magnitude to initiate a new phylum at one fell swoop will be so large, in hundreds of dimensions simultaneously, that it would have to be preposterously lucky to land on another island of viability. Almost inevitably, a megamutation of that magnitude will land in the middle of the ocean of inviability: probably unrecognisable as an animal at all.

Creationists foolishly liken Darwinian natural selection to a hurricane blowing through a junkyard and having the luck to assemble a Boeing 747. They are wrong, of course, for they completely miss the gradual, cumulative nature of natural selection. But the junkyard metaphor is entirely apt to the hypothetical overnight invention of a new phylum. An evolutionary step of the same magnitude as, say, the overnight transition from earthworm to snail, really would have to be as lucky as the hurricane in the junkyard.

We can, then, with complete confidence, reject the third of our three hypotheses, the bonkers one. That leaves the other two, or some compromise between them, and here I find myself agnostic and eager for more data. As we shall see in the epilogue to this tale, it seems to be increasingly accepted that the earliest molecular clock estimates were exaggerating when they pushed the major branch points many hundreds of millions of years back into the Pre-cambrian. On the other hand, the mere fact that there are few, if any, fossils of most animal phyla before the Cambrian should not stampede us into assuming that those phyla must have evolved extremely rapidly. The hurricane in a junkyard argument tells us that all those Cambrian fossils must have had continuously evolving antecedents. Those antecedents had to be there, but they have not been discovered. Whatever the reason, and whatever the timescale, they failed to fossilise, but they must have been there. On the face of it, it is harder to believe that a whole lot of animals could be invisible for 100 million years than that they could be invisible for only 10 million years. This leads some people to prefer the short-fuse Camb-rian Explosion theory. On the other hand, the shorter you make the fuse, the harder it is to believe all that diversification could be crammed into the time available. So this argument cuts both ways and doesn't decisively choose between our two surviving hypotheses.

The fossil record is not completely void of metazoan life before Chengjiang and Sirius Passet. Around 20 million years earlier, almost plumb on the Cambrian/Precambrian boundary, start to appear a variety of microscopic fossils that superficially resemble tiny shells – together they are known as the 'small shelly fauna'. These seem to be allied to disparate animal phyla. The oldest, *Claudina*, might be a part of a segmented worm. Or perhaps the skeleton of a coral-like animal. Or maybe neither. Others are almost certainly fragments of the body armour from various sorts of bilaterians, particular molluscs and – pleasingly for this tale – lobopods, relatives of the velvet worm. That came as a surprise to most palaeontologists because it means the divergences between different groups of protostomes *must* have occurred in the Precambrian, before the visible 'explosion'. And armour plates imply that predators were already hunting in the earliest Cambrian seas.

There are hints of older animal diversity too. Twenty million years before the start of the Cambrian, in the Ediacaran Period of the late Precambrian, there was a worldwide flourishing of a mysterious group of animals called the Ediacaran fauna, named after the Ediacara Hills in South Australia where they were first found. The rough sandstone in which they are usually embedded makes it hard to know quite what most of them were, but they were among the first large animals to be fossilised. Some of them are probably sponges. Some are a bit like jellyfish. Others somewhat resemble sea anemones, or sea pens (feather-like relatives of sea anemones), and one of them, *Haootia*, even seems to have had muscles. Some, such as *Kimberella*, look a bit worm-like or slug-like, and could conceivably represent true Bilateria. Others are just plain mysterious. What are we to make of *Dickinsonia*, the weird wonder in plate 39? Is it a coral? Or a worm? Or a fungus? Or something completely different from anything that survives today? Until we find new fossils, or new techniques for probing the existing ones, it is generally thought that the Ediacarans, though intriguing, don't help us much one way or the other in deducing the ancestry of bilaterally symmetrical animals.

There are also fossil imprints that appear to be the trails and tunnels of Precambrian animals. These traces tell us of the early existence of large crawling or burrowing animals, probably true bilaterians. Unfortunately, they don't give us much detail about what those animals looked like. There are also some even older, mostly microscopic fossils found at Doushantuo in China which appear to be embryos, though it is not clear what kind of animal they might have grown into. Older still are small, disc-shaped impressions from Newfoundland, dated between about 600 and 610 million years ago, but these animals are, if anything, even more enigmatic than the Ediacaran forms.

This book is hung upon a series of 40 rendezvous points and it seemed desirable to make some sort of guess as to the date of each one. Most of the rendezvous points can now be dated with some confidence, using a combination of datable fossils and molecular clocks calibrated by datable fossils. Not surprisingly, the fossils start to let us down when we reach the older rendezvous points. This means that the molecular methods can no longer be reliably calibrated, and we enter a wilderness of undatability. For completeness I have forced myself to put some sort of date on these wilderness concestors, roughly Concestors 23 onwards. The most recently available evidence seems to me to favour, even if only slightly, a view closer to a medium-fuse explosion. This goes against my earlier bias in favour of no real explosion. When more evidence comes in, as I hope it will, I shall not be in the least surprised if we find ourselves pushed the other way again into the deep Precambrian in our quest for the concestors of modern animal phyla. Or we might be pulled back to an impressively short explosion, in which the concestors of most of the great animal phyla are compressed into a period of 20 or even 10 million years around the beginning of the Cambrian. In this case, my strong expectation would be that even if we correctly place two Cambrian animals in different phyla on the basis of their resemblance to modern animals, back in the Cambrian they would have been much closer to each other than the modern descendants of one are to the modern descendants of the other. Cambrian zoologists would not have placed them in separate phyla but only in, say, separate subclasses.

I wouldn't be surprised to see either of the first two hypotheses vindicated. I'm not sticking my neck out. But I'll eat my hat if any evidence is ever found in favour of Hypothesis Three. There is every reason to suppose that evolution in the Cambrian was essentially the same kind of process as evolution today. All that over-excited rhetoric about the mainspring of evolution running down after the Cambrian; all that euphoric shouting about wild, heedless dances of extravagant invention, with new phyla leaping into existence in a blissful dawn of zoological irresponsibility – now here's something I am prepared to stick my neck out for: all that stuff is just plain dotty.

I hasten to say I have nothing against prose-poetry on the Cambrian. But give me Richard Fortey's version, on page 120 of his beautiful book *Life: An Unauthorised Biography*:

I can imagine standing upon a Cambrian shore in the evening, much as I stood on the shore at Spitsbergen and wondered about the biography of life for the first time. The sea lapping at my feet would look and feel much the same. Where the sea meets the land there is a patch of slightly sticky, rounded stromatolite pillows, survivors from the vast groves of the Precambrian. The wind is whistling across the red plains behind me, where nothing visible lives, and I can feel the sharp sting of wind-blown sand on the back of my legs. But in the muddy sand at my feet I can see worm casts, little curled wiggles that look familiar. I can see trails of dimpled impressions left by the scuttling of crustacean-like animals . . . Apart from the whistle of the breeze and the crash and suck of the breakers, it is completely silent, and nothing cries in the wind . . .

EPILOGUE TO THE VELVET WORM'S TALE

For much of this book we have tossed rendezvous dates around with insouciance, and even been rash enough, when introducing many of the concestors, to stick a specific number of 'greats' before 'grand-parent'. Our dates have mostly been based upon fossils which, as we shall see in the Redwood's Tale, can be dated to a precision commensurate with the vast timescales involved. But fossils never helped us much with tracing the ancestry of soft-bodied animals such as flatworms. Coelacanths went missing from the record for the past 70 million years, which was why the discovery of a live one in 1938 was such an exhilarating surprise. The fossil record, even at the best of times, can be a fickle witness. And now, having reached the Cambrian Period, we are sadly running out of fossils. Whatever interpretation we place on 'explosion', everyone agrees that almost all the predecessors of the great Cambrian fauna have, for uncertain reasons, failed to fossilise. As we seek concestors that predate the Cambrian, we find no more help in the rocks. Fortunately, fossils are not our only recourse. In the Chimpanzee's Tale, the Elephant Bird's Tale, the Coelacanth's Tale and other places we have made use of the ingenious technique known as the molecular clock. The time has come to explain the molecular clock properly.

Wouldn't it be wonderful if measurable, or countable, evolutionary changes happened at a fixed rate? We could then use evolution itself as its own clock. And this needn't involve circular reasoning because we could calibrate the evolutionary clock on parts of evolution where the fossil record is good, then extrapolate to parts where it isn't. But how do we measure rates of evolution? And, even if we could measure them, why on earth should we expect that any aspect of evolutionary change should go at a fixed rate like a clock?

There is not the slightest hope that leg length, or brain size, or number of whiskers will evolve at a fixed rate. Such features are important for survival, and their rates of evolution will surely be hideously inconstant. As clocks they are doomed by the very principles of their own evolution. In any case, it is hard to imagine an agreed standard for measuring rates of visible evolution. Do you measure evolution of leg length in millimetres per million years, as percentage change per million years, or what? J. B. S. Haldane proposed a unit of evolutionary rate, the darwin, which is based upon proportional change per generation. Wherever it has been used on real fossils, results vary from millidarwins to kilo-darwins and megadarwins, and nobody is surprised.

Molecular change looks like a much more promising clock. First, because it is obvious what you must measure. Since DNA is textual information written in a four-letter alphabet, there is an entirely natural way to measure its rate of evolution. You just count letter differences. Or, if you prefer, you can go to the protein products of the DNA code and count substitutions of amino acids.* There are reasons to hope that the majority of evolutionary change at the molecular level is neutral rather than being steered by natural selection. Neutral is not the same as useless or functionless – it only means that different versions of the gene are equally good, therefore change from one to the other is not noticed by natural selection. This is good for the clock.

Contrary to my rather ludicrous reputation as an 'ultra-Darwinist' (a slander I would protest more vigorously if the name sounded less of a compliment than it does), I do not think that the majority of evolutionary change at the molecular level is favoured by natural selection. On the contrary, I have always had a lot of time for the so-called neutral theory associated with the great Japanese geneticist Motoo Kimura, or its extension, the 'nearly neutral' theory of his collaborator Tomoko Ohta. The real world has no interest in human tastes, of course, but as it happens I positively *want* such theories to be true. This is because they give us a separate, independent chronicle of evolution, unlinked to the visible features of the creatures around us, and they hold out the hope that some kind of molecular clock might really work.

Just in case the point is misunderstood, I must emphasise that the neutral theory does not in any way denigrate the importance of selection in nature. Natural selection is all-powerful with respect to those visible changes that affect survival and reproduction. Natural selection is the only explanation we know for the functional beauty and apparently 'designed' complexity of living things. But if there are any changes that have no visible effect – changes that pass right

---

* When the molecular clock was first proposed, by Emile Zuckerkandl and the great Linus Pauling, this was the only method available.

under natural selection's radar – they can accumulate in the gene pool with impunity and may supply just what we need for an evolutionary clock.

As ever, Charles Darwin was way ahead of his time with respect to neutral changes. In the first edition of *The Origin of Species*, near the beginning of Chapter 4, he wrote:

> This preservation of favourable variations and the rejection of injurious variations, I call natural selection. Variations neither useful nor injurious would not be affected by natural selection, and would be left a fluctuating element, as perhaps we see in the species called poly-morphic.

By the sixth and last edition, the second sentence had an even more modern-sounding addendum:

> . . . as perhaps we see in certain polymorphic species, or would ultim-ately become fixed . . .

'Fixed' is a genetic technical term and Darwin surely cannot have meant it in quite the modern sense, but it gives me a lovely lead-in to the next point. A new mutation, whose frequency in the population begins near zero by definition, is said to become 'fixed' when it has reached 100 per cent in the population. The rate of evolution that we seek to measure, for purposes of a molecular clock, is the rate at which a succession of mutations of the same genetic locus become fixed in the population. The obvious way for fixation to happen is if natural selection favours the new mutation over the previous 'wild type' allele, and therefore drives it to fixation – it becomes the norm, 'the one to beat'. But a new mutation can also go to fixation even if it is exactly as good as its predecessor – true neutrality. This is nothing to do with selection: it happens by sheer chance. You can simulate the process by tossing pennies, and can calculate the rate at which it will happen. Once a neutral mutation has drifted to 100 per cent, it will become the norm, the so-called 'wild type' at that locus, until another mutation has the luck to drift to fixation.

If there is a strong component of neutrality, we could potentially have a marvellous clock. Kimura himself wasn't particularly con-cerned with the molecular clock idea. But he believed – it now seems rightly – that the majority of mutations in DNA are indeed neutral – 'neither useful nor injurious'. And, in a remarkably neat and simple piece of algebra, which I shall not spell out here, he calculated that, if this is true, the rate at which genuinely neutral genes should 'ultimately become fixed' is exactly equal to the rate at which the variations are generated in the first place: the mutation rate.

You see how perfect this is for anybody who wants to date bifurcation ('rendezvous') points using a molecular clock. As long as the mutation rate at a neutral genetic locus remains constant over time, the fixation or 'substitution' rate will also be constant. You can now compare the same gene in two different animals, say a pangolin and a starfish, whose most recent common ancestor was Concestor 25. Count the number of letters by which the starfish gene differs from the pangolin gene. Assume that half the differences accumulated in the line leading from concestor to starfish, and the other half in the line leading from concestor to pangolin. That gives you the number of ticks of the clock since their genes separated, and (as discussed in the Epilogue to the Chimpanzee's Tale and at the end of the Bonobo's Tale) a reasonable estimate for Rendezvous 25.

But it isn't as simple as that, and the complications are interesting. First, if you listened to the ticking of the molecular clock, it would not be regular like a pendulum clock or a hairspring watch; it would sound like a Geiger counter near a radioactive source. Completely random! Each tick is the fixation of yet another mutation. Under the neutral theory, the interval between successive ticks could be long or it could be short, by chance – 'genetic drift'. In a Geiger counter, the timing of the next tick is unpredictable. But – and this is really important – the *average* interval over a large number of ticks is highly predictable. The hope is that the molecular clock is predictable in the same way as a Geiger counter, and in general this is true.

Second, the tick rate varies from gene to gene within a genome. This was noticed early, when geneticists could look only at the protein products of DNA, not DNA itself. Cytochrome-c evolves at its own characteristic rate, which is faster than histones but slower than globins, which in turn are slower than fibrinopeptides. In the same way, when a Geiger counter is exposed to a very slightly radioactive source such as a lump of granite, versus a highly radioactive source such as a lump of radium, the timing of the next tick is always unpredictable but the average rate of ticking is predictably and dramatically different as you move from granite to radium. Histones are like granite, ticking at a very slow rate; fibrinopeptides are like radium, buzzing like a dementedly randomised bee. Other proteins such as cytochrome-c (or rather the genes that make them) are intermediate. There is a spectrum of gene clocks, each running at its own speed, and each useful for different dating purposes, and for cross-checking with each other.

Why do different genes run at different speeds? What distinguishes 'granite' genes from 'radium' genes? Remember that neutral doesn't mean useless, it means equally good. Granite genes and radium genes are both useful. It is just that radium genes can change at many places along their length and still be useful. Because of the way a gene works, portions of its length can change with impunity without affecting its functioning. Other portions of the same gene are highly sensitive to mutation, and its functioning is devastated if these portions are hit by a mutation. Maybe all genes have a granite portion, which mustn't change much if the gene is to go on working, and a radium portion that can freewheel unchecked as long as the granite portion is not affected. Maybe the cytochrome-c gene has a mixture of granite bits and radium bits; fibrinopeptide genes have a higher proportion of radium bits, while histone genes have a higher proportion of granite bits. There are some problems, or at least complications, with this as an explanation for the differences in tick rates between genes. But what matters for us is that tick rates really do vary between genes, while the rate for any given gene is pretty constant even in widely separated species.

Not completely constant, however, and this brings us to our next problem, which is a serious one. Tick rates are not just vague and sloppy. For any given gene they can be systematically greater in some kinds of creatures than in others, and this introduces a real bias. Bacteria have a much less effective DNA-repair system than our DNA's sophisticated 'proofreading', so their genes mutate at a higher rate and their molecular clocks tick faster. Rodents, too, have slightly sloppy repair enzymes, which might explain why molecular evolution is faster in rodents than in other mammals. Major changes in evolution, like the move to 'warm blood', have the potential to change the mutation rate, which could play havoc with our clock estimates of branch dates. But more sophisticated methods are now available that allow for mutation rates to change in a heritable (so-called 'autocorrelated') manner along lineages, and where possible we have used these for the dating in this book.

Another problem might be anticipated because we expect mutations to happen when DNA is copied, when cells divide. Because animals tend to set aside reproductive cells early in development, vastly different animal species experience surprisingly similar numbers of cell divisions per generation. For example, to go from the egg cell in a mother back to the egg cell from which she was conceived we would count about 31 divisions in humans, 25 in mice and 37 in fruit flies. Such similar figures would make for a fine clock if these animals all had the same generation time. But of course they don't. Nowhere near. Those 30-odd cell divisions occur every 30 days in fruit flies but only every 30 years in humans*. On this argument, we would not expect the mutation rate to be constant per calendar year. To a very rough approximation, we might expect the molecular clock to count numbers of generations. Actually, however, when molecular biologists looked at rates of change in gene sequences, using lineages that happened to have a good fossil record for calibration, this isn't what they found. There really did seem to be a molecular clock that measured time in years, and not in generations. This was convenient, but how to explain it?

One suggestion was that, even though the reproductive turnover in elephants is slow compared to fruit flies, during all the years between reproductive events elephant genes are subject to the same bombardment of cosmic rays and other events that can cause mutation as fruit fly genes. Admittedly, fruit fly genes are hopping into a new fly once a fortnight, but why should cosmic rays care about that? Well, genes sitting in one elephant for ten years are being hit by the same number of cosmic rays as genes hopping through a succession of 250 fruit flies during the same period. There may be something in this theory, but it probably isn't a sufficient explanation. It really is true that most mutations occur when DNA is copied, so we seem to need another explanation for the molecular clock's apparent ability to tell the time in years rather than in generations.

Here's where Kimura's colleague Tomoko Ohta made a clever contribution: her *nearly* neutral theory. Kimura, as I said, calculated from his fully neutral theory that the rate of fixation of neutral genes should equal the mutation rate. This remarkably simple conclusion depended on an elegant piece of 'cancelling out' in the algebra. And

---

* Compared to egg cells, the number of cell divisions in sperm is more variable and shows a slight correlation with calendar time. The male line sees, for example, 40 divisions per generation in fruit flies versus perhaps ten times that in humans. Nevertheless, this is still dwarfed by the difference in generation times. Incidentally, the fact that men's sperm undergoes over ten times as many rounds of cell division than women's eggs means that most human mutations trace back to fathers, more specifically to their testicles, where the majority of germ cell division takes place.

the quantity that cancelled out was the population size. Population size comes into the equation, but it ends up both above and below the line, so it conveniently vanishes in a puff of mathematical smoke, and fixation rate emerges as equal to mutation rate. But *only* if the genes concerned really are completely neutral. Ohta revisited Kimura's algebra but she allowed her mutations to be nearly neutral instead of completely neutral. And this made all the difference. Population size no longer cancelled out.

This is because – as has long been calculated by mathematical geneticists – in a large population, slightly harmful genes are more likely to be eliminated by natural selection before they have a chance to drift to fixation. In a small population, luck is more likely to carry a slightly harmful gene to fixation before natural selection 'notices'. To push to the extreme, imagine a population almost entirely wiped out by some catastrophe, with only half a dozen individuals remaining. It would not be very surprising if, by chance, all six happened to have the slightly deleterious gene. In that case, we have fixation – 100 per cent of the population. That's an extreme, but the mathematics shows the same effect more generally. Small populations favour the drifting to fixation of genes that would be eliminated in a large population.

So, as Ohta pointed out, population size no longer cancels out of the algebra. On the contrary, it stays in just the right place to do the molecular clock theory a bit of good. Now, back to our elephants and fruit flies. Large animals with long life cycles, such as elephants, also tend to have small populations. Small animals with short life cycles, such as fruit flies, tend to have large populations. This isn't just a vague effect, it is a pretty lawful one, and it holds for reasons that are not hard to imagine. So even if fruit flies have short generation times which would tend to speed the clock up, they also have a large population, which slows it down again. Elephants may have a slow clock as far as mutations are concerned, but their small populations speed the clock up again in the fixation department.

Professor Ohta has evidence that truly neutral mutations, as in junk DNA or in 'synonymous' substitutions* seem to tick in generation time as opposed to real time: creatures with short generation times show accelerated DNA evolution if you measure it in real time. Conversely, mutations that actually change something, and therefore

---

* The DNA code being 'degenerate', any one amino acid can be specified by more than one 'synonymous' mutation. A mutational change resulting in an exact synonym makes no difference at all to the final outcome.

fall foul of natural selection, tick away more or less constantly in real time.

Whatever the theoretical reason, it does seem to be the case in practice that, with known exceptions that we can usually allow for (by carefully choosing our clock genes, and avoiding species with exceptional rates of mutation), the molecular clock has proved itself a workable instrument. To use it, we need to draw the evolutionary tree that relates the set of species we are interested in, and estimate the amount of evolutionary change in each lineage. This is not as simple as just counting differences between the genes of two modern species and dividing by two. We need to use the advanced tree-building techniques of maximum likelihood and especially Bayesian phylogenetics that we met in the Gibbon's Tale. Anchored with robust fossil dates for calibration and where possible a measure of uncertainty in those dates, we can then make a good guess at the dates of branching points on the tree.

Molecular clock dates, especially those based on carefully chosen DNA, multiple fossil calibrations, and a good range of species within the groups of interest, are becoming more and more accepted by a wide range of biologists. They have been used successfully across the tree of life, and in most cases they agree with the more limited set of dates that can be deduced from the fossil record. And we have relied extensively on molecular clock dates at every stage in this book so far.

Complacency alert! Listen to those alarm bells ringing.

Molecular clocks ultimately depend on calibration by fossils. Radiometrically calculated dates for fossils are accepted with the respect that biology rightly bestows upon physics (see the Redwood's Tale). One strategically located fossil that confidently places a lower bound on the dating of an important evolutionary branch point can be used to calibrate a whole lot of molecular clocks dotted around the genomes of a range of animals dotted around the phyla. But when we get back to Precambrian territory where the supply of fossils gives out, we have to depend on relatively young fossils to calibrate great-great-grandfather clocks that are then used to estimate much older dates. And that spells trouble.

A commonly chosen date for the junction point between mammals and sauropsids (birds, crocodiles, snakes, etc.) is 310 million years ago. Until relatively recently, this one date provided the master calibration for many molecular clock datings of much older branch points. Now, any date estimate has a certain margin of error, and in their scientific papers scientists try to remember to place 'error bars' on each of their estimates. A date is quoted plus or minus, say, 10 million years (and indeed it now seems as though the mammal/sauropsid date should be more like 320 Ma). Such errors are fine when the dates we seek with the molecular clock are in the same ballpark as the fossil dates used to calibrate it. When there is a great mismatch between ballparks, the error bars can grow alarmingly. The implication of a wide error bar is that if you tweak some small assumption, or slightly alter some small number that you feed into the calculation, the impact on the final result could be dramatic. Not plus or minus 10 million years but plus or minus half a billion years, say. Wide error bars mean that the estimated date is not robust against measurement error.

In the Velvet Worm's Tale itself, we saw various molecular clock estimates that placed important branch points in the deep Precambrian, for example 1,200 million years for the split between vertebrates and molluscs. More recent studies, using sophisticated techniques that allow for possible variations in mutation rates, bring the estimates down to dates in the 600- to 700-million-year range: a dramatic shortening – accommodated in the error bars of the original estimate, but that is small consolation.

Although I am a firm supporter of the molecular clock idea in general, I think its estimates of very early branch points need to be treated with caution. Extrapolating backwards from a 320-million-year-old calibration fossil to a rendezvous point more than twice as old is fraught with danger. For example, it is possible that the rate of molecular evolution in the vertebrates (which enters into our calibration calculation) is not typical of the rest of life. As we saw in the Lamprey's Tale, vertebrates are thought to have undergone two rounds of doubling of their entire genome. The sudden creation of large numbers of duplicate genes may affect the selection pressure on nearly neutral mutations. Some scientists (I am not one of them, as I have already made clear) believe that the Cambrian marked a great shift in the whole process of evolution. If they are right, the molecular clock would need a radical recalibration before it should be let loose in the Precambrian.

In general, as we go back further in time and the supply of fossils peters out, we enter a realm of almost complete conjecture. Nevertheless, I am hopeful of future studies. The dazzling fossils of the Chengjiang and similar formations may greatly extend the range of calibration points into regions of the animal kingdom hitherto off limits.

Meanwhile, recognising that we are wandering in an ancient wilderness of conjecture, we have adopted the following rough strategy in trying to estimate dates beyond the sponges at Rendezvous 31, where reliable fossils become hard to find. We have provisionally placed Rendezvous 35, the junction of animals and fungi, at 1,200 million years ago. This is a compromise between some studies that suggest figures of 900–1,000 million and others that suggest 1,300 or even 1,600 million. The rendezvous date is nevertheless old enough to mesh well with *Bangiomorpha*, a fossilised red alga of approximately the same age. We then space out Concestors 32 to 37 roughly in the ratios indicated by molecular clock studies. However, if we have got Rendezvous 35 badly wrong, then our dates from here on in the pilgrimage could be overestimates by many tens or even hundreds of millions of years. Please bear this in mind as we enter that wilderness of undatability. I am so unconfident of dates in this vicinity that from now on I shall give up the already rather quaint conceit of estimating the number of 'greats' to put before grandparent. That number will soon be getting into the billions. The order of joinings at the successive rendezvous points is more certain, but even that could be wrong too.

ACOELOMORPH FLATWORMS

When we were talking about the protostomes, descendants of Concestor 26, I grouped the flatworms, Platyhelminthes, firmly within them. But now we have an interesting little complication. Recent evidence quite strongly suggests that the Platyhelminthes are a fiction. I'm not saying flatworms themselves don't exist, of course. But they are a heterogeneous collection of worms who should not be united under one name. Most of them are true protostomes and we met them at Rendezvous 26, but a few of them, the acoelomorphs, are quite separate and belong somewhere else. Where exactly they belong is still a matter of dispute, but many authorities place them as 'basal bilaterians', in other words, here at Rendezvous 27. This we are dating at 565 million years ago, although out in these remote reaches of geological time these datings become more and more uncertain.

What all flatworms have in common, apart from their eponymous flatness, is that they lack an anus and they lack a coelom. The coelom of a typical animal, such as you or me or an earthworm, is the body cavity. This doesn't mean the gut: the gut, though a cavity, is topologically part of the outside world, the body being a topological doughnut, the hole in the middle of the ring being the mouth, the anus and the gut that connects them. The coelom, by contrast, is the closed cavity within the body in which the intestines, the lungs, heart, kidneys and so on all sit. Flatworms don't have a coelom. Instead of a body cavity in which the guts slop about, their guts and other internal organs are embedded in solid tissue called parenchyma. This may seem a trivial distinction, but the coelom is embryologically defined and lies deep in the collective unconscious of zoologists.

Lacking an anus, how do flatworms expel wastes? Through the mouth if there is nowhere else. The gut may be a simple sac or, in larger flatworms, it branches into a complicated system of blind alleys, like the air tubes in our lungs. Our lungs, too, could theoretically have had an 'anus' – a separate hole for the air to leave by, with its waste carbon dioxide. Fish sort of do the equivalent, for their respiratory stream of water enters by one hole, the mouth, and leaves by others, the gill apertures. But our lungs are tidal, and so is the

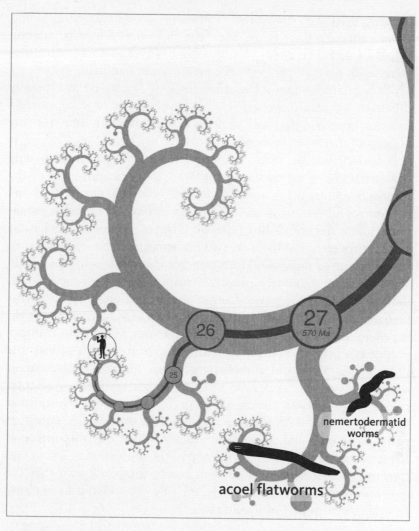

Acoelomorph flatworms join. The vast majority of bilaterally symmetrical animals are deuterostomes or protosomes (including the platyhelminth 'flatworms'). However, some molecular data hives off a pair of flatworm groups as neither protostomes nor deuterostomes but earlier branching lineages. These are the classes Acoela (about 320 described species) and Nemertodermatida (ten described species), together known as the acoelomorph flatworms. Current evidence hints that the Acoela and the Nemertodermatida are sister groups, as shown here. That the acoelomorphs do not belong with the other flatworms is now generally accepted. Nevertheless, controversy still remains, because they have seen much molecular evolution and so are subject to the 'long branch attraction' effect discussed in the Gibbon's Tale. Certain geneticists have recently claimed that they lie within *Xenoturbella* in the deuterostomes (but perhaps this deuterostome affinity is another red herring).

digestive system of flatworms. Flatworms lack lungs or gills and breathe through their skins. They also lack a system of circulating blood, so their branched gut presumably serves to transport nutriment to all parts of the body. There are some free-living flatworms which are large – tens of centimetres long – and often elegantly coloured. These are more likely to have a long or exceptionally complex branched gut, and some have reinvented an anus (or lots of anuses) after a long absence.

Because flatworms lack a coelom and mostly lack an anus, they have always been regarded as primitive – the most primitive of the bilaterally symmetrical animals. It was always assumed that the ancestor of all deuterostomes and protostomes was probably something like a flatworm. But now, as I began by saying, molecular evidence suggests that there are two unconnected kinds of flatworms, and only one of these two kinds stands a chance of being genuinely primitive. The genuinely primitive suspects are the acoelomorph flatworms, comprising two groups: the Acoela and the Nemertodermatida. The Acoela are named for their lack of coelom which, for them and the Nemertodermatida but not the Platyhelminthes proper, is a primitive lack. The main group of flatworms proper, the flukes, tapeworms and turbellarians, seem to have evolved their simplicity later, and may have secondarily lost their anus and their coelom. They passed through a stage of being more like normal Lophotrochozoa, then reverted to being like their earlier ancestors again, *sans* anus and *sans* coelom. They joined our pilgrimage at Rendezvous 26, along with all the rest of the protostomes. I won't go into the detailed evidence, but will accept the conclusion that the Acoela and the Nemertodermatida are different and join us as a tiny incoming stream here at Rendezvous 27.

At this point I should describe these tiny worms that are joining us but, though I hate to say it, at least by comparison with most of the wonders we have seen, there is not a lot to describe. They live in the sea and they not only lack a coelom but lack a proper lined gut too. In the species that have a mouth, food ends up being temporarily crammed into spaces between their body cells, or sometimes engulfed directly into their central tissues. This situation is viable only in animals that are very small, which they are.

Some of them supplement their diet by giving house room to plants, and hence benefiting indirectly from their photosynthesis. Members of the genus *Waminoa* have symbiotic dinoflagellates (unicellular algae) and live off their photosynthesis. Another acoel, *Symsagittifera*, has a similar relationship with a single-celled green alga, *Tetraselmis convolutae*. Symbiotic algae presumably make it possible for these little worms to be less little. The worms seem to take steps to make life easier for their algae, and hence themselves, crowding at the surface to give them as much light as possible. Their colloquial name of 'mint sauce worm' is an apt description – Peter Holland writes to me that *Symsagittifera roscoffensis*

> . . . are amazing animals to see in their natural habitat. They appear as a green 'slime' on certain beaches in Brittany, the slime really being thousands of acoels plus their endosymbiotic algae. And as you creep up on the 'slime' it hides! (by disappearing into the sand). Very strange to see.

Unlike Acoels, their sister group the Nematodermatida have the unusual feature of keeping their nervous system on their surface. Indeed, there is an intriguing hypothesis, due to Nicholas Holland (no relation) that animal brains originally evolved in the skin.

The acoelomorphs are still with us, and therefore must be treated as modern animals, but their form and simplicity suggest that they might not be greatly changed since the time of Concestor 27. Modern acoel worms might be a reasonable approximation to the ancestor of all bilaterally symmetrical animals.

Our gathered pilgrims now include all the phyla recognised as Bilateria, which means the great bulk of the animal kingdom. The name refers to their bilateral symmetry, and is intended to exclude the two main radially symmetrical phyla, previously grouped together as the 'Radiata', who will join our pilgrimage presently: the Cnidaria (sea anemones, corals, jellyfish, etc.) and the Ctenophora (comb jellies). Unfortunately for this simple terminology, starfish and their kin, which zoologists are sure are descended from Bilateria, are also radially symmetrical, at least in the adult phase. Echinoderms are assumed to have become secondarily radial when they took to a bottom-living existence. They have bilaterally symmetrical larvae, and are not closely related to the 'truly' radiate animals such as jellyfish. Reflexively, not all the 'Radiata' are (quite) radially symmetrical. Ctenophores show a combination of bilateral and radial symmetry (known as 'biradial'), and as we shall see at Rendezvous 29 they include a few worm-like species. Recently, Peter Holland's group has even identified a worm-like cnidarian, which tends to support those zoologists who suspect that all these creatures may have had bilaterally symmetrical ancestors.

All in all, Bilateria is an unfortunate name by which to unite the descendants of Concestor 27 and separate them from those pilgrims still to join. Another possible criterion is 'triploblasty' (three layers of cells) versus 'diploblasty' (two). At a crucial stage in their embryology, cnidarians and ctenophores build their bodies out of two main layers of cells ('ectoderm' and 'endoderm'), the Bilateria out of three (they add 'mesoderm' in the middle). Even this is open to dispute, however. Some zoologists believe 'Radiata' also have mesodermal cells. I think the sensible thing is not to worry about whether Bilateria and Radiata are really good words to use, nor diploblastic and triploblastic, but just concentrate on who are the next pilgrims to join.

Even this is subject to dispute. Nobody doubts that the cnidarians are a unitary group of pilgrims who all join up with each other more recently than they join anyone else. And nobody doubts the same of the ctenophores. The question is, in what order do they join each other and join us? A number of possibilities have been supported and, as we shall see, some molecular taxonomists even think, against all zoological intuition, that ctenophores join our pilgrimage further back in time than the sponges. To make matters worse, there is a tiny phylum, the Placozoa, containing only a single genus, *Trichoplax*, and nobody knows for sure where to put *Trichoplax*. I shall follow the school of thought that says the cnidarians are the first to join us at Rendezvous 28, then the ctenophores at Rendezvous 29, then *Trichoplax* at Rendezvous 30, and the sponges, the most distantly related to all other animals, joining at Rendezvous 31. But be warned that the placement of these groups, especially *Trichoplax* and the ctenophores, could change, requiring a renumbering of Rendezvous 28 to 31.

CNIDARIANS

Our pilgrim band of worms and their diverse descendants has now swelled to very large numbers, and we all pass on back to Rendezvous 28 where we are joined by the cnidarians (the c is silent). They include the freshwater hydras and the more familiar marine sea anemones, corals and jellyfish, all very different from worms. Unlike the Bilateria, they are radially symmetrical about a central mouth. They have no obvious head, no front or rear, no left or right, only an up or down.

What is the date of the rendezvous? Well, who knows? There are only a few much older fossils which can be relied upon for molecular clock calibrations, so dates from this technique are enormously variable, ranging from 650 to 1,000 million years ago. Direct evidence comes from 560-million-year-old Ediacaran fossils, some of which have been interpreted as cnidarians,* and from the microscopic Doushantuo embryos, roughly 585 million years old, which may be cnidarian larvae. But, like most Precambrian evidence, these are controversial. Our inclination is to fix a date for the oldest animal rendezvous, when we meet the sponges. There are variably disputed fossil sponge remains between 600 and 630 million years ago, and sponge-like chemical traces from 645, which lead us to place a date of 650 Ma for Rendezvous 31. Relative genetic distances then place the current rendezvous approximately 600 million years ago – but many geneticists would disagree by hundreds of millions of years.

Because they are among our most distant animal cousins (some were once even confused with plants), the cnidarians are often regarded as very primitive. Of course this doesn't follow – they have had the same time to evolve since Concestor 28 as we have. But it is true that they lack many of the features that we regard as advanced in an animal. They have no long-distance sense organs, their nervous system is a diffuse network, not urbanised into brain, ganglia or major nerve trunks, and their digestive organ is a single, usually

---

* Most recently by Latha Menon, the excellent and tireless editor of the first edition of this book.

uncomplicated cavity with only one opening, the mouth, which also does duty as anus.

On the other hand, there aren't many animals who could claim to have redrawn the map of the world. Cnidarians make islands: islands you can live on; islands big enough to need, and accommodate, an airport. The Great Barrier Reef is more than 2,000 kilometres long. It was Charles Darwin himself who worked out how such coral reefs are formed, as we shall see in the Polypifer's Tale. Cnidarians also include the most dangerously venomous animals in the world, the extreme example being the box jellyfish, which oblige prudent Australian bathers to wear nylon bodystockings. The weapon cnidarians use is remarkable for various reasons, in addition to its formidable power. Unlike a snake's fangs, or the sting of a scorpion or a hornet, the jellyfish sting emerges from inside a cell as a miniature harpoon. Well, thousands of cells, called cnidocytes (or sometimes

nematocysts, but this is strictly just one variety of cnidocyte), each with its own cell-sized harpoon called a cnida. *Knide* is Greek for nettle, and it gives the Cnidaria their name. No others are as dangerous to us as box jellyfish, and many are not even painful. When you touch the tentacles of a sea anemone, the 'sticky' feeling on your finger is the clutch of hundreds of tiny harpoons, each on the end of its own little thread, which attaches it to the anemone.

The cnidarian harpoon is probably the most complicated piece of apparatus inside any cell anywhere in the animal or plant kingdoms. In the resting state, waiting to be launched, the harpoon is a coiled tube inside the cell, under pressure (osmotic pressure, if you want the details) waiting to be released. The hair trigger is indeed a tiny hair, the cnidocil, projecting outwards from the cell. When triggered,

Probably the most complicated piece of apparatus inside any cell. Cross-section of a cnidarian harpoon.

the cell bursts open, and the pressure turns the entire coiled mechanism inside out with great force, shooting into the body of the victim and injecting poison. Once triggered in this way, the harpoon

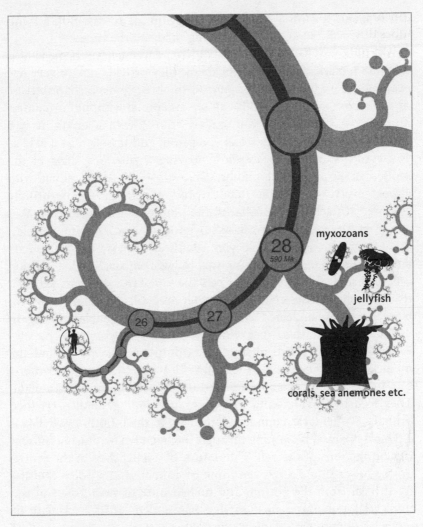

Cnidarians join. The order of branching of the cnidarians (jellyfish, corals, sea anemones and the like) and the ctenophores (comb jellies) is effectively unresolved. Most authors position either one or the other (or sometimes both) as the closest living relative of the bilaterally symmetrical animals. Certain molecular data hints that cnidarians may occupy this position. Unfortunately, the branching order of subgroups within the 9,000 or so cnidarian species is also disputed, but the deep and fundamental division between lineages with an evolved medusa stage in their life cycle (jellyfish) and the others (including corals and sea anemones) is widely accepted. Molecular and morphological data also suggests that a group of almost-unicellular parasites called myxozoans are actually highly derived cnidarians, related to the jellyfish, as shown here.

cell is spent. It cannot be charged up again for re-use. But, as with most kinds of cell, new ones are being made all the time.

All cnidarians have cnidae, and only cnidarians have them. That is the next remarkable thing about them: they provide one of very few examples of an utterly unambiguous, single diagnostic characteristic of any major animal group. If you see an animal without any cnidae, it is not a cnidarian. If you see an animal with a cnida, it is a cnidarian. Actually, there is one exception, and it is as neat a case as you could want of an exception proving a rule. Sea slugs of the molluscan group called nudibranchs (they joined us along with almost everybody else at Rendezvous 26) often have beautifully coloured tentacles on their backs, the kind of coloration that makes would-be predators back off. With good reason. In some species, these tentacles contain cnidocytes, identical to those of true cnidarians. But only Cnidaria are supposed to have cnidae, so what is going on? As I said, the exception proves the rule. The slug eats jellyfish, from which it passes cnidocytes, intact and still working, to its own tentacles. Commandeered weapons, they are still capable of firing, in defence of the sea slug – hence the bright warning coloration.

Cnidarians have two alternative body plans: the polyp and the medusa. A sea anemone or a *Hydra* is a typical polyp: sedentary, mouth uppermost, the opposite end fixed to the ground like a plant. They feed by waving tentacles about, harpooning small prey, then bringing the tentacle, complete with prey, to the mouth. A jellyfish is a typical medusa, swimming through the open sea by pulsing muscular contractions of the bell. The mouth of a jellyfish is in the centre, on the lower side. So you can think of a medusa as a polyp that has freed itself from the bottom and turned over to swim. Or you can think of a polyp as a medusa that has settled on its back with its tentacles uppermost. Many species of cnidarian have both polypoid and medusoid forms, alternating them through the life cycle, a bit like caterpillar and butterfly.

Polyps often reproduce by budding vegetatively, like plants. A new baby polyp grows on the side of a freshwater *Hydra*, eventually breaking off as a separate individual: a clone of the parent. Many marine relatives of *Hydra* do something similar, but the clone doesn't break off and assume a separate existence. It stays attached and becomes a branch, as in a plant. These 'colonial hydrozoans' branch and branch again, making it easy for us to understand why they were thought to be plants. Sometimes more than one kind of polyp grows on the same polyp tree, specialised for different roles, such as feeding, defence, or reproduction. You can think of them as a colony of polyps, but there is a sense in which they are all parts of one individual, for the tree is a clone: all the polyps have the same genes. Food caught by one polyp may be used by others, since their gastric cavities are all continuous. The branches of the tree and its main trunk are hollow tubes that you can think of as a shared stomach – or maybe as a kind of circulatory system playing the role that in us is played by blood vessels. Some of the polyps bud off tiny medusae, which swim away like miniature jellyfish to reproduce sexually and disperse the genes of the parent polyp tree to distant places.

A group of cnidarians called the siphonophores have taken the colonial habit to an extreme. We can think of them as polyp trees which, instead of being fixed to a rock or a piece of seaweed, hang down either from one or a cluster of swimming medusae (which are, of course, members of the clone) or to a float at the surface. The Portuguese man-of-war *Physalia* has a large gas-filled float with a vertical sail on top. A complicated colony of polyps and tentacles dangles beneath. It doesn't swim but gets about through being blown by the wind. The smaller *Velella* is a flat, oval raft with a diagonally placed vertical sail. It too uses the wind to disperse, and its English names are Jack-sails-by-the-wind or by-the-wind-sailor. You often find the dried-up little rafts with their sails on the beach, where they usually lose their blue colour and seem to be made of whitish plastic. *Velella* resembles the true Portuguese man-of-war in that both sail by the wind. *Velella* and its relative *Porpita* are not siphonophore colonies, however, but single, highly modified polyps, hanging down from a float rather than sticking up from a rock (see plate 40).

Many siphonophores can adjust their depth in the water, rather as bony fish do with their swim bladders, by secreting gas into the float, or releasing it. Some have a combination of floats and swimming medusae, and all have polyps and tentacles dangling beneath. The siphonophores are treated by E. O. Wilson, founder of the science of sociobiology, as one of the four pinnacles of social evolution (the others being the social insects, the social mammals and ourselves). This, then, is another superlative that one can attach to the Cnidaria. Except that, since the members of a colony are clones, genetically identical to each other, it is by no means clear that we should call them a colony rather than a single individual.

Hydrozoans see the medusa as a way for their genes to hop

occasionally from one stable living place to another. Jellyfish could be said to take the medusoid form seriously, as what living is all about. Corals, by contrast, take sedentary living to the extreme lengths of building a hard, solid house that is destined to stay there for thousands of years. We shall take their tales in order.

THE JELLYFISH'S TALE

Jellyfish ride the ocean currents as Jack-sails-by-the-wind. They don't pursue their prey, as a barracuda or squid might. Instead, they rely on their long, trailing, armed tentacles to trap planktonic creatures that are unlucky enough to bump into them. Jellyfish do swim, with the languorous heartbeat of the bell, but they are not swimming in any particular direction, at least as we would understand direction. Our understanding, however, is limited by our two dimensional trammels: we crawl over the surface of the land, and even when we take off into the third dimension it is only in order to move a bit faster in the other two. But in the sea, the third dimension is the most salient. It is the dimension in which travelling has the most effect. In addition to the steep pressure gradient with depth, there is a light gradient, complicated by a gradient of colour balance. But the light disappears anyway as day gives way to night. As we shall see, a planktonic animal's preferred depth changes dramatically with the 24-hour cycle.

During the Second World War, sonar operators looking for submarines were puzzled by what seemed to be a false bottom of the sea that rose towards the surface every evening, and sank back down again the next morning. It turned out to be the bulk of the plankton, millions of tiny crustaceans and other creatures, rising to feed near the surface at night, then sinking at morning. Why should they do this? The best guess seems to be that during the hours of daylight they are vulnerable to visually hunting predators such as fish and squids, so they seek the dark safety of the depths by day. Why, then, come to the surface at night, for it is a long journey that must consume a lot of energy? One student of the plankton has compared it to a human daily walking 25 miles each way, just to get breakfast.

The reason for visiting the surface is that food ultimately comes from the sun, via plants. The surface layers of the sea are unbroken green prairies, with microscopic single-celled algae in the role of waving grass. The surface is where the food ultimately is, and that is where the grazers, and those that feed on the grazers, and those that in turn feed on them, must be. But if it is safe to be there only by night because of visually hunting predators, a diurnal migration is exactly what the grazers and their small predators must undertake. And apparently they do. The 'prairie' itself doesn't migrate. If there were any sense in doing so, it should swim against the animal tide, for its whole *raison d'etre* is to catch sunlight at the surface during the day, and avoid being eaten.

Whatever the reason, most of the animals in the plankton migrate down for the day and up for the night. The jellyfish, or many of them, follow the herds, like lions and hyenas tracking the wildebeest across the Mara and Serengeti plains. Although, unlike lions and hyenas, jellyfish don't target individual prey, even blindly trailing tentacles will benefit by following the herds, and this is one of the reasons jellyfish swim. Some species increase their catch rate by zigzagging about, again not individually targeting prey, but increasing the area swept by those tentacles with their batteries of lethal harpoons. Others just migrate up and down.

A different kind of migration has been described for the massed jellyfish of 'Jellyfish Lake' on Mercherchar, one of the Palau Islands (an American colony in the western Pacific). The lake, which communicates underground with the sea and is therefore salty, is named after its huge population of jellyfish. There are several kinds, but the dominant one is *Mastigias*, an estimated 20 million of them in a lake 2.5 kilometres long and 1.5 kilometres wide. All the jellyfish spend the night near the western end of the lake. When the sun rises in the east, they all swim straight towards it and therefore the eastern end of the lake. They stop before they reach the shore, for an interestingly simple reason. The trees fringing the shore cast a deep shadow, cutting off so much of the sun's light that the jellyfish's sun-seeking automatic pilot starts to drive them towards the now brighter west. As soon as they come out from the trees' shadow, however, they turn east again.

This internal conflict traps them around the line of the shadow, with the consequence (which I dare not think is more than co-incidence) of keeping them a safe distance from the dangerously predatory sea anemones – also cnidarians of course – that line the shore itself. In the afternoon, the jellyfish follow the sun back to the western end of the lake, where the whole armada again becomes trapped at the shadow line of the trees (see plate 41). When it becomes dark, they swim vertically up and down at the western end of the lake, until the dawn sun lures their automatic guidance system back towards the east. I don't know what they might gain from this remarkable twice-daily migration. The published explanation satisfies me too little to bear repetition. For now, the lesson of the tale must be that the living world offers much that we don't *yet* understand, and that is exciting in itself.

THE POLYPIFER'S TALE

All evolving creatures track changes in the world: changes in the weather, in temperature, rainfall and – more complicated because they hit back in evolutionary time – changes in other evolving lines such as predators and prey. Some evolving creatures alter, by their very presence, the world in which they live, and to which they must adapt. The oxygen we breathe was not there before photosynthetic creatures put it there. At first a poison, it provided radically changed conditions that most animal lineages were forced first to tolerate, and then to depend upon. On a shorter timescale, the trees in a mature forest inhabit a world that they themselves have created, over hundreds of years – the time it takes to transform bare sand into climax forest. A climax forest is, of course, also a complex and rich environment to which huge numbers of other plant and animal species have become adapted.

Because the word 'coral' is used both for an organism and for the hard material that it builds, I shall indulge a fancy and adopt from Darwin the older word 'polypifer' for the coral organism that tells this tale. Pronounce it 'pol-lip-if-er', with the stress on lip. Coral organisms, or polypifers, transform their world, over a timespan of hundreds of thousands of years, by building on the dead skeletons of their own past generations to construct huge underwater mountains: wave-resisting ramparts. Before they die, corals combine with countless other corals to condition the world in which future corals will live. And not just future corals, but future generations of an enormous and intricate community of animals and plants. The idea of community will be the main message of this tale.

The picture reproduced in plate 42 shows Heron Island, the one island of the Great Barrier Reef that I have visited (twice). The houses dotted around the near end of the little island give an idea of scale. The huge pale area surrounding the island itself is the reef, of which the island is just the highest tip, covered with sand made of crushed coral (much of it having passed through fish guts) in which vegetation of limited variety grows, supporting a similarly limited fauna of land animals. For objects that are entirely made by living creatures, coral reefs are big, and core drillings show some of them to be many hundreds of metres deep. Heron Island is just one of the more than 1,000 islands and nearly 3,000 reefs that constitute the Great Barrier Reef, arcing round the north-east side of Australia for 2,000 kilometres. The Great Barrier Reef is often said – with what veracity I don't know – to be the only evidence of life on our planet that is large enough to be visible from outer space. It is also said to be home to 30 per cent of the world's sea creatures, but again I am not sure quite what that means – what is being counted? Never mind, the Great Barrier Reef is an utterly remarkable object, and it has been entirely built by the small sea anemone-like animals called corals or polypifers. The living polypifers occupy only the surface layers of a coral reef. Beneath them, to a depth of hundreds of metres in some oceanic atolls, are the skeletons of their predecessors, compacted to limestone.

Nowadays only corals build reefs, but in earlier geological eras they had no such monopoly. Reefs have at various times been built by algae, sponges, molluscs and tube worms too. The great success of coral organisms themselves seems to stem from their association with microscopic algae, which live inside their cells and photosynthesise in the sunlit shallows, to the eventual benefit of the corals. These algae, called zooxanthellae, have a variety of different coloured pigments for trapping light, which accounts for the vividly photogenic appearance of coral reefs. It is not surprising that corals were once thought to be plants. They get much of their food in the same way as plants, and they compete for light as plants do. It is only to be expected that they would take on similar shapes. Moreover, their struggles to overshadow, and not be overshadowed, lead to the whole community of corals taking on something of the appearance of a forest canopy. And, like any forest, a coral reef is also home to a large community of other creatures.

Coral reefs hugely increase the 'ecospace' of an area. As my late colleague Richard Southwood put it in his book *The Story of Life*:

> Where there would otherwise be a surface of rock or sand with a column of water above it, the reef provides a complex three-dimensional structure with a great amount of extra surface, with many cracks and small caves.

Forests do the same kind of thing, inflating the effective surface area available for biological activity and colonisation. Increased ecospace is the kind of thing we expect to find in complex ecological communities. Coral reefs are home to a huge variety of animals of all kinds, nestling in every corner and nook of the prodigious ecospace provided.

Something similar happens in the organs of a body. The human brain increases its effective area – and hence its functional capacity – by elaborate folding. It may be no accident that the 'brain coral' so strikingly resembles it.

Darwin himself was the first to understand how coral reefs are formed. His debut scientific book (after his travel book on the *Voyage of the Beagle*) was the treatise on *Coral Reefs* that he published when only 33. Here is Darwin's problem as we would see it today, although he did not have access to most of the information that is relevant either to posing the problem or solving it. Darwin, indeed, was as astoundingly prescient in his theory of coral reefs as he was to be in his more famous theories of natural selection and sexual selection.

Corals can live only in shallow water. They depend upon the algae in their cells, and the algae of course need light. Shallow water is also good for the planktonic prey with which corals supplement their diet. Corals are denizens of shorelines, and you can indeed find shallow 'fringing reefs' around tropical coasts. But what is puzzling about corals is that you can also find them surrounded by very deep water. Oceanic coral islands are the summits of lofty underwater mountains made by generations of dead corals. Barrier reefs are an intermediate category, following the line of a coast, but farther out than fringing reefs, and with deeper water between them and the shore. Even in the case of remote coral islands completely isolated in the deep ocean, the living corals are always in shallow water, close to the light where they and their algae can thrive. But the water is shallow only by courtesy of the generations of earlier corals on which they sit.

Darwin, as I say, didn't have all the information needed to realise the extent of the problem. It is only because people have drilled down into reefs, and found compacted coral to great depths, that we now know that coral atolls are the summits of towering underwater mountains made of ancient coral. In Darwin's time the prevailing theory was that atolls were superficial encrustations of coral on top of submerged volcanoes that lay only just below the surface. On this theory there was no problem to solve. Corals grew only in shallow water, and it was the volcanoes that gave them the perch they needed to find shallow water. But Darwin didn't believe it, even though he had no way of knowing the dead coral was so deep.

Darwin's second feat of prescience was his theory itself. He suggested that the sea bottom was continually subsiding in the vicinity of the atoll (while rising in other places, as he vividly knew from finding marine fossils high in the Andes). This was, of course, long before the theory of plate tectonics. Darwin was inspired by his mentor, the geologist Charles Lyell, who believed that parts of the Earth's crust rose and sank relative to one another. Darwin proposed that as the sea bottom subsided, it took the coral mountain down with it. Corals grew on top of the subsiding undersea mountain, just keeping pace with the subsidence in such a way that the summit was always near the surface of the sea, in the zone of light and prosperity. The mountain itself was just layer upon layer of dead corals which had once prospered in the sun. The oldest corals, at the base of the underwater mountain, probably began as a fringing reef of some forgotten piece of land or long-dead volcano. As the land gradually submerged beneath the water, the corals later became a barrier reef, at an increasing distance away from the receding coastline. With further subsidence the original land disappeared altogether, and the barrier reef became the basis for a prolonged extension of the under-water mountain for as long as the subsidence continued. Remote oceanic coral islands got their start perched on the top of volcanoes, the base of which slowly subsided in the same way. Darwin's idea is still substantially supported today, with the addition of plate tectonics to explain the subsidence.

A coral reef is a textbook example of a climax community, and this shall be the climax of the Polypifer's Tale. A community is a collection of species that have evolved to flourish in each other's presence. A rainforest is a community. So is a bog. So is a coral reef. Sometimes the same kind of community springs up in parallel in different parts of the world where the climate favours it. 'Mediterra-nean' communities have arisen not just around the Mediterranean Sea itself, but on the coasts of California, Chile, south-western Australia and the Cape region of South Africa. The particular species of plants found in these five regions are different, but the plant communities themselves are as characteristically 'Mediterranean' as, say, Tokyo and Los Angeles are recognisably 'urban sprawl'. And an equally characteristic fauna goes with the Mediterranean vegetation.

Tropical reef communities are like that. They vary in detail but are the same in essentials, whether we are talking about the South Pacific, the Indian Ocean, the Red Sea or the Caribbean. There are also temperate-zone reefs, which are somewhat different, but one very particular thing the two have in common is the remarkable phenomenon of cleaner fish – a wonder which epitomises the sort of subtle intimacy that can arise in a climax ecological community.

A number of species of small fish, and some shrimps, ply a prosperous trade, harvesting nutritious parasites, or mucus, off the surfaces of larger fish, and in some cases even entering their mouths, picking their teeth and exiting through the gills. This argues for an astonishing level of 'trust',* but here my interest is more focused on cleaner fish as an example of a 'role' in a community. Individual cleaners typically have a so-called 'cleaning station', to which larger fish come to be serviced. The advantage of this to both parties is presumably a saving of time that might otherwise be spent searching for a cleaner or searching for a client. Site-tenacity also allows repeated meetings between individual cleaners and clients, which allows the all-important 'trust' to build up. These cleaning stations have been compared to barbers' shops (see plate 43). It has been claimed – though the evidence has more recently been disputed – that if all the cleaners are removed from a reef, the general health of the fish on the reef nosedives.

In different parts of the world, the local cleaners have evolved independently, and are drawn from different groups of fish. On the Caribbean reefs, the cleaner trade is mostly filled by members of the goby family, which typically form small groups of cleaners. In the Pacific, on the other hand, the best-known cleaner is a wrasse, *Labroides. L. dimidiatus* runs its 'barber's shop' by day, while *L. bicolor,* so George Barlow, my colleague from Berkeley days told me, services the nocturnal guild of fishes who take refuge in caves during the day. Such divvying-up of a trade among species is typical of a mature ecological community. Professor Barlow's book, *The Cichlid Fishes,* gives examples of freshwater species in the great lakes of Africa that have taken convergent steps towards the cleaner habit.

---

* The evolutionary problems of evolving 'trust' are interesting, but I have already dealt with the matter in *The Selfish Gene,* so must refrain from repeating myself here.

On tropical coral reefs, the almost fantastic levels of co-operation achieved between cleaner fish and 'client' is symbolic of the way an ecological community can sometimes simulate the intricate harmony of a single organism. Indeed, the resemblance is seductive – too seductive. Herbivores depend on plants; carnivores depend on herbivores; without predation, population sizes would spiral out of control with disastrous results for all; without scavengers like burying beetles and bacteria, the world would sate with corpses, and manure would never be recycled into the plants. Without particular 'key-stone' species, whose identity is sometimes quite surprising, the whole community would 'collapse'. It is tempting to see each species as an organ in the super-organism that is the community.

To describe the forests of the world as its 'lungs' does no harm, and it might do some good if it encourages people to preserve them. But the rhetoric of holistic harmony can degenerate into a kind of dotty, Prince Charles-style mysticism. Indeed, the idea of a mystical 'balance of nature' often appeals to the same kind of airheads who go to quack doctors to 'balance their energy fields'. But there are profound differences between the way the organs of a body and the species of a community interact with each other in their respective domains to produce the appearance of a harmonious whole.

The parallel must be treated with great caution. Yet it is not completely without foundation. There is an ecology within the individual organism, a community of genes in the gene pool of a species. The forces that produce harmony among the parts of an organism's body are not wholly unlike the forces that produce the illusion of harmony in the species of a coral reef. There is balance in a rainforest, structure in a reef community, an elegant meshing of parts that recalls co-adaptation within an animal body. In neither case is the balanced unit favoured *as a unit* by Darwinian selection. In both cases the balance comes about through selection at a lower level. Selection doesn't favour a harmonious whole. Instead, harmonious parts flourish in the presence of each other, and the illusion of a harmonious whole emerges.

Carnivores flourish in the presence of herbivores, and herbivores flourish in the presence of plants. But what about the other way around? Do plants flourish in the presence of herbivores? Do herbivores flourish in the presence of carnivores? Do animals and plants need enemies to eat them in order to flourish? Not in the straightforward way suggested by the rhetoric of some ecological activists. No creature normally benefits from being eaten. But grasses that can withstand being cropped better than rival plants really do flourish in the presence of grazers – on the principle of 'my enemy's enemy'. And something like the same story might be told of victims of parasites – and predators, although here the story is more complicated. It is still misleading to say that a community 'needs' its parasites and predators like a polar bear needs its liver or its teeth. But the 'enemy's enemy' principle does lead to something like the same result. It can be right to see a community of species, such as a coral reef, as a kind of balanced entity that is potentially threatened by removal of its parts.

This idea of community, as made up of lower-level units that flourish in the presence of each other, pervades life. Even within the single cell, the principle applies. Most animal cells house communities of bacteria so comprehensively integrated into the smooth working of the cell that their bacterial origins have only recently become understood. Mitochondria, once free-living bacteria, are as essential to the workings of our cells as our cells are to them. Their genes have flourished in the presence of ours, as ours have flourished in the presence of theirs. Plant cells by themselves are incapable of photosynthesis. That chemical wizardry is performed by guest workers, originally bacteria, now re-labelled chloroplasts. Plant eaters, such as ruminants and termites, are themselves largely incapable of digesting cellulose. But they are good at finding and chewing plants (see the Mixotrich's Tale). The gap in the market offered by their plant-filled guts is exploited by symbiotic micro-organisms that possess the biochemical expertise necessary to digest plant material efficiently. Creatures with complementary skills flourish in each other's presence.

What I want to add to that familiar point is that the process is mirrored at the level of every species' 'own' genes. The entire genome of a polar bear or a penguin, of a caiman or a guanaco, is an ecological community of genes that flourish in each other's presence. The immediate arena of this flourishing is the interior of an individual's cells. But the long-term arena is the gene pool of the species. Given sexual reproduction, the gene pool is the habitat of every gene as it is recopied and recombined down the generations.

CTENOPHORES

The ctenophores, who join us at Rendezvous 29, are some of the most beautiful of all the animal pilgrims. They are colloquially known in English as comb jellies, and at first glance they do look a bit like jellyfish. Both used to be placed in the same phylum, which was known as the Coelenterata, celebrating their shared characteristic, the fact that the main body cavity is also the digestive chamber. Likewise, they have a simple nerve net, and their bodies are built from (disputably) only two layers of tissue.

Ctenophore in Greek means 'comb-bearer'. The 'combs' are prominent rows of hair-like cilia, whose beating normally propels these delicate creatures in place of the pulsating muscles that do the same for the superficially similar jellyfish. It is not a fast system of propulsion, but it presumably serves adequately, not least because (unlike jellyfish but like, say, a shark) it allows ctenophores to swim mouth-first. There aren't many species of them – only about 100 – but the total number of individuals is not small, and they beautify, by any standards, all the oceans of the world. Waves of synchronised motion pass up the comb rows in eerie iridescence (see plate 44).

Belying their delicate appearance, ctenophores are voracious predators, and they have evolved a surprising diversity of methods. A few species have huge mouths, the width of their entire body, with which they can perform the seemingly impossible feat of engulfing other ctenophores that are actually larger than themselves. Some of these monsters have even independently invented teeth by gluing thousands of cilia together into hard spikes, which gives them the additional possibility of biting chunks off their prey. To add to this alien weirdness, they can zip their mouths open and shut by breaking or forming cellular connections between their 'lips'.

Other ctenophore species tend to trap prey using long dangling tentacles but, unlike jellyfish, these lack cnidocytes. Instead, they have their own make of 'lasso cells', which discharge a kind of glue instead of sharp, poisonous harpoons. They also include species which are far from bell-shaped. The ravishingly beautiful *Cestum veneris* is one of those rare animals whose English and Latin names mean exactly

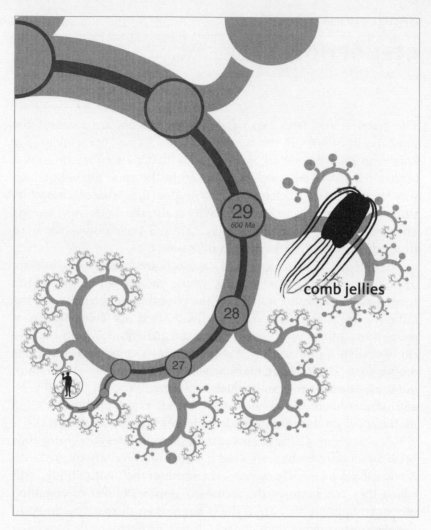

comb jellies

Ctenophores join. The bilaterally symmetrical animals, together with the cnidarians and ctenophores, are sometimes referred to as the 'Eumetazoa'. Following early molecular studies, the 100 known species of ctenophore are placed as the most distant relatives of the rest, but whole-genome sequencing implies that they may actually be the outgroup to all animals, including sponges, which would place them at Rendezvous 31 instead. This is still to be accepted by the majority of zoologists: we reflect a more traditional position.

the same thing, Venus's girdle, and no wonder: the body is a long, shimmering, ethereally beautiful ribbon, too good for a goddess (see plate 45). Notice that, although the Venus's girdle is long and thin like a worm, and even swims by undulating its body, the ends of the 'worm' are not its head and tail. Instead, it is mirrored about the middle, where the mouth is – the 'buckle' of the girdle. It is still radially (or strictly biradially) symmetrical.

Perhaps we should see the ctenophores as a sort of alternative experiment in animal forms, an idea which is reinforced by the fascinating Platyctenida ('flat ctenophores'), which really have evolved a proper wormy body, and quite recently at that. Rather than moving by waving cilia like other ctenophores, they crawl on the sea bottom, using the inside of their mouth as a muscular foot, hence most have lost their diagnostic lines of combs. No wonder that they are often mistaken for marine flatworms, although they are given away by the streaming sticky tentacles that emerge from their 'back'.

Nowadays everyone agrees that ctenophores are not jellyfish, but there is no consensus on what they are. At the time of writing this is one of the hot topics in animal evolution, as a result of a few ctenophore genome sequences becoming available for study. Various analyses of their DNA have kicked off a storm among students of evolutionary developmental biology, because they suggest that ctenophores are the most distant relatives to all other animals, including the sponges. This would mean that instead of joining the other pilgrims here, the ctenophores join further back in time than the sponges, at Rendezvous 31. There has been understandable scepticism at this revolutionary suggestion, which would mean either that ctenophores have independently invented muscle, nerves, cell layers, symmetrical embryology and jelly-filled bodies, or that sponges have lost them all.

You might have thought that having a set of entire DNA sequences would have settled the matter, but ctenophore genomes appear to be rather reduced in size, making comparisons slightly less obvious. For example, it is quite possible that they have lost a number of key genes such as the Hox cluster. Another problem is that, unlike jellyfish, whose two major lineages diverged very deep in time, living ctenophores are all rather closely related to each other. That means a single 'long branch' separates the ctenophores from the other animals, exposing them to the hazard of long branch attraction, the pitfall that we discussed in the Gibbon's Tale.

If we can sequence more species, or develop better genome-wide analytical techniques, or (even better) find earlier-diverging ctenophore groups, we may be able to say decisively where these pilgrims belong on the tree of life. Until then, we are placing ctenophores here as sister to the rest of the 'complex animals' – by which we mean all of them apart from the sponges and the puzzling little blob that we are about to meet.

PLACOZOANS

Here is an enigmatic little animal: *Trichoplax adhaerens*, the only species described in its entire phylum, the Placozoa – which, of course, doesn't necessarily mean it is the only one. I should mention that in 1896 a second placozoan from the Gulf of Naples was described and named *Treptoplax reptans*. It has never been found again, however, and most experts think that particular specimen was *Trichoplax* itself. Molecular evidence from the lab of Bernd Schierwater, the acknowledged authority on *Trichoplax*, implies perhaps 200 genetically distinct lineages exist, but their lack of distinguishing features has hindered conventional taxonomy.

Trichoplax lives in the sea and doesn't look like anything much, not symmetrical in any direction; a bit like an amoeba, except that it is made of lots of cells instead of just one; a bit like a very small flatworm, except that it doesn't have any obvious front or back end, nor left or right side. A tiny, irregularly shaped mat, perhaps two millimetres across, *Trichoplax* creeps over the surface on a little upside-down carpet of beating cilia. It feeds on single-celled creatures, mostly algae, even smaller than itself, which it digests through its lower surface without taking them inside its body. It almost certainly has sex, but no one has yet observed its full life cycle.

There is not much in its anatomy to connect *Trichoplax* with any other kind of animal. It has two main cell layers, like a cnidarian or a ctenophore. Sandwiched between the two main layers are a few contractile cells that work as its nearest approach to muscles. The animal shortens these strings to change its shape. Strictly the two main cell layers should probably not be called dorsal and ventral. The upper layer is sometimes called protective and the lower layer digestive. Some authors claim that the digestive layer invaginates to form a temporary cavity for digestive purposes, but not all observers have seen this and it may not be true. Careful study has revealed that its body is composed of six different cell types, rather than four as previously thought. Nevertheless, that is still rather paltry in comparison to over 200 in us.

Trichoplax has had a somewhat confused history in the zoological

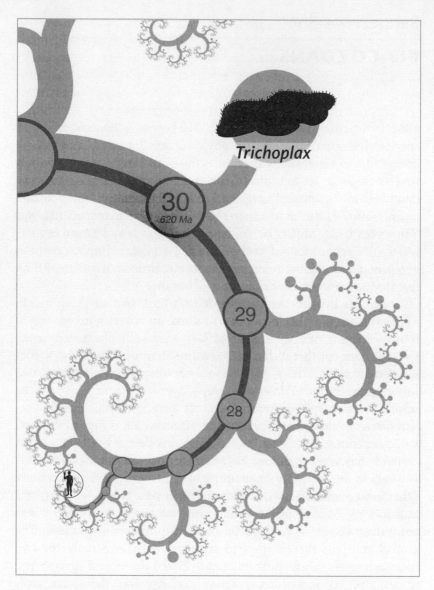

Trichoplax

Placozoans join. As with Rendezvous 28 and 29, the order of Rendezvous 30 and 31 is pretty well unresolved. Rendezvous 30 could be with the placozoans (represented by what is currently treated as a single species, *Trichoplax*), or it could be with the sponges. At present this ordering is essentially arbitrary. It would be entirely unsurprising if Rendezvous 30 and 31 had to swap or merge.

literature, as is recounted by Schierwater and his student Tareq Syed in a 2002 paper. When first described in 1883, *Trichoplax* was thought to be very primitive; it has now recovered that honoured status. Unfortunately, it bears a superficial resemblance to the so-called planula larva of some cnidarians. In 1907 a German zoologist called Thilo Krumbach thought he saw *Trichoplax* where he had previously seen planula larvae, and he regarded the little creatures as modified planulae. That wouldn't have mattered too much but for the death in 1922 of W. Kükenthal, the editor of the authoritative multi-volume *Handbuch der Zoologie*. Unluckily for *Trichoplax*, Kükenthal's under-study as editor was the very same Thilo Krumbach. *Trichoplax* was duly billed as a cnidarian in Kükenthal & Krumbach, and this was copied by the French equivalent, the *Traité de Zoologie*, edited by P. P. Grassé (who, incidentally, retained anti-Darwinian sympathies long after he should have known better). It was also picked up from the *Handbuch* by Libbie Henrietta Hyman, author of the leading American multi-volume work *Invertebrates*.

With such a weight of multi-volumed authority bearing down upon it, what chance had poor little *Trichoplax*, especially given that nobody had looked at the animal itself for more than half a century? It languished as an alleged cnidarian larva until the molecular revolution opened up the possibility of discovering its real affinities. Whatever else it is, it is definitely not a cnidarian. In 2008, full genome sequencing found that *Trichoplax* is a more distant relative than that. Nevertheless it seems to be closer to us than the sponges, hence our guess at Rendezvous 30, something that may change in the future. Given that sponges have a more complex structure than *Trichoplax*, it is quite likely that Concestor 30 was also more complex. In other words, the microscopic animal that we see today is the result of secondary simplification. That is borne out by its genome. *Trichoplax* has about 11,000 genes, fewer than some sponges (18,000) and ourselves (20,000), and there are hints that this is because it has lost many genes over evolutionary time. The homeobox genes, discussed at length in the Fruit Fly's Tale, provide a good example. *Trichoplax* has a single ParaHox gene, identified not only through direct similarity, but also because the surrounding genes are similar to those that surround our own six ParaHox genes. *Trichoplax* also contains the approximate arrangement of genes that surround true Hox genes in most animals, but it seems to have lost the actual Hox cluster itself. To invert the metaphor, the smile of the Cheshire Cat has vanished, but a fair bit of the body remains as a clue to its old location. This 'ghost Hox' observation illustrates another point: most of the genes that remain not only show similarities to other animal genes, they are in very roughly the same order along the genome. *Trichoplax* has a fairly small genome for an animal (about 30 times shorter than our own), but, unlike many other animals with tiny genomes, it has not actually rearranged its DNA that much. In the long term, that should allow comparison of its genome on a grander scale, and finally reveal what this strange little creature really is.

SPONGES

Sponges are the last pilgrims to join us who are members of the Metazoa, the truly multicellular animals. Sponges haven't always been dignified as Metazoa, but were written off as 'Parazoa' – a name for a kind of second-class citizen of the animal kingdom. Nowadays the same class distinction is fostered by placing the sponges in the Metazoa, but coining the word Eumetazoa ('true' Metazoa) for all the rest *except* sponges.*

People are occasionally surprised to learn that sponges are animals rather than plants. Like plants, they don't appear to move. Well, there is movement at the cellular level, but that is true of plants too. Neither plants nor sponges have muscles, so it was rather surprising when a study in 2014 produced a video of sponges 'sneezing': contracting their whole body in a co-ordinated movement which seems to have evolved to clear detritus from within. The reflex is useful because sponges make their living by passing a ceaseless current of water right through the body, from which they filter food particles. Consequently, they are full of holes, which is what makes them so good at holding water in the bath.

Bath sponges, however, don't give a good idea of the typical body form, which is a hollow pitcher with a big opening at the top and lots of smaller holes all round the sides. As is easy to tell by putting a little dye in the water outside the pitcher of a living sponge, water is drawn in through the small holes around the sides, and expelled into the main hollow interior, from which it flows out through the main entrance of the pitcher. The water is driven by special cells called choanocytes, which line the chambers and canals of the walls of a sponge. Each choanocyte has a waving flagellum (like a cilium, only larger) surrounded by a deep collar of tiny hairs. It has been suggested that they are the evolutionary precursors to certain cells in other animals, notably those involved in sense organs. Choanocytes also provide an important clue to the origin of animals, as we shall see towards the end of this chapter.

* Some authors also except *Trichoplax*, the little animal we met at Rendezvous 30.

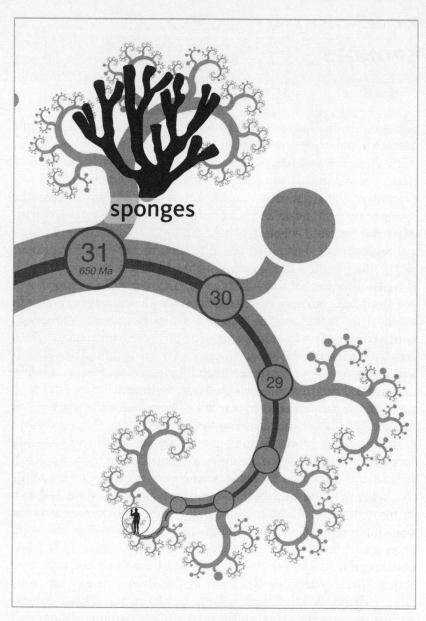

sponges

31
650 Ma

30

29

Sponges join. Since the time of Linnaeus, the animals ('metazoans') have been classified as one of the kingdoms of life. It is usually assumed that there was a very early divide between the approximately 10,000 described species of sponge and the other animals, something that has been confirmed by molecular data. However, a minority of molecular taxonomists think that there are two lineages of sponges, one more closely related to the rest of the Metazoa than the other – this implies that the earliest metazoans really did look like sponges and would have been classified as such.

Although their cells can sense and communicate well enough to co-ordinate a sneeze, sponges have no nervous system and possess only a relatively simple internal structure. They have several different kinds of cells, but those cells don't organise themselves into tissues and organs the way ours do. Sponge cells are 'toti potent', which means that every cell is capable of becoming any of the sponge's repertoire of cell types. This is not true of our cells. A liver cell is not capable of giving rise to a kidney cell or a nerve cell. But sponge cells are so flexible that any isolated cell is capable of growing a whole new sponge (and there's more to it, as we shall see in the Sponge's Tale).

Unsurprisingly, therefore, sponges make no distinction between 'germ line' and 'soma'. In the Eumetazoa, germ-line cells are those that are capable of giving rise to reproductive cells and whose genes are therefore in principle immortal. The germ line is a small minority of cells residing in ovaries or testes, and insulated from the need to do anything else but reproduce. Soma is that part of the body that is not germ line – somatic cells are destined not to pass their genes on indefinitely. In a eumetazoan such as a mammal, a subset of cells is set aside early in embryology as germ line. The rest of the cells, the cells of the soma, may divide a few times to make liver or kidney, bone or muscle, but then their dividing career comes to an end.

Cancer cells are the sinister exception. They have somehow lost the ability to stop dividing. But as Randolph Nesse and George C. Williams, the authors of *The Science of Darwinian Medicine*, point out, we should not be surprised. On the contrary, the surprising thing about cancer is that it is not more common than it is. Every cell in the body, after all, is descended from an unbroken line of billions of generations of germ-line cells that have not stopped dividing. Suddenly being asked to become a somatic cell like a liver cell, and learn the art of *not* dividing, has never happened before in the entire history of the cell's ancestors! Don't be confused. Of course the bodies that housed the cell's ancestors had livers. But germ-line cells – by definition – are not descended from liver cells.

All sponge cells are germ-line cells – all potentially immortal. They have several different cell types, but they are deployed in development in a different way from most multicellular animals. Eumetazoan embryos form cell layers that fold and invaginate in complicated 'origami' ways to build the body. Sponges don't have that kind of embryology. Instead they self-assemble – each of their toti-potent cells has an affinity for hooking up to other cells, as though they were autonomous protozoa with sociable tendencies. Nevertheless, modern zoologists include the sponges as members of the Metazoa, and we follow this trend. They are probably the most primitive living group of multicellular animals, giving us a better idea of the early Metazoa than any other modern animals. Indeed, our reconstruction of Concestor 31 (see plate 46) looks very like the larval stage of some modern sponges. In fact, a minority of molecular taxonomists find evidence that certain classes of sponge are more closely related to us than others, which would not only add another rendezvous point, but also imply that the ancestral animal *was* a sponge.

As with other animals, each species of sponge has its own characteristic shape and colour. The hollow pitcher is only one of many forms. Others are variants on it, systems of hollow cavities connected to one another. Sponges characteristically toughen their structures with collagen fibres (that's what makes bath sponges spongy) and with mineral spicules: crystals of silica or calcium carbonate, the shape of which is often the most reliable diagnostic of the species. Sometimes the spicule skeleton can be intricate and beautiful, as in the glass sponge, *Euplectella* (see plate 47).

Considering their hard spicule skeleton, the fossil record of sponges is surprisingly poor, and the topic of much debate. Some researchers argue that definitive evidence for sponges is only found after the Cambrian explosion. Others claim to have discovered primitive sponges in rocks 600 or even 640 million years old. Following these claims, we place Rendezvous 31 at roughly 650 million years ago. Even this figure, however, is considerably more recent than most molecular clock estimates for Rendezvous 31, and the usual despairing warnings for such ancient datings apply.

It is tempting to link the appearance of sponges, indeed animals in general, with the most extreme episodes of worldwide glaciation ever known, which happened around this time and have led geologists to name this geological period the Cryogenian. At least two such episodes are thought to have occurred – one from 717 to 660 million years ago, and another from 640 to 635 million years ago – and there is ongoing debate about whether the ice entirely covered land and sea, in what has been nicknamed 'Snowball Earth'. Even if not in full deep freeze, the effect on the Earth and its inhabitants must have been both drastic and prolonged. Nevertheless, it is probably no more than coincidence that our estimated rendezvous date falls right in the middle of the interglacial interval.

Whenever and wherever it took place, the evolution of multicellular sponges from single-celled protozoa is one of the landmark events in evolution – the Origin of the Metazoa – and we shall examine it in the next two tales.

THE SPONGE'S TALE

The 1907 issue of the *Journal of Experimental Zoology* contains a paper on sponges by H. V. Wilson of the University of North Carolina. The research was classic, and the paper describing it recalls a golden age when scientific papers were written in a discursive style that you could understand, and at a length that made it possible to visualise a real person doing real experiments in a real laboratory.

Wilson took a living sponge and separated its cells by forcing them through a fine sieve – a piece of 'bolting cloth'. The disassembled cells were passed into a saucer of sea water, where they formed a red cloud, mostly consisting of single cells. The cloud settled down into a sediment at the bottom of the saucer, where Wilson observed them with his microscope. The cells behaved like individual amoebas, crawling over the bottom of the saucer. When these amoeboid crawlers met others of their kind, they joined up to form growing agglomerations of cells. Eventually, as Wilson and others showed in a series of papers, such agglomerations grow to become whole new sponges. Wilson also tried mashing up sponges of two different species and mixing the two suspensions together. The two species were of different colours, so he could easily see what happened. The cells chose to agglomerate with their own species and not the other. Oddly, Wilson reported this result as a 'failure', since he was hoping – for reasons I don't understand, and which perhaps reflect the different theoretical preconceptions of a zoologist of nearly a century ago – that they would form a composite sponge of two different species.

Sociable cells. Portion of sponge wall showing choanocytes, with their distinctive collars and flagella.

The 'sociable' behaviour of sponge cells as exhibited by such experiments perhaps sheds light on the normal embryonic development of individual sponges. Does it also give us some sort of hint of how the first multicellular animals (metazoans) *evolved* from single-celled ancestors (protozoans)? The metazoan body is often called a colony of cells. In keeping with this book's pattern of using some tales as modern re-enactments of evolutionary happenings, could the Sponge's Tale be telling us something about the remote evolutionary past? Could the behaviour of the crawling and agglomerating cells in Wilson's experiments represent some sort of re-enactment of how the first sponge arose – as a colony of protozoans?

Almost certainly it was not the same in detail. But here is a hint. The most characteristic cells of sponges are the choanocytes, which they use for generating currents of water. The picture on the previous page shows a portion of the wall of a sponge, with the inside of the cavity to the right. The choanocytes are the cells that line the cavity of the sponge. 'Choano-' comes from the Greek for 'funnel', and you can see the little funnels or collars, made up of many fine hairs known as microvilli. Each choanocyte has a beating flagellum, which draws water through the sponge, while the collar catches nutrient particles in the stream. Take a good look at those choanocytes, for we shall meet something rather like them at the next rendezvous. And then, in the light of that, the following tale will complete our speculation about the origin of multicellularity.

CHOANOFLAGELLATES

The choanoflagellates are the first protozoans to join our pilgrimage, and they do so at Rendezvous 32 which, with worrying uncertainty, we place at 800 million years ago, at the youngest end of most molecular clock estimates. Look at the picture below. Do the little flagellated cells remind you of anything? Yes, they are very similar to the choanocytes lining the water canals of sponges. It has long been suspected that either they represent a hangover from a common ancestor, or choanoflagellates are the evolutionary descendants of sponges that have degenerated to single cells or very few cells. Molecular genetic evidence suggests the former, which is why I am considering them as separate pilgrims, joining our pilgrimage here.

There are about 140 species of choanoflagellates. Some are free-swimming, propelling themselves along with the flagellum. Others are attached by a stalk, sometimes several together in a colony, as in the picture. They use their flagellum to drive water into the funnel, where food particles such as bacteria are trapped and engulfed. In this respect they are different from the choanocytes of sponges. In a sponge, each flagellum is used not to drive food into the individual funnel of the choanocyte, but in co-operation with other choanocytes to draw a current of water in through holes in the walls of the sponge and out through the sponge's main opening. But anatomically each individual choanoflagellate, whether it is in a colony or not, is suspiciously similar to a

Is this how it was? Haeckel's drawing of colonial choanoflagellates on a stalk, from *Kunstformen der Natur* (Art Forms in Nature) [171].

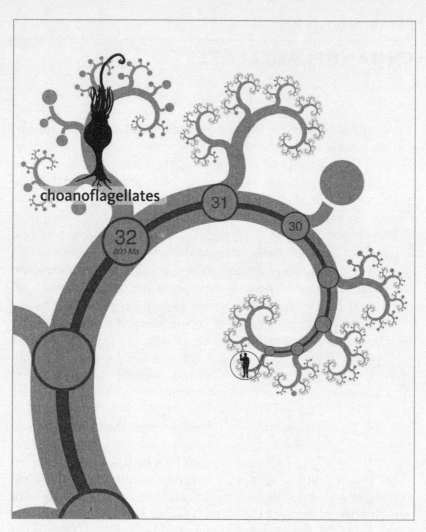

Choanoflagellates join. The 120 or so known species of choanoflagellate are conventionally seen as close relatives of the animals, a position strongly supported by both morphological and molecular data.

sponge choanocyte. This fact will bulk large in the Choanoflagellate's Tale, which resumes the topic begun by the Sponge's Tale: the origin of multicellular sponges.

THE CHOANOFLAGELLATE'S TALE

Zoologists have long enjoyed speculating about how multicellularity evolved from protozoan ancestors. The great nineteenth-century German zoologist Ernst Haeckel was one of the first to propose a theory of the origin of the Metazoa, and some version of his theory is still much favoured today: the first metazoan was a colony of flagellate protozoa.

We met Haeckel in the Hippo's Tale, in connection with his prescient linking of hippos with whales. He was a passionate Darwinian, who made a pilgrimage to Darwin's house (which the great man found irksome). He was also a brilliant artist, a dedicated atheist (he sardonically called God a 'gaseous vertebrate'), and a particular enthusiast for the now unfashionable theory of recapitulation: 'Ontogeny recapitulates phylogeny', or 'The developing embryo climbs up its own family tree.'

You can see the appeal of the idea of recapitulation. The life story of every young animal is a telescoped re-enactment of its (adult) ancestry. We all start as a single cell: that represents a protozoan. The next stage in development is a hollow ball of cells, the blastula. Haeckel suggested that this represents an ancestral stage, which he called the blastaea. Next in embryology, the blastula invaginates, like a ball punched in as a dent from one side, to form a cup lined by a double layer of cells, the gastrula. Haeckel imagined a gastrula-stage ancestor, which he called the gastraea. A cnidarian, such as a hydra or a sea anemone, has two layers of cells, like Haeckel's gastraea. In Haeckel's recapitulationist view, cnidarians stop climbing up their family tree when they reach the gastrula stage, but we soldier on. Subsequent stages in our embryology resemble a fish with gill slits and a tail. Later we lose our tail. And so on. Each embryo stops climbing up its family tree when it reaches its appropriate evolutionary stage.

Appealing as it is, the recapitulation theory has become unfashionable – or rather it is now regarded as a small part of what is sometimes but not always true. The whole matter is thoroughly discussed in Stephen Gould's book *Ontogeny and Phylogeny*. We must leave it there, but it is important for us to see where Haeckel was coming from. From the point of view of the origin of the Metazoa, the interesting stage in Haeckel's theory is the blastaea: the hollow ball of cells that, in his view, was the ancestral stage now reprised in embryology as the blastula. What modern creature can we find that resembles a blastula? Where shall we find an adult creature that is a hollow ball of cells?

Setting aside the fact that they are green and photosynthesise, the group of colonial algae called the Volvocales seemed almost too good to be true. The eponymous member of the group is the largest, *Volvox* itself, and Haeckel could hardly have wished for a neater model blastaea than *Volvox*. It is a perfect sphere, hollow like a blastula, with a single layer of cells, each resembling a unicellular flagellate (which happens to be green).

Haeckel's theory did not have the field to itself. In the mid-twentieth century a Hungarian zoologist called Jovan Hadzi proposed that the first metazoan was not round at all, but elongated like a flatworm. His contemporary model for the first metazoan was an acoelomorph worm of the kind we met at Rendezvous 27. He derived it from a ciliate protozoan (we shall meet them at Rendezvous 38) with many nuclei (which some of them have to this day). It crawled along the bottom with its cilia, as some small flatworms do today. Cell walls appeared between the nuclei, turning an elongate protozoan with one cell but many nuclei (a 'syncitium'), into a creeping worm with many cells, each with its own nucleus – the first metazoan. On Hadzi's view the round metazoans such as cnidarians and ctenophores secondarily lost their elongated worm shape and became radially symmetrical, while most of the animal kingdom continued to expand upon the bilateral worm shape in ways that we see all around us.

Hadzi's ordering of the rendezvous points would, therefore, be very different from ours. The rendezvous with the cnidarians and ctenophores would come earlier in the pilgrimage than the rendezvous with the acoelomorph flatworms. Unfortunately for Hadzi, modern molecular evidence goes against his ordering. Most zoologists today support some version of the Haeckel 'colonial flagellate' theory against the Hadzi 'syncitial ciliate' theory. But attention has switched away from the Volvocales, elegant as they are, and to the group whose tale this is, the choanoflagellates.

One type of colonial choanoflagellate is so sponge-like it even used to be called *Proterospongia*. The individual choanoflagellates (or should we stick our necks out and call them choanocytes?) are embedded in a matrix of jelly. The colony is not a ball, which would not have pleased Haeckel, although he appreciated the beauty of the choanoflagellates, as his wonderful drawings of them show. *Proterospongia* is a colony of cells of a type almost indistinguishable from those that dominate the interior of a sponge. The choanoflagellates marginally get my vote as the most plausible candidates for a recent re-enactment of the origin of the sponges, and ultimately of the whole group of Metazoa.

Proterospongia. The choanocyte-like cells facing outwards use their flagella to move the colony through the water.

The choanoflagellates would once have been lumped in the phylum 'Protozoa', together with many teeming bands of single-celled organisms who have not yet joined our pilgrimage. But 'Protozoa' is now considered a misleading name. As we have discovered more of these little creatures and probed their secrets, we have come to realise their enormous variability, both in their natural history and their relationships to us. Nevertheless, we shall continue to use the word protozoan as an informal name for a single-celled eukaryote, among whom are two newcomers we are about to meet.

FILASTEREANS

Unless you are a zoologist, you will be forgiven for not having heard of the choanoflagellates, the group that we met at the previous rendezvous. But even if you *are* a zoologist, chances are you won't have heard of the long-lost relatives who join us at Rendezvous 33. *Ministeria* and *Capsaspora* are a pair of single-celled protozoans which became known to science in 1997 and 2002 respectively. Only in 2008 did a team from Oslo University, together with the ever-industrious Tom Cavalier-Smith here in Oxford, collect the genetic evidence to show that they are each other's closest relatives, and form the next band to join our pilgrimage.*

Both creatures extrude fine, thread-like filaments from the surface of their cell, giving rise to the name for the new class chosen to describe the two species: Filasterea. Moreover, these slender tentacles have been suggested as precursors to the hairs that form the 'collar' of the choanocyte cells we met in the last two chapters. Hence all the descendants of Concestor 33, including ourselves, now go under the name 'Filozoa' – the filament-animals.

Filaments and genetics apart, the two organisms live quite different lives. *Ministeria vibrans*, to use its full name, was discovered floating freely in sea water just off the south coast of England, near Southampton. Its filaments ensnare bacteria from the surrounding sea water, which it then engulfs whole. *Capsaspora owczarzaki* (now there's a mouthful) is certainly not free-living. It lives within the circulatory system of a freshwater snail. Not just any species of snail, but one that hosts the parasitic flatworms responsible for schisto-somiasis. This debilitating disease is a significant burden on hundreds of millions of humans worldwide; you may have heard of it under its alternative name of bilharzia. *Capsaspora* helps the snails, and in doing so happens to benefit us too. It uses its filamentous tentacles to kill and then eat the flatworm larvae, giving snails a somewhat limited resistance to the disease.

* The first edition of this book was published in 2004, before the true relationship of these creatures was known. Hence this is a new rendezvous point, and the rendezvous numbers from 33 onwards have changed accordingly.

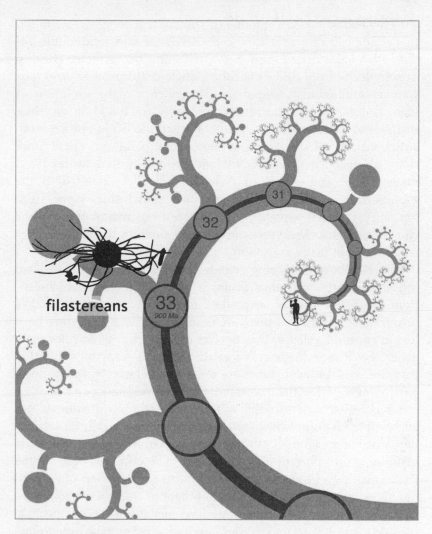

filastereans

32

31

33
900 Ma

Filasterans join. This rendezvous has only been known since 2008, and the two species in the group only a little longer. The silhouette shows *Ministeria vibrans*, catching two bacteria in its filaments.

Compared to the dazzling mix of animals that we have met so far, this arcane pair of protozoans seem, dare it be said, a little dull. Yet their DNA belies their appearance. First, a little background. The leap in complexity from undistinguished single-celled protozoan to gargantuan multicellular animal was dependent on the evolution of many new capabilities: the ability to identify and bind cells together, methods to send signals and resources between cells, and processes to switch on and off specialised developmental pathways. These functions are encoded in our DNA, and geneticists have worked out many of the genes involved. We are naturally interested in the evolutionary origin of these genes, and initial clues came from hunting for related sequences in the closest relatives of animals, the choanoflagellates. Their genomes contain some of the gene families concerned, but many are absent.

Several important missing gene families, especially those involved in cell signalling, were thus assumed to be uniquely animal inventions. Until, that is, they were found in the *Capsaspora* genome. We now realise that many of them are more ancient, and have since been lost in choanoflagellates. It is not yet entirely clear what roles these genes play in our single-celled relatives, but it seems possible that they are used by protozoans to sense and interact with the local environment, rather than necessarily to form colonies.

An increased understanding of these close relatives of animals also allows us to pinpoint which sorts of genetic changes really do seem to be particular to animals (those involving control genes and genes involved in transportation, it seems). More broadly, zoologists of the future may even be able to model the form and function of most of our ancestral genes: the genomic equivalent of concestor reconstruction. The obscure little creatures we have met at this rendezvous may provide the means to understand how and why animals came to be.

As we travel back in time from here, more esoteric members of the group formerly known as Protozoa will continue to join our pilgrimage in DRIPs and drabs, alongside major contingents of multicellular creatures such as fungi and plants. Our continual encounters with many otherwise obscure organisms, and the possibility that there are more yet to be discovered, should not be seen as disheartening. Rather, it should be a cause for celebration that among the uncharted backwaters of the natural world lie yet more rendezvous points. For in these deep recesses of geological time, they are practically our only clues to the emergence of the major groups of life on Earth.

DRIPs

There is a small group of single-celled parasites known as either Mesomycetozoea or Ichthyosporea, mostly parasites of fish and other freshwater animals. The name Mesomycetozoea* suggests an association with both fungi and animals, and it is true that on our backwards journey their rendezvous with us is immediately followed by the rendezvous where we all meet the fungi. This fact is now known from molecular genetic studies, which unite what had hitherto been a rather miscellaneous set of single-celled parasites, both with each other and (at a grander scale) with animals and their kin. This huge grouping, all the creatures we have met so far, has an appropriately inclusive title, the Holozoa, and it is the sister group to the fungi, who we will meet at the next rendezvous.

Both 'Mesomycetozoea' and 'Ichthyosporea' are quite hard to remember, and there is disagreement over which of them to prefer. This may be why they are also referred to by the nickname DRIPs – an acronym from the initial letters of the only four genera known to the discoverers of the group. The genera that provide the D, the I and the P are *Dermocystidium*, *Ichthyophonus* and *Psorospermium*. The R was always a bit of a cheat because it is not a Latin name. It stood for 'Rosette agent', a commercially important parasite of salmon, now formally named *Sphaerothecum destruens*. So I suppose the acronym should really have been amended to DIPS, or DIPSs in the plural. But DRIPs with an s for plural was what stuck. Then, with what seems like the workings of nomenclatural providence, another organism, whose name happened to begin with R, was subsequently discovered to be a DRIP too. This was *Rhinosporidium seeberi*, a parasite of human noses. So we can redesign the name DRIPS, with all five letters being comfortably acronymical, and try to ignore the embarrassing question of whether it is singular or plural.

---

* Confusingly (that's putting it mildly), the name Mesomycetozoa, as opposed to Mesomycetozoea (can you spot the difference?), has been used for a more inclusive group. This seems positively designed to confuse, like the Hominoidea, Hominidae, Homininae, Homimini complex of names for our own relatives, and I prefer to boycott them all.

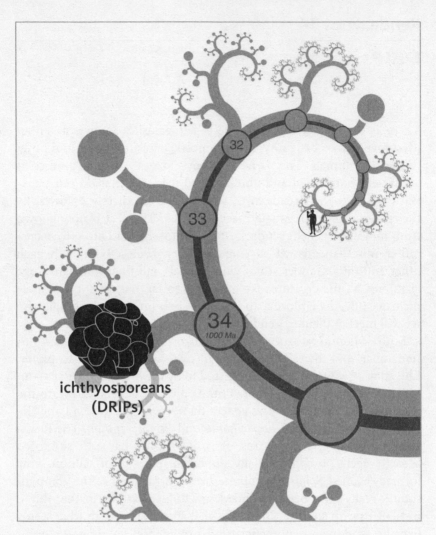

ichthyosporeans
(DRIPs)

32

33

34
1000 Ma

DRIPs join. This small band of mostly single-celled parasites, also known as the Ichthyosporea, is now well established as the last band to join the animals before we meet the fungi. Originally known from only a couple of species, DNA sequencing has added a few more obscure creatures to the group, and it now numbers around 50 species.

Rhinosporidium seeberi was first discovered in 1890, and it has long been known as the cause of rhinosporidiosis, an unpleasant disease of the human, indeed mammalian, nose, but its affinities were a mystery. At different times it has been moved from protozoan pillar to fungal post, but molecular studies now show it to be the fifth DRIP. Fortunately for pun-haters, *R. seeberi* doesn't seem to cause the nose to drip. On the contrary, it blocks the nostril with polyp-like growths. Rhinosporidiosis is mainly a disease of the tropics, and doctors have long suspected that people catch it by bathing in freshwater rivers or lakes. Since all other known DRIPs are parasites of freshwater fish, crayfish or amphibians, it seems likely that freshwater animals constitute the primary host of *R. seeberi* too. The discovery that it is a DRIP might be helpful to doctors in other ways. For example, attempts to treat it with antifungal agents have failed, and we can now get an inkling as to why: it is not a fungus.

Dermocystidium appears as cysts in the skin or gills of carp, salmonids, eels, frogs and newts. *Ichthyophonus* causes economically important systemic infections of more than 80 species of fish. *Psorospermium*, which, incidentally, was originally discovered by our old friend Ernst Haeckel, infects crayfish (not fish at all, of course, but crustaceans), and again has economically important effects on crayfish stocks. And *Sphaerothecum*, as we have seen, infects salmon. With our new ability to sequence genomes from obsure creatures, a few other genera that cluster with the DRIPs have recently been found, and the band that joins us here is now known to contain in excess of 50 species.

DRIPs organisms themselves would be dismissed as unremarkable, but for their evolutionarily aristocratic status – their branch point, after all, is the deepest in the Holozoa, their rendezvous with us the oldest. We don't know what Concestor 34 looked like, except in so far as single-celled organisms all look pretty much alike to our jaded multicellular eyes. It was not a parasite like a DRIP – not of fish, amphibians, crustaceans or humans, that's for sure, for they all still lay unimaginably far in the future.

The one adjective that is always applied to DRIPs is 'enigmatic', and who am I to break with tradition? If a DRIP were to tell its enigmatic tale, I suspect that it would be a tale of how, now that we have reached such ancient rendezvous points, it is nearly arbitrary which of our single-celled cousins happen to have survived. Not accidentally, it is also pretty arbitrary which single-celled organisms scientists have chosen to examine at the level of molecular genetics. People have looked at DRIPs because some of them are commercially important parasites of fish, and others, we now know, bung up our noses. There could be single-celled organisms that are just as pivotal on the family tree of life, but which nobody has bothered to look at because they parasitise, say, Komodo dragons rather than salmon or people.

Nobody, however, could overlook the fungi. We are about to greet them.

FUNGI

At Rendezvous 35 we animals are joined by the second of the three great multicellular kingdoms, the fungi. The third consists of the plants. It might at first be surprising that fungi, which seem so plant-like, are more closely related to animals than they are to plants, but molecular comparison leaves little doubt. Perhaps that is not too surprising. Plants import energy from the sun into the biosphere. Animals and fungi, in their different ways, are parasites on the plant world. And parasitism pays. There seem to be far more ways of being an animal or fungus than being a plant. Only about 400,000 plant species are though to exist, but estimates of the true number of fungi reach to ten times that number (and animals perhaps two or three times that again).

Diverse and important though fungi are, of these three kingdoms they are certainly the most hidden. Of the doubtless millions of species, only 99,000 have been described, substantially less than the tip of an iceberg. Even the familiar species are mostly concealed from us, for mushrooms and toadstools are merely the spore-producing reproductive organs of a vast underground organism. The conspicuous plant-like structures above ground have been pushed into the light by the main body of the fungus: a spreading subterranean network of threads called hyphae. The collection of hyphae belonging to one individual fungus is called the mycelium. The total length of mycelium of an individual fungus may be measured in kilometres, and may spread through a substantial area of soil.

A single mushroom is like a flower growing on a tree. But the 'tree', instead of being a tall, vertical structure, is spread out like the strings of a giant tennis racket underground, in the surface layers of the soil. Fairy rings are a vivid reminder of this. The circumference of the ring represents the extent of growth of a mycelium, spreading outwards from a central starting point, perhaps originally a single spore. The circular leading edge – feeding edge – of the expanding mycelium, the frame of the racket, is where the broken-down products of digestion are richest. These are a source of nutriment for the grass, which consequently grows more luxuriantly around the

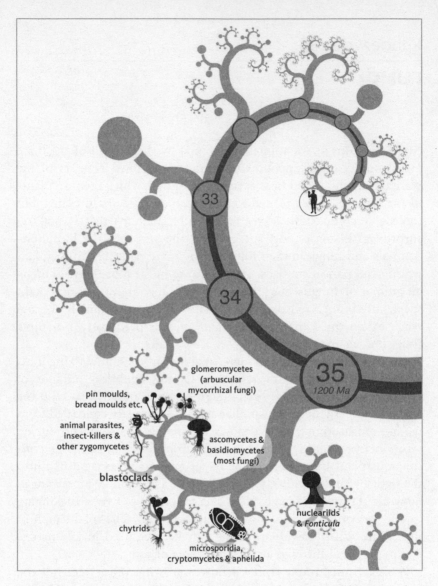

Fungi join. Fungi are closer to animals than to plants, and taxonomists place us together in a huge group called the opisthokonts. Here we use 'fungi' in a loose sense, to describe what are technically known as the Holomycota: the true fungi plus their sister group (a few single-celled amoeboid creatures together with the slime mould *Fonticula*). Most of the initial fungal divergences split various single-celled lineages from the rest, and the exact details are unsure. The parasitic microsporidia (silhouette pictured) are among the most divergent, and may cluster with some other cryptic fungal groups, as shown here. The familiar mushrooms nest deep in the tree, in two large sister groups: the Ascomycota (64,000 species known in 2008, now probably many more) and the Basidiomycota (about half that number). Their nearest relatives are the 160 species of arbuscular mycorrhizal fungi, which form intimate relations with plant roots. The relationships of the remaining 3,000 or so fungal species are only just being sorted out: we show a provisional arrangement.

ring. Where there are fruiting bodies (mushrooms, or any of dozens of species of related fungi) they tend to grow up in the ring too.

Hyphae may be divided into cells by cross-walls. But sometimes they are not, and the nuclei containing the DNA are dotted along the hypha in a syncitium, meaning a tissue with many nuclei not divided into separate cells (we met other syncitia in the early development of *Drosophila*, and in Hadzi's theory of the origin of the Metazoa). Not all fungi have a thread-like mycelium. Some, such as yeasts, have reverted to single cells which divide and grow in a diffuse mass. What the hyphae (or yeast cells) are doing is digesting whatever it is they are burrowing through: dead leaves and other decaying matter (in the case of soil fungi), curdled milk (in the case of cheese-making fungi), grapes (in the case of wine-making yeasts), or the grape-treader's toes (if he happens to suffer from athlete's foot).

The key to efficient digestion is to expose a large area of absorptive surface to the food. We achieve that by chewing the food into small pieces and passing the fragments through a long coiled gut whose already large area is compounded by a forest of tiny projections, or villi, covering its lining. Each villus in turn has a brush border of hair-like micro-villi, so the total absorptive area of an adult human intestine is millions of square centimetres. A fungus such as the well-named *Phallus* (see plate 48) or the field mushroom *Agaricus campestris* spreads its mycelium over a similar area of soil, secreting digestive enzymes and digesting the soil material where it lies. The fungus doesn't walk about devouring food and digesting it inside its body as a pig or a rat would. Instead it spreads its 'intestines', in the form of thread-like mycelia, right through the food and digests it on the spot. From time to time hyphae come together to form a single solid structure with recognisable form: a mushroom (or toadstool, or bracket). This structure manufactures spores that float high and far on the wind, spreading the genes for making new mycelium and, eventually, new mushrooms.

As you'd expect of a new influx of millions of pilgrims, they have already joined up with each other in large sub-contingents which reflect their kinship. All the major subgroups of fungi end in 'mycete', the Greek for 'mushroom', which sometimes turns into 'mycota'. We have already met 'mycete' in Mesomycetozoea, the name for DRIPs that implies some sort of intermediate status between animals and fungi. Fungi too have a number of obscure single-celled groups which split off early in evolution. These include the appropriately named cryptomycetes, and the chytridiomycetes to which belongs *Batra-chochytrium dendrobatidis*, a fungal parasite blamed for recent and calamitous declines in many frog and toad species. But the two largest and most important of the sub-contingents of the fungus pilgrims are the ascomycetes (Ascomycota) and the basidiomycetes (Basidio-mycota).

The ascomycetes include some famous and important fungi, such as *Penicillium*, the mould from which the first antibiotic was acci-dentally discovered and largely ignored by Fleming until Florey, Chain and their colleagues rediscovered it 13 years later. Incidentally, it is rather a pity that the name antibiotic has stuck. These agents attack bacteria, not viruses, and if only they had been called anti-bacterials instead of antibiotics, patients might stop demanding that doctors prescribe them (uselessly and even counterproductively) for viral infections. Another Nobel Prize-winning ascomycete is *Neuro-spora crassa*, the mould with which Beadle and Tatum developed the 'one gene one enzyme hypothesis'. Then there are the human-friendly yeasts that make bread, wine and beer, and the unfriendly *Candida* from which we get unpleasant diseases, such as vaginitis (thrush). Edible morels and the highly prized truffles are asco-mycetes. Truffles are traditionally found with the aid of female pigs, who are strongly attracted by the smell of what appears to be alpha-androstenol, a male sex pheromone secreted by boars. It isn't clear why truffles produced this dead giveaway, but it may – in some interesting way yet to be worked out – account for their gastronomic appeal to us.

Most of the edible and notoriously inedible or hallucinogenic fungi are basidiomycetes: mushrooms, chanterelles, boletuses, shii-takes, ink caps, death caps, stinkhorns, bracket fungi, toadstools and puffballs. Some of their spore-producing bodies can reach impressive sizes. Basidiomycetes are also of economic importance as causes of the plant diseases known as rusts and smuts. Some basidiomycetes and ascomycetes, as well as all members of a specialised group called the glomeromycetes, collaborate with plants to supplement their root hairs with mycorrhizae, a most remarkable story, which I'll briefly relate.

We saw that the villi in our intestines and the mycelium threads of a fungus are thin, to increase the surface area for digestion and absorption. In just the same way, plants have numerous fine root hairs to increase the surface area for absorption of water and nutrients from the soil. But it is an amazing fact that most of what appear to be root hairs are no part of the plant itself. Instead, they are provided by symbiotic fungi, whose mycelium both resembles and works like true root hairs. These are the mycorrhizae, and close examination reveals that there are several independently evolved ways in which the mycorrhizal principle has been implemented. Much of plant life on our planet is utterly dependent on mycorrhizae.

In an even more impressive feat of symbiotic co-operation, basidiomycetes and – independently evolved again – ascomycetes form associations with algae or cyanobacteria to create lichens, those remarkable confederacies which can achieve so much more than either partner on its own, and can produce body forms so dramatically different from the body form of either partner. Lichens (pronounced LIE-kins) are sometimes mistaken for plants, and that isn't so far from the truth – for plants too, as we shall see at the Great Historic Rendezvous, originally made a compact with photosynthetic micro-organisms for their food production. The fungus could almost be said to 'farm' captured crops of photosynthesisers. The metaphor gains from the fact that in some lichens the partnership is largely co-operative, and in others the fungus is more exploitative. Evolutionary theory predicts that the lichens in which the reproduction of the fungus and the photosynthesiser go hand-in-hand generally form co-operative relationships. Lichens in which the fungus just captures available photosynthetic organisms from the environment are predicted to have more exploitative relationships. And this seems to be the case.

What especially fascinates me about lichens is that their phenotypes (see the Beaver's Tale) look nothing like a fungus – nor indeed like an alga. They constitute a very special kind of 'extended phenotype', wrought of a collaboration of two sets of gene products. In my vision of life, explained in other books, such a collaboration is not in principle different from the collaboration of an organism's 'own' genes. We are all symbiotic colonies of genes – genes co-operating to weave phenotypes about them.

UNCERTAIN

The Microbe is so very small
You cannot make him out at all,
But many sanguine people hope
To see him through a microscope.
His jointed tongue that lies beneath
A hundred curious rows of teeth;
His seven tufted tails with lots
Of lovely pink and purple spots,
On each of which a pattern stands,
Composed of forty separate bands;
His eyebrows of a tender green;
All these have never yet been seen –
But Scientists, who ought to know,
Assure us that they must be so . . .
Oh! let us never, never doubt
What nobody is sure about.

HILAIRE BELLOC (1870–1953)
From *More Beasts for Worse Children* (1897)

Hilaire Belloc was a brilliant versifier but a prejudiced man. If there is an element of anti-scientific prejudice above, let us not play up to it. There is much that we are unsure about in science. Where science scores over alternative world views is that we know our uncertainty, we can often measure its magnitude, and we work optimistically to reduce it.

Rendezvous 36 is a case in point. In the first edition of this book, the tiny handful of pilgrims we meet here were not even mentioned in our journey. Indeed, we used the word 'uncertain' for a different rendezvous, from which the mists have now partially cleared. The three groups opposite were microscopic gliding cells of uncertain affinity, unceremoniously dumped in with any number of other protozoans. To some extent that is still true. Although they are now thought to group separately from the other major branches of the tree of life, and have been given their own phylum, the 'Apusozoa',

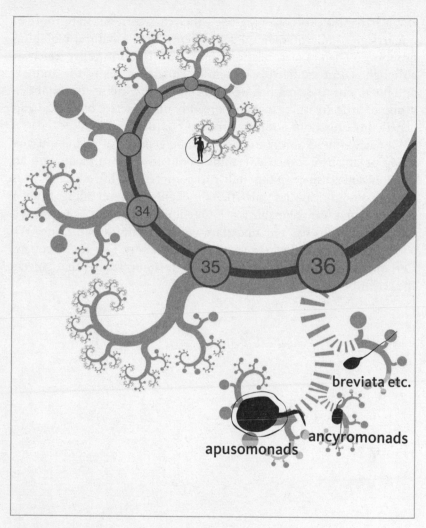

Uncertain. The three 'protist' groups shown here are poorly known, and probably do not join our pilgrimage as a single group, hence the dashed lines used here. They are grouped into a single rendezvous more for convenience than anything else.

this is in truth a protozoan rag-bag which may turn out to be more of a marriage of microbiological convenience than a natural biological grouping. Their exact relationship to us is also rather unclear, although recent studies place them as close relatives of the animals and fungi. For these reasons we label this rendezvous as 'uncertain', a group of little creatures that are currently under active biological and genetic investigation, not least here in Oxford.

Our new-found ability to peer into the genome of just about any living organism is, when you think about it, truly miraculous – an unparalleled advance in our ability to comprehend life on Earth. It is raising obscure species into the limelight, and revealing hidden diversity on a unimaginable scale. Our understanding has increased, but inevitably so has our uncertainty. That is not a bad thing. We should perhaps take Belloc's verse as a Cautionary Tale for Scientists. Certainty is not all it is cracked up to be. It is uncertainty that powers the scientific endeavour.

AMOEBOZOANS

Joining us at Rendezvous 37 is a little creature that once had the distinction of being, in the popular and even scientific imagination, the most primitive of all, little more than naked 'protoplasm': *Amoeba proteus*. On this view, Rendezvous 37 would be the final encounter of our long pilgrimage. Well, we still have a way to go, and *Amoeba* has, when compared to bacteria, quite an advanced, elaborate structure. It is also surprisingly large, visible to the naked eye. The giant amoeba *Pelomyxa palustris* can be as much as half a centimetre across.

Amoebas famously have no fixed shape – hence the species name *proteus*, after the Greek god who could change his form. They move by streaming their semi-liquid interior, either as a more or less coherent single blob, or by thrusting out pseudopodia. Sometimes they 'walk' on those temporarily extruded 'legs'. They eat by engulfing prey, throwing pseudopodia around it and enclosing it in a spherical bubble of water. Being engulfed by an amoeba would be a nightmarish experience, if you weren't too small to have nightmares. The spherical bubble or vacuole can be thought of as part of the outside world, lined by a portion of the amoeba's 'outer' wall. Once in the vacuole, the food is digested.

Some amoebas live inside animal guts. For example, *Entamoeba coli* is extremely common in the human colon. It is not to be confused with the (much smaller) bacterium *Escherichia coli* on which it probably feeds. It is harmless to us, unlike its near relative *Entamoeba histolytica*, which destroys the cells lining the colon and causes amoebic dysentery, familiarly known (in British English) as Delhi Belly, or (in American English) as Montezuma's Revenge.

Three rather different groups of amoebozoans are called slime moulds because they have independently evolved similar habits (plus another unrelated group of 'slime moulds', the acrasids, which will join us at the next rendezvous). Of the amoebozoan ones, the best known are the cellular slime moulds or dictyostelids. They have been the life work of the distinguished American biologist J. T. Bonner, and what follows is largely drawn from his scientific memoir *Life Cycles*.

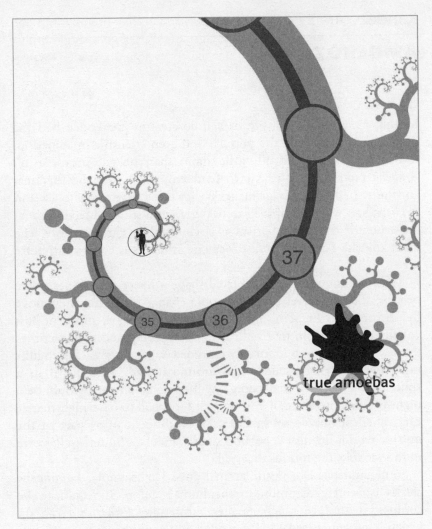

true amoebas

Amoebozoans join. The word 'amoeba' is a description rather than a strict classification because many unrelated eukaryotes exhibit an amoeboid form. Nevertheless, the Amoebozoa include many of the classic amoebas, such as *Amoeba proteus* shown here, as well as most of the slime moulds – a few thousand known species in total. Recent molecular analysis [106] tentatively places some obscure single-celled organisms (including *Collodictyon* and *Malawimonas*) at the base of the Amoebozoa. We show this arrangement here, although it has yet to be fully confirmed.

Cellular slime moulds are social amoebas. They literally blur the distinction between a social group of individuals and a single multicellular individual. In part of their life cycle, separate amoebas creep through the soil, feeding on bacteria and reproducing, as amoebas will, by dividing in two, feeding some more, then dividing again. Then, rather abruptly, the amoebas switch into 'social mode'. They converge on aggregation centres, from which chemical attractants radiate outwards. As more and more amoebas stream in on an attraction centre, the more attractive it becomes, because more of the beacon chemical is released. It is a bit like the way planets form from aggregating debris. The more debris accumulates in a given attraction centre, the more its gravitational attraction. So after a while, only a few attraction centres remain, and they become planets. Eventually the amoebas in each major attraction centre unite their bodies to form a single multicellular mass, which then elongates into a multicellular 'slug'. About a millimetre long, it even moves like a slug, with a definite front and back end, and is capable of steering in a coherent direction – for example towards light. The amoebas have suppressed their individuality to forge a whole organism.

After crawling around for a while, the slug initiates the final phase of its life cycle, the erection of a mushroom-like 'fruiting body'. It begins the process by standing on its 'head' (the front end as defined by its crawling direction), which becomes the 'stalk' of the miniature mushroom. The inner core of the stalk becomes a hollow tube made of swollen cellulose carcasses of dead cells. Now cells around the top of the tube pour into the tube like, in Bonner's simile, a fountain flowing in reverse. The result is that the tip of the stalk rises into the air, with the originally posterior end of the stalk at the top. Each of the amoebas in the originally posterior end now becomes a spore encased in a thick protective coat. Like the spores of a mushroom, they are now shed, each one bursting out of its coat a free-living, bacteria-devouring amoeba, and the life cycle begins again.

Bonner gives an eye-opening list of such social microbes – multicellular bacteria, multicellular ciliates, multicellular flagellates and multicellular amoebas, including his beloved slime moulds. These creatures might represent instructive re-enactments (or pre-enactments) of our kind of metazoan multicellularity. But I suspect that they are all completely different, and the more fascinating because of it.

LIGHT HARVESTERS AND THEIR KIN

By this point in our pilgrimage, our band includes both the animal and the fungal kingdoms, as well as some of their occasionally single-celled amoeboid relatives. Together we are known as the Amorphea, a recent coining to reflect the particularly malleable form of our cells. At Rendezvous 38 we amorpheans – animals, fungi and fellow travellers – meet the last of the traditional kingdoms of large-bodied creatures, the plants. But plants are not the only pilgrims to join here.

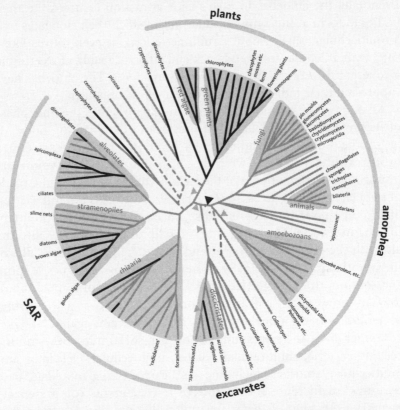

The tree of eukaryotic life modified from a diagram by Fabien Burki in the light of further studies [106]. Light harvesters are shown as dark branches. Names mostly refer to groups mentioned in the text; names for unlabelled branches can be found in [60]. Broken lines show uncertain relationships; triangles suggest positions for the root.

The diagram opposite depicts the family tree of complex life on this planet, as we currently understand it. Before we move on to the main point, notice with proper humility the tiny blob labelled 'animals'. If you can't find them, look to the right, where you'll see us sandwiched between the fungi and the amoebozoans. That is where you and I belong, together with the entire populace of pilgrims who joined us up to and including Rendezvous 31.

The plant kingdom (including the red algae as well as the traditional green forms) occupies the major branch at the top. That leaves at least five branches we have yet to mention, distinct enough that they too probably merit the title of kingdom. At the bottom are a group known as the Excavates. On the left of the diagram, three potential 'kingdoms' cluster together; their acronym, SAR, is easier to remember than their separate names (Stramenopiles, Alveolates and Rhizaria). Finally, between SAR and the plants are a miscellaneous set of creatures of uncertain affinity, as indicated by the dotted lines.

According to the scheme we are following here, these unfamiliar 'kingdoms' are more closely related to the plants than they are to ourselves. That would make Rendezvous 38 a momentous one, marking the deepest split among all complex life forms (that is, excluding the various sorts of bacterial forms who will occupy the last lap of our journey). As my circumspect language implies, this hypothesis is open to revision; it already marks an alteration from the first edition of this book. The main problem is rooting. We met this in the Gibbon's Tale. A star diagram such as this is compatible with many different evolutionary trees, that is, different orders and numbers of rendezvous points. We would normally root a tree like this using a distantly related outgroup. The problem is that even the most closely related outgroup in this case, the Archaea (whom we will meet at Rendezvous 39), are very distant indeed from any of the complex life on our star diagram. With few directly comparable DNA sequences, there are many places where we could reasonably sling the root; the triangles on the diagram indicate suggestions from different groups of experts. We are choosing one popular option, which is to root this tree at the position of the black triangle. Another common option is to place the root within the Excavates, causing them to join separately, at an additional rendezvous point. And indeed there are few notable features which unite the plants with the Excavates, or any of the SARs either.

Among the band who join at Rendezvous 38 there is, however, one recurrent vocation which we have not encountered hitherto. The pilgrims who ply this trade have been marked with darker lines in the

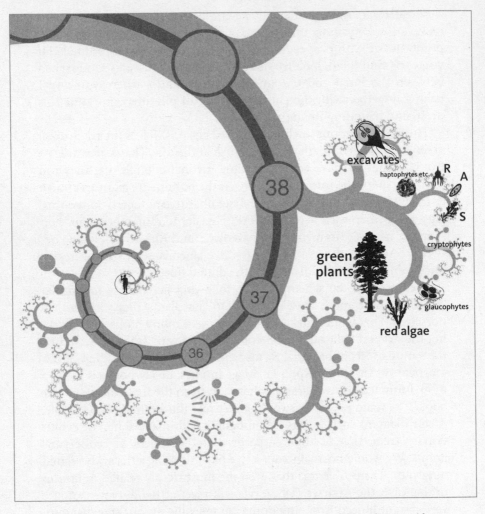

Remaining eukaryotes (including plants) join. This fractal phylogeny shows an identical branching pattern to the eukaryotic tree depicted on page 574, with the root of the eukaryotes placed at the dark arrow on that diagram. If we were to choose a different root, the branching order could change drastically, as discussed in the Gibbon's Tale. Most uncertainty concerns the position of the 1,000 or so excavate species (including *Giardia*, seen here as a silhouette). In contrast, the SAR group of about 15,000 species, shown here on the upper right, is now well established. The plants, which dominate this branch of the tree of life, are shown at the bottom. They include about 20 species of single-celled glaucophytes, over 4,000 species of red algae, and many hundreds of thousands of species of green plants. The branching order of these three groups is generally accepted.

plot, which reveal it to be habitual in the plants and also common among the rest. Unlike animals or fungi, who feed off other creatures or their decaying remains, members of the joining band tend not to exploit others for food. Most are self-sufficient. And most do it by feeding off energy from the sun. Broadly, we might dub them the light harvesters.

Life could get along without animals and without fungi. But abolish the light harvesters, and life would rapidly cease. They sit, indispensably, at the base – the very foundation – of nearly every food chain. They form the most visible creatures on our planet, the first living things any visiting Martian would remark upon. By far the weightiest single organisms that ever lived exist to harvest light, and the creatures we greet at Rendezvous 38 account for an impressive percentage of the world's biomass. This doesn't just happen to be so. A high proportion follows necessarily from the fact that almost* all biomass is put together by energy from the sun, by photosynthesis carried out in plants on land and algae in the sea, and the transaction at every link of the food chain is only about 10 per cent efficient. The surface of the land is green because of plants, and the surface of the sea is blue-green wherever there are enough nutrients to sustain light-harvesting plankton. It is as though these creatures are going out of their way to cover every square centimetre with green, leaving none uncovered. And that is pretty much what they are doing, for a very sensible reason.

A finite number of photons reaches the planet's surface from the sun, and every last photon is precious. The total number of photons that can be garnered from its star by a planet is limited by its surface area, with the complication that only one side is facing its star at any one time. For a light harvester, every square centimetre of the Earth's surface that is anything but green amounts to a negligently wasted opportunity to sweep up photons. Leaves are solar panels, as flat as possible to maximise photons caught per unit expenditure. There is a premium on placing your leaves in such a position that they are not overshadowed by other leaves, especially somebody else's leaves. This is why forest trees grow so tall. Tall trees that are not in a forest are out of place, probably because of human interference. It is a complete waste of effort to grow tall if you are the only tree around. It is much better to spread out sideways like grasses because that way you trap more photons per unit of effort put into growing. As for forests, it is no accident that they are so dark. Every photon that makes it to the ground represents failure on the part of the leaves above.

* The reason for this little hedge will emerge when we get to Canterbury.

Terrestrial ourselves, our thoughts are preoccupied with the green and pleasant land. But about half of global photosynthesis occurs in the sea. The two environments pose quite different problems for a light harvester.

Air is transparent to visible light, meaning an organism can live at the bottom of a thick blanket of air and still harvest photons. Outcompeting other light harvesters means growing upwards which, due to the thinness of air, requires expensive material support. So individual land plants are large, and in total account for most of the biomass, even though photosynthesis originated in the sea and the sea occupies a much larger surface area.

Most sea-dwelling light harvesters are floating plankton, supported by water and so exempted from the need to invest in structures to reach for the light. Not only is there no need for stems or trunks, even leaves are unhelpful: it's more economical for single green cells to float free. Competition manifests itself in other ways, such as in their prodigious rates of reproduction (hence the occurrence of smothering algal blooms) and intricate physical and biochemical weaponry.

A major problem in the sea is that nutrients fall into the sunless depths, hence life exists primarily where nutrients are near the surface, for example at the edges of continents. This has led a few of our pilgrims, 'seaweeds' of various sorts, to evolve large bodies – not so much to grow up, as to reach down and hold fast to the bottom, preventing pulverisation by the surf or stranding on the unforgiving land. Despite their convergent similarity, many seaweeds are not true plants. In fact, marine photosynthesis is carried out by creatures in all six major groups that join us at this rendezvous. This seems as good a point as any to provide a brief description of these 'kingdoms', ending with the most successful, the plants.

The Excavates are the weirdest of the lot. We are placing them with the light harvesters and their kin, but to be honest they are more 'kin' than light harvesters. Most are single cells which swim about using flagella, and they commonly eschew photosynthesis. In some, the mitochondria have been so reduced as to be practically unrecognisable. This lack of functioning energy factories may explain their parasitic inclinations: they include the nasty gut parasite *Giardia* and the sexually transmitted vaginal microbe *Trichomonas*. Not all cause disease though. There are some fascinatingly complicated inhabitants of insect guts, whose story we shall meet in due course.

Excavates with identifiable (if modified) mitochondria are sometimes placed into a sort of sub-kingdom of their own, the discicristates.

These include photosynthetic forms such as *Euglena*, but also parasites such as *Trypanosoma*, which causes sleeping sickness. Also in this group are the acrasid slime moulds, which are not closely related to the dictyostelid slime moulds whom we met at the previous rendezvous. As so often in this long pilgrimage, we marvel at the capacity of life to reinvent similar body forms for similar ways of life. 'Slime moulds' pop up in two or even three different pilgrim bands; so do 'flagellates', so do 'amoebas'. Probably we should think of 'the amoeba' as a way of life, like 'the tree'. 'Trees', meaning very large plants stiffened with wood, pop up in many separate plant families.* It looks as though the same is true of 'amoebas' and 'flagellates'. It is certainly true of multicellularity, which has arisen in animals, fungi, plants, brown algae and various other places, such as slime moulds.

Within the SAR grouping of three kingdoms, the Rhizaria include various groups of single-celled forms, some green and photosynthesising, most not. In the latter category come the exquisite marine foraminiferans and radiolarians. While mostly microscopic, some foraminiferan species grew large enough that their fossilised calcite shells were used as currency in ancient Egypt. Foraminiferans are frequently the main component of limestone, and they tower above us in the shape of the limestone blocks used to build the Egyptian pyramids. Radiolarians are notable for their beautiful miniature glass skeletons, never better captured than in the drawings of Ernst Haeckel, the eminent German zoologist who seems to keep cropping up through this book (see overleaf). Continuing the architectural theme, Haeckel's work inspired the extraordinary Byzantine design of the main entrance to the Paris Exposition Universelle in 1900 (see photo overleaf). Although radiolarians do not harvest light themselves, many can form symbiotic relationships with major aquatic photosynthesisers, in particular the dinoflagellates.

Dinoflagellates belong to the Alveolates, the 'A' in SAR. They too are unicellular. A number of marine species give out tiny flashes of light when disturbed, which, it has been suggested, may attract fish to eat smaller animals that would otherwise prey on the dinoflagellates. Whatever the explanation, the effects can be spectacular at night, producing luminous blue surf and enveloping nocturnal swimmers in a haze of glowing water. Within the alveolates lie the

---

* It is intriguing to note that although trees have independently emerged in many lineages of plants, evolutionary relationships reveal that all flowering plants have some – indeed many – concestors that were woody trees or shrubs. In other words, the tiniest weed or flower in your garden has spent a substantial amount of its ancestry as a woody plant.

Notable for their beautiful miniature glass skeletons. Ernst Haeckel's drawing of radiolarians, from his 1904 treatise *Kunstformen der Natur* (Art Forms in Nature) [171].

Extraordinary Byzantine design. Early photographic postcard of the radiolarian-inspired entrance to the 1900 Paris Exposition Universelle, designed by architect René Binet.

parasitic apicomplexans, photosynthesisers gone to the bad. They include *Plasmodium*, the malaria parasite, which due to its light-harvesting past is sensitive to certain anti-plant compounds – a source of hope for new antimalarial drugs. Also to be found here is *Toxoplasma*, a mind-controlling parasite of rodents and cats, which is also commonly found in human brains, and may influence our psychology and behaviour. Another major group of alveolates are the predatory gliding ciliates, known among geneticists for their bizarre habit of cutting up multiple copies of their genome into tens of thousands of pieces and putting them into a separate nucleus for everyday use. Among the ciliates, or so it would seem, is *Mixotricha paradoxa*, whose tale we shall hear later. The 'so it would seem' and the *'paradoxa'* form the substance of the tale, whose thunder we shall not steal here.

The Stramenopiles, sister to the Alveolates, and completing the SAR acronym, are another mixed bag. They include more beautiful unicellular creatures, such as the photosynthetic diatoms, which encase themselves in delicate glass capsules, again memorably illustrated by Haeckel. On some estimates, diatoms account for almost half of the light-harvesting in the sea. Stramenopiles also include non-photosynthetic forms such as the slime nets, or labyrinthulids, who take an unusual approach to multicellularity, laying down filamentous tracks on the seaweed that they eat, which provides a network for multiple cells to whizz around, akin to a public transport system or Pac-Man game.

But Stramenopiles have also independently discovered true multi-cellularity, in the form of the brown algae. These are the largest and most prominent of all seaweeds, with giant kelps reaching 100 metres in length. The brown algae include the wracks of the genus *Fucus*, the various species of which segregate themselves in strata up the beach, each being best suited to a particular zone of the tide cycle. *Fucus* might well be the genus upon whom the leafy sea dragon (see its tale) is modelled.

The penultimate 'kingdom' comprises the taxonomically homeless creatures at the top of our star diagram. Most important among those are the photosynthesising haptophytes, whose geometric armoured shield plates are seen under the microscope to be the main component of the White Cliffs of Dover and the huge beds of chalk throughout Western Europe, after which the Cretaceous Period is named. They may also include the photosynthesising freshwater algae known as cryptophytes; alternatively, this group may be sister to the plants, to whom we finally turn.

It is an unfortunate side-effect of our backwards journey that the 300,000-odd species of plant – such a fundament of life on land – receive so little focus in this book.* The criticism, levelled at the first edition of this book, must be compounded now that plants have become shoehorned into a shared rendezvous, rather than occupying their own chapter as previously. Incidentally, mycologists might have an even stronger case: fungi probably account for ten times more species than plants yet get even shorter shrift, not only in this book, but more importantly in the taxonomic community at large. By way of rectification for plants, if not for fungi, we will hear some tales shortly. But first, a formal introduction.

Among all the light harvesters at this rendezvous, plants are the original and most important. They account for 95 per cent of the joining species. Moreover, all the other photosynthesisers (with the possible exception of one rhizarian genus, *Paulinella*) have acquired their light-harvesting organs by stealing them from plants. This is not as surprising as it might sound, given the various symbiotic associations between the light harvesters and other unlikely creatures (lichens and coral, to name but two). In fact, even plants originally stole their photosynthetic ability from . . . no, that story must wait until the next chapter, the Great Historic Rendezvous.

---

* See *The Emerald Planet* by David Beerling, which provides the summary of land plant evolution that we have no room to include here.

Not all plants are large and visible, and not all are green. They include the smallest of all the free-living organisms in our journey so far, the single-celled green alga *Ostreococcus*, which is smaller than a typical bacterium. They also include the heaviest single organism, the giant redwood, whose forthcoming tale is of ancient time. Unlike in animals, plant multicellularity appears to have arisen many times over, for example in the red algae, repeatedly in the chlorophyte algae (including *Volvox* which made a cameo appearance in the Choano-flagellate's Tale), and in the most familiar and impressive group, the land plants. The closest relatives of the land plants are another group of green algae, the freshwater charophytes, which implies that plants probably didn't move directly from the sea onto the land but, like animals, went via freshwater. Fossils indicate this happened during the Ordovician, with arthropods probably following soon after. The order is almost obvious, for without plants to eat, what would it profit an animal to be there?

With few exceptions, such as Venus flytraps, plants don't move. With few exceptions, such as sponges, animals do. Why the difference? It must have to do with the fact that plants eat photons while animals (ultimately) eat plants. We need that 'ultimately', of course, because the plants are sometimes eaten at second or third hand, via animals eating other animals. But what is it about eating photons that makes it a good idea to sit still with roots in the ground? What is it about eating plants, as opposed to being a plant, that makes it a good idea to move? Well, I suppose, given that plants stay still, animals have got to move in order to eat them. But why do plants stay still? Maybe it has something to do with the need to be rooted in order to suck nutrients out of the soil. Maybe there is too unbridgeable a distance between the best shape to be if you want to move (solid and compact), and the best shape to be if you want to expose yourself to lots of photons (high surface area, hence straggly and unwieldy). I'm not sure. Whatever the reason, of the three great groups of megalife that have evolved on this planet, two of them, the fungi and the plants, stay mostly still as statues, while the third group, the animals, do most of the scurrying about, most of the active go-getting. Plants even make use of animals to do their scurrying for them, and flowers, with their beauteous colours, shapes and scents, are the instruments of this manipulation.*

---

* I would have included a tale about this if I had not already done it in two chapters of *Climbing Mount Improbable*, 'Pollen Grains and Magic Bullets' and 'A Garden Inclosed'.

Plants, perhaps because they don't move, have evolved a modular approach to body-building. A leaf is one sort of module repeated at the tip of each twig. Another is a stem, which can divide to give rise to two smaller stems, and so on, forming a recursive pattern of branches. This encourages exquisitely economical repetition of form, an example of the 'Fractal Geometry of Nature' as per the title of Benoit Mandelbrot's seminal work. The elegant self-similarity of fractals underlies the first of our tales at this rendezvous, and we will also return to the idea of modularity in the final chapter of this book. Here, I simply wish to note the similarities between the plant architecture and the branching process of gene reproduction. Coupled with the isolation of genes into separate species, this allows us to talk of an evolutionary 'tree of life', the metaphor which has been our guide throughout our pilgrimage.

Linking the tree of life to the mathematical geometry of fractals results in the beautiful visualisations that adorn the start of each rendezvous. These are based on a computer program called OneZoom, developed by James Rosindell of Imperial College London, which in principle allows millions of species to be displayed on a single, navigable evolutionary tree. It is a lovely idea, and provides a brilliantly intuitive way to explore the diversity of life on Earth. The OneZoom website* provides many trees to explore (and we fully expect a tree of all species of life on Earth to be constructed soon). You can explore these fractal trees as far as you like: they seem to go on for ever. Like Miranda in *The Tempest*, the enchantment is palpable: 'Oh, wonder! How many goodly creatures are there here!' And as you climb through the branches, hand over hand as it seems, like a Darwinian monkey in evolutionary tree heaven, remember that each fork you encounter represents a true rendezvous point in exactly the sense of this book. What a delight it would be to demonstrate to Charles Darwin.

THE CAULIFLOWER'S TALE

The tales in this book are intended to be about more than the private concerns of the teller. Like Chaucer's, they are supposed to reflect upon life in general – in his case human life, in our case life. What has the cauliflower to tell the huge assemblage of pilgrims at the great get-together after Rendezvous 36 when the plants join the animals?

* www.onezoom.org

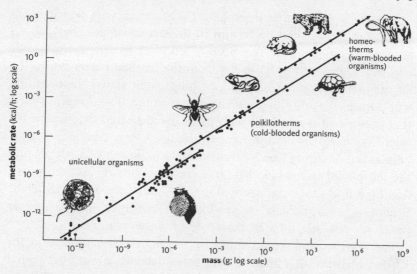

Law holds over 20 orders of magnitude. Kleiber's Law plot, adapted from West, Brown and Enquist [438].

An important principle that applies to every plant and every animal. It could be seen as a continuation of the Handyman's Tale.

The Handyman's Tale was about brain size, and it made great play with the logarithmic way of making scatter plots to compare different species. Larger animals seemed to have proportionally smaller brains than small animals. More specifically, the slope of a log-log graph of body mass against brain mass was pretty much exactly ¾. This fell, you will remember, between two intuitively comprehensible slopes: ¼ (brain mass simply proportional to body mass) and ⅔ (brain area proportional to body mass). The observed slope for log brain mass against log body mass turned out to be not just vaguely higher than ⅔ and lower than ¼. It was exactly ¾. Such precision of data seems to demand equal precision from theory. Can we think up some rationale for the ¾ slope? It isn't easy.

To add to the problem, or perhaps give us a hint, biologists have long noticed that lots of other things besides brain size follow this precise ¾ relationship. In particular, the energy use of various organisms – the metabolic rate – follows a ¾ rule, and this was raised to the status of a natural law, Kleiber's Law, even though there was no known rationale for it. The graph above plots log metabolic rate against log body mass (the Handyman's Tale spells out the rationale for log-log plots).

The really astonishing thing about Kleiber's Law is that it holds good from the smallest bacterium to the largest whale. That's about 20 orders of magnitude. You need to multiply by ten 20 times – or add 20 noughts – in order to get from the smallest bacterium to the largest mammal, and Kleiber's Law holds right across the board. It works for plants and single-celled organisms too. The diagram shows that the best fit is obtained with three parallel lines. One line is for micro-organisms, a second for cold-blooded large creatures ('large' here means anything heavier than about a millionth of a gram!) and the third is for warm-blooded large creatures (mammals and birds). All three lines have the same slope ($\frac{3}{4}$) but they are of different height: not surprisingly, warm-blooded creatures have a higher metabolic rate, size for size, than cold-blooded creatures.

For years no one could think of a really convincing reason for Kleiber's Law, until a piece of brilliant collaborative work between a physicist, Geoffrey West, and two biologists, James Brown and Brian Enquist. Their derivation of the precise $\frac{3}{4}$ law is a piece of mathematical magic that is hard to translate into words, but it is so ingenious and important that it is worth the effort.

The theory of West, Enquist and Brown, henceforth WEB, takes off from the fact that the tissues of large organisms have a supply problem. That is what blood systems in animals and vascular tubing in plants are all about: transporting 'stuff' to and from tissues. Small organisms don't face the problem to the same extent. A very small organism has such a large surface area compared to its volume that it can get all the oxygen it needs through its body wall. Even if it is multicellular, none of its cells are very far from the outside body wall. But a large organism has a transport problem because most of its cells are far away from the supplies they need. They need to pipe stuff from place to place. Insects literally pipe air into their tissues in a branching network of tubes called tracheae. We too have richly branched air tubes, but they are confined to special organs, the lungs, which have a correspondingly richly branched blood network to take the oxygen from the lungs to the rest of the body. Fish do a similar thing with gills: area-intensive organs designed to increase the interface between water and blood. The placenta does the same kind of thing for maternal blood and foetal blood. Trees use their richly dividing branches to supply their leaves with water drawn from richly branched roots in the ground, and pump sugars back from the leaves to the trunk.

The cauliflower opposite, freshly bought from a local greengrocer

Tissues have a supply problem.
The complex supply system of a cauliflower.

and cut in half, shows what a typical stuff-transporting system looks like. You can see how much effort a cauliflower puts in to provide a supply network for its surface covering of 'flower buds' – buds that have been grotesquely modified in this case by artificial selection under domestication, but the principle still stands.

Now, we could imagine that such supply networks – tubes of air, pipes of blood or sugar solution, or whatever they are – might compensate perfectly for increased body size. If they did, a typical cell of a modest cauliflower would be exactly as well supplied as a typical cell of a giant redwood, and the metabolic rate of the two cells would be the same. Since the number of cells in an organism is proportional to its mass or volume, the scatter plot of total metabolic rate against body mass, with both axes on a log scale, would fall on a line with a slope of 1. Yet what we actually observe is a slope of $\frac{3}{4}$. Small organisms have a higher metabolic rate than they 'should' have, for their mass, compared with large organisms. What this means is that the metabolic rate of a cauliflower cell is higher than the metabolic rate of an equivalent cell in a redwood, and the metabolic rate of a mouse is higher than the metabolic rate of a whale.

At first sight, this seems strange. A cell is a cell is a cell, and you might think there is an ideal metabolic rate which would be the same for a cauliflower as for a redwood, a mouse as for a whale. Perhaps there is. But the WEB theory suggests that the difficulty of delivering water, or blood, or air, or whatever 'stuff' it is, places a limit on achieving that ideal. There has to be a compromise. The theory explains the compromise, and why it ends up delivering exactly a slope of $\frac{3}{4}$, and it does so in precise, quantitative detail.

There are three key points to the WEB theory. The first is that the pipes, which deliver stuff to the cells, are most economically laid out as a fractal-like, branching network, the smallest being roughly standard-sized capillaries, or their equivalent. We see this with the cauliflower, but it is also true of our own circulatory system, or indeed our lungs. Given that Kleiber's Law extends down to single-celled organisms, WEB suggest that a similar network might also be true for the cytoskeleton within a cell too. The second point is that this supply network itself occupies some volume, competing for space with the cells that it is supplying. Towards the tips of the supply network, the pipes occupy a substantial space in their own right. And if you double the number of cells that need to be supplied, the volume of piping more than doubles because a new, large pipe is needed to plumb the network into the main system, a pipe which itself occupies space. If you want to double the number of supplied cells whilst only doubling the space occupied by the pipes, you need a more sparsely distributed plumbing network. The third key point is that, whether you are a mouse or a whale, the most efficient transport system – the one that wastes the least energy in moving stuff around – is one that takes up a fixed percentage of the volume of your body. That's how the mathematics works out, and it is also an empirically observed fact.* For example, mammals, whether mice, humans or whales, have a volume of blood (i.e. the size of the transport system) which occupies between 6 and 7 per cent of their body.

Taking these three points together, it means that if we wish to double the volume of cells to be supplied, but still keep the most efficient transport system, we need a more sparsely distributed supply network. And a more sparse network means that less stuff is supplied per cell, meaning that the metabolic rate must go down. But by how much, precisely, must it go down?

WEB calculated the answer to this question. Wonderful to say, the mathematics predicts a straight line with a slope of exactly $\frac{3}{4}$ for the graph of log metabolic rate against log body size! The exact details of their theory, and the ubiquity of Kleiber's Law, have seen several recent challenges. But the explanatory power of the basic idea remains compelling. It seems that Kleiber's Law – whether in plants, animals, or even at the level of transport within a single cell – follows from the physics and geometry of supply networks.

THE REDWOOD'S TALE

People argue about the one place in the world that you must visit before you die. My candidate is Muir Woods, just north of the Golden Gate Bridge. Or, if you leave it too late, I can't imagine a

---

* The actual percentage might differ a little depending on, say, whether you are warm-blooded or cold-blooded.

better place to be buried (except I doubt that it's allowed, nor should it be). It is a cathedral of greens and browns and stillness, the nave lofted by the world's tallest trees, *Sequoia sempervirens*, the Pacific coast redwoods, whose padded bark damps out the echoes that would fill a man-made cathedral. The related species *Sequoiadendron giganteum* (see plate 50), found inland on the foothills of the Sierra Nevada range, is typically slightly shorter but more massive. The largest single living creature in the world, the General Sherman tree, is a *giganteum* over 30 metres in circumference and over 80 metres tall, with an estimated weight of 1,260 tonnes. Its age is not known for certain but the species is known to survive more than 3,000 years. General Sherman's age could be ascertained exactly to the nearest year if we cut it down – a major undertaking, for the bark alone is about a metre thick.* Let us hope this will never happen, in spite of Ronald Reagan's notorious opinion, when he was Governor of California: 'If you've seen one, you've seen them all.'

How is it that we can know the age of a large tree, even one as old as General Sherman, accurately to the nearest year? We count the rings in its stump. Ring counting, in a more sophisticated form, has given rise to the elegant technique of dendrochronology, by which archaeologists working on a timescale of centuries can precisely date any wooden artefact.

It has been left to this tale to explain how, throughout our pilgrimage, we have been able to date historical specimens on an absolute timescale. Tree rings are very accurate but only for the nearest reaches of the historical record. Fossils are dated by other methods, mostly involving radioactive decay, and we shall come to them, along with other techniques, in the course of the tale.

Annual rings in a tree result from the unsurprising fact that a tree puts on more growth in some seasons than in others. But, by the same token, whether in summer or winter, trees grow more in a good year than in a poor year. Good years are two a penny, and so are bad years, so one tree ring is no good for identifying a particular year. But a sequence of years has a fingerprint pattern of wide and narrow rings, which labels that sequence in different trees over a wide area. Dendrochronologists compile catalogues of these labelled signature patterns. Then a fragment of wood, perhaps from a Viking longship

---

* Without cutting it down, a core sample would give a good estimate, although it might miss the smallest rings.

buried in mud, can be dated by matching its ring pattern against previously collected libraries of signatures.

The same principle is used in dictionaries of melodies. Suppose you have a tune in your head and you can't remember its name. How might you look it up? Various principles are used, of which the simplest is the Parsons code. Turn your tune into a series of ups and downs (the first note is ignored because, obviously, it can be neither up nor down). Here, for instance, is the pattern of a favourite tune, the Londonderry Air, or Air from County Derry, which I have just typed into the Melodyhound website:

UUUDUUDDDDDUUUUUDDDUD

Melodyhound correctly sleuthed my tune (calling it 'Danny Boy' – the name by which Americans know it because of some twentieth-century words that were set to it). At first it seems surprising that a tune should be identified by such a short sequence of symbols, telling only the direction of movement, not the distance, and with no indication of durations of notes. But it really works. For the same kind of reason, a fairly short consecutive pattern of tree rings suffices to identify a particular sequence of yearly growth rings.

In a newly felled tree, the outer ring represents the present. The past can be exactly reckoned by counting inwards. So absolute dates can be put upon ring pattern signatures in recent trees whose date of felling is recorded. By looking for overlaps – signature patterns near the core of a young tree that match the pattern in the outside layers of an older tree – we can put absolute dates on ring patterns in older trees too. By daisy-chaining the overlaps backwards, it is in principle possible to put absolute dates on very old wood indeed – in principle even from the Petrified Forest of Arizona, if only there were a continuous series of petrified intermediates – if only! By this tech-nique of overlapping jigsaws, libraries of fingerprint patterns can be built up and consulted to recognise wood that is older than the oldest tree we ever see alive. The changing thickness of tree rings can also, incidentally, be used not just for dating wood but for reconstructing year-to-year climate and ecological patterns dating from long before meteorological records were kept.

Dendrochronology is limited to the relatively recent time domains inhabited by archaeologists. But tree growth is not the only process that spurts and slows on an annual cycle, or on some other regular or even irregular cycle. Any such process can in principle be used for dating, aided by the same ingenious trick of daisy-chaining over-lapped patterns. And some of these techniques work over a longer period than dendrochronology itself. Sediments are laid down on the sea bottom at an uneven rate, and in stripes which we can think of as equivalent to tree rings. These stripes can be counted, and signatures recognised, in core samples extracted by deep cylindrical probes.

Another example, which we encountered in the Epilogue to the Sloth's Tale, is palaeomagnetic dating. As we saw there, the Earth's magnetic field reverses from time to time. What had been magnetic north suddenly becomes magnetic south for some thousands of years, then flips again. This has happened 282 times during the last 10 million years. Although I say 'flips' and 'suddenly', it is sudden only by geological standards. Entertaining as it might be if a polar flip today turned every plane and ship around in its tracks, that isn't the way things work. The 'flip' actually takes a few thousand years, and is much more complicated than the flip word suggests. The magnetic North Pole in any case seldom coincides exactly with the true, geographic North Pole (around which the Earth spins). It wanders around the polar region over the years. At present the magnetic North Pole is about 300 miles from the true North Pole, heading rapidly from Canada towards Russia. During a 'flip' there is an interregnum of magnetic confusion, with large and complicated variations in field strength and direction, sometimes involving the temporary appearance of more than one magnetic north and more than one magnetic south. Eventually the confusion stabilises again, and when the dust settles it may turn out that the previous magnetic north is now near the true South Pole and vice versa. Stability, with wandering, then resumes for perhaps a million years until the next flip.

A thousand years in geology's sight is but an evening gone. The time spent 'flipping' is negligible compared to the time spent in the rough vicinity of either the true North or the true South Pole. Nature, as we saw earlier, keeps an automatic record of such events. In molten volcanic rock, certain minerals behave like little compass needles. When the molten rock solidifies, these mineral needles constitute a 'frozen' record of the Earth's magnetic field at the moment of solidification (by a rather different process, palaeomag-netism can be observed in sedimentary rock, too). After a 'flip', the miniature compass needles in the rocks point in the opposite dir-ection compared with before the flip. It's like tree rings all over again, except that the stripes are not a year apart but of the order of a million years. Once again, patterns of stripes can be matched up with other patterns, and a continuous chronology of magnetic flips can be daisy-chained together. Absolute dates can't be calculated by counting stripes because, unlike tree rings, the stripes represent unequal durations. Nevertheless, the same signature pattern of stripes can be picked up in different places. This means that if some other method of absolute dating is available for one of the places, magnetic stripe patterns, like the Parsons code for a melody, can be used to recognise the same time zone in other places. As with tree rings and other dating methods, the full picture is built up from fragments gathered in different places.

Tree rings are good for dating recent relics to the nearest year. For older dates, with inevitably less fine pinpointing, we exploit the well-understood physics of radioactive decay. To explain this, we begin with a digression.

All matter is made of *atoms*. There are more than 100 types of atoms, corresponding to the same number of *elements*. Examples of elements are iron, oxygen, calcium, chlorine, carbon, sodium and hydrogen. Most matter consists not of pure elements but of *compounds:* two or more atoms of various elements bonded together, as in calcium carbonate, sodium chloride, carbon monoxide. The binding of atoms into compounds is mediated by *electrons*, which are tiny particles orbiting (a metaphor to help us understand their real behaviour, which is much stranger) the central *nucleus* of each atom. A nucleus is huge compared to an electron but tiny compared to an electron's orbit. Your hand, consisting mostly of empty space, meets hard resistance when it strikes a block of iron, also consisting mostly of empty space, because forces associated with the atoms in the two solids interact in such a way as to prevent them passing through each other. Consequently iron and stone seem solid to us because our brains most usefully serve us by constructing an illusion of solidity.

It has long been understood that a compound can be separated into its component parts, and recombined to make the same or a different compound with the emission or consumption of energy. Such easy-come easy-go interactions between atoms constitute chemistry. But, until the twentieth century, the atom itself was thought to be inviolate. It was the smallest possible particle of an element. A gold atom was a tiny speck of gold, qualitatively different from a copper atom, which was a minimal particle of copper. The modern view is more elegant. Gold atoms, copper atoms, hydrogen atoms and so on are just different arrangements of the same fundamental particles, just as horse genes, lettuce genes, human genes and bacterial genes have no essential 'flavour' of horse, lettuce, human or bacteria but are just different combinations of the same four DNA letters. In the same way as chemical compounds have long been understood to be arrangements put together from a finite repertoire of 100 or so atoms, so each atomic nucleus turns out to be an arrangement of two fundamental particles, the *protons* and *neutrons*. A gold nucleus is not 'made of gold'. Like all other nuclei, it is made of protons and neutrons. An iron nucleus differs from a gold nucleus, not because it is made of a qualitatively different kind of stuff called iron, but simply because it contains 26 protons and some neutrons, instead of gold's 79 protons and some neutrons. At the level of a single atom there is no 'stuff' that has the properties of gold or iron. There are just different combinations of protons, neutrons and electrons. Physicists go on to tell us that protons and neutrons are themselves composed of yet more fundamental particles, the quarks, but we shall not follow them to such depths.

Protons and neutrons are almost the same size as each other, and much larger than electrons. Unlike a neutron, which is electrically neutral, each proton has one unit of electric charge (arbitrarily designated positive), which exactly balances the negative charge of one electron 'in orbit' around the nucleus. A proton can be transformed into a neutron if it absorbs an electron, whose negative charge neutralises the proton's positive one. Conversely, a neutron can transform itself into a proton by expelling a unit of negative charge – one electron. Such transformations are examples of nuclear reactions, as opposed to chemical reactions. Chemical reactions leave the nucleus intact. Nuclear reactions change it. They usually involve much larger exchanges of energy than chemical reactions, which is why nuclear weapons are so much more devastating, weight for weight, than conventional (i.e. chemical) explosives. The alchemists' quest to change one metallic element into another failed only because they tried to do it by chemical rather than nuclear means.

Each element has a characteristic number of protons in its atomic nucleus, and the same number of electrons in 'orbit' around the nucleus: one for hydrogen, two for helium, six for carbon, 11 for sodium, 26 for iron, 82 for lead, 92 for uranium. It is this number, the so-called atomic number, which (acting via the electrons) largely determines an element's chemical behaviour. The neutrons have little effect on an element's chemical properties, but they do affect its mass and they do affect its nuclear reactions.

A nucleus typically has roughly the same number of neutrons as protons, or a few more. Unlike the proton count, which is fixed for any given element, the neutron count varies. Normal carbon has six protons and six neutrons, giving a total 'mass number' of 12 (since the mass of electrons is negligible and a neutron weighs approximately the same as a proton). It is therefore called carbon 12. Carbon 13 has one extra neutron, and carbon 14 two extra neutrons, but they all have six protons. Such different 'versions' of an element are called 'isotopes'. The reason all three of these isotopes have the same name, carbon, is that they have the same atomic number, 6, and therefore all have the same chemical properties. If nuclear reactions had been discovered before chemical reactions, perhaps the isotopes would have been given different names. In a few cases, isotopes are different enough to earn different names. Normal hydrogen has no neutrons. Hydrogen 2 (one proton and one neutron) is called deuterium. Hydrogen 3 (one proton and two neutrons) is called tritium. All behave chemically as hydrogen. For example, deuterium combines with oxygen to make a form of water called heavy water, famous for its use in the manufacture of hydrogen bombs.

Isotopes, then, differ only in the number of neutrons they have, along with the fixed number of protons that characterise the element. Among the isotopes of an element, some may have an unstable nucleus, meaning it has an occasional tendency to change at an unpredictable instant, though with predictable probability, into a different kind of nucleus. Other isotopes are stable: their probability of changing is zero. Another word for unstable is radioactive. Lead has four stable isotopes and 25 known unstable ones. All isotopes of the very heavy metal uranium are unstable – all are radioactive. Radioactivity is the key to the absolute dating of rocks and their fossils: hence the need for this digression to explain it.

What actually happens when an unstable, radioactive element changes into a different element? There are various ways in which this can happen, but the two best known are called alpha decay and beta decay. In alpha decay the parent nucleus loses an 'alpha particle', which is a pellet consisting of two protons and two neutrons stuck together. The mass number therefore drops by four units, but the atomic number drops by only two units (corresponding to the two protons lost). So the element changes, chemically speaking, into whichever element has two fewer protons. Uranium 238 (with 92 protons and 146 neutrons) decays into thorium 234 (with 90 protons and 144 neutrons).

Beta decay is different. One neutron in the parent nucleus turns into a proton, and it does so by ejecting a beta particle, which is a single unit of negative charge or one electron. The mass number of the nucleus remains the same because the total number of protons plus neutrons remains the same, and electrons are too small to bother with. But the atomic number increases by one because there is now one more proton than before. Sodium 24 transforms itself, by beta decay, into magnesium 24. The mass number has remained the same, 24. The atomic number has increased from 11, which is uniquely diagnostic of sodium, to 12, which is uniquely diagnostic of magnesium.

A third kind of transformation is neutron-proton replacement. A stray neutron hits a nucleus and knocks one proton out of the nucleus, taking its place. So, as in beta decay, there is no change in the mass number. But this time the atomic number has decreased by one because of the loss of one proton. Remember that the atomic number is simply the number of protons in the nucleus. A fourth way in which one element can turn into another, which has the same effect on atomic number and mass number, is electron capture. This is a kind of reversal of beta decay. Whereas in beta decay a neutron turns into a proton and expels an electron, electron capture transforms a proton into a neutron by neutralising its charge. So the atomic number drops by one, while the mass number remains the same. Potassium 40 (atomic number 19) decays to argon 40 (atomic number 18) by this means. And there are various other ways in which nuclei can be radioactively transformed into other nuclei.

One of the cardinal principles of quantum mechanics is that it is impossible to predict exactly when a particular nucleus of an unstable element will decay. But we can measure the statistical likelihood that it will happen. This measured likelihood turns out to be utterly characteristic of a given isotope. The preferred measure is the half-life. To measure the half-life of a radioactive isotope, take a lump of the stuff and count how long it takes for exactly one half of it to decay into something else. The half-life of strontium 90 is 28 years. If you have 100 grams of strontium 90, after 28 years you'll have only

50 grams left. The rest will have turned into yttrium 90 (as it happens, which in turn changes into zirconium 90). Does this mean that after another 28 years you'll have no strontium left? Emphatically no. You'll have 25 grams left. After another 28 years the amount of strontium will have halved again, to 12.5 grams. Theoretically, it never reaches zero but only approaches it by successively halved steps. This is the reason we have to talk about the half-life rather than the 'life' of a radioactive isotope.

The half-life of carbon 15 is 2.4 seconds. After 2.4 seconds you'll be left with half of your original sample. After another 2.4 seconds you'll have only a quarter of your original sample. After another 2.4 seconds you are down to an eighth, and so on. The half-life of uranium 238 is nearly 4.5 billion years. This is approximately the age of the solar system. So, of all the uranium 238 that was present on Earth when it first formed, about half now remains. It is a wonderful and very useful fact about radioactivity that half-lives of different elements span such a colossal range, from fractions of seconds to billions of years.

We are approaching the point of this whole digression. The fact that each radioactive isotope has a particular half-life offers an opportunity to date rocks. Volcanic rocks often contain radioactive isotopes, such as potassium 40. Potassium 40 decays to argon 40 with a half-life of 1.3 billion years. Here, potentially, is an accurate clock. But it's no use just measuring the amount of potassium 40 in a rock. You don't know how much there was when it started! What you need is the ratio of potassium 40 to argon 40. Fortunately, when potassium 40 in a rock crystal decays, the argon 40 (a gas) remains trapped in the crystal. If there are equal amounts of potassium 40 and argon 40 in the substance of the crystal, you know that half the original potassium 40 has decayed. It is therefore 1.3 billion years since the crystal was formed. If there's, say, three times as much argon 40 as potassium 40, only one quarter (half of a half) of the original potassium 40 remains, so the age of the crystal is two half lives or 2.6 billion years.

The moment of crystallisation, which in the case of volcanic rocks is the moment when the molten lava solidified, is the moment when the clock was zeroed. Thereafter, the parent isotope steadily decays and the daughter isotope remains trapped in the crystal. All you have to do is measure the ratio of the two amounts, look up the half-life of the parent isotope in a physics book, and it is easy to calculate the age of the crystal. As I said earlier, fossils are usually found in sedimentary rocks, while dateable crystals are usually in volcanic rocks, so fossils themselves have to be dated indirectly by looking at volcanic rocks that sandwich their strata.

A complication is that often the first product of the decay is itself another unstable isotope. Argon 40, the first product of decay of potassium 40, happens to be stable. But when uranium 238 decays it passes through a cascade of no fewer than 14 unstable intermediate stages, including nine alpha decays and seven beta decays, before it finally comes to rest as the stable isotope lead 206. By far the longest half-life of the cascade (4.5 billion years) belongs to the first transition, from uranium 238 to thorium 234. An intermediate step in the cascade, from bismuth 214 to thallium 210, has a half-life of only 20 minutes, and even that is not the fastest (i.e. most probable). The later transitions take negligible time compared to the first, so the observed ratio of uranium 238 to the finally stable lead 206 can be set against a half-life of 4.5 billion years to calculate the age of a particular rock.

The uranium/lead method and the potassium/argon method, with their half-lives measured in billions of years, are useful for dating fossils of great age. But they are too coarse for dating younger rocks. For these, we need isotopes with shorter half-lives. Fortunately a range of clocks is available with a wide selection of isotopic half-lives. You choose your half-life to give best resolution for the rocks with which you are working. Better yet, the different clocks can be used as checks on each other.

The fastest radioactive clock in common use is the carbon 14 clock, and this brings us full circle to the teller of this tale, for wood is one of the main materials subjected to carbon 14 dating by archaeologists. Carbon 14 decays to nitrogen 14 with a half-life of 5,730 years. The carbon 14 clock is unusual in that it is used to date the actual dead tissues themselves, not volcanic rocks sandwiching them. Carbon 14 dating is so important for relatively recent history – much younger than most fossils, and spanning the range of history normally called archaeology – that it deserves special treatment.

Most of the carbon in the world consists of the stable isotope carbon 12. About one million-millionth part of the world's carbon consists of the unstable isotope carbon 14. With a half-life measured in only thousands of years, all the carbon 14 on Earth would long since have decayed to nitrogen 14 if it were not being renewed.

Fortunately, a few atoms of nitrogen 14, the most abundant gas in the atmosphere, are continually being transformed, by bombardment of cosmic rays, into carbon 14. The rate of creation of carbon 14 is approximately constant. Most of the carbon in the atmosphere, whether carbon 14 or the more usual carbon 12, is chemically combined with oxygen in the form of carbon dioxide. This gas is sucked in by plants, and the carbon atoms used to build their tissues. To plants, carbon 14 and carbon 12 look almost the same (plants are only 'interested' in chemistry, not the nuclear properties of atoms). The two varieties of carbon dioxide are imbibed approximately in proportion to their availability. Plants are eaten by animals, which may be eaten by yet other animals, so carbon 14 is dispersed in a known proportion relative to carbon 12 throughout the food chain during a time which is short compared to the half-life of carbon 14. The two isotopes exist in all living tissues in approximately the same proportion as in the atmosphere, one part in a million million. To be sure, they occasionally decay to nitrogen 14 atoms. But this constant rate is offset by their continuous exchange, via the links of the food chain, with the ever-renewed carbon dioxide of the atmosphere.

All this changes at the moment of death. A dead predator is cut off from the food chain. A dead plant no longer takes in fresh supplies of carbon dioxide from the atmosphere. A dead herbivore no longer eats fresh plants. The carbon 14 in a dead animal or plant continues to decay to nitrogen 14. But it is not replenished by fresh supplies from the atmosphere. So the ratio of carbon 14 to carbon 12 in the dead tissues starts to drop. And it drops with a half-life of 5,730 years. The bottom line is that we can tell when an animal or plant died by measuring the ratio of carbon 14 to carbon 12. This method is gaining more and more precision as researchers, not least Tom Higham here in Oxford, develop techniques to extract individual ancient molecules, which avoids the problem of contamination. Carbon 14 dating is how it was proved that the Turin Shroud cannot have belonged to Jesus – its date is medieval. It is a wonderful tool for dating the relics of relatively recent history, but it is of no use for more ancient dating because almost all the carbon 14 has decayed to nitrogen 14, and the residue is too tiny to measure accurately.

There are other methods of absolute dating, and new ones are being invented all the time. The beauty of having so many methods is partly that they collectively span such an enormous range of timescales. It is also that they can be used as a cross-check on each other.

It is extremely hard to argue against datings that are corroborated across different methods.

THE HUMPED BLADDERWORT'S TALE

The humped bladderwort, *Utricularia gibba*, is an aquatic plant with yellow flowers that look somewhat like those of a garden snapdragon. The small 'bladders' that adorn its roots may seem odd at first glance – noteworthy only because they contribute to its entertaining name. But their purpose is more sinister. Like other bladderworts in the same genus, the plant is carnivorous. Each bladder is a lethal trap, with (literally) a hair trigger. It is a sealed low-pressure chamber, rigged to open suddenly when a nearby hair is brushed. Passing animals catch the hair, the trap is sprung, and they are sucked into the chamber, where they are imprisoned then gradually digested.

Organisms exploiting one another are a cornerstone of natural history documentaries. The humped bladderwort captivates us because of the reversal of fortunes between plant and animal. But it also provides a natural history lesson on a different level, a lesson concering exploitation not between species in an ecosystem, but between genes in a genome. To grasp the message of this odd little plant, we must temporarily park it to one side. It will return to centre stage at the end of this tale, where its endearingly small genome will disclose what most of our DNA actually does.

Since the 1970s, we have known that the amount of DNA in a creature's genome, called the 'C-value', tells us very little about its overall complexity. This even has a name, the 'C-value paradox'. We have already met the marbled lungfish, with its bloated 133-billion-letter genome. Another example that cries for attention is advocated by Ryan Gregory, one of the leading researchers in this field. While you prepare dinner, tearfully ponder on the fact that onions (*Allium cepa*) have a genome five times larger than humans. Perhaps this is because plants and animals arrange their genetic instructions in different ways? No. The same disparity is also seen between closely related species. *Plethodon*, a genus of salamander, contains some species with genomes four times larger than others. It's also true for onion relatives. *Allium altyncolicum*, a chive-like species, has a genome 'only' twice as large as ours (the C-value of chives themselves is a fraction more). *Allium ursinum*, the wild garlic, has a genome fully nine times larger than our own. An even starker example is maize (sweetcorn), where genomes can vary by 50 per cent in size within a single species.

The C-value paradox can't simply be explained by whole genome duplication, the process that quadrupled the vertebrate genome in our distant past (see the Lamprey's Tale). It's true that the resulting condition, known as polyploidy, is fairly common in plants, especially crops like wheat and ornamental plants such as varieties of daffodil. In these cases the duplication is recent, and it is very clear under the microscope that the chromosomes come in (say) four or six near-identical copies, rather than the more conventional two. It's also true that we can detect more ancient doubling by looking at full genome sequences, and hunting for multiple copies of similar genes (we saw this in the Hox genes of vertebrates). Although something similar can clearly be seen in most groups of flowering plants, it is not evident in the case of onions. And more importantly, it does not explain the fourfold differences in genome size *within* the different species of *Allium*, or the different species of plethodontid salamanders.

Most of the difference in genome size is not down to the number of genes either, if we use 'gene' in the technical sense to mean a protein-coding sequence. Onion genomes have not yet been sequenced, but we know that in humans, protein-coding regions account for only just over 1 per cent of the DNA, and we actually expect that figure to be less in onions. By comparing known genomes, it is clear that the answer to the C-value paradox lies in the vast expanse of non-coding DNA in our pilgrims' genomes.

'Non-coding DNA' is an unfortunate term. It sounds like nonsense sequence which does not do anything at all. That is not necessarily the case. Coding DNA is so named because it uses the 'genetic code' (the three-letter mapping of DNA letters to amino acids in a protein). Non-coding DNA does not, but can still contain functional instructions. It's confusing terminology, and – as pointed out by Francis Crick, no less – the translation from DNA to protein is not really a code anyway. Technically it is a cipher: a method for converting one length of information into another. It would be better to axe the term 'non-coding DNA' from the genetic lexicon and use something less misleading, perhaps 'untranslated DNA'. As we saw in the Mouse's Tale, these untranslated regions contain the vital switches that turn genes on and off. But as we also mentioned there, such controls account for less than a tenth of our untranslated DNA. What is the rest for?

What is anything for? Here's a story from the entomologist George McGavin. At an open day in the museum where both authors of this book used to work, he was accosted by a lady who demanded, 'What are wasps *for*?' As he recounts it, he patiently explained to her the basics of evolution, culminating in the conclusion that, from a human point of view, a creature need not be *for* anything at all. Nodding in a satisfied way, she thought a little, and then came back with the next question: 'What are *worms* for?'

Dr McGavin says that now his retort is simply 'wasps are for making other wasps'. That pithy description is even more accurate at the level of the DNA. The function of a DNA sequence, inasmuch as we read one into it, is ultimately to make more copies of itself. This is the message of *The Selfish Gene*, and it holds the key to the C-value paradox. Of course, in hindsight we can look at the effects of a DNA sequence and propose that they give some advantage to the organism (which thus indirectly benefits the sequence). If it is reasonable to say that our eyes have evolved 'for seeing', then at the molecular level it is reasonable to say that red, green and blue opsin genes have evolved 'for seeing' too. We ascribe them a function: to produce proteins that absorb certain wavelengths of light, allowing us not just to see, but to see in colour (see the Howler Monkey's Tale). But this is shorthand. There is only one ultimate reason that both red and green opsin genes exist: because they have been copied more times than the bodies that they are in have died. Looking back, there might be many reasons for this. Perhaps it was just chance. More likely is that they confer a statistical benefit on the body as a whole, allowing an ancestral monkey to distinguish red from green, perhaps pick riper fruit, and have more children – children which themselves are likely to inherit copies of the opsin genes. Most protein-coding genes exist for the same historical reason. They have co-operated with other genes to ensure that more copies are made of the entire genome – a succession of entire genomes through populations and through generations. That is what gives the illusion both of function and of coherent design.

But there are other ways that a DNA sequence can make copies of itself. The most obvious is to move independently, between different bodies. The most familiar example is a virus, and it is worth introducing viruses here, both because they are relevant to our tale, and also because they will not formally join us elsewhere on our pilgrimage, as it is unclear how they are related to other life forms.

Viruses are short stretches of DNA or RNA, wrapped up in a protective protein coat. At least some sorts of virus are likely to be pieces of the genomes of other organisms, which have evolved the ability to move between cells. These are genes that have literally gained a life of their own.*

Viruses are too short to encode all the proteins required for DNA replication (the shortest known virus, *Porcine circovirus*, is only 1,768 letters long). Rather than arranging its own replication, a virus hijacks host cells to do so. The host's cellular machinery is used to copy the viral genome, and also to synthesise the other parts of the virus such as the outer coat and the machinery that allows the virus to sneak into cells in the first place.

Your body contains viruses as you are reading this. *Varicella zoster* virus causes chickenpox and then sits dormant in your nerve cells for the rest of your life. If you had chickenpox as a child, chances are the virus is still in your body. We might well ask what the virus is for, why it is there? Clearly, because it can arrange for its own reproduction and because your immune system, for whatever reason, cannot get rid of it. Viruses, in other words, exist to make more viruses.

Everyone would accept this conclusion in the case of *Varicella*, a virus whose genome always stays isolated from the rest of the DNA in the cell. But the implication gets more uncomfortable in the case of some viruses, known as retroviruses, which permanently paste their genome into the main DNA of the cell, and so can become part of the host genome – part of ourselves. As you might expect, depending on the insertion point, this can cause serious problems for the host, cancer being a particular problem (a distressingly common example is leukaemia in cats). A greater fear for humans is a retrovirus that integrates into certain cells of the immune system, the infamous human immunodeficiency virus, HIV. Although HIV is the best known, it is not the most common human retrovirus, for immune system cells are ultimately a dead end. When we die, they die too. There is a related virus, in fact a set of viruses, that get around this problem. All of us are infected with them, because they have managed to infiltrate the human germ line, and so are passed from parent

---

* Arguments arise over whether viruses are 'alive', although everyone agrees that they evolve. When we reach our final Canterbury, the division into life and non-life will be seen to be another case of the tyranny of the discontinuous mind. It is an irrelevant semantic point whether we choose to label viruses as alive. We know how they work. That is enough.

to child just like the rest of our DNA. These are called endogenous retroviruses, and they have been infecting us and our ancestors for millions of years.* There are many different types, which have infected us at different times in evolution, and they have spread not just to additional cells but (more frequently) to additional places in the genome. By spreading around our DNA, viral-derived sequences – including a number of non-retroviruses that have also managed to sneak into the germ line – have come to make up about 10 per cent of the human genome.

Here, then, is an explanation for some untranslated DNA, and why it might vary in quantity. It simply reflects a history of ancient infections. Gkikas Magiorkinis and colleagues have recently shown that in the apes (but not in the Old World monkeys) the amount of endogenous viral replication in our genome has slowed. One major family of endogenous viruses has even gone extinct. That doesn't mean that their DNA has been removed, however. It has simply mutated so that the sequence can no longer replicate. The bodies of these extinct viruses still litter our genome.

We can draw a reasonable analogy between a genome and a computer hard disk. Like a genome, a hard disk is full of random old clutter, which is not visible from the outside. When you delete an outdated file on your computer, the file itself is not erased: it is the pointer to it which is removed (which is why, in an emergency, you can recover previously deleted files). Similarly in a genome, unused genes are usually not removed, but gradually diverge from their originally functional sequence due to accumulated mutations. Viruses take the analogy one step further: we even speak of 'computer viruses', pieces of code that self-replicate, sometimes with the malign intent of the programmer. As with an infected computer, biological viruses, or the remains of them, can persist undetected in our genome as long as they do not interfere too drastically with normal function. In fact, it turns out that endogenous viruses are not the main malware in our genomes. They are merely one player in a much larger cast of actors, which go under the general name of transposable elements.

Transposable elements, or transposons, are united by their ability

---

* This leads to an interesting suggestion for why animals separate the germ line from the rest of the body early on in life: to stop viral infections acquired during adulthood spreading into the germ line. It is a specific example of a more general reason for separate germ lines: to prevent selfish replicators evolving methods which help them infiltrate the germ line but are deleterious for the body as a whole.

to replicate within a genome: they spread by jumping into new locations along the DNA of a single cell. Transposons account for around half of all human DNA yet, like computer malware, we see little of their activities because their continued survival depends on stealth. Careful detective work in the 1940s demonstrated their existence in maize, but it took decades to be accepted, despite transposons making up a staggering 85 per cent of the maize genome (Barbara McClintock, their discoverer, was finally awarded the Nobel Prize in 1983). As we have seen, some elements are viral, and can jump between organisms. Far more common, however, are the unseen multitude of transposons that are specialised for replication purely within a single genome. DNA sequencing reveals that the variable amounts of genetic material found in onions, salamanders, maize and all other complex life forms are largely due to differences in the number of transposons.

It's time to return to the C-value paradox. As long ago as 1976 I made the following suggestion in *The Selfish Gene*: 'The simplest way to explain the surplus DNA is to suppose that it is a parasite, or at best a harmless but useless passenger, hitching a ride in the survival machines created by the other DNA.' Four years later a seminal pair of research papers were published by Ford Doolittle and Carmen Sapienza on one hand, and Leslie Orgel and Francis Crick on the other. The first was entitled 'Selfish genes, the phenotype paradigm, and genome evolution', the second 'Selfish DNA: the Ultimate Parasite'. I am, of course, extremely gratified, and the term 'selfish DNA' is now commonly used to describe all transposable elements. But the word selfish here is, I think, infelicitous. In the original sense of the book title *The Selfish Gene, all* pieces of DNA ('genes' in the loose sense) are selfish: selfish by default although in practice the conventional system of inheritance encourages them to co-operate in building bodies. Transposons are a particular sort of selfish gene – we might call them 'ultra-selfish' – which have hit upon another way of replication, by spreading within the genome.

By the way, transposons are not the only piece of DNA deserving of the name ultra-selfish. There are others which are positively malevolent. According to the normal 'Mendelian' rules of inheritance, a normal gene has a 50 per cent chance of finding itself in any one sperm. Genes called 'segregation distorters' bias this chance in their own favour; the so-called t-gene in mice, for example, kills sperms which don't contain it. There are many other examples of genetic sequences which have found ways of gaming the Mendelian system. Ultra-selfish genes of all kinds are the subject of a very comprehensive book by Austin Burt and Robert Trivers, called *Genes in Conflict: The Biology of Selfish Genetic Elements.*

Of all the types of ultra-selfish DNA, transposons are, however, the most prevalent. We know much more about them than we did 30 years ago. We know that, like the endogenous viruses, most of them are inactive – usually because they have been hit by an incurable mutation or have jumped into an unfavourable spot in the genome. We also know, for instance, that perhaps the most common parasitic element in the human genome is a short sequence called *Alu*, roughly 300 letters long and repeated over a million times. Not much functionality can be contained in 300 letters, and *Alu* can only replicate by using parts of other transposons: in other words it is a 'hyperparasite', a parasite of other genomic parasites. *Alu* is specific to primates, where it presumably arose some time between Concestors 9 and 8, starting out as a single copy. Its origin is revealing. Parts of its DNA sequence match a vital gene for a molecule called 7SL RNA, which is found in all living cells and which helps move proteins around the cell. *Alu* seems to have been created by a deletion in the middle of the gene, which accidentally endowed it with the ability to transpose. This is a salutary lesson in how natural selection works. Thinking of genes as 'wanting to replicate' is misleading. It is purely that sequences that happen to acquire this property are – for obvious reasons – the ones that are now found in our genome.

We also have a number of other transposable elements, mostly longer than *Alu* and many of which paste themselves into the genome in a similar way to retroviruses. A persuasive theory is that retroviruses such as HIV evolved when these elements managed to 'go viral' by acquiring coat proteins from other viruses. More generally, the types and numbers of transposons within a genome vary between species, depending on their evolutionary history. That is what mostly accounts for the C-value paradox. Why is this DNA there? As Doolittle and Sapienza put it: 'When a given DNA, or class of DNAs, of unproven phenotypic function can be shown to have evolved a strategy (such as transposition) which ensures its genomic survival, then no other explanation for its existence is necessary.'

Some people, tempted by the vice of human perfectionism, object to this idea. 'Functions' for transposable sequences are often sought, without acknowledging that this may be unnecessary. Possibly, these people are misled by the large number of important benefits to the genome which can be traced back to mutations caused by transposons. We saw this in the Howler Monkey's Tale where we found that *Alu* caused the gene duplication that gave us three-colour vision. Transposons are also known to pick up the 'switches' that turn genes on and off, and move these around the genome, changing the activity of existing genes. This deep involvement with the evolution of our genome is unsurprising, since ultra-selfish elements not only make up a huge proportion of our genome but have also evolved to chop and change our DNA. We might even expect some of this DNA to have been later co-opted into roles that help the entire organism to reproduce. But in the main, the mutation-causing activities of these parasitic pieces of DNA are deleterious, simply because the vast majority of all mutations are deleterious. And to repeat the caution of the Howler Monkey's Tale, however much parasitic elements are involved with the long-term restructuring of our genomes (and the evidence suggests they are the main driving force), this is not the *reason* that they are kept in our genomes. Natural selection doesn't keep pieces of DNA on the off-chance that they might provide useful mutations one day: 'might come in handy in the Cretaceous', as Sydney Brenner witheringly put it.

The search for benefits provided by ultra-selfish DNA has been given a misguided boost by an otherwise useful international collaboration called ENCODE, whose laudable aim was to map which parts of the genome are used by the cell. Recently, this team has been widely reported as stating that 80 per cent of the genome in humans had a 'biochemical function'. Given that half of our genome is parasitic, most of it dead or decaying, this seems a little odd. But as we have seen, function is a rather slippery word. The ENCODE project seem to be using it in an even more slippery manner than normal, defining a DNA sequence to have 'biochemical function' if it interacts with the cell in almost any repeatable way. That of course includes the activities of viruses and transposons (a 'biochemical function', although not necessarily of benefit to the whole organism). It also includes defunct elements whose sequence inadvertently interacts with the cellular machinery.

A more meaningful picture comes from taking an explicitly evolutionary approach. If a particular DNA sequence provides a useful function for the organism, we expect it to be conserved over geological time. This is a relatively easy thing to measure. For example, a recent study by Chris Rands and a team here at Oxford looked for regions of our genome that showed very few insertion or deletion mutations. These 'indels' are disruptive mutations that are unlikely to be tolerated in DNA which serves a useful function – they are likely to be removed by natural selection. By this measure, undisruptable regions total 7 to 9 per cent of our DNA. That 7 to 9 per cent contains the basic instructions for making a human. The rest apparently provides little useful function. Or rather, human beings seem unaffected by major mutations in these areas, despite much of this DNA having some sort of biochemical activity. Infamously, these 'useless' areas are often termed 'junk DNA', although, as the case of transposons shows, such DNA can nevertheless have a 'function' – just not necessarily one that benefits the organism as a whole.* The function of any DNA is to replicate itself: we are accustomed to dealing with that subset of DNA which achieves this end by the roundabout route of collaborating in the embryonic construction of an organism.

Whatever method we use to assign function to our DNA requires subtle and indirect calculations, and there is still much we don't understand. For example, it may be that the sequence itself is unimportant, but the DNA structure still has a role, for instance as some sort of padding. The real test would be to remove the 90 per cent of our DNA that we suspect is junk, and see if it still makes a human. Quite aside from being technically impossible, such an experiment would rightly fail any ethics committee on Earth. Fortunately a similar experiment has been done for us. We can finally turn to the teller of this tale, the humped bladderwort.

The Humped Bladderwort's Tale could be subtitled 'the case of the incredible shrinking genome'. Its genome is about 40 times smaller than our own, even though it actually has a substantially greater number of protein coding genes. Its genome is tiny even when compared to other flowering plants with small genomes, such as the closely related monkey flower and its slightly more distant relative the tomato. The humped bladderwort's genome was sequenced by an international team of researchers in 2013, revealing how it has pulled off this rather spectacular feat. As you will no doubt have guessed, it has done it by deleting the vast majority of its ultra-selfish DNA.

* A excellent and freely available online summary is the 'quick guide' of questions and answers recently written by Sean Eddy for the journal *Current Biology*, entitled 'The C-value paradox, junk DNA and ENCODE'.

Transposable elements now make up less than 3 per cent of its genome.

It is still unclear quite how the plant has managed to do this, but, counterintuitively, it seems to have actually *increased* its genome first, in three successive rounds of whole genome duplication. One suggestion is that having spare gene copies enabled the plant to survive when very large chunks of the genome were subsequently deleted; that logic seems unclear to us. More research is needed before we can properly work out how this plant has managed to rid itself of its freeloading parasitic DNA.

Why has the humped bladderwort, in particular, done this? Evolutionary botanists are not sure, but the answer may lie in the fact that like other carnivorous plants it lives in very nutrient-poor conditions. In particular the humped bladderwort is usually chronically short of phosphorus, a major component of DNA. It may be that it has seen extremely strong selection pressure to economise on DNA production.

Even in other organisms, such as ourselves, the amount of ultra-selfish DNA must be under *some* sort of constraint, otherwise it would fill almost the entire genome. One way to keep it under control is to evolve transposon-suppressing mechanisms. The 2006 Nobel Prize in Medicine was awarded to Andrew Fire and Craig Mello for the discovery in 1998 of a previously unknown mechanism of gene control. This mechanism, called RNA interference (RNAi), acts rather like a genomic immune system. It recognises and shuts down any gene which leaves (double-stranded) RNA floating around the cell – as transposons and retroviruses do. The descendants of Concestor 38 have since evolved to use the RNAi mechanism to regulate all sorts of genes, not just the parasitic ones, and lately it has become a staple of biotechnology, but it quite clearly originated as means to rein in the pernicious activity of transposons. More recently, other sets of control genes have been found that specifically shut down endogenous retroviruses. And the humped bladderwort shows that, *in extremis*, it is possible to delete some of this genomic junk. It represents an instructive extreme outcome – probably a temporary one – in the long-running molecular arms race between the merely selfish and the ultra-selfish components of eukaryotic genomes. The amount of DNA in our own bodies, and in all organisms, reflects the past winners and losers in this cellular civil war.

THE MIXOTRICH'S TALE

Mixotricha paradoxa means 'unexpected combination of hairs', and we shall see why in a moment. It is a micro-organism that lives in the gut of an Australian termite, 'Darwin's termite', *Mastotermes darwiniensis*. Pleasingly, though not necessarily for the human inhabitants, one of the main places where it flourishes is the town of Darwin in northern Australia.

Termites bestride the tropics like a distributed colossus. In tropical savannahs and forests, they reach population densities of 10,000 per square metre, and are estimated to consume up to a third of the total annual production of dead wood, leaves and grass. Their biomass per unit area is double that of migrating herds of wildebeest on the Serengeti and Masai Mara, but is spread across the entire tropics.

If you ask the source of the termites' alarming success, it is twofold. First, they can eat wood, which includes cellulose, lignin and other matter that animal guts normally can't digest. I'll return to this. Second, they are highly social and gain great economies from division of labour among specialists. A termite mound has many of the attributes of a single large and voracious organism, with its own anatomy, its own physiology and its own mud-fashioned organs, including an ingenious ventilation and cooling system. The mound itself stays in one place, but it has a myriad mouths and six myriad legs, and these range over a foraging area the size of a football pitch.

Termites' legendary feats of co-operation are possible, in a Darwinian world, only because the majority of individuals are sterile but closely related to a minority who are very fertile indeed. Sterile workers act like parents towards their younger siblings, thereby freeing the queen to become a specialised egg factory, and a grotesquely efficient one at that. Genes for worker behaviour are passed to future generations via the minority of the workers' siblings who are destined to reproduce (helped by the majority of their siblings who are destined to be sterile). You will appreciate that the system works only because it is a strictly non-genetic decision whether a young termite shall become a worker or a reproducer. All young termites have a genetic ticket to enter an environmental lottery which decides whether they become reproductives or workers. If there were genes for being unconditionally sterile, they obviously could not be passed on. Instead, they are *conditionally* switched-on genes. They are passed on when they find themselves in queens or kings because copies of the very same genes cause workers to labour for that end and forgo reproduction themselves.

The analogy of insect colony to human body is often made, and it is not a bad one. The majority of our cells subjugate their individuality, devoting themselves to assisting the reproduction of the minority that are capable of it: 'germ-line' cells in the testes or ovaries, whose genes are destined to travel, via sperms or eggs, into the distant future. But genetic relatedness is not the only basis for subjugation of individuality in fruitful division of labour. Any sort of mutual assistance, where each side corrects a deficiency in the other, can be favoured by natural selection on both. To see an extreme example, we dive inside the gut of an individual termite, that seething and, as I assume, noisome chemostat which is the world of the mixotrich.

Termites, as we have seen, enjoy an additional advantage over bees, wasps and ants: their prodigious feats of digestion. There is almost nothing that termites can't eat, from houses to billiard balls to priceless First Folios. Wood is potentially a rich food source but it is denied to almost all animals because cellulose and lignin are so indigestible. Termites and certain cockroaches are the outstanding exception. Termites are, indeed, related to cockroaches, and Darwin's termite, like other so-called 'lower' termites, is a sort of living fossil. One could imagine it halfway between cockroaches and advanced termites.

In order to digest cellulose, you need enzymes called cellulases. Most animals can't make cellulases, but some micro-organisms can. As Taq's Tale will explain, bacteria and archaea are biochemically more versatile than the rest of the living kingdoms put together. Animals and plants perform a fraction of the biochemical mix of tricks available to bacteria. For digesting cellulose, herbivorous mammals all rely upon microbes in their guts. Over evolutionary time, they have entered into a partnership in which they make use of chemicals such as acetic acid which, to the microbes, are waste products. The microbes themselves gain a safe haven with plenty of raw materials for their own biochemistries, preprocessed and ready-chopped into small, manageable pieces. All herbivorous mammals have bacteria in the lower gut, which the food reaches after the mammal's own digestive juices have had a go at it. Sloths, kangaroos, colobus monkeys and especially cud-chewing ruminants have independently evolved the trick of also keeping bacteria in the upper portion of the gut, which precedes the mammal's own main digestive efforts.

Unlike mammals, termites are capable of manufacturing their own cellulases, at least in the case of the so-called 'advanced' termites. But up to one-third of the net weight of a more primitive (i.e. more cockroach-like) termite, such as Darwin's termite, consists of its rich gut fauna of microbes, including eukaryotic protozoa as well as bacteria. The termites locate and chew the wood into small, manageable chips. The microbes live on the wood chips, digesting them with enzymes unavailable to the termites' own biochemical toolkit. Or you could say the microbes have become tools in the termites' toolkit. As with the cattle, it is the waste products of the microbes that the termites live on. I suppose we could say that Darwin's termite and the other primitive termites farm micro-organisms in their guts.* And this brings us, eventually, to the mixotrich, whose tale this is.

Mixotricha paradoxa is not a bacterium. Like many of the microbes in termite guts, it is a large protozoan, half a millimetre long or more, and large enough to contain hundreds of thousands of bacteria inside itself – as we shall see. It lives nowhere except in the gut of Darwin's termite, where it is a member of the mixed community of microbes that thrive on the wood chips milled by the termite's jaws. Micro-organisms populate the termite's gut as richly as the termites themselves populate the mound, and as termite mounds populate the savannah. If the mound is a town of termites, each termite gut is a town of micro-organisms. We have here a two-level community. But – and now we come to the crux of the tale – there is a third level, and the details are utterly remarkable. *Mixotricha* itself is a town. The full story was revealed by the work of L. R. Cleveland and A. V. Grimstone, but it is especially the American biologist Lynn Margulis who drew our attention to *Mixotricha*'s significance for evolution.

When J. L. Sutherland first examined *Mixotricha* in the early 1930s she saw two kinds of 'hairs' waving on its surface. It was almost completely carpeted by thousands of tiny hairs, beating to and fro.

---

* There are two main processes by which energy is extracted from food fuel: anaerobic (without oxygen) and aerobic (with oxygen). Both are chemical sequences in which fuel, rather than being burned, is coaxed into trickling out its energy in a way that can be efficiently used. The most common anaerobic sequence yields pyruvate as its major product, and this is the starting point of the most common aerobic cascade. Termites go out of their way to deprive their guts of free oxygen, thereby forcing their microbes to use only the anaerobic process, using wood fuel to produce pyruvate that the termite can then use for aerobic energy release.

She also saw a few very long, thin, whip-like structures at the front end. Both seemed familiar to her, the small ones as 'cilia', the large ones as 'flagella'. Cilia are common in animal cells, for instance in our nasal passages, and they cover the surface of those protozoans called, not surprisingly, ciliates. Another traditionally recognised group of protozoans, the flagellates, have much longer, whip-like 'flagella' (singular 'flagellum' and, unlike cilia, they often are). Cilia and flagella share an identical ultrastructure. Both are like multi-stranded cables, and the strands have exactly the same signature pattern: nine pairs in a ring surrounding one central pair.

Cilia, then, can be seen as just smaller and more numerous flagella, and Lynn Margulis went so far as to abandon the separate names and call them all by her own name of 'undulipodia', reserving 'flagella' for the very different appendages of bacteria. Nevertheless, according to the taxonomy of Sutherland's day, protozoans were supposed to have either cilia or flagella but not both.

This is the background to Sutherland's naming of *Mixotricha paradoxa*: 'unexpected combination of hairs'. *Mixotricha*, or so it seemed to Sutherland, has *both* cilia and flagella. It violates protozoological protocol. It has four large flagella at the front end, three pointing forwards and one backwards, in the manner characteristic of a particular, previously known group of flagellates called the Parabasalia. But it also has a dense coat of waving cilia. Or so it seemed.

As it has turned out, *Mixotricha*'s 'cilia' are even more unexpected than Sutherland realised, and they don't violate precedent in the way she feared. It's a pity she didn't get the chance to see *Mixotricha* alive, instead of fixed on a slide. Mixotrichs swim too smoothly to be swimming with their own undulipodia. In the words of Cleveland and Grimstone, flagellates normally 'swim at varying speeds, turning from side to side, changing direction, and sometimes coming to rest'. The same is true of ciliates. *Mixotricha*, however, glides along smoothly, usually in a straight line, never stopping unless physically blocked. Cleveland and Grimstone concluded that the smooth glid-ing movement is caused by the waving of the 'cilia' but – a far more exciting conclusion, this – they demonstrated with the electron microscope that the 'cilia' are not cilia at all. They are bacteria. Each one of the hundreds of thousands of tiny hairs is a single spirochaete – a bacterium whose entire body is a long, wiggling hair. Some important diseases, such as syphilis, are caused by spirochaetes. They

Arrangement of pill bacteria (b), brackets (br) and spirochaetes (s) on the surface of the mixotrich. From Cleveland and Grimstone [70].

normally swim freely, but *Mixotricha*'s spirochaetes are stuck to its body wall, exactly as though they were cilia.

They don't move like cilia, however: they move like spirochaetes. Cilia move with an actively propulsive rowing stroke, followed by a recovery stroke in which they bend so as to present less resistance to the water. Spirochaetes undulate in a completely different and very characteristic manner, and that is just what *Mixotricha*'s 'hairs' do. Amazingly, they seem to be co-ordinated with each other, moving in waves that begin at the front end of the body and travel backwards. Cleveland and Grimstone measured the wavelength (the distance between wave-crests) as about a hundredth of a millimetre. This suggests that the spirochaetes are somehow 'in touch' with each other. Probably they are literally in touch: responding directly to the movement of neighbours, with a delay that determines the wavelength. I don't think it is known why the waves pass from front to back.

What is known is that the spirochaetes are not just jammed haphazardly into the mixotrich's skin. On the contrary, the mixotrich has, in a repeat pattern all over its surface, a complicated apparatus for holding spirochaetes and, what's more, pointing them in a posterior direction so that their undulating movements drive the mixotrich forwards. If these spirochaetes are parasites, it is hard to think of a more remarkable example of a host being 'friendly' to its parasites. Each spirochaete has its own little emplacement, called a 'bracket' by Cleveland and Grimstone. Each bracket is tailor-made to hold one spirochaete, or sometimes more than one. No cilium could ask for more. It becomes quite tricky to draw the line between 'own' body and 'alien' body in such cases. And that, to anticipate, is one of the main messages of this tale.

The resemblance to cilia goes further. If you look with a powerful microscope into the very fabric of a ciliate protozoan, such as *Paramecium*, you'll find that every cilium has a so-called basal body at its root. Now, amazingly, although the 'cilia' of *Mixotricha* are not cilia at all, they do appear to have basal bodies. Each spirochaete-toting bracket has at its base one basal body, shaped rather like a vitamin pill. Except that . . . well, having learned about *Mixotricha*'s idiosyncratic way of doing things, what would you guess those 'basal bodies' actually are? Yes! They too are bacteria. A completely different kind of bacteria – not spirochaetes but oval, pill-shaped bacteria.

Over large parts of the body wall there is a one-to-one relationship between bracket, spirochaete and basal bacterium. Each bracket has one spirochaete stuck through it, and one pill bacterium at its base. Looking at this, it is easy to understand why Sutherland saw 'cilia'. She naturally expected to see basal bodies wherever there are cilia . . . and when she looked, lo and behold, there were the 'basal bodies'. Little could she know, both 'cilia' and 'basal bodies' were hitch-hiking bacteria. As for the four 'flagella', the only true undulipodia the mixotrich possesses, they seem to be used not for propulsion at all, but as rudders for steering the craft as it is propelled by the thousands of spirochaete 'galley slaves'. Much as I'd like to claim it, by the way, that evocative phrase is not my own. It was coined by S. L. Tamm, who found, after Cleveland and Grimstone's work on *Mixotricha*, that other termite-gut protozoa do the same trick, but instead of spirochaetes, their galley slaves are ordinary bacteria with flagella.

Now for the other bacteria in the mixotrich, the pill-shaped ones that look like basal bodies – what are they doing? Are they contributing to the economy of their host? Are they getting something out of the relationship themselves? Probably yes, but it has not been shown definitely. They may well be making cellulases that digest wood. For of course, the mixotrichs subsist on the tiny chips of wood in the termite's gut, originally broken up by the powerful jaws of the termite. We have here a triple-decker dependency, reminiscent of Jonathan Swift's verse:

> So, naturalists observe, a flea
> Has smaller fleas that on him prey;
> And these have smaller still to bite 'em;
> And so proceed ad infinitum.
> Thus every poet, in his kind,
> Is bit by him that comes behind.

By the way, Swift's scansion in the middle lines is (surprisingly) so ungainly that we can understand why Augustus De Morgan came behind for another bite, giving us the rhyme in the form that most of us know today:

> Great fleas have little fleas upon their backs to bite 'em,
> And little fleas have lesser fleas, and so ad infinitum.
> And the great fleas themselves, in turn, have greater fleas to go on;
> While these again have greater still, and greater still, and so on.

And finally we come to the strangest part of the Mixotrich's Tale, the climax towards which the narrative has been directed. This whole story of vicarious biochemistry, the borrowing by greater creatures of the biochemical talents of lesser ones inside them, is charged with evolutionary *déjà vu*. The message of the mixotrich to the rest of the pilgrims is this: *This has all happened before.* We have reached the Great Historic Rendezvous.

THE GREAT HISTORIC RENDEZVOUS

Rendezvous, in this book, has a special meaning, following the central metaphor of a backward pilgrimage. But there is one cataclysmic event, arguably the most decisive event in the history of life, which really was a rendezvous, literally a historic rendezvous that actually took place in the true, forward direction of history. This was the origin of the eukaryotic cell: the high-tech, miniature machine that is the microfoundation of all large-scale and complex life on this planet. To distinguish it from all the other metaphorical backwards rendezvous points, I call it the Great Historic Rendezvous. The word 'historic' has here a double meaning: it means 'of major importance', and it also means 'forward chronology' as opposed to backward.

Eukaryotic cells base their sophistication on internal compartments, something that can clearly be seen in our reconstruction of Concestor 38 in plate 49. The word eukaryote itself refers to the 'true kernel' or nucleus of the cell (*eu*, 'true', and *karyon*, 'kernel' or 'nut'), the compartment which houses the DNA. As in a large office or factory, other compartments exist to serve different purposes: energy production, storage areas, assembly lines, and so forth. And of course there are complex shuttling arrangements to move stuff between them. This delicate world is hard to visualise, limited as we are by our macroscopic vision. But it possesses a natural history which is, to my eyes, even more spectacular than the sprinting of a cheetah or the co-operative intricacies of a beehive. If only David Attenborough could be shrunk so that his whispering tones could do justice to this magical world of the very small.*

Most astonishing of all is the story of where this all came from. I have referred to the Great Historic Rendezvous as a single event, because of what now appears to be its single momentous consequence, but it was actually two or three events, perhaps widely spaced in time. Each one of these historic rendezvous events was a merging with bacterial cells to form a larger cell. The Mixotrich's

---

* One of the best attempts so far, although now almost ten years old, is an animated video called 'The Inner Life of the Cell', sponsored by Harvard University. It is full of wonder – literally wonderful – and is freely viewable online.

Tale, as a recent re-enactment, has prepared us to see the kind of thing that happened.

Perhaps 2 billion years ago, an ancient single-celled organism, some kind of proto-protozoan, entered into a strange relationship with a bacterium: a relationship similar to that between *Mixotricha* and its bacteria. As with *Mixotricha*, the same thing happened more than once, involving different bacteria, the events possibly separated by hundreds of millions of years. All our cells are like individual mixotrichs, stuffed with bacteria which have become so transformed by generations of co-operation with the host cell that their bacterial origins are almost lost to sight. As with the mixotrich, only more so, the bacteria have become so intimately enmeshed in the life of the eukaryotic cell that it was a major scientific triumph to detect that they were there at all. I am fond of the Cheshire Cat simile, used by Sir David Smith, one of our leading experts on this internal, or endosymbiosis, to describe the co-operative living together of once-distinct elements in cells.

> In the cell habitat, an invading organism can progressively lose pieces of itself, slowly blending into the general background, its former existence betrayed only by some relic. Indeed, one is reminded of Alice in Wonderland's encounter with the Cheshire Cat. As she watched it, 'it vanished quite slowly, beginning with the tail, and ending with the grin, which remained some time after the rest of it had gone'. There are a number of objects in a cell like the grin of the Cheshire Cat. For those who try to trace their origin, the grin is challenging and truly enigmatic.

What are the biochemical tricks that these once-free bacteria brought into our lives: tricks that they perform to this day, tricks without which life would instantly cease? The two most important ones are photosynthesis, which uses solar power to synthesise organic compounds, and oxygenates the air as a by-product; and oxidative metabolism, which reverses the process, using oxygen (ultimately from photosynthesis) to slow-burn the organic compounds and redeploy the energy that originally came from the sun.* These chemical technologies were developed before the Great Historic Rendezvous by (different) bacteria – and, in a sense, bacteria are still the only game in

---

* Bacteria (including Archaea) also have a monopoly (apart from lightning strikes and human industrial chemists) on nitrogen fixation.

town. All that has changed is that they now practise their biochemical arts in the purpose-built factories called eukaryotic cells.

Photosynthetic bacteria used to be called blue-green algae, a terrible name since most of them aren't blue-green and none of them are algae. Most are green, and it is better to call them green bacteria, although some are reddish, yellowish, brownish, blackish or, yes, in some cases bluish-green. 'Green' also is sometimes used as a word for photosynthetic, and in that sense, too, green bacteria is a good name. Their scientific name is cyanobacteria. They are true bacteria rather than Archaea, and they seem to be a good monophyletic group. In other words, all of them (and nothing else) are descended from a single ancestor which would, itself, have been classified as a cyanobacterium.

The green colour of algae, and of cabbages, pine trees and grasses, comes from small green bodies called chloroplasts within their cells. Chloroplasts are distant descendants of once free-living green bacteria. They still have their own DNA, and they still reproduce by asexual division, building up to a substantial population within each host cell. As far as a chloroplast is concerned, it is a member of a reproducing population of green bacteria. The world in which it lives and reproduces is the interior of a plant cell. From time to time its world suffers a minor upheaval when the plant cell divides into two daughter cells. Roughly half the chloroplasts find themselves in each daughter cell, and they soon resume their normal existence of reproducing to populate their new world with chloroplasts. All the while, the chloroplasts use their green pigment to trap photons from the sun and channel the sun's energy in the useful direction of synthesising organic compounds from carbon dioxide and water supplied by the host plant. The oxygen wastes are partly used by the plant and partly exhaled into the atmosphere through holes in the leaves called stomata (singular 'stoma'). The organic compounds synthesised by the chloroplasts are ultimately made available to the host plant cell.

I previously made the point that many light harvesters stole their photosynthetic ability from plants. As with *Mixotricha*, we can see recent re-enactments in the many examples of photosynthesisers being incorporated in the tissues of fungi and animals, for example in corals. In algae such as the brown seaweeds, a striking proof of theft comes from the membranes inside their cells. Normal plant chloroplasts have two membranes, an inner one that used to be the outer membrane of the bacteria, and an outer one from the host cell. But non-plant algae often have triple or quadruple layered membranes, evidently a result of engulfing an algal plant cell, which itself had engulfed a green bacterium. In some cases, there is even a vestigial plant nucleus, sandwiched halfway between the membrane layers. It doesn't stop there. Dinoflagellates have gained photosynthesis by at least three successive rounds of endosymbiosis: they swallowed an alga that had swallowed a plant that had swallowed a green bacterium. Some scientists believe that their story involves yet another round of engulfing, quadruple endosymbiosis. It is not surprising that the analogy of Russian dolls is often made, although in some cases, the exact number of organisms nested inside one another is hard to deduce, because not all the membranes have been retained – unsurprising given that each extra membrane will absorb a little life-giving light.

All the free oxygen in the atmosphere comes from green bacteria, whether free-living or in the form of chloroplasts. And, as mentioned before, when it first appeared in the atmosphere oxygen was a poison. Indeed, some people colourfully say it still is a poison, which is why doctors advise us to eat 'anti-oxidants'. In evolution, it was a brilliant chemical coup to discover how to use oxygen to extract (originally solar) energy from organic compounds. This discovery, which can be seen as a sort of reverse photosynthesis, was entirely made by bacteria, but a different kind of bacteria. As with photosynthesis itself, bacteria still have a monopoly on the technology. Again as with photosynthesis, eukaryotic cells like ours give house room to these oxygen-loving bacteria, who now travel under the name of mitochondria. We have become so dependent on oxygen, via the biochemical wizardry of mitochondria, that the statement that it is a poison makes sense only when uttered in a tone of self-conscious paradox. Carbon monoxide, the deadly poison in car exhausts, kills us by competing with oxygen for the favours of our oxygen-carrying haemoglobin molecules. Depriving somebody of oxygen is a swift way to kill them. Yet our own cells, unaided, wouldn't know what to do with oxygen. It is only mitochondria, and their bacterial cousins, that do.

As with chloroplasts, molecular comparison tells us the particular group of bacteria from which mitochondria are drawn. Mitochondria sprang from the so-called alpha-proteobacteria and they are therefore related to the rickettsias that cause typhus and other nasty diseases. Most of the original genome of mitochondria has been moved to the nucleus, or lost entirely, and they have become completely adapted to life inside eukaryotic cells. But, like chloroplasts, they still reproduce autonomously by division, making populations within each eukaryotic cell. Although mitochondria have lost most of their genes, they haven't lost all of them, and this is fortunate for evolutionary historians, as we have seen throughout this book.

Lynn Margulis, who was largely responsible for promoting the idea – now all but universally accepted – that mitochondria and chloroplasts are symbiotic bacteria, tried to do the same thing with cilia. Inspired by possible re-enactments such as we saw in the Mixotrich's Tale, she traced cilia back to spirochaete bacteria. Unfortunately, in view of the beauty and persuasiveness of the mixotrich parallel, the evidence that cilia (undulipodia) are symbiotic bacteria is found unpersuasive by almost everybody who was persuaded by Margulis's evidence in the case of mitochondria and chloroplasts.

For early eukaryotes, the engulfing of (or was it invasion by?) mitochondria was to have far-reaching consequences. One seductive theory holds them responsible for both the origin of the nucleus and the peculiar organisation of eukaryotic genes. Unlike bacteria, our genes are split up into pieces along the genome: the useful protein-coding chunks (exons) are separated by ignored rubbish (introns). There is good evidence that these eukaryotic introns trace back to a particular type of selfish genetic element (called a group II self-splicing intron), which is common in the bacterial relatives of mitochondria. The theory has it that mitochondria, while providing a useful service to early eukaryotes, also infected their genomes with ultra-selfish DNA parasites. Our original genes, each a continuous string of useful information, would become interrupted by sections of parasitic DNA (which later evolved into our meaningless introns). Many features of our cells can be seen as an evolutionary response to this problem. Parasitic sequences would somehow have to be cut out before the gene was translated into protein. Hence, the suggestion goes, the evolution of the nucleus, to isolate our genetic material from the site of protein production, allowing introns to be spliced out as the sequence is sent from one to the other. Extensions of the theory, which we will not detail here, can be used to explain why eukaryotes evolved multiple linear chromosomes, rather than keeping the standard, circular bacterial chromosome.

Other theorists have fingered mitochondria as responsible for the evolution of two sexes, later in eukaryotic evolution. There is in principle no reason why we shouldn't have a much larger number of cross-compatible sexes, similar to the large number of 'mating types' that are seen in many species of fungi. Why stop at two, given that it drastically reduces the pool of potential mates? The suggested reason is that one sex – usually but not exclusively the female – is given the role of transmitting mitochondria between generations. As explained in Mark Ridley's provocative book *Mendel's Demon*, this diktat resolves the conflict between mitochondria from different parents, a conflict which could easily undermine mitochondrial function and destroy the cell as a whole.

Both theories, and indeed other hypotheses for the evolution of eukaryotic features, are speculative – necessarily so given the antiquity of eukaryotes and (more importantly) the lack of surviving intermediates. But it is undoubtedly the case that the curious natural history of the eukaryotic cell requires evolutionary explanation. These ideas provide excellent stimulus for further work.

Because the Great Historic Rendezvous is a true rendezvous in the forward historical direction, our pilgrimage, from now on, should strictly be a split pilgrimage. We should follow the separate backward pilgrimages of the various participants in the eukaryotic compact until they are finally reunited in the deep past, but that would make for a gratuitously complicated journey. Both chloroplasts and mitochondria have their affinities with the eubacteria, not the other prokaryotic group, the Archaea. But our nuclear genes, at least those that have not been imported from mitochondria, are slightly closer to Archaea. The next rendezvous in our backwards story is with them.

ARCHAEA

For the first time in our journey, we eukaryotes meet the prokaryotes – commonly known as bacteria. We have chosen to number this as Rendezvous 39, although the exact number is still up for grabs (or, rather, up for research in the next few years). It could be number 40, 41, or even more, depending on the root of the star diagram on page 574.

No matter. The rendezvous numbers are starting to become dubious anyway, because from hereon back the idea of single rendezvous points starts to disintegrate. For a start, the odd circumstances of the Great Historic Rendezvous mean that some bits of our cells have a different ancestry from the rest of the genome. But it's not just that. Because of the way they exchange genetic material, prokaryotes have wantonly promiscuous genes, which can jump between 'species' with relative impunity. Indeed, the very concept of a species starts to lose its meaning in these life forms. As we go deeper and deeper in time, we might therefore expect to start tracing more and more genes back through radically different routes.

Nevertheless, it has been argued that there is a core group of genes, especially those involved in DNA replication, which tend not to jump about, but remain in the same bacterial cell as it goes forth and multiplies. Whether this is true or not, it certainly seems that the majority of our genes do not find their closest relatives among the 'true' (*eu-*)bacteria such as the gut inhabitant *E. coli*, or the typhus-causing kin of the mitochondrion. Instead, the immediate ancestry of most of our body lies among another major group of prokaryotes, the Archaea.

It was the great American microbiologist Carl Woese of the University of Illinois who discovered and defined the Archaea (then called Archaebacteria) in the late 1970s. The deep separation from other bacteria was controversial at first because it was so different from previous ideas. But it is now very widely accepted, and Woese was justly honoured with prizes and medals, including the highly prestigious Crafoord Prize and the Leeuwenhoek Medal.

The Archaea include species that thrive in different kinds of

DPANN group   TACK group

Lokiarchaeum

Euryarchaeota

39b

39c

39a

38

39d

Archaeans join. Most experts believe the archaeans are the sister group to the eukaryotes, on the basis of most of our nuclear DNA, as well as certain details of biochemistry and cellular morphology. However, if mitochondrial DNA were used, our closest relatives would be the α-proteobacteria, from whose ranks mitochondria emerged (see the Great Historic Rendezvous). As we develop tools for exploring the microscopic world around us, we are finding more and more microbial 'species', and more branches on the tree of life. One very recent example is the Lokiarchaeota, described on page 624 and represented by a single species of *Lokiarchaeum* above. Analyses of the Archaea are starting to suggest that we are more closely related to some than to others, as we show here. However, this idea is still controversial, as is the order of branching. To reflect this state of flux, the various rendezvous points with Archaea are labelled 39a, 39b, 39c, etc. and gathered into a single chapter. Most species of bacteria (including the Archaea) look similar in silhouette, so we have omitted images from the remaining fractal diagrams.

extreme conditions, whether it is very high temperatures, or very acid, alkaline or salty water. The archaeans as a group seem to 'push the envelope' of what life can tolerate. Nobody knows whether Concestor 39 was such an extremophile, but it is an intriguing possibility.

My Oxford colleague Tom Cavalier-Smith, whose view of the early evolution of life is informed by his great knowledge of microbial diversity, thinks the archaeans evolved their peculiar biochemical features within the bacteria as an adaptation to thermophily. Thermophily comes from the Greek for 'love of heat', which in practice usually means living in hot springs. He believes that these heat-loving bacteria – 'thermophiles' – then split into two. Some became hyperthermophiles (liking it very hot indeed) and gave rise to the modern Archaea. Others left the hot springs and, under cooler conditions, became the eukaryotes by absorbing other prokaryotes and making use of them, in the manner of the Mixotrich's Tale. If he is right, we know the conditions in which Rendezvous 38 takes place: in a hot spring, or perhaps in a volcanic upwelling from the bottom of the sea. But of course he may not be right, and it has to be said that his view is far from the consensus.

Indeed, there is little overall consensus in this field. Some researchers find evidence that the different groups of archaeans gained their various abilities by stealing different sets of genes from the eubacteria. Others state that eukaryotes are not simply a sister group to the Archaea; we are Archaea, and nest deep within its branches. According to this increasingly popular viewpoint, our core genes are more closely related to some sorts of Archaea than others. As this chapter was being rewritten for the second edition, a Scandinavian research team announced they had discovered our closest archaeal relatives yet, the mischievously named Lokiarchaeota, extracted remotely from a deep-sea hydrothermal vent called Loki's Castle, hundreds of miles north of Norway. More discoveries are likely, particularly now that we have astounding technology which can sort individual cells from sea water, then sequence each genome separately (unfortunately destroying the cells in the process). This reveals many new prokaryotic branches on the tree of life, and has led to an unprecedented situation for naturalists. There are creatures for which we now know the entire genome sequence, but nothing else about them, not even what they look like! Strange times indeed.

Regardless of whether our closest relatives are the Lokiarchaeota or any other newly discovered prokaryote group, it seems plausible that eukaryotes do indeed nest within the Archaea. If so, Rendezvous 39 is a fig leaf concealing many true rendezvous points. This is the reason for the fractal phylogeny which heads this chapter, which shows various archaeal groups joining at Rendezvous 39a (Lokiarchaeota), 39b (the 'TACK' group, acronym for its component four phyla), 39c (a newly united group including the tiny Nanoarchaeota, acronym 'DPANN') and 39d (Euryarchaeota). We have not gone the whole hog and given these groups different numbers and chapters, partly because their position is highly debatable, partly because of all undiscovered Archaea out there, but mostly because of their rampant swapping of DNA – 'horizontal' gene transfer – which means that any particular order may apply to only a minority of genes in our genome. For the purpose of this book, our solution is to visualise a family tree as best we can, focusing on core genes, but flagging up the issue by labelling all the rendezvous under the umbrella of a single Rendezvous 39. It is our necessary compromise.

Despite all this confusion, it does seem that the Archaea evolved from within the eubacteria, and it is those impressive creatures that define the final rendezvous in this book.

EUBACTERIA

When the pilgrimage began, our time machine ground away in bottom gear and we thought in terms of tens of thousands of years. We changed up through the gears, upgrading our imaginations to cope with millions, then hundreds of millions of years as we accelerated back to the Cambrian, picking up animal pilgrims along the way. But the Cambrian is alarmingly recent. For the great majority of its career on this planet life has been nothing but prokaryotic life. We animals are a recent afterthought. For the home stretch to Canterbury, our time machine has to go into hyperdrive to save the book from intolerable *longueur*. With what may seem almost indecent haste, our pilgrims, now including the eukaryotes and the archaeans, speed backwards to what I am assuming is one last rendezvous – Rendezvous 40 with the eubacteria. But it might be more than one, and we might be closer to some eubacteria than others. In this final stage of life's backward journey, we have no other organisms for comparison, no outgroup, and no easy way to root the tree. That is why the fractal tree opposite is drawn as it is.

Bacteria, as we have already seen and as Taq's Tale will agree, are supremely versatile chemists. They are also the only non-human creatures known to me who have developed that icon of human civilisation, the wheel. *Rhizobium* tells the tale.

THE *RHIZOBIUM*'S TALE

The wheel is the proverbial human invention. Take apart any machine of more than rudimentary complexity and you'll find wheels. Ship and aeroplane propellors, spinning drills, lathes, potters' wheels – our technology runs on the wheel and would seize up without it. The wheel may have been invented in Mesopotamia during the fourth millennium BC. We know it was elusive enough to *need* inventing because the New World civilisations still lacked it by the time of the Spanish conquest. The alleged exception there – children's toys – seems so bizarre as to prompt suspicion. Could it be

Eubacteria join. Three factors conspire to make bacterial relationships particularly uncertain. First, bacteria exchange DNA with each other, so different genes can show radically different trees. Second, their ancient divergences result in a star-like phylogeny which is difficult to resolve. Third, we have no outgroup to root the tree (other methods of rooting are sometimes suggested). Our snowflake-like diagram is intended to show that the point of final rendezvous is not known. Traditionally the root of life is placed between the archaea and the eubacteria, marked by the line upwards from 40. But there are suggestions that it lies within the bacteria, as marked for example by the lower line to the right, which would mean our final rendezvous point could be (say) Rendezvous 44.

There is no well-established set of relationships between the 50 or more bacterial 'phyla', and many undiscovered groups are thought to exist. So although the five eubacterial branches here are technically based on the diagram on page 638, the diagram should be seen more as an impression of bacterial diversity than a comprehensive and accurate evolutionary tree. Labels for the different branches can be found in the endnotes, on page 707.

one of those myths that spreads purely because it is so memorable, like the Inuit having 50 words for snow?

Whenever humans have a good idea, zoologists have grown accustomed to finding it anticipated in the animal kingdom. Examples pervade this book, including echo-ranging (bats), electro-location (the Duckbill's Tale), the dam (the Beaver's Tale), the lens and even the parabolic reflector (limpets focus an image with a parabolic mirror), the infrared heat-seeking sensor (some snakes), the hypodermic syringe (wasps, snakes and scorpions), the harpoon (cnidarians) and jet propulsion (squids). Why not the wheel?

It is possible that the wheel impresses us only by contrast with our rather undistinguished legs. Before we had engines driven by fuels (fossilised solar energy), we were easily outpaced by animal legs. No wonder Richard III offered his kingdom for four-footed transportation out of his predicament. Perhaps most animals wouldn't benefit from wheels because they can already run so fast on legs. After all, until very recently, all our wheeled vehicles have been pulled by leg power. We developed the wheel not to go faster than a horse, but to enable a horse to transport us at its own pace – or a bit less. To a horse, a wheel is something that slows you down.

Here's another way in which we risk overrating the wheel. It is dependent for maximum efficiency on a prior invention – the road (or other smooth, hard surface). A car's powerful engine enables it to beat a horse or a dog or a cheetah on a hard, flat road. But run the race over wild country or ploughed fields, perhaps with hedges or ditches in the way, and it is a rout: the horse will leave the car wallowing.

Well then, perhaps we should change our question. Why haven't animals developed the road? There is no great technical difficulty. The road should be child's play compared with the beaver dam or the bower-bird's ornamented arena. There are even some digger wasps that tamp soil hard, picking up a stone tool to do so. Presumably the same skills could be used by larger animals to flatten a road.

But it raises an unexpected problem. Even if roadbuilding is technically feasible, it is a dangerously *altruistic* activity. If I as an individual build a good road from A to B, you may benefit from the road just as much as I do. Why should this matter? Because Darwinism is a selfish game. Building a road that might help others will be penalised by natural selection. A rival individual benefits from my road just as much as I do, but he does not pay the cost of building. Freeloaders, who use my road and don't bother to build their own, will be free to concentrate their energy on out-reproducing me, while I slave away on the road. Unless special measures are taken, genetic tendencies towards lazy, selfish exploitation will thrive at the expense of industrious roadbuilding. The upshot will be that no roads get built. With the benefit of foresight, we can see that everybody will be worse off. But natural selection, unlike us humans with our big, recently evolved brains, has no foresight.

What is so special about humans that we have managed to overcome our antisocial instincts and build roads that we all share? Oh, there is so much. No other species comes remotely close to a welfare state, to an organisation that takes care of the old, that looks after the sick and the orphaned, that gives to charity. On the face of it these things present a challenge to Darwinism, but this is not the place to go into that. We have governments, police, taxation, public works to which we all subscribe whether we like it or not. The man who wrote, 'Sir, You are very kind, but I'd prefer not to join your Income Tax Scheme', heard back, we may be sure, from the Inland Revenue. Unfortunately, no other species has invented the tax. They have, however, invented the (virtual) fence. An individual can secure his exclusive use of a resource if he actively defends it against rivals.

Many species of animals are territorial, not just birds and mammals, but fish and insects too. They defend an area against rivals of the same species, often so as to sequester a private feeding ground, or a private courtship bower or nesting area. An animal with a large territory might benefit by building a network of good, flat roads across the territory from which rivals were excluded. This is not impossible, and we saw something of the sort with elephant shrew runs, but such animal roads would be too local for long-distance, high-speed travelling. Roads of any quality would be limited to the small area that an individual can defend against genetic rivals. Not an auspicious beginning for the evolution of the wheel.

But now, finally, we come to the teller of this tale. There is one revealing exception to my premise. Some very small creatures *have* evolved the wheel in the fullest sense of the word. The wheel may even have been the first locomotor device ever evolved, given that for most of its first 2 billion years, life consisted of nothing but bacteria. Many bacteria, of which *Rhizobium* is typical, swim using thread-like spiral propellors, each driven by its own continuously rotating propellor shaft. It used to be thought that these 'flagella' were wagged like tails, the appearance of spiral rotation resulting from a wave of motion passing along the length of the flagellum, as in a wriggling snake. The truth is much more remarkable. The bacterial flagellum[*] is attached to a shaft that rotates freely and indefinitely in a hole that runs through the cell wall. This is a true axle, a freely rotating hub. It is driven by a tiny molecular motor which uses the same biophysical principles as a muscle. But a muscle is a reciprocating engine, which, after contracting, has to lengthen again to prepare for a new power stroke. The bacterial motor just keeps on going in the same direction: a molecular turbine.

The fact that only very small creatures have evolved the wheel suggests what may be the most plausible reason why larger creatures have not. It's a rather mundane, practical reason, but nonetheless important. A large creature would need big wheels which, unlike man-made ones, would have to grow *in situ* rather than being separately fashioned out of dead materials and then mounted. For a large, living organ, growth *in situ* demands blood or something equivalent, and probably something equivalent to nerves too. The problem of supplying a freely rotating organ with blood vessels (not to mention nerves) that don't tie themselves in knots is too vivid to need spelling out. There might be a solution, but we need not be surprised that it has not been found.

Human engineers might suggest running concentric ducts to carry blood through the middle of the axle into the middle of the wheel. But what would the evolutionary intermediates have looked like? Evolutionary improvement is like climbing a mountain. You can't jump from the bottom of a cliff to the top in a single leap. Sudden, precipitous change is an option for engineers, but in nature the summit of the evolutionary mountain can be reached only via a gradual ramp upwards from the starting point. The wheel may be one of those cases where the engineering solution can be seen in plain view, yet be unattainable in evolution because it lies on the other side of a deep valley: unevolvable by large animals but within the reach of bacteria because of their small size.

In an imaginative piece of lateral thinking, Philip Pullman, in his epic of childhood fiction *His Dark Materials*, solves the problem for big animals in a completely unexpected but very biological way. He

---

[*] The bacterial flagellum, as we have seen, is completely different in structure from the eukaryotic (or protozoan) flagellum or 'undulipodium' that we met in the Mixotrich's Tale. Unlike the eukaryotic arrangement of 9+2 microtubules, the bacterial flagellum is a hollow tube made of the protein flagellin.

A true axle, a freely rotating hub . . . driven by a tiny molecular motor.

invents a species of benevolent, trunked animal, the mulefa, who have evolved symbiotically with a species of gigantic tree that sheds hard, circular, wheel-like seed pods. The feet of the mulefa have a horny, polished spur which fits into a hole in the centre of a seed pod, which then works as a wheel. The trees gain from the arrangement because whenever – as eventually must happen – a wheel wears out and has to be discarded, the mulefa disperse the seeds inside. The trees have evolved to return the favour by making the pods perfectly circular, with a suitable hole for the mulefan axle right in the centre, into which they secrete a high-grade lubricating oil. The mulefa's four legs are placed in a diamond pattern. The fore and aft legs are in the midline, and they are the ones that slot into the wheels. The other two legs, halfway along the body and to the sides, have no wheels and are used to punt the animal along like an old-fashioned boneshaker bicycle without pedals. Pullman cleverly notes that the whole system is made possible only by a geological peculiarity of the world on which these creatures live. Basalt happens to form in long, ribbon-like lines over the savannah, which serve as unmade but hard roads.

Short of Pullman's ingenious symbiosis, we may provisionally accept the wheel as one of those inventions that, even if it were a good idea in the first place, cannot evolve in large animals: either because of the prior need for a road, or because the problem of the twisted blood vessels could never be solved, or because the intermediates to a final solution would never be good for anything. Bacteria were able to evolve the wheel because the world of the very small is so very different and presents such different technical problems.

As it happens, the bacterial flagellar motor itself has recently, in the hands of a species of creationists who call themselves 'Intelligent Design Theorists', been elevated to the status of icon of alleged unevolvability. Since it manifestly exists, the conclusion of their argument is different. Whereas I proposed unevolvability as an explanation for why large animals like mammals don't grow wheels, creationists have seized upon the bacterial flagellar wheel as something that cannot exist and yet does – so it must have come about by supernatural means!

This is the ancient 'Argument from Design', also called the 'Argument from Paley's Watchmaker', or the 'Argument from Irreducible Complexity'. I have less kindly called it the 'Argument from Personal Incredulity' because it always has the form: 'I personally cannot imagine a natural sequence of events whereby X could have come about. Therefore it must have come about by supernatural means.' Time and again scientists have retorted that if you make this argument, it says less about nature than about the poverty of your imagination. The 'Argument from Personal Incredulity' would lead us to invoke the supernatural every time we see a good conjuror whose tricks we cannot fathom.

It is perfectly legitimate to propose the argument from irreducible complexity as a possible explanation for the lack of something that doesn't exist, as I did for the absence of wheeled mammals. That is very different from evading the scientist's responsibility to explain something that *does* exist, such as wheeled bacteria. Nevertheless, to be fair, it is possible to imagine validly using some version of the argument from design, or the argument from irreducible complexity. Future visitors from outer space, who mount archaeological digs of our planet, will surely find ways to distinguish designed machines such as planes and microphones, from evolved machines such as bat wings and ears. It is an interesting exercise to think about how they will make the distinction. They may face some tricky judgements in the messy overlap between natural evolution and human design. If the alien scientists can study living specimens, not just archaeological relics, what will they make of fragile, highly strung racehorses and greyhounds, of snuffling bulldogs who can scarcely breathe and can't be born without Caesarian assistance, of blear-eyed Pekinese baby surrogates, of walking udders such as Friesian cows, walking rashers such as Landrace pigs, or walking woolly jumpers such as Merino sheep? Molecular machines – nanotechnology – crafted for human benefit on the same scale as the bacterial flagellar motor, may pose the alien scientists even harder problems.

Francis Crick, no less, has speculated semi-seriously in *Life Itself* that bacteria might not have originated on this planet but been seeded from elsewhere. In Crick's fantasy, they were sent in the nose-cone of a rocket by alien beings, who wanted to propagate their form of life, but shrank from the technically harder problem of transporting themselves and relied, instead, upon natural evolution to finish the job once the bacterial infection had taken root. Crick, and his colleague Leslie Orgel, who originally suggested the idea with him, supposed that the bacteria had originally evolved by natural processes on the home planet, but they could equally, while in the mood for science fiction, have added a touch of nano-technological artifice to the mix, perhaps a molecular gearwheel like the flagellar motor which we see in *Rhizobium* and many other bacteria.

Crick himself – whether with regret or relief it is hard to say – finds little good evidence to support his own theory of Directed Panspermia. But the hinterland between science and science fiction constitutes a useful mental gymnasium in which to wrestle with a genuinely important question. Given that the illusion of design conjured by Darwinian natural selection is so breathtakingly powerful, how do we, in practice, distinguish its products from deliberately designed artefacts? Another great molecular biologist, Jacques Monod, began his *Chance and Necessity* in similar terms. Could there be genuinely persuasive examples of irreducible complexity in nature: complex organisation made of many parts, the loss of any one of which would be fatal to the whole? If so, might this suggest genuine design by a superior intelligence, say from an older and more highly evolved civilisation on another planet?

It is possible that an example of such a thing might eventually be discovered. But the bacterial flagellar motor, alas, is not it. Like so many previous allegations of irreducible complexity, from the eye on, the bacterial flagellum turns out to be eminently reducible. Kenneth Miller of Brown University deals with the whole question in a *tour de force* of clear exposition. As Miller shows, the allegation that the component parts of the flagellar motor have no other functions is simply false. As one example, many parasitic bacteria have a mechanism for injecting chemicals into host cells called the TTSS (Type Three Secretory System). The TTSS makes use of a subset of the very same proteins that are used in the flagellar motor. In this case they are used not for providing rotatory motion of a circular hub, but for making a circular hole in a host's cell wall. Miller summarises:

> Stated directly, the TTSS does its dirty work using a handful of proteins from the base of the flagellum. From the evolutionary point of view, this relationship is hardly surprising. In fact, it's to be expected that the opportunism of evolutionary processes would mix and match proteins to produce new and novel functions. According to the doctrine of irreducible complexity, however, this should not be possible. If the flagellum is indeed irreducibly complex, then removing just one part, let alone 10 or 15, should render what remains 'by definition nonfunctional.' Yet the TTSS is indeed fully functional, even though it is missing most of the parts of the flagellum. The TTSS may be bad news for us, but for the bacteria that possess it, it is a truly valuable biochemical machine.
>
> The existence of the TTSS in a wide variety of bacteria demonstrates that a small portion of the 'irreducibly complex' flagellum can indeed carry out an important biological function. Since such a function is clearly favored by natural selection, the contention that the flagellum must be fully assembled before any of its component parts can be useful is obviously incorrect. What this means is that the argument for intelligent design of the flagellum has failed.

Miller's indignation at 'Intelligent Design Theory' receives a boost from an interesting source: his deep religious convictions, which are more fully articulated in *Finding Darwin's God*. Miller's God (if not Darwin's) is the God revealed in – or perhaps synonymous with – the deep lawfulness of nature. The creationist quest to demonstrate God through the negative route of the Argument from Personal Incredulity turns out, as Miller shows, to assume that God capriciously *violates* his own laws. And this, to those – like Miller – of a thoughtfully religious disposition, is a cheap and demeaning sacrilege.

As a non-religious person I can sympathetically buttress Miller's argument with a parallel one of my own. If not sacrilegious, the intelligent design style of argument from personal incredulity is *lazy*. I have satirised it in an imagined conversation between Sir Andrew Huxley and Sir Alan Hodgkin, both sometime presidents of the Royal Society, who shared the Nobel Prize for working out the molecular biophysics of the nerve impulse.

'I say, Huxley, this is a terribly difficult problem. I can't see how the nerve impulse works, can you?'

'No, Hodgkin, I can't, and these differential equations are fiendishly hard to solve. Why don't we just give up and say that the nerve impulse propagates by nervous energy?'

'Excellent idea, Huxley, let's write the letter to *Nature* now: it'll only take one line, then we can turn to something easier.'

Andrew Huxley's elder brother Julian made a similar point when, long ago, he satirised vitalism, then usually epitomised by Henri Bergson's name of *élan vital*, as tantamount to explaining that a railway engine was propelled by *élan locomotif.** My censure of laziness, and Miller's of sacrilege, do not apply to the hypothesis of directed panspermia. Crick was talking about superhuman, not supernatural, design. The difference really matters. On Crick's world view, superhuman designers of bacteria, or of the means to seed Earth with them, would themselves have originally evolved by some local equivalent of Darwinian selection on their own planet. Crucially, Crick would always seek what Daniel Dennett calls a 'crane': would never resort – as Henri Bergson would – to a 'skyhook'.

The main objection to the irreducible complexity argument amounts to a demonstration that the allegedly irreducible complex entity, the flagellar motor, the blood-clotting cascade, the Krebs cycle, or whatever it might be, is actually reducible. The personal incredulity was simply wrong. To this we add the reminder that, even if we can't *yet* think of a step-by-step pathway by which the complexity might have evolved, the eager slide to assuming that it is therefore supernatural is either sacrilegious or lazy, according to taste.

But there is another objection that needs to be mentioned: the 'arch and scaffolding' of Graham Cairns-Smith. Cairns-Smith was writing in a different context, but his point works here too. An arch is irreducible in the sense that if you remove part of it, the whole collapses. Yet it is possible to build it gradually by means of scaffolding. The subsequent removal of the scaffolding, so that it no longer appears in the visible picture, does not entitle us to a mystified and obscurantist attribution of supernatural powers to the masons.

---

* It is depressing to reflect that Henri Bergson – a vitalist – represents the nearest approach to a scientist in the entire list of over 100 winners of the Nobel Prize in Literature. The nearest competitor is Bertrand Russell, but he won it for his humanitarian writings.

The flagellar motor is common among bacteria. *Rhizobium* was chosen to tell the tale because of a second claim to impress us with the versatility of bacteria. Farmers sow plants of the pea family, Leguminosae, as a part of most good crop rotation schemes for one very good reason. Leguminous plants can use raw nitrogen straight out of the air (it is by far the most abundant gas in our atmosphere) rather than having to suck up nitrogen compounds from the soil. But it isn't the plants themselves that fix atmospheric nitrogen and turn it into usable compounds. It is symbiotic bacteria – specifically *Rhizobium* – housed for the purpose in special nodules provided for them, with every indication of inadvertent solicitude, on the roots of the plants.

Such contracting out of ingenious chemical tricks to chemically much more versatile bacteria is an extremely common pattern throughout animals and plants. It is the main message of Taq's Tale.

TAQ'S TALE

Having reached our most ancient rendezvous, having gathered into our pilgrimage all of life as we know it, we are in a position to survey its diversity. At the deepest level, the diversity of life is chemical. The trades plied by our fellow pilgrims span a range of skills in the arts of chemistry. And, as we have seen, it is the bacteria, including archaeans, who display the fullest spread of chemical skills. Bacteria taken as a group are the master chemists of this planet. Even the chemistry of our own cells is largely borrowed from bacterial guest workers, and it represents a fraction of what bacteria are capable of. Chemically, we are more similar to some bacteria than some bacteria are to other bacteria. At least as a chemist would see it, if you wiped out all life except bacteria, you'd still be left with the greater part of life's range.

The particular bacterium that I choose to tell the tale is *Thermus aquaticus*, known fondly to molecular biologists as Taq. Different bacteria seem alien to us for different particular reasons. *Thermus aquaticus*, as its name suggests, likes to be in hot water. *Very* hot water. As we saw at Rendezvous 39, many of the archaeans are thermophiles and hyperthermophiles, but the archaeans don't have a monopoly on this way of life. Thermophiles and hyperthermophiles are not taxonomic categories, but something more like trades or guilds, like Chaucer's Clerk, Miller and Physician.

They make their living in places where nothing else can: the scalding-hot springs of Rotorua and Yellowstone Park, or the volcanic vents on mid-ocean ridges. *Thermus* is a eubacterial hyperthermophile. It can survive with little problem in near-boiling water – although it prefers a more balmy 70°C, or so. It doesn't quite hold the world temperature record – there are deep-sea archaeans that thrive at up to 115°C, well above the normal boiling point of water.*

Thermus is famous in molecular biology circles for being the source of the DNA duplication enzyme known as Taq polymerase. Of course, all organisms have enzymes to duplicate DNA, but *Thermus* has had to evolve one that can withstand near-boiling temperatures. This is useful for molecular biologists because the easiest way to ready DNA for duplication is to boil it, separating it into its two constituent strands. Repeated boiling and cooling of a solution containing both DNA and Taq polymerase duplicates – or 'amplifies' – even the most minute quantities of original DNA. The method is called the 'polymerase chain reaction', or PCR, and it is brilliantly clever.

Traditionally, and understandably, life's tale was told from the point of view of big animals – us. Life was divided into the animal kingdom and the vegetable kingdom, and the difference seemed pretty clear. The fungi counted as plants because the more familiar of them are rooted to the spot and don't walk away while you try to study them. We didn't even know about bacteria until the nineteenth century, and when they were first seen through powerful microscopes people didn't know where to put them in the scheme of things. Some thought of them as miniature plants, others as miniature animals. Yet others put the light-trapping bacteria in the plants (as 'blue-green algae') and the rest in the animals. Much the same was done with the 'protists' – single-celled eukaryotes that are not bacteria and are much larger than bacteria. The green ones were called Protophyta and the rest Protozoa. A familiar example of a protozoan is *Amoeba*, once thought to be close to the grand ancestor of all life – how wrong we were, for an *Amoeba* is scarcely distinguishable from a human when viewed through the 'eyes' of bacteria.

All that was in the days when living organisms were classified by their visible anatomy. Bacteria look much less diverse than animals

---

* Again, if it seems surprising that water can be found so far above its normal boiling point, remember that water boils hotter at high pressure.

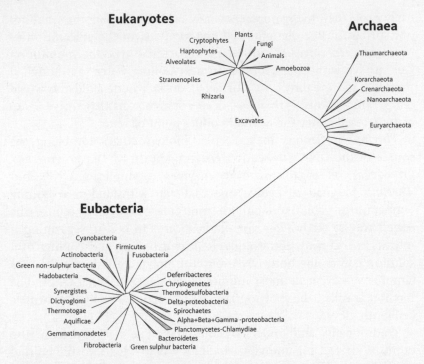

Eukaryotes

Cryptophytes
Plants
Haptophytes
Fungi
Alveolates
Animals
Stramenopiles
Amoebozoa
Rhizaria
Excavates

Archaea

Thaumarchaeota
Korarchaeota
Crenarchaeota
Nanoarchaeota
Euryarchaeota

Eubacteria

Cyanobacteria
Firmicutes
Actinobacteria
Fusobacteria
Green non-sulphur bacteria
Hadobacteria
Deferribacteres
Chrysiogenetes
Synergistes
Thermodesulfobacteria
Dictyoglomi
Delta-proteobacteria
Thermotogae
Spirochaetes
Aquificae
Alpha+Beta+Gamma -proteobacteria
Gemmatimonadetes
Planctomycetes-Chlamydiae
Bacteroidetes
Fibrobacteria
Green sulphur bacteria

The deepest divisions of life. Tree of life, showing division into three major domains, based on recent analysis of ribosomal proteins. The exact relationships within the eubacteria and archaea vary slightly, depending on the method of analysis. Drawn from data provided by Lasek-Nesselquist and Gogarten [237, fig. 5].

or plants and it was pardonable to put them down as primitive animals and plants. It was another matter entirely when we began to classify creatures using the much richer information provided by their molecules, and when we looked at the range of chemical 'trades' perfected by microbes. The diagram above shows approximately how things look today.

If animals and plants are treated as a pair of kingdoms, by the same standards there are dozens of microbial 'kingdoms', whose uniqueness entitles them to the same status as animals and plants. And the diagram shows only the tip of the iceberg. Not only have some long branches been omitted, but I've shown only those that live in accessible places and can be cultured in a laboratory. Indeed, while we were in the very last stages of revising this tale, researchers from Berkeley claim to have identified a previously unknown radiation of ultra-small bacteria which belong on the long branch separating eubacteria from

Archaea in the diagram. Impossible to grow in the laboratory, these have been identified by sequencing whole bacterial genomes directly from groundwater. Too recent to include on this diagram, who knows what other microbial kingdoms also await discovery?

Animals, plants and fungi constitute just three small branches of the tree of life. What distinguishes these three familiar kingdoms from the others is that the organisms in them are large, being built of many cells. The other kingdoms are almost entirely microbial. Why do we not unite them into one microbial kingdom, on a par with the three great multicellular kingdoms? One reason, and a sound one, is that, at the biochemical level, many of the microbial kingdoms are as different from each other, and from the big three, as the three familiar kingdoms are from each other.

It would be worthless to argue in detail whether there 'really' are, say, 20 kingdoms on this scale of difference, or 25 or 100 – indeed bacteriologists tend to avoid trouble by calling them all phyla and eschewing the 'kingdom' moniker entirely. But what is clear from the diagram is that these dozens fall into three main super-kingdoms – 'domains' in the terminology of Carl Woese, already mentioned as the originator of this new view of life. The three domains are first our own, the eukaryotes, in whose company we have been travelling for most of our Pilgrims' Way. Second, the Archaea – the microbes we met at Rendezvous 39 – who, on the old view of life, would be lumped in with the third domain, the true (or eu-) bacteria. Tempting though it is to think of Archaea and eubacteria as just different sorts of bacteria, they are not close relatives at all. We eukaryotes – most of our genes, at least – are much closer to the Archaea. We may be even closer to some sorts of Archaea than others, and there have been calls to recognise only two (not three) major domains of life: the eubacteria, and everything else. We do not go so far off piste here. Nevertheless, the eubacteria are clearly distinct from the rest, and the last leg of our pilgrimage is their domain. It is a privilege to share these final steps with the most ubiquitous and efficient DNA-propagators that have ever existed.

The star diagram itself is not, of course, based upon the sorts of features we can see and touch. If you want to compare organisms, you have to choose features that they all approximately share. You can't compare legs if most species don't have legs. Legs, heads, leaves, collar bones, roots, hearts, mitochondria – each is restricted to a subset of creatures. But DNA is universal, and there's a handful of particular genes that all living creatures share with each other, with only minor, countable differences. These are what we must use for large-scale comparison. Perhaps the best example is provided by the codes that go to make ribosomes.

Ribosomes are cellular machines that read the RNA message 'tapes' sent from the genome, and churn these messages out as proteins. They are miniature protein factories, vital to all cells and universally present. They are themselves made of many specialised proteins wrapped around an RNA core – this core is completely separate from the RNA messages that the ribosomes read and translate. Not only is the ribosome ubiquitous, but the 'genes' that encode both the ribosomal core and its surrounding proteins change very little over time, meaning that even ourselves and bacteria have recognisably similar sequences, which we can use to deduce evolutionary relationships via the methods of the Gibbon's Tale.

We must take care. These sequences are fully vulnerable to 'long branch attraction' and other such pitfalls. But tentative attempts can be made to draw a tree. The diagram we have just seen is one of the most recent efforts, based on many proteins in the ribosome, backed up by analysis of the ribosomal RNA core. Certainly, some branches in this tentative tree are uncertain, particularly within the bacteria, and this may reflect their tendency to exchange DNA between themselves – a problem we have not met in any eukaryotes. Nevertheless, the ribosome is such an integral part of a cell's life that we might expect its genetic code to be very rarely swapped about, if at all. Inasmuch as we can represent deep ancestral history as a single branching tree, rather than an intermeshed network, the diagram on page 638 is our current view of the universal tree of life.

Taxonomic distance, as measured by comparing genomes, is one way of looking at diversity. Another is to look at the range of ways of life, the range of 'trades' that our pilgrims ply. At first sight, different bacteria might seem more similar in this regard than, say, a lion is to a buffalo, or a mole to a koala. To big animals like us, burrowing underground for worms seems very different, as a way of life, from chewing leaves up a gum tree. But from the chemical point of view of our bacterial storyteller, all moles, koalas, lions and buffalos are doing much the same thing. All are deriving their energy by breaking down complex molecules ultimately put together by energy from the sun captured by plants. Koalas and buffalos eat the plants directly. Lions and moles get their solar energy at one remove, by eating other animals that (ultimately) eat plants.

The primary source of outside energy is the sun. The sun, through symbiotic green bacteria inside plant cells, is the only begetter of energy for all the life we can see with the unaided eye. Its energy is trapped by green solar panels (leaves) and used to drive uphill the synthesis of organic compounds, such as sugar and starch in plants. In a series of energy-coupled downhill and uphill chemical reactions, the rest of life is then powered by the energy originally trapped from the sun by plants. Energy flows through the economy of life, from the sun to plants to herbivores to carnivores to scavengers. At every step of the way, not only between creatures but within them, every transaction in the energy economy is wasteful. Inevitably, some of it is dissipated as heat and never recovered. Without the massive inflow of energy from the sun, life would, or so the textbooks used to say, grind to a halt.

That is still mostly true. But those textbooks reckoned without the bacteria and archaeans. If you are a sufficiently ingenious chemist, it is possible to dream up alternative schemes of energy flow on this planet, which do not start with the sun. And if a useful piece of chemistry can be dreamed up, the chances are that a bacterium got there first: maybe even before they discovered the solar energy trick, and that was more than 3 billion years ago. There has to be some kind of external source of energy, but it doesn't have to be the sun. There is chemical energy locked up in lots of substances, energy that can be released by the right chemical reactions. Sources economically worth mining by living creatures include hydrogen, hydrogen sulphide, and some iron compounds. We will revisit the mining way of life at Canterbury.

Although our tales are not, in the main, told in the first person, let us make an exception for the last word of all our tales, and give it to *Thermus aquaticus:*

> Look at life from our perspective, and you eukaryotes will soon cease giving yourselves such airs. You bipedal apes, you stump-tailed tree-shrews, you desiccated lobe-fins, you vertebrated worms, you Hoxed-up sponges, you newcomers on the block, you eukaryotes, you barely distinguishable congregations of a monotonously narrow parish, you are little more than fancy froth on the surface of bacterial life. Why, the very cells that build you are themselves colonies of bacteria, replaying the same old tricks we bacteria discovered a billion years ago. We were here before you arrived, and we shall be here after you are gone.

CANTERBURY

As befits the destination of a 4-billion-year pilgrimage, our Canterbury has a patina of mystery. It is the singularity known as the origin of life, but we could better call it the origin of heredity. Life itself is not clearly defined, a fact that contradicts intuition and traditional wisdom. Ezekiel, Chapter 37, in which the prophet was commanded down into the valley of the bones, identifies life with breath. I cannot resist quoting the passage ('bone to his bone' – such wonderful economy of language).

> So I prophesied as I was commanded: and as I prophesied, there was a noise, and behold a shaking, and the bones came together, bone to his bone.
> And when I beheld, lo, the sinews and the flesh came up upon them, and the skin covered them above: but there was no breath in them.
> Then said he unto me, Prophesy unto the wind, prophesy, son of man, and say to the wind, Thus saith the Lord God: Come from the four winds, O breath, and breathe upon these slain, that they may live.

And, of course, the winds did. A great army breathed and stood up. Breath, for Ezekiel, defines the difference between dead and alive. Darwin himself implied the same in one of his more eloquent passages, the concluding words of *The Origin of Species* (emphasis added):

> Thus, from the war of nature, from famine and death, the most exalted object which we are capable of conceiving, namely, the production of the higher animals, directly follows. There is grandeur in this view of life, with its several powers, *having been originally breathed* by the Creator into a few forms or into one; and that, whilst this planet has gone cycling on according to the fixed law of gravity, from so simple a beginning endless forms most beautiful and most wonderful have been, and are being, evolved.

Darwin rightly reversed Ezekiel's order of events. The breath of life came first and created the conditions under which bones and sinews, flesh and skin would eventually evolve. Incidentally, the phrase 'by the Creator' is not present in the first edition of the *Origin*. It was added in the second edition, probably as a sop to the religious lobby. Darwin later regretted this in a letter to his friend Hooker:

> I have long since regretted that I truckled to public opinion and used the Pentateuchal term of creation, by which I really meant 'appeared' by some wholly unknown process. It is mere rubbish thinking at present of the origin of life; one might as well think of the origin of matter.

Darwin probably (and in my view rightly) saw the origin of primitive life as a relatively (and I stress relatively) easy problem compared with the one he solved: how life, once begun, developed its amazing diversity, complexity and powerful illusion of good design. Nevertheless, Darwin did later (in another letter to Hooker) venture a guess about the 'wholly unknown process' that started it all. He was led to it through wondering why we don't see life originating again and again.

> It is often said that all the conditions for the first production of a living organism are now present, which could ever have been present. But if (and oh! what a big if!) we could conceive in some warm little pond, with all sorts of ammonia and phosphoric salts, light, heat, electricity, &c., present, that a proteine compound was chemically formed ready to undergo still more complex changes, at the present day such matter would be instantly absorbed, which would not have been the case before living creatures were found.

The doctrine of spontaneous generation had only lately been experimentally attacked by Pasteur. It had long been believed that rotting meat spontaneously generated maggots, that goose barnacles spontaneously generated goslings and even that dirty laundry placed with wheat generated mice. Perversely, the spontaneous generation theory was supported by the Church (following Aristotle in this as in so much else). I say perversely because, at least with hindsight, spontaneous generation was as direct a challenge to divine creation as evolution would ever be. The idea that flies or mice could spring spontaneously into existence hugely underestimates the stupendous achievement that the creation of flies or mice would be: an insult to the Creator, one might have thought. But the science-free mindset fails to grasp how complex and inherently improbable a fly or a mouse is. Darwin was perhaps the first to appreciate the full magnitude of this mistake.

As late as 1872 in a letter to Wallace, the co-discoverer of natural selection, Darwin could still find it necessary to express his scepticism about 'Rotifers and Tardigrades being spontaneously generated', as had been suggested in a book, *The Beginnings of Life*, which he otherwise admired. His scepticism was on target as usual. Rotifers and tardigrades are complex life forms beautifully fitted to their respective ways of life. For them to be spontaneously generated would imply that they became fit and complex 'by a happy accident, and this I cannot believe'. Happy accidents of such magnitude were anathema to Darwin, as they should have been to the Church for a different reason. The whole rationale of Darwin's theory was, and is, that adaptive complexity comes about by slow and gradual degrees, step by step, no single step making too large a demand on blind chance as explanation. The Darwinian theory, by rationing chance to the small steps needed to supply variation for selection, provides the only realistic *escape* from sheer luck as the explanation of life. If rotifers could spring into existence just like that, Darwin's life-work was unnecessary.

But natural selection itself had to have a beginning. In this sense alone, some kind of spontaneous generation must have happened, if only once. The beauty of Darwin's contribution was that the single spontaneous generation we must postulate did not have to synthesise anything complicated like a maggot or a mouse. It only had to make . . . well, now we approach the heart of the problem. If not breath, what was the vital ingredient that first enabled natural selection to get going and lead eventually, after epics of cumulative evolution, to maggots, mice and men?

The details lie buried, perhaps beyond recovery, at our ancient Canterbury, but we can give the key ingredient a minimalist name to express the *kind* of thing it must have been. That name is heredity. We should be seeking not the origin of life, which is vague and undefined, but the origin of heredity – true heredity, and that means something very precise. I have previously invoked fire to help explain it.

Fire rivals breath as imagery for life. When we die, the fire of life goes out. Our ancestors who first tamed it probably thought fire a living thing, a god even. Staring into flames or embers, especially at night when the campfire warmed and protected them, did they commune in imagination with a glowing, dancing soul? Fire stays alive as long as you feed it. Fire breathes air; you can suffocate it by cutting off its oxygen supply, you can drown it with water. Wild fire devours the forest, driving animal prey before it with the speed and ruthlessness of a pack of wolves in (literally) hot pursuit. As with wolves, our ancestors could capture a fire cub as a useful pet, tame it, feed it regularly and clear away its ashy excreta. Before the art of firemaking was discovered, society would have prized the lesser art of husbanding a captured fire. Perhaps a live scion of the home fire was carried in a pot for barter to a neighbouring group whose own fire had unfortunately died.

Wild fires would have been observed giving birth to daughter fires, spitting sparks and live cinders up on the wind, like dandelion puffs, to land and seed the dry grass at a distance. Did ergastrine philosophers theorise that fire cannot spontaneously generate, but must always be born of a parent fire, either wild fire out on the plains, or domestic fire fenced in by hearthstones? And did the first firemaking sticks therefore rub out a world view?

Our ancestors might even have imagined a population of reproducing wild fires, or a pedigree of descent among domestic fires traced from a glowing ancestor bought from a distant clan and traded on to others. But still there was no true heredity. Why not? How can you have reproduction and a pedigree, yet no heredity? This is the lesson fire has for us here.

True heredity would mean the inheritance not of fire itself but of *variations* among fires. Some fires are yellower than others, some redder. Some roar, some crackle, some hiss, some smoke, some spit. Some have tinges of blue or green amongst the flames. Our ancestors, if they had studied their domesticated wolves, would have noticed a telling difference between dog pedigrees and fire pedigrees. With dogs, like begets like. At least some of what distinguishes one dog from another is handed down by its parents. Of course some comes in sideways too: from food, disease and accident. With fires, all the variation comes from the environment, none descends from the progenitive spark. It comes from the quality and dampness of the fuel, from the lie and strength of the wind, from the drawing qualities of the hearth, from the soil, from traces of copper and potassium that add touches of blue-green and lilac to sodium's yellow flame. Unlike a dog, nothing about the quality of an adult fire arrives via the spark that gave it birth. Blue fires don't beget blue fires. Crackling fires don't inherit their crackle from the parent fire that threw up their initiating spark. Fires exhibit reproduction without heredity.

The origin of life was the origin of true heredity; we might even say the origin of the first gene. By first gene, I hasten to insist, I don't mean first DNA molecule. Nobody knows whether the first gene was made of DNA, and I bet it wasn't. By first gene I mean first replicator. A replicator is an entity, for example a molecule, that forms lineages of copies of itself. There will always be errors in copying, so the population will acquire variety. The key to true heredity is that each replicator resembles the one from which it was copied more than it resembles a random member of the population. The origin of the first such replicator was not a probable event, but it only had to happen once. Thereafter, its consequences were automatically self-sustaining and they eventually gave rise, by Darwinian evolution, to all of life.

A length of DNA or, under certain conditions, the related molecule RNA is a true replicator. So is a computer virus. So is a chain letter. But all these replicators need a complicated apparatus to assist them. DNA needs a cell richly equipped with pre-existing biochemical machinery highly adapted to read and copy the DNA code. A computer virus needs a computer with some sort of data link to other computers, all designed by human engineers to obey coded instructions. A chain letter needs a good supply of idiots, with evolved brains educated at least enough to read. What is unique about the first replicator, the one that sparked life, is that it had no ready supply of anything evolved, designed or educated. The first replicator worked *de novo, ab initio*, without precedent, and without help other than from the ordinary laws of chemistry.

A powerful source of help to a chemical reaction is a catalyst, and catalysis in some form was surely involved in the origin of replication. A catalyst is an agent that speeds up a chemical reaction while not being consumed by it. All biological chemistry consists of catalysed reactions, the catalysts usually being the large protein molecules called enzymes. A typical enzyme offers the shaped cavities of its three-dimensional form as receptacles for the ingredients of one chemical reaction. It lines them up for each other, enters into temporary chemical liaison with them, matchmakes with an aimed precision that they would be unlikely to discover in open diffusion.

Catalysts, by definition, are not consumed in the chemical reaction they boost, but they may be produced. An autocatalytic reaction is a reaction that manufactures its own catalyst. As you can imagine, an autocatalytic reaction is reluctant to start but, once started, it takes off on its own – like wild fire indeed, for fire has some of the properties of an autocatalytic reaction. Fire is not strictly a catalyst but it is self-generating. Chemically, it is an oxidation process that gives off heat, and needs heat to push it over a threshold to start. Once started, it continues and spreads as a chain reaction because it generates the heat needed to restart itself. Another famous chain reaction is an atomic explosion, in this case not a chemical reaction but a nuclear one. Heredity began as a lucky initiation of an auto-catalytic, or otherwise self-regenerating, process. It immediately took off and spread like a fire, eventually leading to natural selection – and all that was to follow.

We too oxidise carbonaceous fuel to generate heat, but we don't burst into flames because we do our oxidation in a controlled way, step by step, trickling the energy into useful channels instead of dissipating it as undisciplined heat. Such controlled chemistry, or metabolism, is as universal a feature of life as heredity. Theories of the origin of life need to account for both heredity and metabolism, but some writers have mistaken the priority. They have sought a theory of metabolism's spontaneous origin, and somehow hoped that heredity would follow, like other useful devices. But heredity, as we shall see, is not to be thought of as a useful device. Heredity has to be first on the scene because, before heredity, usefulness itself had no meaning. Without heredity, and hence natural selection, there would have been nothing to be useful for. The very idea of usefulness cannot begin until the natural selection of hereditary information does.

The earliest theories of the origin of life that are taken seriously today are those of A. I. Oparin in Russia and J. B. S. Haldane in England, writing in the 1920s in ignorance of each other. Both emphasised metabolism rather than heredity. Both realised, im-portantly, that the atmosphere of Earth lacked free oxygen. Organic compounds (compounds of carbon), when there is free oxygen around, are vulnerable to being burned or otherwise oxidised to carbon dioxide. It seems odd to us, who die within minutes if deprived of oxygen, but as I've already explained, oxygen would have been a deadly poison to our earliest ancestors. Everything we know about other planets makes it almost certain that Earth's original atmosphere lacked free oxygen. It accumulated later, as a polluting waste product of green bacteria, who at first swam free and were later incorporated into plant cells. At some point our ancestors evolved the ability to cope with oxygen, and then came to depend upon it.

Incidentally, having said that oxygen is produced by green plants and algae, it is an oversimplification to leave it at that. It is true that plants give off oxygen. But when a plant dies, its decay, in chemical reactions equivalent to burning all its carbonaceous materials, would use up an amount of oxygen equal to all the oxygen released by that plant during its lifetime. There would therefore be no net gain in atmospheric oxygen, but for one thing. Not all dead plants decay completely. Some parts of them are laid down as coal (or equivalents), other parts are consumed and bits of the consumers themselves may become locked into rocks.* The net effect is to store energy-rich compounds underground and leave some oxygen free to circulate. Releasing some of that stored energy by burning fossil fuels turns this oxygen back into carbon dioxide, taking us back towards the ancient status quo. Fortunately we are unlikely to return the atmosphere all the way to our suffocating Canterbury. But we should not forget that the oxygen we breathe exists only because of compounds tied up underground, which include coal and oil. We burn them at our peril.

Oxygen atoms were always present in that early Earth, but they were not free as oxygen gas. They were tied up in compounds: in solids such as silica (sand), in liquids such as water, and in gas such as carbon dioxide. Carbon today is mostly locked up in living bodies or – a far greater proportion – in rocks, such as chalk, limestone and coal, many of which themselves come from the remains of once-living bodies. In Canterbury times, far more of those same carbon atoms would have been in the atmosphere, mainly as carbon dioxide.

Oparin and Haldane realised that a similar atmosphere would have been friendly to the spontaneous synthesis of simple organic compounds. Here are Haldane's own words, which I quote for the sake of his famous concluding phrase:

> Now, when ultraviolet light acts on a mixture of water, carbon dioxide,
> and ammonia, a vast variety of organic substances are made, including
> sugars and apparently some of the materials from which proteins are

---

* An important example is fool's gold, iron pyrite, an energy-rich sulphide compound which indirectly comes from the plant-digesting activities of sulphur bacteria.

built. This fact has been demonstrated in the laboratory by Baly of Liverpool and his colleagues. In this present world such substances, if left about, decay – that is to say, they are destroyed by micro-organisms.* But before the origin of life they must have accumulated till the primitive oceans reached the consistency of hot dilute soup.

This was written in 1929, more than 20 years before the much cited experiment of Miller and Urey, which, one would think, from Haldane's account, to have been a kind of repeat of Baly's. However, E. C. C. Baly was not concerned with the origin of life. His interest was photosynthesis, and his achievement was to synthesise sugars by beaming ultraviolet rays into water containing dissolved carbon dioxide in the presence of a catalyst, such as iron or nickel. It was Haldane, rather than Baly himself, who, with characteristic brilliance,† was anticipating something remarkably like the Miller–Urey experiment and reading it backwards into Baly's work.

What Miller, under Urey's direction, did was take two flasks, one above the other, connected by two tubes. The lower flask contained heated water to represent the primaeval ocean. The upper flask housed the mocked-up primordial atmosphere (then thought to be methane, ammonia, water vapour and hydrogen). Through one of the two tubes, vapour rose from above the heated 'ocean' in the lower flask and was fed into the top of the 'atmosphere' in the upper flask. The other tube returned downwards from 'atmosphere' to 'ocean'. On the way it passed through a spark chamber ('lightning') and a cooling chamber, where vapour condensed to form 'rain' which replenished the 'ocean'.

After only a week in this recycling simulacrum, the ocean had turned yellow-brown and Miller analysed its content. As Haldane would have predicted, it had become a soup of organic compounds, including no fewer than seven amino acids, the essential building blocks of proteins. Among the seven were three – glycine, aspartic acid and alanine – from the list of 20 found in living things.

Miller's dramatic result has been influential. Perhaps it is not strictly relevant any more, as geologists now tell us that the early atmosphere was otherwise (primarily nitrogen and carbon dioxide). Nevertheless, it established a general principle: complex organic

* This was the point Darwin was making in his 'warm little pond' letter.
† Sir Peter Medawar, no slouch himself, described Haldane as the cleverest man he ever knew.

molecules can form without living things to build them. Most recently, attention has turned to reactions involving simple molecules such as hydrogen sulphide and hydrogen cyanide, calculated to be produced by meteor bombardment. With the addition of ultraviolet light and certain common minerals, researchers led by John Sutherland in Cambridge have created not just proto amino acids (eight in this case), but also the complex precursors to fats and, significantly (as we shall see), RNA. Their particular method requires different reactions to happen in specific environmental conditions, with rainwater merging the results together later.

Whether or not these reactions exactly recreate the early stages of life, we know that natural, abiotic reactions do create complex organic chemicals (including amino acids) because we find them in meteorites. It is not certain how these compounds were formed in outer space, but formed they were, and they were presumably available in the prebiotic environment of Earth. It almost makes one think that Miller needn't have bothered. The riddle remains of how the first self-replicating molecule arose, but at least the first stage – the one simulated by Miller – is now seen to be commonplace.

For Oparin, the vital step was the origin of the first cell. And, to be sure, cells, like organisms, have the important property that they never arise spontaneously but always from other cells. It was pardonable to see the origin of the first 'cell' (metaboliser) as a synonym for the origin of life, rather than the first 'gene' (replicator), as I would. Among more modern theorists with the same metabolism-first bias, the distinguished theoretical physicist Freeman Dyson is aware of it and defends it, and the concept has been extensively developed by Mike Russell at NASA's Jet Propulsion Laboratory. The majority of theorists, including Leslie Orgel in California, Manfred Eigen and his colleagues in Germany, and Graham Cairns-Smith in Scotland – more of a lone maverick, but by no means to be written off – give priority to self-replication, both chronologically and in terms of centrality: rightly so, in my opinion.

What would heredity without a cell look like? Don't we have a chicken-and-egg problem? Certainly so if we take heredity to demand DNA, for DNA can't replicate without a large supporting cast of molecules, including proteins that can only be made by DNA-coded information. But just because DNA is the main self-replicating molecule we know, it doesn't follow that it is the only one we could imagine, or the only one that has ever existed in nature. Cairns-

Smith has persuasively argued that the original replicators were inorganic mineral crystals, with DNA a late usurper, stepping into the starring role when life had evolved to the point where such a *Genetic Takeover* became possible. I won't expound his case here, partly because I have already given it my best shot in *The Blind Watchmaker*, but also for a larger reason. Cairns-Smith makes the clearest case I have read that replication was primary, and that DNA must have had a forerunner of some kind whose nature is unknown, save that it exhibited true heredity. I think it is a shame if this unassailable part of his case becomes tied in people's minds to his more controversial and speculative case for mineral crystals as the forerunner.

I have nothing against the mineral crystals theory, which is why I expounded it before, but what I really want to emphasise is the primacy of replication, and the strong likelihood that there was a late takeover by DNA from some forerunner. I can make the point most forcefully by deliberately switching in this book to a different *particular* theory of what that forerunner might have been. Whatever its ultimate merits as the original replicator, RNA is certainly a better candidate than DNA, and it has been cast as forerunner by a number of theorists in their so-called 'RNA World'. To introduce the RNA World theory, I need to digress on enzymes. If the replicator is the star of life's show, the enzyme is the co-star, more than just supporting cast.

Life depends utterly on the virtuoso ability of enzymes to catalyse biochemical reactions in a very fussy way. When I first learned about enzymes at school, the conventional (and in my view mistaken) wisdom that science should be taught by homely example meant that we spat into water to demonstrate the power of the salivary enzyme amylase to digest starch and make sugar. From this we gained the impression that an enzyme is like a corrosive acid. Biological washing powders, which use enzymes to digest dirt out of clothes, give the same impression. But these are destructive enzymes, working to dismember large molecules into their smaller components. Constructive enzymes are involved in synthesising large molecules from smaller ingredients, and they do so by behaving as 'robotic matchmakers', as I shall explain.

The interior of a cell contains a solution of thousands of molecules, atoms and ions of many different kinds. Pairs of these could combine with each other in almost infinitely varied ways, but on the whole they do not. So there is a huge repertoire of potential chemistry waiting to happen in a cell, but most of it doesn't happen. Hold that in mind while reflecting on the following. A chemistry lab has hundreds of bottles on its shelves, all securely stoppered so their contents don't meet each other unless a chemist desires it, in which case a sample from one bottle is added to a sample from another. You could say that the shelves in a chemistry lab also house a huge repertoire of potential chemistry waiting to happen. And again most of it doesn't happen.

But imagine taking all the bottles off all the shelves and tipping them into a single vat full of water. A preposterous act of scientific vandalism, yet such a vat is pretty much what a living cell is, although admittedly with a lot of internal membranes that complicate the picture. The hundreds of ingredients of thousands of potential chemical reactions are not kept in separate bottles until required to react together. Instead, they are all mixed up together in the same shared space, all the time. But still they wait, largely unreactive, until required to react, as though separated in virtual bottles. There are no virtual bottles, but there are enzymes working as robotic match-makers, or we might even call them robotic lab assistants. Enzymes discriminate, much as a radio tuner does when it puts a particular wireless set in touch with a particular transmitter while ignoring the hundreds of other signals simultaneously bombarding its aerial with a babel of carrier frequencies.

Suppose there is an important chemical reaction in which in-gredient A combines with ingredient B to yield product Z. In a chemistry lab we achieve this by taking the bottle labelled A off the shelf, and the bottle labelled B from another shelf, mixing their contents in a clean flask, and providing other necessary conditions, such as heat or stirring. We achieve the specific reaction we want by taking only two bottles off the shelf. In the living cell lots of A molecules and lots of B molecules are among the huge variety of molecules floating around in the water, where they may meet, but seldom combine even if they do. In any case, they are no more likely to meet than thousands of other possible combinations. Now we introduce an enzyme called abzase, which is specifically shaped to catalyse the A+B=Z reaction. There are now millions of abzase molecules in the cell, each one acting as a robotic lab assistant. Each abzase lab assistant grabs one A molecule, not off a shelf but floating free in the cell. It then grabs a B molecule as it drifts by. It holds the A firmly in its grip so that it faces in a particular direction. And it holds the B equally firmly so that it abuts the A, in just the right position and orientation to bond with the A and make a Z. The enzyme may do other things too – the equivalent of the human lab assistant wielding a stirrer or lighting a Bunsen burner. It may form a temporary chemical alliance with A or B, exchanging atoms or ions that will eventually be paid back, so the enzyme ends up as it started, thus qualifying as a catalyst. The result of all this is that a new Z molecule forms in the shaped 'grip' of the enzyme molecule. The lab assistant then releases the new Z into the water and waits for another A to come by, whereupon it grabs it and the cycle resumes.

If there were no robotic lab assistant, a drifting A would occasionally bump into a drifting B under the right conditions to bond. But this lucky occurrence would be rare, no more common than the occasional chance-matched encounters that either A or B might make with lots of other potential partners. A might bump into C and make Y. Or B might bump into D and make X. Small amounts of Y and X are being made all the time by lucky drift. But it is the presence of the lab assistant enzyme abzase that makes all the difference. In the presence of abzase, Z is churned out in (from the cell's point of view) industrial quantities: an enzyme typically multiplies the spontaneous rate of reaction by a factor varying between a million and a trillion. If a different enzyme, acyase, were introduced, A would be combined with C instead of B, again at racing conveyor-belt speed, to make a lavish supply of Y. It is the very same A molecules we are talking about, not confined to a bottle but free to combine with either B or C, depending on which enzyme is present to grab them.

The production rates of Z and Y will therefore depend on, among other things, how many of each of the two rival lab assistants, abzase and acyase, are floating about in the cell. And *that* depends on which of two genes in the nucleus of the cell is turned on. It is, however, a little more complicated than that: even if a molecule of abzase is present, it may be inactivated. One way this can happen is that another molecule comes and sits in the active 'cavity' of the enzyme. It is as though the lab assistant's robotic arms were temporarily handcuffed. The handcuffs remind me, by the way, to issue the ritual warning that, as always with metaphors, there is a risk that 'robotic lab assistant' might mislead. An enzyme molecule doesn't actually have arms to reach out and seize ingredients such as A, let alone submit to handcuffs. Instead, it has special zones in its own surface for which A, say, has an affinity, either because of a snug physical fit to a shaped cavity, or due to some more recondite chemical property. And this affinity can be temporarily negated in ways that resemble the calculated throwing of an off-switch.

Most enzyme molecules are special-purpose machines which make only one product: a sugar, say, or a fat; a purine or a pyrimidine (building blocks of DNA and RNA), or an amino acid (20 of them are building blocks of natural proteins). But some enzymes are more like programmable machine tools that take in a punched paper tape to determine what they do. Outstanding among these is the ribosome, briefly explained in Taq's Tale, a large and complicated machine tool constructed from both protein and RNA, which makes proteins themselves. Amino acids, the building blocks of proteins, have already been made by special-purpose enzymes and are floating around in the cell, available to be picked up by the ribosome. The punched paper tape is RNA, specifically 'messenger RNA' (mRNA). The messenger tape, which itself has copied its message from DNA in the genome, feeds into the ribosome and, as it passes through the 'reading head', the appropriate amino acids are assembled into a protein chain in the order specified by the tape using the genetic code.

How this specification works is known, and it is unspeakably wonderful. There is a set of small 'transfer RNAs' (tRNA), each about 70 building blocks long. Each of the tRNAs attaches itself selectively to one, and only one, of the 20 kinds of natural amino acids. At the other end of the tRNA molecule is an 'anti-codon', a triplet precisely complementing the short mRNA sequence (codon) that specifies the particular amino acid according to the genetic code. As the tape of mRNA moves through the reading head of the ribosome, each codon of the mRNA binds to a tRNA with the right anti-codon. This causes the amino acid dangling off the other end of the tRNA to be brought into line, in the 'matchmaking' position, to attach to the growing end of the newly forming protein. Once the amino acid is attached, the tRNA peels off in search of a new amino acid molecule of its preferred type, while the mRNA tape inches forward another notch. So the process continues and the protein chain is extruded step by step. Amazingly, one physical tape of mRNA can cope with several ribosomes at once. Each of these ribosomes moves its reading head along a different portion of the tape's length, and each extrudes its own copy of the newly minted protein chain.

As each new protein chain is completed, when the mRNA feeding its ribosome has completely gone through that ribosome's reading head, the protein detaches itself. It coils up into a complicated three-dimensional structure whose shape is determined, through the laws of physics and chemistry, by the sequence of amino acids in the protein chain. That sequence was itself determined by the order of code symbols along the length of the mRNA. And that order was, in turn, determined by the complementary sequence of symbols along the DNA, which constitutes the master database for the cell.

The coded sequence of DNA therefore controls what goes on in the cell. It specifies the sequence of amino acids in each protein, which determines the protein's three-dimensional shape, which in turn gives that protein its particular enzymatic properties. Importantly, the control may be indirect in that, as we saw in the Mouse's Tale, genes determine which other genes shall be turned on and when. Most genes in any one cell are not switched on. This is why of all the reactions that could be going on in the 'vat full of mixed ingredients', only one or two actually do go on at any one time: the ones whose specific 'lab assistants' are active in the cell.

After that digression on catalysis and enzymes, we now turn from ordinary catalysis to the special case of autocatalysis, some version of which probably played a key role in the origin of life. Think back to our hypothetical example of molecules A and B combining to make Z under the influence of the enzyme abzase. What if Z itself is its own abzase? I mean, what if the Z molecule happens to have just the right shape and chemical properties to seize one A and one B, bring them together in the correct orientation, and combine them to make a new Z, just like itself? In our previous example we could say that the amount of abzase in the solution would influence the amount of Z produced. But now, if Z actually is one and the same molecule as abzase, we need only a single molecule of Z to seed a chain reaction. The first Z grabs As and Bs and combines them to make more Zs. Then these new Zs grab more As and Bs to make still more Zs and so on. This is autocatalysis. Under the right conditions the population of Z molecules will grow exponentially – explosively. This is the kind of thing that sounds promising as an ingredient for the origin of life.

But it is all hypothetical. Julius Rebek and his colleagues at the Scripps Institute in California made it real. They explored some fascinating examples of autocatalysis in real chemistry. In one of their examples, Z was amino adenosine triacid ester (AATE), A was amino adenosine and B was pentafluorophenyl ester, and the reaction took place not in water but in chloroform. Needless to say, none of these particular chemical details, and certainly not the long names, need to be remembered. What matters is that the product of the chemical reaction is its own catalyst. The first molecule of AATE is reluctant to form but, once formed, an immediate chain reaction is set in train as more and more AATE synthesises itself by serving as its own catalyst. As if that weren't enough, this brilliant series of experiments went on to demonstrate true heredity in the sense defined here. Rebek and his team found a system in which more than one variant of the autocatalysed substance existed. Each variant catalysed the synthesis of itself, using its preferred variant of one of the ingredients. This raised the possibility of true competition in a population of entities showing true heredity, and an instructively rudimentary form of Darwinian selection.

Rebek's chemistry is highly artificial. Nevertheless, his chloroform-based toy world beautifully illustrates the principle of autocatalysis, according to which the product of a chemical reaction serves as its own catalyst. It is something like autocatalysis that we need for the origin of life. Could RNA, or something like RNA under the conditions of the early Earth, have autocatalysed its own synthesis Rebek-style, and in water instead of chloroform?

The problem is a formidable one, as explained by the German Nobel Prize-winning chemist Manfred Eigen. He pointed out that any self-replication process is subject to degradation by copying error – mutation. Imagine a population of replicating entities in which there is a high probability of error in every copying event. If a coded message is to hold its own against the ravages of mutation, at least one member of the population in any one generation must be identical to its parent. If there are ten code units ('letters') in an RNA chain, for example, the average error rate per letter must be less than one in ten: we can then expect that at least some members of the offspring generation will have the full complement of ten correct code letters. But if the error rate is greater, there will be a relentless degradation as the generations go by, simply because of mutation alone, no matter how strong the selection pressure. This is called an error catastrophe. Mark Ridley's excellent book *Mendel's Demon*, which I have mentioned before, argues that sex evolved to prevent error catastrophes in advanced genomes. Here we are concerned with the simplest possible genomes, and the error catastrophe that threatened the origin of life itself.

Short chains of RNA and, indeed, DNA can spontaneously self replicate without an enzyme. But the error rate per letter is far higher than when an enzyme is present. And this means that long before a sufficient length of gene could be built up to make the protein for a working enzyme, the fledgling gene would have been destroyed by mutation. That is the catch-22 of the origin of life. A gene big enough to specify an enzyme would be too big to replicate accurately without the aid of an enzyme of the very kind that it is trying to specify. So the system apparently cannot get started.

The solution to the catch-22 that Eigen offers is the theory of the hypercycle. It uses the old principle of divide and rule. The coded information is subdivided into sub-units small enough to lie below the threshold for an error catastrophe. Each sub-unit is a mini replicator in its own right, and it is small enough for at least one copy to survive in each generation. All the sub-units co-operate in some important larger function, large enough to suffer an error catastrophe if catalysed by a single large chemical rather than being subdivided.

As I have so far described the theory, there is a danger that the whole system would be unstable because some sub-units would self-replicate faster than others. This is where the clever part of the theory kicks in. Each sub-unit flourishes in the presence of the others. More specifically, the production of each is catalysed by the presence of another, such that they form a cycle of dependency: a 'hypercycle'. This automatically prevents any one element from racing ahead. It cannot do so because it depends on its predecessor in the hypercycle.

John Maynard Smith pointed out the similarity of a hypercycle to an ecosystem. Fish numbers depend on the population of *Daphnia* (waterfleas) on which they feed. In turn, fish numbers affect the population of fish-eating birds. The birds provide guano, which assists blooms of algae on which the *Daphnia* flourish. The whole cycle of dependency is a hypercycle. Eigen and his colleague Peter Schuster propose some kind of molecular hypercycle as the solution to the catch-22 riddle of the origin of life.

More recently, Nilesh Vaidya and colleagues in the United States have managed to produce a three-component chemical hypercycle in the lab, using an RNA molecule split into three pieces, A, B, and C, in which A helps make B, B helps make C, and C helps make A again. They chose to base their experiment on RNA for much the same reason that the RNA world is a strong theoretical contender for an early stage in the origin of life. To understand this, we need to look at why proteins are good at being enzymes but bad at being replicators; at why DNA is good at replicating but bad at being an enzyme; and finally why RNA might just be good enough at both roles to break out of the catch-22.

Three-dimensional shape is largely what matters for enzyme activity. Proteins are good at being enzymes because they can assume almost any shape you want in three dimensions, as an automatic consequence of their amino acid sequence in one dimension. It is the chemical affinities of amino acids for other amino acids in different parts of the chain that determine the particular knot into which the protein chain ties itself. So the three-dimensional shape of a protein molecule is specified by the one-dimensional sequence of amino acids, and that is itself specified by the one-dimensional sequence of code letters in a gene. In principle (practice is a different matter, and formidably difficult) it should be possible to write down a sequence of amino acids that would spontaneously coil itself up into almost any shape you like: not just shapes that make good enzymes, but any arbitrary shape you choose to specify.* It is this protean talent that qualifies proteins to act as enzymes. There is a protein capable of selecting any one out of the hundreds of potential chemical reactions that could go on in a cell full of jumbled ingredients.

Proteins, then, make wonderful enzymes, capable of tying themselves into knots of any desired shape. But they are lousy replicators. Unlike DNA and RNA, whose component elements have specific pairing rules (the 'Watson–Crick pairing rules' discovered by those two inspired young men), amino acids have no such rules. DNA, by contrast, is a splendid replicator but a lousy candidate for the enzyme role in life. This is because, unlike proteins with their near infinite variety of three-dimensional shapes, DNA has only one basic shape, the famous double helix itself. The double helix is ideally suited to replication because the two sides of the stairway peel easily away from one another, each being then exposed as a template for new letters to join, following the Watson–Crick pairing rules. It is not much good for anything else.

RNA has some of the virtues of DNA as a replicator and some of the virtues of protein as a versatile shaper of enzymes. Both DNA and RNA can form double helices, and the four letters of RNA are

* Indeed, there are lots of different amino acid sequences that will yield the same shape, which is one reason to doubt naive calculations of the astronomical 'improbability' of a particular protein chain, obtained by raising 20 to the power of its length.

sufficiently similar to the four letters of DNA that either set can serve as a template for the other. On the other hand, compared to DNA, RNA breaks down more easily into single letters, its double helix form is less flexible, and the pairings between the bases are stronger. These three properties make it somewhat inferior to DNA as a replicator, and only simple organisms with small genomes, such as some viruses, can get away with using RNA as their primary replicator.

But the last two properties of RNA conspire to provide an upside as well as a downside. Its relative inflexibility means that RNA does not comfortably form long lengths of double helix. Yet the strong base pairing means that an RNA strand will readily form short lengths of double helix – moreover, it will tend to do so with other bits of itself, using the ordinary Watson–Crick base-pairing rules, the same ones as are used to make copies of RNA. RNA frequently finds small stretches of itself with which it can pair, and that leads to it tying itself in knots rather like a protein. The pairing rules insist that the paired stretches have to be going in opposite directions, so a knotted RNA chain therefore tends to be made up of a series of hairpin bends (see plate 51).

The repertoire of three-dimensional shapes into which an RNA molecule is capable of throwing itself may not be as great as the repertoire of large protein molecules. But it is large enough to encourage the thought that RNA might furnish a versatile armoury of enzymes. And, to be sure, many RNA enzymes, called ribozymes, have been discovered. The conclusion is that RNA has some of the replicator virtues of DNA and some of the enzyme virtues of proteins. Maybe, before the coming of DNA, the arch-replicator, and before the coming of proteins, the arch-catalysts, there was a world in which RNA alone had enough of both virtues to stand in for both experts. Perhaps an RNA fire ignited itself in the original world, and then later started to make proteins that turned around and helped synthesise RNA, and later DNA too, which took over as the dominant replicator. That is the hope of the RNA World theory. It receives indirect support from a lovely series of experiments initiated by Sol Spiegelman of Columbia University, and repeated in various forms by others over the years. Spiegelman's experiments use a protein enzyme, which might be thought to be cheating, but they produce such spectacular results, illuminating such important links in the theory, that you can't help feeling it was worth it anyway.

First, the background. There is a virus called Qβ. It is an RNA virus, which means that, instead of DNA, its genes are entirely made of RNA. It uses an enzyme to replicate its RNA, called Qβ replicase. In the wild state, Qβ is a bacteriophage (phage for short) – a parasite of bacteria, specifically of the common gut bacterium *Escherichia coli*. The bacterial cell 'thinks' the Qβ RNA is a piece of its own messenger RNA, and its ribosomes process it exactly as though it were, but the proteins that it manufactures are good for the virus instead of for the host bacterium. There are four such proteins: a coat protein to protect the virus; a glue protein to stick it to the bacterial cell; a so-called replication factor, which I'll mention again in a moment; and a bomb protein to destroy the bacterial cell when the virus has finished replicating, thereby releasing some tens of thousands of viruses, each to travel in its little protein coat until it bumps into another bacterial cell and renews the cycle. I said I would return to the replication factor. You might think that this must be the enzyme Qβ replicase, but actually it is smaller and simpler. All that the little viral gene itself does is make a protein that sews together three other proteins which the bacterium is making anyway for its own (completely different) purposes. When these are stitched together by the virus's own little protein, the composite so formed is the Qβ replicase.

Spiegelman was able to isolate from this system just two components, Qβ replicase and Qβ RNA. He put them together in water with some small-molecule raw materials – the building blocks for making RNA – and watched what happened. The RNA seized small molecules and built copies of itself using Watson–Crick pairing rules. It managed this feat without any bacterial host, and without the protein coat or any other part of the virus. That in itself was a nice result. Notice that protein synthesis, which is part of the normal action of this RNA in the wild, has been completely taken out of the loop. We have a stripped-down RNA replication system making copies of itself without bothering to make protein.

Then Spiegelman did something wondrous. He set a form of evolution in motion in this wholly artificial test-tube world, with no cells involved at all. Imagine his set-up as a long row of test tubes, each containing Qβ replicase and raw building blocks but no RNA. He seeded the first tube with a small amount of Qβ RNA, and it duly replicated lots more copies of itself. He then drew out a small sample of the liquid, and put a drop of it into the second tube. This seed RNA now set about replicating in the second tube, and when this had been going on for a while Spiegelman drew out a drop from the second tube and seeded the third virgin tube. And so on. This is like the spark from our fire seeding a new fire in the dry grass, and the new fire seeding another, and so on in a chain of seedings. But the result was very different. Whereas fires don't inherit any of their qualities from the seed, Spiegelman's RNA molecules did. And the consequence was . . . evolution by natural selection in its most basic and stripped-down form.

Spiegelman sampled the RNA in successive tubes as the 'generations' went by and monitored its properties, including its potency in infecting bacteria. What he found was fascinating. The evolving RNA became physically smaller and smaller and, at the same time, less and less infective when bacteria were offered to samples of it. After 74 generations* the typical RNA molecule in a tube had evolved to a small fraction of the size of its 'wild ancestor'. The wild RNA had been a necklace about 3,600 'beads' long. After 74 generations of natural selection, the average inhabitant of a test tube had reduced itself down to a mere 550: no good at infecting bacteria but brilliant at infecting test tubes. What had happened was clear. Spontaneous mutations in the RNA had occurred all along the line, and the mutants that survived were well fitted to do so in the test-tube world, as opposed to the natural world of bacteria waiting to be parasitised. The main difference was presumably that the RNA in the tube world could dispense with all the coding devoted to making the four proteins needed to make the coat, the bomb and the other requirements for survival of the wild virus as a working parasite of bacteria. What was left was the bare minimum required to replicate in the featherbedded world of test tubes full of Qβ replicase and raw materials.

This bare minimum survivor, less than a tenth the size of its wild ancestor, has become known as Spiegelman's Monster. Being smaller, the streamlined variant reproduces more rapidly than its competitors, and therefore natural selection gradually increases its representation in the population (and population, by the way, is exactly the right word, even though we are talking about free-floating molecules, not viruses or organisms of any kind).

Amazing to relate, almost the same Spiegelman monster repeatedly evolves when the experiment is run over again. Moreover,

* That's tube generations, of course: the number of RNA generations would be more, because RNA molecules are replicating many times within each tube generation.

Spiegelman and Leslie Orgel, one of the leading figures in research on the origin of life, performed further experiments in which they added a nasty substance, such as ethidium bromide, to the solution. Under these conditions, a different monster evolves, one that is resistant to ethidium bromide. Different chemical obstacle courses foster evolution towards different specialist monsters.

Spiegelman's experiments used natural 'wild type' Qβ RNA as a starting point. M. Sumper and R. Luce, working in the laboratory of Manfred Eigen, obtained a truly stunning result. Under some conditions, a test tube containing *no RNA at all*, just the raw materials for making RNA plus the Qβ replicase enzyme, can spontaneously generate self-replicating RNA which, under the right circumstances, will evolve to become similar to Spiegelman's Monster. So much, incidentally, for creationist fears (or hopes, we might rather say) that large molecules are too 'improbable' to have evolved. Such is the simple power of cumulative natural selection (so far is natural selection from being a process of blind chance) Spiegelman's Monster takes only a few days to build itself up from scratch.

These experiments are still not direct tests of the RNA World hypothesis of the origin of life. In particular, we still have the 'cheat' of Qβ replicase being present throughout. The RNA World hypothesis pins its hopes on RNA's own catalytic powers. If RNA can catalyse other reactions, as it is known to do, might it not catalyse its own synthesis? Sumper and Luce's experiment dispensed with RNA but provided the Qβ replicase. What we need is a new experiment that dispenses with the Qβ replicase too. Research continues, and I expect exciting results. But now I want to switch another line of thought, fully compatible with the RNA World, and with many others among the current theories of the origin of life. What is interesting is the suggested location in which the crucial events first took place. Not 'warm little pond' but 'hot deep rock'. The main inspirer of the theory was another maverick, Thomas Gold, originally an astronomer but versatile enough to deserve the now-rare accolade 'general scientist', and distinguished enough to have been elected to both the Royal Society of London and the American National Academy of Sciences.

Gold believed that our emphasis on the sun as energetic prime mover of life may be misplaced. Perhaps we have yet again been misled by what happens to be familiar: yet again assigned to ourselves and our kind of life a centrality in the scheme of things that we do not deserve. There was a time when textbooks asserted that all life depended ultimately on sunlight. Then, in 1977, the startling discovery was made that volcanic vents on the floors of deep oceans support a strange community of creatures, living without benefit of sunlight. Heat from red-hot lava raises the water temperature to more than 100°C, still well below boiling point at the colossal pressures of those depths. The surrounding water is very cold, and the temperature gradient drives various kinds of bacterial metabolism. These thermophile bacteria, including sulphur bacteria who make use of hydrogen sulphide streaming from the volcanic vents, constitute the base of elaborate food chains, higher links of which include blood-red tube worms up to three metres long, together with limpets, mussels, starfish, barnacles, white crabs, prawns, fish and other annelid worms capable of thriving at 80°C. There are bacteria, as we have seen, which can take such Hadean temperatures in their stride, but almost no animals are known to do so, and these annelids have accordingly been dubbed Pompeii worms. Some of the sulphur bacteria are given house room by animals, for example by mussels, and by the huge tube worms, who take special biochemical steps, using haemoglobin (hence their blood-red colour) to feed sulphide to their own bacteria. These colonies of life, based on bacterial extraction of energy from hot volcanic vents, astonished everybody, first by their very existence, and then by their abundant richness, which contrasted startlingly with the near-desert conditions of the surrounding sea bottom.

Even after this sensational discovery, most biologists continue to believe that life is centred on the sun. The creatures of the deep-sea smoker communities, fascinating though they might be, are assumed by most of us to be a rare and unrepresentative aberration. Gold suggested otherwise. He believed hot, dark, high-pressure depths are where life fundamentally belongs and where it originated. Not necessarily in the deep, smoking sea, although that is another popular theory, but in the rocks, deep underground. We who live at the surface, in the light and the cool and the fresh air, we are the anomalous aberrations! He pointed out that 'hopanoids', organic molecules made in bacterial cell walls, are ubiquitous in rocks, quoting an authoritative estimate of between 10 trillion and 100 trillion tonnes of hopanoids in the rocks of the world. This comfortably exceeds the trillion tonnes or so of organic carbon in surface-dwelling life.

Gold noted that the rocks are seamed with cracks and fissures, which, though small to our eyes, provide more than a billion trillion cubic centimetres of hot, wet space suitable for life on the bacterial scale of existence. Heat energy, and the chemicals of the rocks themselves, would be enough to sustain bacteria in huge numbers. Many bacteria thrive at temperatures up to 110°C, and this would permit them to live down to depths of between 5 and 10 kilometres, a distance that would take them less than 1,000 years to travel. It is impossible to verify his estimate, but he calculated that the biomass of bacteria in the hot, deep rocks might exceed the biomass of the surface sun-based life with which we are familiar.

Turning to the question of the origin of life, Gold and others have pointed out that thermophily – love of high temperatures – is not a rare oddity among bacteria and archaeans. It is common: so common, and so widely distributed around bacterial family trees, that it might well be the primitive state from which our familiar cool forms of life have evolved. With respect to both chemistry and temperature, the conditions on the surface of the primitive Earth – some scientists call it the Hadean Age – were more like those in Gold's hot deep rocks than they were like today's surface conditions. A persuasive case can indeed be made that when we dig down into the rocks we are digging backwards in time, and rediscovering something like the conditions of life's scalding Canterbury.

The idea has been further championed recently by the physicist Paul Davies, whose book *The Fifth Miracle* summarises new evidence discovered since Gold's paper of 1992. Various drilling samples have been found to contain hyperthermophile bacteria, alive and reproducing, amid scrupulous precautions to preclude contamination from the surface. Some of these bacteria have been successfully cultured . . . in a modified pressure cooker! Davies, like Gold, believes life may have originated deep underground, and that the bacteria which still live there may be relatively unchanged relics of our remote ancestors. This idea is especially appealing for our pilgrimage because it offers us the hope of meeting something like the earliest bacteria, rather than the more familiar bacteria, modified for modern conditions of light, cold and oxygen. Having endured ridicule at first, the hot deep rock theory of the origin of life is now verging on the positively fashionable. Whether it will turn out to be right must await more research, but I confess to hoping that it will.

There are many other theories that I have not gone into. Maybe one day we shall reach some sort of definite consensus on the origin of life. If so, I doubt if it will be supported by direct evidence because I suspect that it has all been obliterated. Rather, it will be accepted because somebody produces a theory so elegant that, as the great American physicist John Archibald Wheeler said in another context:

> . . . we will grasp the central idea of it all as so simple, so beautiful, so compelling that we will say to each other, 'Oh, how could it have been otherwise! How could we all have been so blind for so long!'

If that isn't how we finally realise we know the answer to the riddle of life's origin, I don't think we ever shall know it.

THE HOST'S RETURN

The genial host, having guided Chaucer and the other pilgrims from London to Canterbury and stood impresario to their tales, turned around and led them straight back to London. If I now return to the present, it must be alone, for to presume upon evolution's following the same forward course twice would be to deny the rationale of our backward journey. Evolution was never aimed at any particular endpoint. Our backwards pilgrimage has been a series of swelling mergers, as we were swallowed up in ever more inclusive groupings: the apes, the primates, the mammals, the vertebrates, the deuterostomes, the animals and so on back to the arch ancestor of all life. If we turn around and move forward now, we cannot retrace our steps. That would imply that evolution, were it to be rerun, would follow the same course, putting those same mergers into reverse gear in the form of splits. The stream of life would branch in all the 'right' places. Photosynthesis and an oxygen-based metabolism would be rediscovered, the eukaryotic cell would reconstitute itself, cells would club together in neometazoan bodies. There would be a new split between plants on the one hand, and animals plus fungi on the other; a new split between protostomes and deuterostomes; the backbone would be rediscovered, and so would eyes, ears, limbs, nervous systems . . . Eventually a swollen-brained biped would emerge, with skilled hands guided by forward-looking eyes, culminating in the proverbial cricket team to beat the Australians.

My disavowal of aimed evolution underlay my original choice to do history backwards. And yet in my opening lines I confessed to an ear for a rhyme that would lead me into cautious flirtation with recurring patterns, with lawfulness and forward directionality in evolution. So although my return as host will not be a retracing of steps, I shall be publicly wondering whether something a little bit like a retracing might not be appropriate.

Rerunning Evolution

The American theoretical biologist Stuart Kauffman put the question well in a 1985 article:

One way to underline our current ignorance is to ask, if evolution were to recur from the Precambrian when early eukaryotic cells had already been formed, what organisms in one or two billion years might be like. And, if the experiment were repeated myriads of times, what properties of organisms would arise repeatedly, what properties would be rare, which properties were easy for evolution to happen upon, which were hard? A central failure of our current thinking about evolution is that it has not led us to pose such questions, although the answers might in fact yield deep insight into the expected character of organisms.

I especially like Kauffman's statistical proviso. He envisages not just one thought experiment but a statistical sample of thought experiments in quest of general laws of life, as opposed to local manifestations of particular lifes. The Kauffman question is akin to the science fiction question of what life on other planets might be like – except that on other planets the starting and prevailing conditions would be different. On a large planet, gravity would impose a whole new set of selection pressures. Animals the size of spiders could not have spidery limbs (they'd break under the weight) but would need the support of stout, vertical columns, like the tree trunks on which our elephants stand. Conversely, on a smaller planet, animals the size of elephants but of gossamer build could skitter and leap over the surface like jumping spiders. Those expectations about body build will apply to the whole statistical sample of high-gravity worlds and the whole statistical sample of low-gravity worlds.

Gravity is a given condition of a planet, which life cannot influence. So is its distance from its central star. So is its speed of rotation, which determines day length. So is the tilt of its axis, which, on a planet like ours with its near-circular orbit, is the main determinant of seasons. On a planet with a far-from-circular orbit like Pluto, the dramatically changing distance from the central star would be a much more significant determiner of seasonality. The presence, distance, mass and orbit of a moon or moons exert a subtle but strong influence on life via the tides. All these factors are givens, uninfluenced by life and therefore to be treated identically in successive reruns of the Kauffman thought experiment.

Earlier generations of scientists would have treated the weather and the chemical composition of the atmosphere as givens too. Now we know that the atmosphere, especially its high oxygen and low carbon content, is conditioned by life. So our thought experiment must allow for the possibility that in successive reruns of evolution the atmosphere might vary under the influence of whatever life forms evolve. Life could thereby influence the weather, and even major climatic episodes, such as ice ages and droughts. My late colleague W. D. Hamilton, who was right too many times to be laughed away, suggested that clouds and rain are themselves adaptations manufactured by micro-organisms for their own dispersal.

So far as we know, the innermost workings of the Earth remain unaffected by the froth of life on its surface. But thought experiments in the rerunning of evolution should acknowledge possible differences in the course of tectonic events, and hence the histories of continental positions. It is an interesting question whether episodes of volcanism and earthquakes, and bombardments from outer space, should be assumed to be the same on successive Kauffman reruns. It is probably wise to treat tectonics and celestial collisions as important variables that can be averaged out if we imagine a sufficiently large statistical sample of reruns.

How shall we set about answering the Kauffman question? What would life be like if the 'tape' were rerun a statistical number of times? Immediately we can recognise a whole family of Kauffman questions, of steadily increasing difficulty. Kauffman chose to reset the clock at the moment when the eukaryotic cell was assembled from its bacterial components. But we could imagine restarting the process two or three eons earlier, with the origin of life itself. Or, at the other extreme, we can restart the clock much later, say, at Concestor 1, our split from the chimpanzees, and ask whether the hominids would, in a statistically significant number of reruns given that life had reached Concestor 1, have evolved bipedality, brain enlargement, language, civilisation and baseball. In between, there is a Kauffman question for the origin of the mammals, for the origin of the vertebrates, and any number of other Kauffman questions.

Short of pure speculation, does the history of life, as it actually happened, provide anything approaching a natural Kauffman experiment to guide us? Yes it does. We met several natural experiments throughout our pilgrimage. By happy accidents of prolonged geographical isolation, Australia, New Zealand, Madagascar, South America, even Africa, furnish us with approximate reruns of major episodes of evolution.

These land masses were isolated from each other, and from the rest of the world, for significant parts of the period after the dinosaurs disappeared, when the mammal group displayed most of its evolutionary creativity. The isolation was not total, but was sufficient to foster the lemurs in Madagascar, and the ancient and diverse radiation of Afrotheria in Africa. In the case of South America, we have distinguished three separate foundations of mammals, with long periods of isolation in between. Australinea provides the most perfect conditions for this kind of natural experiment – its isolation was nearly perfect for much of the period in question, and it began with a very small, possibly single, inoculum of marsupials. New Zealand is different, for – alone among these revealing natural experiments – it was colonised not by mammals but by birds during the period in question.

As I look at these natural experiments, mostly I am impressed by how similarly evolution turns out when it is allowed to run twice. We have seen how similar the marsupial *Thylacinus* is to a dog, *Notoryctes* to a mole, *Petaurus* to flying squirrels, *Thylacosmilus* to the sabretooths (and to various 'false sabretooths' among the placental mammals). The differences are instructive too. Kangaroos are hopping antelope-substitutes. Bipedal hopping, when perfected at the end of a line of evolutionary progression, may be as impressively fast as quadrupedal galloping. But the two gaits are radically different from each other, in ways that have wrought major changes in the whole anatomy. Presumably, at some ancestral parting of the ways, either of the two 'experimental' lineages could have followed the route of perfecting bipedal hopping, and either could have perfected quadrupedal galloping. As it happens – possibly for almost accidental reasons originally – the kangaroos hopped one way and the antelopes galloped the other. We now marvel at the downstream divergences between the end-products.

The mammals underwent their disparate evolutionary radiations at roughly the same time as each other, on different land masses. The vacuum left by the dinosaurs freed them to do so. But the dinosaurs in their time had similar evolutionary radiations, although with notable omissions – for example, I can't get an answer to my question of why there seem to have been no dinosaur 'moles'. And before the dinosaurs there were yet other multiple parallels, notably among the mammal-like reptiles, and these too culminated in similar ranges of types.

When I give public lectures I always try to answer questions at the end. The commonest question by far is, 'What might humans evolve into next?' My interlocutor always seems touchingly to imagine it is a fresh and original question, and my heart sinks every time. For it is

Reapperarance of Ichthyosauri. This cartoon by Henry de la Beche lampooned a view expressed by Charles Lyell that the Earth's periodic changes in climate and associated wildlife might lead to a future world in which *Iguanodon* would once more roam the woods and ichthyosaurs reappear in the seas. Note the human skull under Professor Ichthyosaurus's lectern.

a question that any prudent evolutionist will evade. You cannot, in detail, forecast the future evolution of any species, except to say that statistically the great majority of species have gone extinct. But although we cannot forecast the future of any species, say, 20 million years hence, we can forecast the general range of ecological types that will be around. There will be herbivores and carnivores, grazers and browsers, meat eaters, fish eaters and insect eaters. These dietary forecasts themselves presuppose that in 20 million years there will still be foods corresponding to the definitions. Browsers presuppose the continued existence of trees. Insectivores presuppose insects, or anyway small, leggy invertebrates – doodoos, to borrow that useful technical term from Africa. Within each category, herbivores, carnivores and so on, there will be a range of sizes. There will be runners, fliers, swimmers, climbers and burrowers. The species won't be exactly the same as the ones we see today, or the parallel ones that evolved in Australia or South America, or the dinosaur equivalents, or the even earlier mammal-like reptile equivalents. But there will be a similar range of types, making their livings in a similar range of ways.

If, during the next 20 million years, there is a major catastrophe and a mass extinction comparable to the end of the dinosaurs, we can expect the range of ecotypes to be drawn from new ancestral starting points, and – notwithstanding my speculation about rodents at Rendezvous 10 – it might be quite hard to guess which of today's animals will provide those starting points. A Victorian cartoon (see opposite) shows Professor Ichthyosaurus discoursing upon a human skull from some remote recycling past. If, in the time of the dinosaurs, Professor Ichthyosaurus had mooted their catastrophic end, it would have been quite hard for him to forecast that their place would be taken by the descendants of the mammals, which were then small, insignificant, nocturnal insectivores.

Admittedly, all this concerns quite recent evolution, not so prolonged a rerun as Kauffman imagined. But these recent reruns can surely teach us some lessons about the inherent reproducibility of evolution. If early evolution ran along similar lines to later evolution, those lessons might amount to general principles. My hunch is that the principles we learn from recent evolution since the decease of the dinosaurs probably hold good at least back to the Cambrian, and probably back to the origin of the eukaryotic cell. I have a hunch that the parallelism of mammal radiations in Australia, Madagascar, South America, Africa and Asia may provide a sort of template for answering Kauffman questions for much older starting points, such as the one he chose, the origin of the eukaryotic cell. Earlier than that landmark event, confidence evaporates. My colleague Mark Ridley, in *Mendel's Demon*, suspects that the origin of eukaryotic complexity was a massively improbable event, perhaps even more improbable than the origin of life itself. Influenced by Ridley, my bet is that most rerun thought experiments that start with the origin of life will not make it into the eukaryocracy.

We don't have to rely on geographical separation as in the Australian natural experiment to study convergence. We can think of the experiment of evolution being rerun, not from the same starting point in different geographical areas, but from different starting points – very possibly in the same geographical area: convergence in animals so unrelated to each other that what they tell us has nothing to do with geographical separation. It has been estimated that 'the eye' has evolved independently between 40 and 60 times

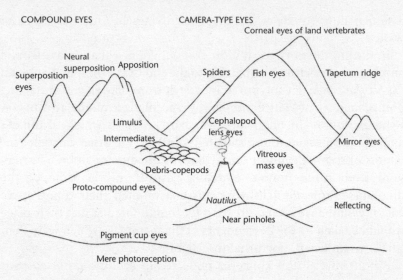

COMPOUND EYES   CAMERA-TYPE EYES

The Fortyfold Path to Enlightenment. Landscape of eye evolution by Michael Land.

around the animal kingdom. This inspired my chapter called 'The Fortyfold Path to Enlightenment' in *Climbing Mount Improbable*. In that book I quoted my friend Professor Michael Land of Sussex University, our leading expert on the comparative zoology of eyes, as recognising nine independent principles of optical mechanism, each of which has evolved more than once. He was kind enough to prepare for that book the landscape reprinted above, in which separate peaks represent independent evolutions of eyes. Where on his landscape would Professor Land have placed the example opposite? It's almost too good to be true, like something out of a zoologist's dream. In 2015 the journal *Nature* published detailed pictures of a tiny structure which looks tantalisingly like the camera eye of a vertebrate or an octopus but is possessed by a single-celled organism, a dinoflagellate. We met the dinoflagellates at Rendezvous 38, you may remember them as the source of bewitchingly luminous nocturnal tides. The ones that have the beautiful 'eyes' are a subgroup of dinoflagellates called the warnowiids.

Conventional zoological wisdom says that a respectable single-celled organism has no business making an elaborate eye, endowed with a lens (L) and a retina (r). Retinas are supposed to be made of cells, each cell stimulated by a different part of the image. How could a single-celled organism get a retina together? Well, remember the Mixotrich's Tale? *Mixotricha* incorporates thousands of bacteria into its cell wall, each one acting as a swimming hair. Other hijacked (or 'volunteer') bacteria perform other functions within the cell. As we noted in the Great Historic Rendezvous, all our eukaryotic cells have one such bacterial stowaway: the mitochondrion. We saw that plants have others too, in the form of chloroplasts which do their photosynthesis for them. And you may recall that dinoflagellates photosynthesise using a hijacked plant cell (a red alga) that itself had previously hijacked a bacterium, in a manner akin to an endosymbiotic Russian doll. According to the *Nature* paper, these are the building materials for the warnowiid eye. The 'cornea' which bounds the 'lens' is said to be made of mitochondria. And the 'retina' is put together as a conglomerate of their hijacked cargo of algal cells.

No business in making an elaborate eye. The reported eye-like structure, formed from symbiotic components, in *Erythropsidinium*, a single-celled warnowiid dinoflagellate. The claim is that the structure marked 'L' may act as a lens, with 'r' acting as the retina. From Govelis *et al.* [147]

There doesn't seem to be any strong evidence that these tiny creatures can actually see images. If the retina is indeed what it purports to be, its image would need to be processed by something equivalent to a nervous system, and nervous systems as we know them are made of many cells. But single celled creatures such as *Mixotricha* and the warnowiids have already surprised us mightily so . . . watch this space.

It seems that life, at least on this planet, is almost indecently eager to evolve eyes (see plate 52). We can confidently predict that a statistical sample of Kauffman reruns would culminate in eyes. And not just eyes, but compound eyes like those of an insect, a prawn or a trilobite, and camera eyes like ours or a squid's, with colour vision and with mechanisms for fine-tuning the focus and the aperture. Also very probably parabolic reflector eyes like those of a limpet, and pinhole eyes like those of *Nautilus*, the latter-day ammonite-like mollusc in its floating coiled shell, whom we met at Rendezvous 26. And if there is life on other planets around the universe, it is a good bet that there will also be eyes, based on the same range of optical principles as we know on this planet. There are only so many ways to make an eye, and life as we know it may well have found them all.

We can do the same kind of count for other adaptations. Echolocation – the trick of emitting sound pulses and navigating by accurate timing of the echoes – has evolved at least four times: in toothed whales, oilbirds, cave swiftlets and most famously in bats (where it may have evolved twice independently). Not as many times as the eye, but still often enough to make us think it not too unlikely that, if the conditions are right, it will evolve. Very probably, too, reruns of evolution would rediscover the same specific principles: the same tricks for confronting difficulties. Once again, I shall not repeat my exposition from a previous book,* but will simply summarise what we might predict for reruns of evolution. Echolocation should repeatedly evolve using very high-pitched cries (for better resolution of detail than low-pitched). The cries in at least some species are likely to be frequency-modulated, sweeping down or up in pitch during the course of each cry (accuracy is improved because early parts of each echo are distinguishable from late parts by their pitch). The computational apparatus used to analyse the echoes might very well make (subconscious) calculations based on Doppler shifts in frequency of echoes, for the Doppler effect is certainly universally present on any planet where there is sound, and bats make sophisticated use of it.

How do we know that something like the eye or echolocation has evolved independently? By looking at the family tree. Relatives of oilbirds and of cave swiftlets don't do echolocation. Oilbirds and cave swiftlets have separately taken up life in caves. We know they have evolved the technology independently of bats and whales, since nothing else in the surrounding family tree does it. Different groups of bats may have evolved echolocation more than once independently. We don't know how many more times echolocation evolved. Some shrews and seals have a rudimentary form of the skill (and some blind human individuals have learned it). Did pterodactyls do it? Since there is a good living to be made flying at night, and since bats weren't around in those days, it is not unlikely. The same goes for ichthyosaurs. They looked very like dolphins, and presumably

---

* *The Blind Watchmaker* in this case.

made their living in a similar way. Since dolphins make heavy use of echolocation, it is reasonable to wonder whether ichthyosaurs did too, in the days before dolphins. There is no direct evidence, and we must remain open-minded. One point against: ichthyosaurs had extraordinarily big eyes – it is one of their most conspicuous features – which might suggest that they relied upon vision instead of echolocation. Dolphins have relatively small eyes, and one of *their* most conspicuous features, the rounded bump or 'melon' above the beak, acts as an acoustic 'lens', focusing sound into a narrow beam projected in front of the animal like a searchlight.

Convergence doesn't just happen at the visible level. As you might expect, there are also instances of molecular convergence. In some cases that might mean two different molecules evolving to provide a similar chemical function. But in a growing number of examples the actual sequence of a gene or protein has been found to converge. The classic case is a digestive enzyme, lysozyme, which shows unexpected similarity between ruminating cows and leaf-eating monkeys. But a more comprehensive example is again provided by echolocation. The mammalian inner ear has a fascinating sound-amplifying mechanism, in which special hairs are actively vibrated at the same frequency as the incoming sound. The motor that drives this vibration is a protein called prestin. Yang Liu and colleagues for Shanghai, along with collaborators from Queen Mary University in London, inferred an evolutionary tree of prestin molecules from different mammals. It placed the dolphins and other toothed whales apparently within the bats.

Notwithstanding the surprise of the Hippo's Tale, taxonomists are absolutely sure that dolphins aren't a sort of bat. These animals appear to group together because both whales and bats have independently changed parts of the prestin molecule in the same way, and for the same reason: to work better with the high-frequency sounds necessary for sonar. In total, 14 separate amino acids in the prestin molecule show this sort of convergence.

It's not just prestin either. The same group from Queen Mary University examined a selection of 2,000 genes across 22 mammal species, looking for ones that misleadingly group bats and dolphins together. They found convergent protein sequences in 200 places in the genome, predominantly in hearing-related genes but also in vision-related genes too. This somewhat surprising finding hints at an intriguing evolutionary link between vision and hearing. It might also be interesting to look outside the protein-coding parts, at regions of DNA that switch genes on and off. We might expect these regulatory sequences to show convergence too; indeed Jason Gallant, Lindsay Traeger and their colleagues recently revealed something of the sort in the genomes of the convergently electric fish that we mentioned in the Duckbill's Tale (eels, catfish, rays, and so on). With so many new genomes coming online, and with DNA comparison trivially carried out on computers, this sort of research is surely buzzing with potential.

Like any zoologist, I can search my mental database of the animal kingdom and come up with an estimated answer to questions of the form: 'How many times has X evolved independently?' It would make a good research project, to do the counts more systematically. Presumably some Xs will come up with a 'many times' answer, as with eyes, or 'several times', as with echolocation. Others 'only once' or even 'never', although I have to say it is surprisingly difficult to find examples of these. And the differences could be interesting. I suspect that we'd find certain potential evolutionary pathways which life is 'eager' to go down. Other pathways have more 'resistance'. In *Climbing Mount Improbable*, I developed the analogy of a huge museum of all life, both real and conceivable, with corridors going off in many dimensions and representing evolutionary change, again both real and conceivable. Some of these corridors were wide open, almost beckoning. Others were blocked off by barriers that were hard or even impossible to surmount. Evolution repeatedly races down the easy corridors, and just occasionally, and unexpectedly, leaps one of the hard barriers. I'll return to the idea of 'eagerness' and 'reluctance' to evolve when I discuss the 'evolution of evolvability'.

Let's now go quickly through some more examples where it might be worth making a systematic count of how many times X has evolved. The venomous sting (injecting poison hypodermically through a sharp-pointed tube) has evolved at least ten times independently: in jellyfish and their relatives, spiders, scorpions, centipedes, insects,* molluscs (cone shells) snakes, the shark group (stingrays), bony fish (stonefish), mammals (male platypus) and plants (stinging nettles). It's a good bet that venom, including hypodermic injection, would evolve in reruns.

Sound production for social purposes has evolved independently

---

* In the bees, wasps and ants, the hypodermic is a modified egg-laying tube, and only females sting.

in birds, mammals, crickets and grasshoppers, cicadas, fish and frogs. Electrolocation, the use of weak electric fields for navigation, has evolved several times, as we saw in the Duckbill's Tale. So has the – probably subsequent – use of electric currents as weapons. The physics of electricity is the same on all worlds, and we could bet with some confidence on repeated evolution of creatures that exploit electricity for both navigational and offensive purposes.

True flapping flight, as opposed to passive gliding, or parachuting, seems to have evolved four times: in insects, pterosaurs, bats and birds. Parachuting and gliding of various kinds have evolved many times, maybe hundreds of times independently, and may well be the evolutionary precursor to true flight. Examples include lizards, frogs, snakes, 'flying' fish, squids, colugos, marsupials and rodents (twice). I'd put a lot of money on gliders turning up in Kauffmanian reruns, and a reasonable sum on true flapping fliers.

Jet propulsion may have evolved twice. Cephalopod molluscs do it, at high speed in the case of squids. The other example I can think of is also a mollusc, but it is not high speed. Scallops mostly live on the sea bottom, but occasionally they swim. They rhythmically open and close their two shells, like a pair of snapping castanets. You'd think (I would) that this would propel them 'backwards' in a direction opposite to the snapping. In fact they move 'forwards', as though biting their way into the water. How can this be? The answer is that the snapping movements pump water through a pair of apertures *behind* the hinge. These two jets propel the animal 'forwards'. The effect is so counter-intuitive it is almost comical.

How about things that have evolved only once, or not at all? As we learned from the *Rhizobium*'s Tale, the wheel, with a true, freely rotating bearing, seems to have evolved only once, in bacteria, before being finally invented in human technology. Language, too, has apparently evolved only in us: that is to say at least 40 times less often than eyes. It is surprisingly hard to think of 'good ideas' that have evolved only once.

I put the challenge to my Oxford colleague the entomologist and naturalist George McGavin, and he came up with a nice list, but still a short one compared with the list of things that have evolved many times. Bombardier beetles of the genus *Brachinus* are unique, in Dr McGavin's experience, in mixing chemicals to make an explosion. The ingredients are made and held in separate glands. When danger threatens, they are squirted into a chamber near the rear end of the beetle, where they explode, forcing noxious (caustic and searing-hot) liquid out through a directed nozzle at the enemy. The case is well known to creationists, who love it. They think it is self-evidently impossible to evolve by gradual degrees because the intermediate stages would all explode. I enjoyed demonstrating the error of this argument during my Royal Institution Christmas Lectures for Children, shown on BBC television in 1991. Donning a Second World War helmet, and inviting nervous members of the audience to leave, I mixed hydroquinone and hydrogen peroxide, the two ingredients of the bombardier explosion. Nothing happened. It didn't even get warm. The explosion requires a catalyst. I raised the concentration of catalyst gradually, which steadily increased the hot whoosh to a satisfactory climax. In nature, the beetle provides the catalyst, and would have had no difficulty in gradually, and safely, increasing the dose over evolutionary time.

Next in the McGavin list is the archer fish, family Toxotidae, which may be unique in shooting a missile to knock prey down from a distance. It comes to the surface of the water and spits a mouthful of water at a perched insect, knocking it down into the water, where it eats it. The other possible candidate for a 'knocking down' predator might be an ant lion. Ant lions are insect larvae of the order Neuroptera. Like many larvae, they look nothing like their adults. With their huge jaws, they would be good casting for a horror film. Each ant lion lurks in sand, just below the surface at the base of a conical pit trap which it digs itself. It digs by flicking sand vigorously outwards from the centre – this causes miniature landslides down the sides of the pit, and the laws of physics do the rest, neatly shaping the cone. Prey, usually ants, fall into the pit and slide down the steep sides into the ant lion's jaws. The possible point of resemblance to the archer fish is that prey don't fall only passively. They are sometimes knocked down into the pit by the particles of sand. These are not, however, aimed with the precision of an archer fish's spit, which is guided, with devastating accuracy, by binocularly focused eyes.

Spitting spiders, family Scytodidae, are a bit different again. Lacking the fleetness of a wolf spider or the net of a web spider, the spitting spider chucks a venomous glue some distance towards its prey, pinning it to the ground until the spider arrives and bites it to death. This is different from the archer fish technique of knocking prey down. Various animals, for example venom-spitting cobras, spit defensively, not to catch prey. The bolas spider, *Mastophora*, is different again, and is probably another unique case. It could be said to throw a missile at prey (moths, attracted by the fake sexual scent of a female moth, which the spider synthesises). But the missile, a blob of silk, is attached to a thread of silk which the spider whirls around like a lasso (or bola) and reels in. Chameleons could be said to spit a missile at prey. The missile is the heavy thickening at the tip of the tongue, and the (much thinner) rest of the tongue is a bit like the rope with which a harpoon is retrieved. The tip of the tongue is technically ballistic, meaning it is hurled free, unlike the tip of your tongue. Chameleons are not unique in this respect, however. Some salamanders also hurl the end of the tongue ballistically at prey, and their missile (but not the chameleon's) contains part of the skeleton. It is fired like a melon pip squeezed between your fingers.

McGavin's next candidate for an evolutionary one-off is a beauty. It is the diving bell spider, *Argyroneta aquatica*. This spider lives and hunts entirely underwater but, like dolphins, dugongs, turtles, freshwater snails and other land animals that have returned to water, it needs to breathe air. Unlike all those other exiles, however, *Argyroneta* constructs its own diving bell. It spins it of silk (silk is the universal solution to any spider problem) attached to an underwater plant. The spider goes to the surface to collect air, which it carries in the same way as some water bugs, in a layer trapped by body hairs. But unlike the bugs, which just carry the air like a scuba cylinder wherever they go, the spider takes it to its diving bell, where it unloads it to replenish the supply. The spider sits in the diving bell watching for prey, and it stores and eats prey there, once caught.

But George McGavin's champion example of a one-off is the larva of an African horsefly called *Tabanus*. Predictably in Africa, the pools of water in which the larvae live and feed eventually dry up. Each larva buries itself in the mud and pupates. The adult fly emerges from the baked mud and flies off to feed on blood, eventually to complete the cycle by laying eggs in pools of water when the rains return. The buried larva is vulnerable to a predictable danger. As mud dries out, it cracks, and there is a risk that a crack will tear right across the grub's refuge. It could theoretically save itself if it could somehow engineer a way for any crack that approaches it to be diverted around it instead. And it does indeed achieve this in a truly wonderful and probably unique manner. Before burying itself in its own pupation chamber, it first corkscrews its way down into the mud in a spiral. It then corkscrews its way back to the surface in an opposite spiral. Finally, it dives into the mud straight down the centre between the two spirals, and that is its resting place through the bad times until water returns. Now, you see what this means? The larva is encased in a cylinder of mud whose circular boundary has been weakened in advance by the preliminary spiral burrowing. This means that when a crack snakes across the drying mud, if it hits the edge of the cylindrical column, instead of cutting straight across the middle it goes instead in a curved bypass around the edge of the cylinder, and the larva is spared. It is just like the perforations around a stamp that stop you tearing the stamp across. Dr McGavin believes this ingenious trick is literally unique to this one genus of horsefly.*

But are there any good ideas that have *never* evolved by natural selection? As far as I know, no animal on this planet has ever evolved an organ for the transmission or reception of radio waves for long-distance communication. The use of fire is another example. Human experience shows how powerfully useful it can be. There are some plants whose seeds need fire to germinate, but I don't think that constitutes using it in the same sense as, for example, electric eels use electricity. The use of metal for skeletal purposes is another example of a good idea that has never evolved except in human artefacts. Presumably it is hard to achieve without fire.

This kind of comparative exercise, counting up which things evolve often, which seldom, when we do it alongside the geographical comparisons discussed earlier, might enable us to predict things about life outside this planet, as well as to guess the likely outcome of Kauffman-style thought experiments in rerunning evolution. We positively expect eyes, ears, wings and electric organs, but perhaps not bombardier beetle explosions or archer fish water bullets.

Those biologists who could be said to take their lead from the late Stephen Jay Gould regard all of evolution, including post-Cambrian evolution, as massively contingent – lucky, unlikely to be repeated in a Kauffman rerun. Calling it 'rewinding the tape of evolution', Gould independently evolved Kauffman's thought experiment. The chance of anything remotely resembling humans on a second rerun is widely seen as vanishingly small, and Gould voiced it persuasively in *Wonderful Life.* It was this orthodoxy that led me to the cautious self-denying ordinance of my opening chapter; led me, indeed, to undertake my backwards pilgrimage, and now leads me to forsake my pilgrim companions at Canterbury and return alone. And yet . . . I

---

* The habit was first described by W. A. Lambourn [235].

have long wondered whether the hectoring orthodoxy of contingency might have gone too far. My review of Gould's *Full House* (reprinted in *A Devil's Chaplain*) defended the unpopular notion of progress in evolution: not progress towards humanity – Darwin forfend! – but progress in directions that are at least predictable enough to justify the word. As I shall argue in a moment, the cumulative build-up of complex adaptations like eyes strongly suggests a version of progress – especially when coupled in imagination with some of the wonderful products of convergent evolution.

Convergent evolution also inspired the Cambridge geologist Simon Conway Morris, whose provocative book *Life's Solution: Inevitable Humans in a Lonely Universe* presents exactly the opposite case to Gould's 'contingency'. Conway Morris means his subtitle in a sense which is not far from literal. He really thinks that a rerun of evolution would result in a second coming of man: or something extremely close to man. And, for such an unpopular thesis, he mounts a defiantly courageous case. The two witnesses that he repeatedly calls are convergence and constraint.

Convergence we have met again and again through this book, including in this chapter. Similar problems call forth similar solutions, not just twice or three times but, in many cases, dozens of times. I thought I was pretty extreme in my enthusiasm for convergent evolution, but I have met my match in Conway Morris, who presents a stunning array of examples, many of which I had not met before. But whereas I usually explain convergence by invoking similar selection pressures, Conway Morris adds the testimony of his second witness, constraint. The materials of life, and the processes of embryonic development, allow only a limited range of solutions to a particular problem. Given any particular evolutionary starting situation, there is only a limited number of ways out of the box. So if two reruns of a Kauffman experiment encounter anything like similar selection pressures, developmental constraints will enhance the tendency to arrive at the same solution.

You can see how a skilled advocate could deploy these two witnesses in defence of the daring belief that a rerun of evolution would be positively likely to converge on a large-brained biped with two skilled hands, forward-pointing camera eyes and other human features. Unfortunately, it has only happened once on this planet, but I suppose there has to be a first time. I admit that I was impressed by Conway Morris's parallel case for the predictability of the evolution of insects.

Among the defining features of insects are the following: an articulated exoskeleton; compound eyes; a characteristic six-legged gait, whereby three of the six walking legs are always on the ground and thereby define a triangle (two legs on one side, one leg on the other) which keeps the animal stable; respiratory tubes known as tracheae that serve to bring oxygen into the interior of the animal via special openings (spiracles) along the side of the body; and, to complete the list of evolutionary peculiarities, the repeated (at least 12 times independently!) evolution of complex eusocial colonies, as in the honey bees. All pretty strange? All one-offs in the great lottery of life? On the contrary, all are convergent.

Conway Morris goes through his list, showing that each item has evolved more than once in different parts of the animal kingdom, in many cases several times, including several times independently in insects themselves. If nature finds it so easy to evolve the component parts of insecthood separately, it is not all that implausible that the whole collection should evolve twice. I am tempted by Conway Morris's belief that we should stop thinking of convergent evolution as a colourful rarity to be remarked and marvelled at when we find it. Perhaps we should come to see it as a norm, exceptions to which are occasions for surprise. For example, true syntactic language seems to be unique to one species, our own. Perhaps this is one thing that a re-evolved brainy biped would lack?

In my opening chapter, The Conceit of Hindsight, I listened to warnings against seeking patterns, rhymes or reasons in evolution, but said that I would cautiously flirt with them. The Host's Return has provided an opportunity to sweep over the whole course of evolution in the forward direction and see what patterns we can descry. The idea that all evolution was aimed at producing *Homo sapiens* was certainly well rejected, and nothing we have seen on our backwards journey reinstates it. Even Conway Morris claims only that something *approximately* similar to our *kind* of animal is one of several outcomes – others being insects, for example – that we would expect to see recurring if evolution were rerun again and again.

Value-free and Value-laden Progress

What other patterns or rhymes do we discern if we look over our long pilgrimage? Is evolution progressive? There is at least one reasonable definition of progress under which I would defend it. I need to work up to this. To begin with, progress can be defined in a weak, minimalist sense with no value judgement, as the predictable continuation into the future of trends from the past. The growth of a child is progressive in that whatever trends we observe in weight, height and other measurements over one year continue in the next. There is no value judgement in this weak definition of progress. Growth of a cancer is progressive in exactly the same weak sense. So also is shrinkage of a cancer under therapy. What, then, would *not* be progressive in the weak sense? Random, aimless fluctuation: the tumour grows a little, shrinks somewhat, grows a lot, shrinks a bit, grows a little, shrinks a lot, and so on. A progressive trend is one in which there are no reversals; or if there are reversals, they are outnumbered and outweighed by movement in the dominant direction. In a sequence of dated fossils, progress in this value-neutral sense would mean simply that whatever anatomical trend you see as you go from early to intermediate, is continued as a trend from intermediate to late.

I now need to clarify the distinction between value-neutral progress and value-laden progress. Progress in the weak sense just defined is value-neutral. But most people think of progress as value-laden. The doctor, reporting shrinkage of the tumour in response to chemotherapy, announces with satisfaction, 'We are making progress.' Doctors do not, on the whole, look at an X-ray of a swelling tumour with numerous secondaries and announce that the tumour is making progress, though they easily could. That would be value-laden but with negative value. 'Progressive', in human political or social affairs, usually refers to a trend in a direction which the speaker considers desirable. We look over human history and regard the following trends as progressive: abolition of slavery; widening of the franchise; reduction of discrimination by sex or race; reduction of disease and poverty; increase in public hygiene; reduction in atmospheric pollution; increase in education. A person of certain political views might see at least some of these trends as negatively value-laden, and yearn nostalgically for the days before women had the vote or were allowed into the club dining room. But the trends are still progressive in more than just the weak, minimalist, value-neutral sense we first defined. They are progressive according to some specified value system, even if it is not a value-system we would all share.

Astonishingly, it is only a hundred years since the Wright brothers first achieved powered flight in a heavier-than-air machine. The history of aviation since 1903 has been unmistakeably progressive, and at amazing speed. Only 42 years later, in 1945, Hans Guido Mutke of the Luftwaffe broke the sound barrier in a Messerschmitt jet fighter.* Only 24 years further on, men walked on the moon. The fact that they no longer do so, and the fact that the only supersonic passenger service has now been discontinued, are economically dictated temporary reversals in an overall trend that is unquestionably progressive. Aircraft are getting faster, and they are progressing in all sorts of other ways at the same time. Much of this progress does not chime with everyone's values – for example those unlucky enough to live under a flightpath. And much of the progress in aviation is driven by military needs. But nobody would deny the existence of a coherently expressible set of values, which at least some sane people might hold, according to which even fighters, bombers and guided missiles have progressively improved over the whole century since the Wrights. The same could be said of all other forms of transport, indeed other forms of technology, including, more than anything else, computers.

I must repeat that in calling this value-laden progress, I am not saying the values necessarily have a positive sign, either for you or for me. As I just remarked, much of the technological progress we are talking about is driven by, and contributes towards, military purposes. We could reasonably decide that the world was a better place before such inventions were made. In this sense 'progress' is value-laden with a negative sign. But it is still value-laden in an important sense, over and above my original value-free minimalist definition of progress as any trend into the future continued from the past. The development of weapons, from the stone to the spear through the longbow, the flintlock, musket, rifle, machine gun, shell, atomic bomb, through hydrogen bombs of ever increasing megatonnage, represents progress according to *somebody's* value system, even if not yours or mine – otherwise the research and development to produce them would not have been done.

Evolution exhibits progress not just in the weak, value-free sense. There are episodes of progress that are value-laden, according to at least some entirely plausible value systems. Since we are talking armaments, it is a good moment to note that the most familiar examples come out of arms races between predators and prey.

---

* Mutke's claim is disputed. Either way, the first time it was done wasn't when USAF Major Chuck Yeager set the official record in 1947, as patriotic Americans are taught. An American civilian, George Welch, did it two weeks before him.

The first use of the 'arms race' listed in the *Oxford English Dictionary* is from *Hansard* (written proceedings of the House of Commons) for 1936:

> This House cannot agree to a policy which in fact seeks security in national armaments alone and intensifies the ruinous arms race between the nations, inevitably leading to war.

The *Daily Express* in 1937, under the headline 'Arms Race Worry', said, 'All were worried at the armament race.' It was not long before the theme found its way into the literature of evolutionary biology. Hugh Cott, in his classic *Adaptive Coloration in Animals*, published in 1940, deep in the Second World War, wrote:

> Before asserting that the deceptive appearance of a grasshopper or butterfly is unnecessarily detailed, we must first ascertain what are the powers of perception and discrimination of the insects' natural enemies. Not to do so is like asserting that the armour of a battle-cruiser is too heavy, or the range of her guns too great,* without inquiring into the nature and effectiveness of the enemy's armament. The fact is that in the primeval struggle of the jungle, as in the refinements of civilized warfare, we see in progress a great evolutionary armament race – whose results, for defence, are manifested in such devices as speed, alertness, armour, spinescence, burrowing habits, nocturnal habits, poisonous secretions, nauseous taste, and procryptic, aposematic, and mimetic coloration; and for offence, in such counter-attributes as speed, surprise, ambush, allurement, visual acuity, claws, teeth, stings, poison fangs, and anticryptic and alluring coloration. Just as greater speed in the pursued has developed in relation to increased speed in the pursuer; or defensive armour in relation to aggresssive weapons; so the perfection of concealing devices has evolved in response to increased powers of perception.

My Oxford colleague John Krebs and I took up the whole matter of evolutionary arms races in a paper given at the Royal Society in 1979. We pointed out that the improvements to be seen in an animal arms race are improvements in equipment to survive, not generally improvements in survival itself – and for an interesting reason. In an arms race between attack and defence, there may be episodes during

---

* Tom Lehrer, probably the all-time wittiest composer of comic songs, included the following musical direction at the head of one of his piano scores: 'A little too fast.'

which one side or the other temporarily pulls ahead. But in general, improvements on one side cancel out improvements on the other. There is even something a bit paradoxical about arms races. They are economically costly for both sides, yet there is no net benefit to either, because potential gains on one side are neutralised by gains on the other. From an economic point of view, both sides would be better off coming to an agreement to call off the arms race. As a ludicrous extreme, prey species might sacrifice a tithe of their number in exchange for secure and untroubled grazing for the rest. Neither predators nor prey need to divert valuable resources into muscles for fast running, sensory systems for detection of enemies, vigilance and prolonged hunts that are time-wasting and stressful for both sides. Both sides would benefit if such a trades union agreement could be reached.

Unfortunately, Darwinian theory knows no route by which this could happen. Instead, both sides pour resources into competing with their own side to outrun the other, and individuals of both are forced into difficult economic trade-offs within their own bodily economies. If there were no predators, rabbits could devote all their economic resources, and all their valuable time, to feeding and reproducing more rabbits. Instead, they are forced to devote substantial time to looking out for predators, and substantial economic resources into building up escape equipment. In turn, this forces predators to shift the balance of their economic investment away from the central business of reproducing, and into improving their weaponry for catching prey. Arms races, in animal evolution and human technology alike, show themselves not in improved performance but in increased shifting of economic investment away from alternative aspects of life and into servicing the arms race itself.

Krebs and I recognised asymmetries in arms races that might result in one side shifting more economic resources into the arms race than the other. One such imbalance we dubbed the 'Life/Dinner Principle'. It takes its name from the Aesop Fable in which the rabbit runs faster than the fox because the rabbit is running for his life, while the fox is only running for his dinner. There is an asymmetry in the cost of failure. In the arms race between cuckoos and hosts, every individual cuckoo can confidently look back on an unbroken line of ancestors who literally never failed to fool a foster parent. An individual of the host species, on the other hand, can look back on ancestors, many of whom never even met a cuckoo, and many of whom met one and were fooled by it. Plenty of genes for failing to detect and kill cuckoos have passed successfully down the generations of the host species. But genes that cause cuckoos to fail in fooling hosts have a much more hazardous ride down the generations. This asymmetry of risk fosters another: an asymmetry in resources devoted to the arms race as opposed to other parts of life's economy. To repeat this important point, the cost of failure is harder on the cuckoos than on the hosts. This leads to asymmetries in how the two sides set their balance between competing calls on their time and other economic resources.

Arms races are deeply and inescapably progressive in a way that, for example, evolutionary accommodation to the weather is not. For an individual of any one generation, predators and parasites just make life harder in pretty much the same way as bad weather does. But over evolutionary time there is a crucial difference. Unlike the weather, which fluctuates aimlessly, predators and parasites (and prey and hosts) are themselves evolving in a systematic direction, getting systematically worse from their victims' point of view. Unlike evolutionary tracking of ice ages and droughts, arms race trends from the past can be extrapolated into the future, and those trends are value-laden in the same kind of way as technological improvements in planes and weapons. Predators' eyes get sharper, though not necessarily more effective, because prey get harder to see. Running speeds increase progressively on both sides, though again the benefits are in general cancelled out by parallel improvements on the other side. Sabre teeth get sharper and longer as hides get tougher. Toxins get nastier as biochemical tricks for neutralising them improve.

With the passing of evolutionary time, the arms race progresses. All the features of life that a human engineer would admire as complex and elegant become more complex, more elegant, and more redolent of the illusion of design.* In *Climbing Mount Improbable* I distinguished designed from 'designoid' (pronounced design-oid, not dezzig-noid). Spectacular feats of designoid engineering, such as the eye of a buzzard, the ear of a bat, the musculo-skeletal apparatus of a cheetah or a gazelle, are all climactic products of evolutionary arms races between predators and prey. Parasite/host arms races culminate in even more finely meshed, co-adaptive designoid climaxes.

And now for an important point. The evolution of any complex designoid organ in an arms race must have come about in a large

---

* Hume said: 'All these various machines, and even their most minute parts, are adjusted to each other with an accuracy which ravishes into admiration all men who have ever contemplated them.'

number of steps of progressive evolution. Such evolution qualifies as progressive by our definition because each change tends to continue the direction of its predecessors. How do we know there are many steps rather than just one or two? By elementary probability theory. The parts of a complex machine, such as a bat's ear, could be rearranged at random in a million ways before you hit another arrangement that could hear as well as the real thing. It is statistically improbable, not just in the boring sense that any particular arrangement of parts is as improbable, with hindsight, as any other. Very few permutations of atoms are precision auditory instruments. A real bat's ear is one in a million. It works. Something so statistically improbable cannot sensibly be explained as the result of a single stroke of luck. It has to be constructed by some sort of improbability-generating process, ratcheted up by what the philosopher Daniel Dennett calls a 'crane' (as opposed to a 'skyhook': the analogy is to the man-made lifting machine, not the bird). The only such cranes known to science (and I would bet the only cranes there have ever been, or ever will be, in the universe) are design and selection. Design explains the efficient complexity of microphones. Natural selection explains the efficient complexity of bat ears. Ultimately, selection explains microphones and everything designed too because the designers of microphones are themselves evolved engineers generated by natural selection. Ultimately, design cannot explain anything because there is an inevitable regression to the problem of the origin of the designer.

Design and natural selection are both processes of gradual, step-by-step, progressive improvement. Natural selection, at least, could not be anything else. In the case of design it may or may not be a matter of principle, but it is an observed fact. The Wright brothers did not have a blinding flash of inspiration and promptly build a *Concorde* or a *Stealth* bomber. They built a creaking, rickety crate that barely lifted off the ground and lurched into a neighbouring field. From Kitty Hawk to Cape Canaveral, every step of the way was built on its predecessors. Improvement is gradual, step by step in the same continued direction, fulfilling our definition of progressive. We could, with difficulty, conceive of a Victorian genius designing a Sidewinder missile fully formed within his Zeusian, side-whiskered head. The notion defies all common sense and all history, but it does not instantly fall foul of the laws of probability in the way we would have to say of the spontaneous evolution of a flying, echolocating, modern bat.

A single macromutational leap from ground-dwelling ancestral shrew to flying, echolocating bat is ruled out just as safely as we can rule out luck when a conjuror successfully guesses the complete order of a shuffled pack of cards. Luck is not literally impossible in either case. But no good scientist would advance such prodigious luck as an explanation. The card-guessing feat has to be a trick – we've all seen tricks that appear just as baffling to the uninitiated. Nature does not set out to fool us, as a conjuror does. But we can still rule out luck, and it was the genius of Darwin to rumble nature's sleight of hand. The echo-ranging bat is the result of an inching series of minor improvements, each adding cumulatively to its predecessors as it propels the evolutionary trend on in the same direction. That is progress, by definition. The argument applies to all complex biological objects that project the illusion of design and are therefore statistically improbable in a specified direction. All must have evolved progressively.

The returning host, now unabashedly sensitive to major themes in evolution, notes progress as one of them. But progress of this kind is not a uniform, inexorable trend from the start of evolution all the way to the present. Rather, to take up the initial quotation from Mark Twain on history, it rhymes. We notice an episode of progress during the course of an arms race. But that particular arms race comes to an end. Perhaps one side is driven extinct by the other. Or both sides go extinct, maybe in the course of a mass catastrophe of the kind that did for the dinosaurs. Then the whole process starts again, not from scratch, but from some discernibly earlier part of the arms race. Progress in evolution is not a single upward climb but has a rhyming trajectory more like the teeth of a saw. A sawtooth plunged deeply at the end of the Cretaceous, when the last of the dinosaurs abruptly gave way to the mammals' new and spectacular climb of progressive evolution. But there had been lots of smaller sawteeth during the long reign of the dinosaurs. And since their immediate post-dinosaur rise, the mammals too have had smaller arms races followed by extinctions, followed by renewed arms races. Arms races rhyme with earlier arms races in periodic spurts of many-stepped progressive evolution.

Evolvability

That's all I want to say about arms races as drivers of progress. What other messages from the past does the returning host carry back to the present? Well, I must mention the alleged distinction between macroevolution and microevolution. I say 'alleged' because my own view is that macroevolution (evolution on the grand scale of millions of years) is simply what you get when microevolution (evolution on the scale of individual lifetimes) is allowed to go on for millions of years. The contrary view is that macroevolution is something qualitatively different from microevolution. Neither view is self-evidently silly. Nor are they necessarily contradictory. As so often, it depends on what you mean.

Again we can use the parallel of the growth of a child. Imagine an argument about an alleged distinction between macrogrowth and microgrowth. To study macrogrowth, we weigh the child every year. Every birthday we stand her up against a white doorpost and draw a pencil line to record her height. More scientifically, we could measure various parts of the body, for example the diameter of the head, the width of the shoulders, the length of the major limb bones, and plot them against each other, perhaps logarithmically transformed for the reasons given in the Handyman's Tale. We also note significant events in development such as the first appearance of pubic hair, or the first sign of breasts and menstruation in girls, and of facial hair in boys. These are the changes that constitute macrogrowth, and we measure them on a timescale of years or months. Our instruments are not sensitive enough to pick up the daily and hourly changes in the body – microgrowth – which, when summed over months, constitute macrogrowth. Or, oddly, they may be too sensitive. A very accurate weighing machine could in theory pick up hourly growth, but the delicate signal is swamped by blundering increases in weight with every meal, and decreases with every act of elimination. The acts of microgrowth itself, which all consist of cell divisions, make no immediate impact on weight at all, and an undetectable impact on gross body measurements.

So, is macrogrowth the sum of lots of small episodes of microgrowth? Yes. But it is also true that the different timescales impose completely different methods of study and habits of thought. Microscopes looking at cells are not appropriate for the study of child development at the whole-body level. And weighing machines and measuring tapes are not suitable for the study of cell multiplication. The two timescales in practice demand radically different methods of study and habits of thought. The same could be said of macroevolution and microevolution. If the terms are used to signify differences in how best to study them, I have no quarrel with a working distinction between microevolution and macroevolution. I do have a quarrel with those people who elevate this rather mundane practical distinction into one of almost – or more than almost – mystical import. There are those who think Darwin's theory of evolution by natural selection explains microevolution, but is in principle impotent to explain macroevolution, which consequently needs an extra ingredient – in extreme cases a *divine* extra ingredient!

Unfortunately, this hankering after skyhooks has been given aid and comfort by real scientists whose intentions are innocent of any such thing. I have discussed the theory of 'punctuated equilibrium' before, too often and too thoroughly to repeat myself in this book,* so I shall only add that its advocates usually go on to propose a fundamental 'decoupling' between microevolution and macroevolution. This is an unwarranted inference. No extra ingredient needs to be added at the micro level to explain the macro level. Rather, an extra level of explanation *emerges* at the macro level as a *consequence* of events at the micro level, extrapolated over geological timespans.

The working distinction between micro- and macroevolution is similar to one that we meet in many other situations. The changes in the map of the world over geological time are due to the effects, summed over millions of years, of plate tectonic events occurring on a timescale of minutes, days and years. But, as with the growth of a child, there is practically no overlap between the methods of study that serve the two timescales. The language of voltage fluctuations is not useful for discussing how a large computer program, such as Microsoft Excel, works. No sensible person denies that computer programs, however complicated, are entirely executed by temporal and spatial patterns of changes between two voltages. But no sensible person attends to that fact while writing, debugging, or using a large computer program.

I have never seen any good reason to doubt the following proposition: macroevolution is lots of little bits of microevolution joined end to end over geological time, and detected by fossils instead of by genetic sampling. Nevertheless, there could be – and I believe are – major events in evolutionary history after which the very nature of evolution itself changes. Evolution itself might be said to evolve. So far in this chapter, progress has meant individual organisms becoming better over evolutionary time at doing what individuals do, which is survive and reproduce, or otherwise promote their genes. But we can

* My view is that it is an interesting empirical question, which is likely to have a different answer in different particular cases, and which does not deserve its elevation to the status of major principle.

also countenance changes in the phenomenon of evolution itself. Might evolution itself become better at doing something – what evolution does – as history goes by? Is late evolution some kind of improvement on early evolution? Do creatures evolve to improve not just their capacity to survive and reproduce, but the lineage's capacity to evolve? Is there an evolution of evolvability?

I coined the phrase 'Evolution of Evolvability' in a paper published in the *Proceedings of the 1987 Inaugural Conference on Artificial Life*. Artificial life was a newly invented merger of other disciplines, notably biology, physics and computer science, founded by the visionary physicist Christopher Langton, who edited the *Proceedings*. Since my paper, but probably not because of it, the evolution of evolvability has become a much discussed topic among students of both biology and artificial life. Long before I invented the phrase, others had proposed the idea. For example, the American ichthyologist Karel F. Liem in 1973 used the phrase 'prospective adaptation' for the revolutionary jaw apparatus of cichlid fishes which enabled them, as their tale describes, so suddenly and explosively to evolve hundreds of species in all the great African lakes. Liem's suggestion goes beyond the idea of pre-adaptation, I should say. A pre-adaptation is something that originally evolves for one purpose and is co-opted to another. Liem's prospective adaptation and my evolution of evolvability carry the suggestion not just of co-option to a new function but of unshackling a new outburst of divergent evolution. I am suggesting a permanent and even progressive trend towards becoming better at evolving.

In 1987, the idea of the evolution of evolvability was somewhat heretical, especially for me as the alleged 'ultra-Darwinist'. I was placed in the odd situation of advocating an idea at the same time as apologising for it to people who couldn't see why it needed any apology. It is now a much discussed topic, and others have taken it further than I ever envisaged, for example, the cell biologists Marc Kirschner and John Gerhart, and the evolutionary entomologist Mary Jane West-Eberhard in her magisterial book *Developmental Plasticity and Evolution*.

What makes an organism good at evolving, over and above being good at surviving and reproducing its genes? An example, first. We have already met the idea that island archipelagos are workshops of speciation. If the islands are near enough to each other to allow occasional immigrations, but far enough apart to allow time for evolutionary divergence between immigrations, we have a recipe for speciation, which is the first step towards evolutionary radiation. But how near is near enough? How far is far enough? That depends on the locomotive powers of the animals. For woodlice, a separation of a few yards is equivalent to a separation of many miles for a flying bird or bat. The Galapagos Islands are spaced just right for the divergent evolution of small birds such as Darwin's finches, not necessarily for divergent evolution generally. For this purpose, separation of islands should be measured not in absolute units but in units of travelability calibrated to the kind of animal we are talking about – as with the Irish boatman who, when my parents asked him the distance of the Great Blasket Island, replied, 'About three miles in fine weather.'

It follows that a Galapagos finch which either decreased *or increased* its flight range in evolution might thereby *decrease* its evolvability. Shortening the range lowers the chance of initiating a new race of descendants on another island. That way round, it is easily understood. Lengthening the range has a less obvious effect in the same direction. Descendants are seeded onto new islands so frequently that there is no time for separate evolution before the next immigrant arrives. To push to the extreme, birds whose flight range is large enough to render the distance between islands trivial, no longer see the islands as separate at all. As far as gene flow is concerned, the whole archipelago counts as one continent. So once again speciation is not fostered. High evolvability, if we choose to measure evolvability as speciation rate, is an inadvertent consequence of intermediate locomotor range, where what counts as intermediate, as opposed to too short or too long, depends upon the spacing of the islands concerned. Of course 'island' in this sort of argument does not have to mean land surrounded by water. As we saw in the Cichlid's Tale, lakes are islands for aquatic animals, and reefs can be islands within lakes. Mountain tops are islands for land-bound animals that cannot easily tolerate low altitudes. A tree can be an island for an animal with a short range. For the AIDS virus, every man is an island.

If an increase or a decrease in travelling range results in an increase in evolvability, would we want to call this an evolved 'improvement'? My ultra-Darwinist hackles start to quiver at this point. My heresy litmus starts to blush. It sounds uncomfortably like evolutionary foresight. Birds evolve an increase or a decrease to their flying range because of natural selection for individual survival. Future effects on evolution are an irrelevant consequence. Nevertheless, we might find with hindsight that the species that fill the world tend to be descended from ancestral species with a talent for evolution. You could say, therefore, that there is a kind of high-level, between-lineage selection in favour of evolvability – an example of what the great American evolutionist George C. Williams called clade selection. Conventional Darwinian selection leads to individual organisms being finely tuned survival machines. Could it be that, as a consequence of clade selection, life itself has increasingly become a set of finely tuned evolving machines? If this is so, we might expect that in Kauffmanian reruns of evolution, the same progressive improvements in evolvability might be rediscovered.

When I first wrote about the evolution of evolvability, I proposed a number of 'watershed events' in evolution, after which evolvability suddenly improved. The most illustrative example of a watershed event I could think of was segmentation. Segmentation, you remember, is the train-like modularisation of the body, in which parts and systems are repeated serially along the body. It seems to have been independently invented, in its full form, in arthropods, vertebrates and annelid worms (although the universality of Hox genes argues for some sort of fore and aft serial organisation as a predecessor). The origin of segmentation is one of those evolutionary events that cannot have been gradual. Bony fish typically have about 50 vertebrae, but eels have as many as 200. Caecilians (worm-like amphibians) vary between 95 and 285 vertebrae. Snakes differ hugely in vertebral number: the record known to me is 565 for an extinct snake.

Every vertebra of a snake represents one segment with its own pair of ribs, its own muscle blocks, its own nerves sprouting from the spinal cord. You can't have fractional numbers of segments, and the evolution of variable segment numbers must include numerous instances in which a mutant snake differed from its parents in some whole number of segments: at least one, possibly more, in one fell swoop. Similarly, when segmentation originated, there must have been a mutational transition straight from unsegmented parents to a child with two (at least) segments. It is hard to imagine such a freak surviving, let alone finding a mate and reproducing, but it evidently happened because segmented animals are all around us. Very probably the mutation involved Hox genes, like those of the Fruit Fly's Tale. In my 1987 evolvability paper I guessed that

> . . . the individual success, or otherwise, of the first segmented animal
> during its own lifetime is relatively unimportant. No doubt many other

new mutants have been more successful as individuals. What is important about the first segmented animal is that its descendant lineages were champion *evolvers*. They radiated, speciated, gave rise to whole new phyla. Whether or not segmentation was a beneficial adaptation during the individual lifetime of the first segmented animal, segmentation represented a change in embryology that was pregnant with evolutionary potential.

The ease with which whole segments can be added or subtracted from the body is one thing that contributes to enhanced evolvability. So is differentiation among segments. In animals such as millipedes and earthworms, most of the segments are the same as each other. But there is a recurrent tendency, especially among arthropods and vertebrates, for particular segments to become specialised for particular purposes, and hence different from other segments (compare a lobster with a centipede). A lineage that manages to evolve a segmented body plan is immediately able to evolve a whole range of new animals by altering segmental modules, all along the body.

Segmentation is an example of modularity, and modularity in general is a main ingredient in the thinking of more recent writers on the evolution of evolvability. Of the many meanings of module listed in the *Oxford English Dictionary*, the relevant one is:

> One of a series of production units or component parts that are standardized to facilitate assembly or replacement and are usually prefabricated as selfcontained structures.

Modular is the adjective describing an assemblage of modules, and modularity is the corresponding abstract noun, the property of being modular. Other examples of modular construction include many plants (leaves and flowers are modules). But perhaps the best examples of modularity are to be found at the cellular and biochemical level. Cells themselves are modules *par excellence*, and within cells so are protein molecules and, of course, DNA itself.

So the invention of multicellularity is another important watershed event that almost certainly enhanced evolvability. It preceded segmentation by hundreds of millions of years, and segmentation is itself a kind of large-scale reenactment of it, another leap in modularity. What other watersheds have there been? The dedicatee of this book, John Maynard Smith, collaborated with his Hungarian colleague Eörs Szathmàry on *The Major Transitions in Evolution*. Most of their 'major transitions' would fit under my heading of 'watershed events' – major improvements in evolvability. This obviously includes the origin of replicating molecules, for without them there could be no evolution at all. If, as Cairns-Smith and others have suggested, DNA usurped the key role of replicator from some less proficient predecessor, bridged by intermediate stages, each one of those stages would constitute a leap forward in evolvability.

If we accept the RNA World theory, there would have been a major transition or watershed, when a world of RNA serving as both replicator and enzyme gave over to a separation between DNA in the replicator role and proteins in the enzyme role. Then there was the clubbing together of replicating entities ('genes') in cells with walls, which prevented the gene products leaking away and kept them together with the products of other genes with which they could collaborate in cellular chemistry. A very major transition, and very probably a watershed of evolvability, was the birth of the eukaryotic cell by the commingling of several prokaryotic cells. So was the origin of sexual reproduction, which gives birth to the concept of the species itself, with its own gene pool, and all that that implies for future evolution. Maynard Smith and Szathmàry go on to list the origin of multicellularity, the origin of colonies such as ant and termite nests, and the origin of human societies with language. There is a rhyming similarity between at least several of these major transitions: they often involve the coming together of previously independent units in a larger grouping at a higher level, with concomitant loss of independence at the lower level.

To their list I have already added segmentation, and I would stress another, which I call bottlenecking. Once again, to spell it out in full would be to repeat from previous books (especially the final chapter, 'Rediscovering the Organism', of *The Extended Phenotype*). Bottlenecking refers to a type of life history in multicellular organisms. In bottlenecking, the life cycle regularly returns to a single cell, from which a multicellular body is grown anew. The alternative to a bottlenecked life cycle might be a hypothetical straggling water plant that reproduces by breaking off small, multicellular chunks of itself which drift off, grow and then break off more small chunks. Bottlenecking has three important consequences, all of which are certainly good candidates for improvements in evolvability.

First, evolutionary innovations can be reinvented from the bottom up, rather than as remouldings of existing structures – the equivalent of beating swords into ploughshares. An improvement in, say, a heart, has a better chance of being a clean improvement if genetic changes can alter the whole course of development from a single cell. Imagine the alternative: take the existing heart and modify it by differential tissue growth within its continuously beating fabric. This on-the-trot remodelling would impair the working of the heart and compromise the would-be improvement.

Second, by continually resetting to a consistent starting point in a recurrent life cycle, bottlenecking provides a 'calendar' by which embryological events may be timed. Genes may be turned on or off at key points in the growth cycle. Our hypothetical straggling chunkextruder lacks a recognisable timetable to regulate such switchings on and off.

Third, without bottlenecking, different mutations would accumulate in different parts of the straggling chunk-extruder. The incentive among cells to co-operate would be reduced. In effect, sub-populations of cells would be tempted to behave as cancers, to increase their chance of contributing genes to the extruded chunks. With bottlenecking, since every generation starts out as a single cell, the whole body has a good chance of being made of a uniform genetic population of co-operating cells, all descended from that single cell. Without bottlenecking, the cells of the body might have, from a genetic point of view, 'divided loyalties'.

Related to bottlenecking is another important landmark event in evolution, and one that may well have contributed to evolvability and might be rediscovered in Kauffman reruns. This is the separation of the germ line from the soma, first clearly understood by the great German biologist August Weismann. As we saw at Rendezvous 31, what happens in the developing embryo is that a portion of the cells are set aside for reproduction (germ-line cells) while the rest are destined to make the body (somatic cells). Germ-line genes are potentially immortal, with a prospect of direct descendants millions of years into the future. Somatic genes are destined for a finite, if not always predictable, number of cell divisions to make the body tissues, after which their line will come to an end and the organism will die. Plants often violate the separation, most obviously when they practise vegetative reproduction. This could constitute an important difference between the ways plants and animals evolve. Before the evolutionary invention of the separate soma, all living cells were potentially the ancestors of an indefinite line of descendants, as sponge cells still are.

The invention of sex is a major watershed, which is superficially confusable both with bottlenecking and with the separation of the germ line, but is logically distinct from both. In its most general form, sex is the partial mixing of genomes. We are familiar with a particular, highly regimented version of it in which every individual gets 50 per cent of its genome from each of two parents. We are used to the idea that there are two kinds of parent, female and male, but this is not a necessary part of sexual reproduction. Isogamy is a system in which two individuals, not distinguished as male and female, combine half their genes to make a new individual. As we saw in the Great Historic Rendezvous, the divide into two genders is best seen as a further watershed event, which came after the origin of sex itself. Regimented sex of this kind is accompanied in every generation by a 'reduction division', in which each individual donates 50 per cent of its genome to each offspring. Without this reduction, genomes would double in size with every generation.

Bacteria practise a haphazard form of sexual donation that is some-times described as sex but which is really very different, having more in common with the cut-and-paste, or copy-and-paste, functions of a computer program. Fragments of a genome are copied or cut from one bacterium and pasted into another, which does not have to be a member of the same 'species' (though the very meaning of species is in doubt for bacteria for that very reason). Because genes are software subroutines that perform cellular operations, a 'pasted' gene can immediately go to work in its new environment, doing the same task as it did before.*

What is in it for the donating bacterium? That may be the wrong question. The right question might be, what is in it for the donated gene? And the answer is that genes that successfully get themselves donated, and then successfully help the recipient bacterium to survive and pass them on, thereby increase the number of copies of them-selves in the world. It is not clear whether our regimented eukaryotic sex has evolved from bacterial 'cut-and-paste' sex, or whether it was an entirely new watershed event. Both must have had a huge impact on subsequent evolution and are candidates for discussion under the

---

* This is why transgenic manipulation in modern agricultural breeding works, for example, the legendary importing of 'antifreeze' genes from Arctic fish into tomatoes. It works for the same reason as a computer subroutine, copied from one program into another, can be relied upon to deliver the same result. The case of GM crops isn't quite so straightforward. But the example serves to allay fears of the 'unnaturalness' of importing, say, fish genes into tomatoes, as though some kind of fishy 'flavour' goes too. A subroutine is a subroutine, and DNA's language of programming is identical in fish and tomatoes.

heading of the evolution of evolvability. Regimented sex, as we saw in the Rotifer's Tale, has a dramatic effect upon future evolution because it makes possible the very existence of species with their gene pools.

The positioning of the apostrophe in *The Ancestor's Tale* indicates a singular. I admit that the motive was partly stylistic. Nevertheless, through the millions – probably billions – of individual ancestors whose lives we could have touched along our Pilgrims' Way, one singular hero has recurred in the minor, like a Wagnerian leitmotiv: DNA. It is our DNA that links us to the rest of the natural world, whose ancestral journey – or should we say journeys – we have been following all the while. We saw it in Eve's Tale, when we traced the human story contained in our genomes. We saw it in the Bonobo's Tale, as genes clamoured to assert their different views of ancestral history. We saw it in the Howler Monkey's and the Lamprey's Tales, where genes and genomes duplicated to give us our colour vision and our sophisticated vertebrate bodies, and in so doing produced family trees analogous to the conventional family tree of species. We saw it in the Humped Bladderwort's Tale, where we followed the replicating DNA parasites that clutter our own genomes. Finally, we saw it in the last stages of our journey, as DNA passed horizontally from cell to cell, and the very concept of a single tree of life crumbled to pieces. The leitmotiv in genetic history echoes, but is distinct from, the major theme of the 'selfish gene' in the understanding of natural selection. And it is at the heart of our modern understanding of the natural world.

The Host's Farewell

If, as returning host, I reflect on the whole pilgrimage of which I have been a grateful part, my overwhelming reaction is one of amazement. Amazement not only at the extravaganza of details that we have seen; amazement, too, at the very fact that there are any such details to be had at all, on any planet. The universe could so easily have remained lifeless and simple – just physics and chemistry, just the scattered dust of the cosmic explosion that gave birth to time and space. The fact that it did not – the fact that life evolved out of nearly nothing, some 10 billion years after the universe evolved (by some physicists' accounts) out of literally nothing – is a fact so staggering that I would be mad to attempt words to do it justice. And even that is not the end of the matter. Not only did evolution happen: it eventually led to beings capable of comprehending the process, and even of comprehending the process by which they comprehend it.

This pilgrimage has been a trip, not just in the literal sense but in the countercultural sense I met when a young man in California in the 1960s. The most potent hallucinogen on sale in Haight or Ashbury or Telegraph Avenue would be tame by comparison. If it's amazement you want, the real world has it all. Not to stray outside the covers of this book, think of Venus's girdle, migrating jellyfish and tiny harpoons; think of the platypus's radar and the electric fish; of the horsefly larva with the apparent foresight to pre-empt cracks in the mud; think redwood; think peacock; think starfish with its piped hydraulic power; think cichlids of Lake Victoria, evolving *how* many orders of magnitude faster than *Lingula, Limulus* or *Latimeria*? Think on the fact that while your own genome is chock full of ultra-selfish DNA parasites, it also contains the history of humanity. It is not pride in this book but reverence for life itself that encourages me to say, if you want a justification for the latter, open the former anywhere, at random. And reflect on the fact that although this book has been written from a human point of view, another book could have been written in parallel for any of 10 million starting pilgrims. Not only is life on this planet amazing, and deeply satisfying, to all whose senses have not become dulled by familiarity: the very fact that we have evolved the brain power to understand our evolutionary genesis redoubles the amazement and compounds the satisfaction.

'Pilgrimage' implies piety and reverence. I have not had occasion here to mention my impatience with traditional piety, and my disdain for reverence where the object is anything supernatural. But I make no secret of them. It is not because I wish to limit or circumscribe reverence; not because I want to reduce or downgrade the true reverence with which we are moved to celebrate the universe, once we understand it properly. 'On the contrary' would be an understatement. My objection to supernatural beliefs is precisely that they miserably fail to do justice to the sublime grandeur of the real world. They represent a narrowing-down from reality, an impoverishment of what the real world has to offer.

I suspect that many who call themselves religious would find themselves agreeing with me. To them I would only quote a favourite remark that I overheard at a scientific conference. A distinguished elder statesman of my subject was having a long argument with a colleague. As the altercation came to an end, he twinkled and said, 'You know, we really do agree. It's just that you *say* it wrong!'

I feel I have returned from a true pilgrimage.

FURTHER READING

The first edition of *The Ancestor's Tale* included a short list of recommended books, over and above the extensive bibliography of works referred to in the text. When considering how we might revise the list for this edition, we decided that a list of our favourite biological authors would be less likely to go out of date than a list of specific books. The following are authors who have influenced or inspired us, who write for non-specialist readers and who, in our opinion, can be relied upon to keep on giving!

Sean B. Carroll (e.g. *The Serengeti Rules: The Quest to Discover How Life Works and Why It Matters.* Princeton University Press, Princeton, N.J. 2016)

Jared Diamond (e.g. *Guns, Germs and Steel: A Short History of Everybody for the Last 13,000 Years.* Chatto & Windus, London 1997)

Richard Fortey (e.g. *Life: An Unauthorised Biography.* HarperCollins, London 1997)

Steve Jones (e.g. *Almost Like A Whale: The Origin Of Species Updated.* Doubleday, London 1999)

Nick Lane (e.g. *Life Ascending: The Ten Great Inventions of Evolution.* Profile Books, London 2010)

Mark Ridley (e.g. *Mendel's Demon: Gene Justice and the Complexity of Life.* Weidenfeld & Nicolson, London 2000)

Matt Ridley (e.g. *Genome: The Autobiography Of Species In 23 Chapters.* Fourth Estate, London 1999)

Adam Rutherford (e.g. *Creation: The Origin of Life / The Future of Life.* Viking, London 2013)

Neil Shubin (e.g. *Your Inner Fish: The Amazing Discovery of our 375-million-year-old Ancestor.* Allen Lane, London 2008)

Edward O. Wilson (e.g. *The Diversity of Life.* Harvard University Press, Cambridge, Mass. 1992)

Carl Zimmer (e.g. *Evolution: The Triumph of an Idea.* Roberts & Co., Colorado 2001)

In addition, the following titles give comprehensive, more advanced treatments of the areas covered in this book.

Benton, M.J. (2015) *Vertebrate Palaeontology.* 4th Edn. Wiley-Blackwell, Hoboken, New Jersey.

Brusca, R.C. & Brusca, G.J. (2003) *Invertebrates.* 2nd Edn. Sinauer Associates, Sunderland, Mass.

Macdonald, D.W. (ed.). (2009) T*he Princeton Encyclopedia of Mammals.* Princeton University Press, Princeton, NJ.

Maynard Smith, J. & Szathmáry, E. (1998) *The Major Transitions in Evolution.* Oxford University Press, Oxford.

Pääbo, S. (2014) *Neanderthal Man: In Search of Lost Genomes.* Basic Books, New York.

Tudge, C. (2000) *The Variety of Life: A Survey and a Celebration of all the Creatures that Have Ever Lived.* Oxford University Press, Oxford.

Watson, J.D. (ed.). (2014) *Molecular Biology of the Gene.* 7th Edn. Pearson, Boston.

NOTES TO THE PHYLOGENIES AND RECONSTRUCTIONS

Numbers in square brackets refer to sources listed in the Bibliography.

FRACTAL PHYLOGENY DIAGRAMS

There is an ongoing and ambitious effort to create a single tree of all life on Earth, combining published phylogenies with standard taxonomic classification. This 'Open Tree of Life' [192] can be accessed at www.opentreeoflife.org. The fourth version of the tree (November 2015) has been used to fill out many of our fractal phylogenies to the species level, with polytomies randomly resolved into a sequence of dichotomous branches. However, the Open Tree does not yet incorporate many studies relevant to this book, nor does it provide dates for branch points. For this reason, many additional sources have been used to construct and date specific parts of the phylogenies and rendezvous points in this book. These are listed below. As well as using molecular clock dates, we have relied on [28] for a comprehensive list of fossil maximum and minimum dates to inform each rendezvous.

RENDEZVOUS 1 For a discussion of great ape branch order and dates, see the Chimpanzee's Tale and the Bonobo's Tale. Branching patterns for primates (Rendezvous 1–8) are mostly based on a recent comprehensive primate tree [400, AUTOsoft tree], with dates recalibrated using pathd8 software to fix rendezvous points at the dates described below.

RENDEZVOUS 2 Date controversial: two recent molecular clock studies quote 10 Ma [400] and 15 Ma [373] for the date of gene divergence. The human/gorilla 'speciation' date is certainly more recent than this: analysis of the gorilla genome suggests 9–11 Ma or so [373]. The date of the concestor may be slightly more recent still, hence our rendezvous estimate of 8 Ma.

RENDEZVOUS 3 Gene divergences dated from 15–18 Ma [193, 400], with separation dates of 9–13 Ma [193]. These do not account for slower mutation rates (see the Chimpanzee's Tale), hence Rendezvous 3 is set at 14 Ma, compatible with analysis of the fossil record [28] and with the fossil ranges shown in the diagram on page 147.

RENDEZVOUS 4 Dates and branching order of the four gibbon genera taken from whole genome studies [56, fig. 4b]. Note that this disagrees with the general primate tree [400] which has nevertheless been used for species relationships within each genus. Rendezvous date is compatible with 12–34 Ma from fossils [28], and also a gene divergence date of 20.9 Ma calculated using an 'autocorrelated molecular clock' [400] (see page 512).

RENDEZVOUS 5 & 6 Dated phylogenies based on [400]. Dates for Rendezvous 5 and 6 are deliberately set marginally earlier than estimated gene divergences, which are 28.8/25.1 and 40.6/44.5 Ma respectively (AUTOsoft/mean dates), placing the date of Rendezvous 5 just within the minimum of 24.44 Ma expected from the fossil record [28].

RENDEZVOUS 7 The overall position of tarsiers in the tree of life was previously controversial but has been fully resolved by molecular data [e.g. 184]. Dated phylogeny from [400] gives an initial genetic divergence of 61–63 Ma, which is compatible with a placement of the rendezvous point at 60 Ma, and *Archicebus* at 55 Ma [303].

RENDEZVOUS 8 Dated phylogeny from [400]. Molecular clock studies [113, 400] place human/lemur genetic divergences at 67 or 68 Ma, corresponding to a rendezvous point of perhaps 66 or 65 Ma, just within the range of 56–66 Ma suggested by fossils [28]. Nevertheless, the timing is much debated [114] – some genetic analyses place the divergence as much as 20 million years earlier, with a consequent increase in the age of Concestors 9, 10 and 11.

RENDEZVOUS 9 & 10 Rendezvous dates are not well established: we have followed molecular clock calculations which place them approximately halfway between Rendezvous 8 and 11 [113]. Conflict between gene trees implies Rendezvous 9 and 10 occurred close in time [207, 304]. Dates within colugos are from a recent reanalysis of colugo taxonomy [206]. Within tree shrews, branches and dates are from an older analysis of all mammals [32], rescaled to a 70 Ma rendezvous point.

RENDEZVOUS 11 Rodent phylogeny is now reasonably well established: the tree shown here, from [32], agrees with more recent studies of the group [129]. Dates have been scaled to a 75 Ma rendezvous, on the basis of a recent molecular clock study [112] which gives 69, 88 or 75 Ma depending on the calibration strategy.

RENDEZVOUS 12 Phylogeny and dates are from [32], but with odd-toed ungulates and Carnivores (but not odd-toed ungulates and whales) moved to be the closest relative of the bats, as in a recent phylogenomic study [113], which has also been used to fix divergence dates between the six laurasiathere orders: dates within orders have consequently been rescaled. As with Rendezvous 10–13, the exact rendezvous date is rather uncertain. We have chosen a pragmatic compromise of various dates from [112].

RENDEZVOUS 13 The grouping 'Atlantogenata' (afrotheres plus xenathrans) is advocated by [295], although the divergence between them is probably very close in time to Rendezvous 13 [307]. Other studies [e.g. 362] place afrotheres as the outgroup to all other placental mammals, as in the first edition of this book. Phylogeny and dates are based on [32], but rescaled and updated from [336], which places the afrothere/xenarthran divergence at 89.1 Ma, and the elephants as sister to the dugongs and manatees. The rendezvous date is close to a recent gene divergence estimate of 89 Ma [113].

Rendezvous 14 Phylogeny and divergence dates of the seven marsupial orders largely taken from [292], apart from the basal divergence of opposums, which follows both [32] and [306]. Within-order branching taken from [32], with dates rescaled to fit. Rendezvous date is compatible with *Juramaia* (see page 273 and [260]), and with molecular clocks that place the marsupial divergence close to Rendezvous 15 [112].

Rendezvous 15 Dates and phylogeny within monotremes from [32]. Rendezvous date from [113], which agrees with fossil constraints of 162.5–191.1 Ma [28].

Rendezvous 16 Sauropsid phylogeny taken from OneZoom tetrapod tree, which is a synthesis of multiple sources: birds from www.birdtree.org, lizards from [29], snakes from [342], turtles from [205] and crocodiles from [309]. Turtle placement agrees with [67]. Sauropsid/mammal divergence is a classic calibration point, previously set at 310 Ma [28], but recent fossil reanalysis suggests 320–325 Ma, hence the rendezvous is likely to be around 320 Ma.

Rendezvous 17 Phylogeny from [204], with divergence dates between major groups from [341]. Rendezvous date accords well with fossil constraints of 330.4–350.1 Ma [28] and molecular clock estimates of 330–345 Ma from [45].

Rendezvous 18 Lungfish supported as sister to tetrapods on the basis of [248], with rendezvous date tightly constrained to 408–419 Ma [298]. Lungfish phylogeny from [78], with divergence dates from [189].

Rendezvous 19 Rendezvous date close to that of Rendezvous 18 [411], hence set at 420 Ma, with divergence between the two coelacanth species about the same as between humans and chimps, or fractionally older [8].

Rendezvous 20 Dated phylogeny of ray-finned fish orders from www.deepfin.org (version 3) [45], which gives a molecular clock date of 427 Ma for the lobefin/ray-fin divergence – a slightly more recent date than the one shown here. However, correction for their too-young lungfish divergence date (which is incompatible with fossils [305]) pushes Rendezvous 20 back towards 430–435 Ma. This range is within the fossil constraints of 445–420 Ma [28]. Within fish orders, most species-level phylogenies have been filled out using the Open Tree of Life, apart from sturgeons [327], bichirs [408], and gars [45].

Rendezvous 21 Recent molecular clock estimates give shark/tetrapod gene divergences of 465 Ma [45], agreeing with fossil constraints of 421–462 Ma [28], and just compatible with the discovery of shark-like scales from ~460 Ma [370]. High-level holocephalan dated phylogeny from [203], batoids (rays etc.) from [15], with families filled out to species level using the Open Tree of Life.

Rendezvous 22 Lampreys and hagfish are now accepted as sister groups [188]. Hagfish phylogeny from [132] and divergence from lampreys from [233], which also provides dates and relationships between lamprey genera. Species details from the Open Tree of Life.

RENDEZVOUS 23 DNA reveals tunicates not lancelets to be closest relatives to vertebrates [103]. For rendezvous date, Chengjiang stem vertebrates provide minimum age of ~515 Ma, possibly a little later [28]. Earliest reliable trace fossils place Rendezvous 26 not much earlier than 555 Ma [49, fig. 1.1], constraining this rendezvous date to a fair bit more recent than 555 Ma. Given the uncertainty of tunicate/lancelet branching order, the rendezvous date is likely to be close to Rendezvous 24, hence 535 Ma seems reasonable. Dates within tunicates are uncertain due to accelerated rates of evolution [30], although a long initial branch [419] hints that divergences within tunicates occurred over 50 Ma after Rendezvous 23, hence the 500 Ma contour line must pass through the tunicate stem. The entire phylogeny has been taken from the Open Tree of Life.

RENDEZVOUS 24 The youngest molecular clock dates set this around 540 Ma [114, 329]. Divergences within lancelets are generally accepted to be relatively recent, and the 500 Ma contour line certainly passes through the lancelet stem, as shown. The entire phylogeny has been taken from the Open Tree of Life.

RENDEZVOUS 25 Rendezvous date placed evenly between Rendezvous 26 and 24, on the basis of [329], which also places echinoderm/hemichordate divergence at 530 Ma. Relationships between echinoderm classes and dates of their divergences taken from [334], and phylogenies of each class filled out from the Open Tree of Life. The placement of *Xenoturbella* is highly controversial – here it is based on [39]. It is possible that acoelomorph flatworms belong here too [1].

RENDEZVOUS 26 Order of branching within protostome phyla primarily based on [119, 123, 190]. Phylogenies within phyla taken from the Open Tree of Life, apart from within Arthropoda, which has been further broken down (but crustacean paraphyly retained) on the basis of [244]. The order of branching within 'Spiralia' (lophotrochozoa and platyzoa) is still highly uncertain; we have followed [151]. For discussion of the rendezvous date, see the Velvet Worm's Tale. We take trace fossils as a reasonably accurate record [49], hence set this rendezvous only 35 Ma before the Cambrian, rather than hundreds of millions of years before, as implied by older molecular clock studies.

RENDEZVOUS 27 Acoelomorphs are strongly argued to be basal bilaterians by [1] and morphological data, as well as in some summary trees [123] but this is still not fully resolved. Species-level phylogenies of Acoela and Nemertodermatida taken from the Open Tree of Life.

RENDEZVOUS 28 Higher-level phylogeny of cnidarian classes and subclasses largely based on [71] (also see the cnidarian tree of life, www.cnidarian.info and [424]), with myxozoa as sister group on the basis of [301]. Within classes, phylogenies are from the Open Tree of Life. Rendezvous date discussed on page 523: our choice conforms to fossil constraints of approximately 550–636 Ma [28], but note that even the youngest molecular clocks tend to give greater ages, e.g. 709–645 Ma, for the initial genetic divergence [329, table 1].

RENDEZVOUS 29 Dates from this point on are extremely uncertain, and the placement of ctenophores is controversial (see page 539), so the date for this rendezvous is rather arbitrary. We place it close to the cnidarian split on the basis of an uncertain order of Concestors 28 and 29. High-level phylogeny of eight ctenophore orders from [335]; lower levels from the Open Tree of Life.

RENDEZVOUS 30 Reference [401] places *Trichoplax* at about 45 per cent of the distance between cnidarians and sponges, with sponges at 650 Ma and cnidarians at 600 Ma. This places the rendezvous at around 620 Ma.

RENDEZVOUS 31 Sponge phylogeny from the World Porifera Database (www.marinespecies.org/porifera), but note that many papers are now starting to suggest or assume sponge paraphyly, with the silicaceous sponges (demosponges and hexactinellida) diverging first (for a contradictory view, see [10]). For divergence date discussion, see page 523.

RENDEZVOUS 32, 33 & 34 Branch points from [384], with rough timescale deduced from branch length ratios in fig. 1, assuming fungal divergence at 1,200 Ma [321, fig. 1B]. Species-level phylogenies for Rendezvous 32 and 34 from the Open Tree of Life.

RENDEZVOUS 35 Early-branching fungal phylogeny mostly from [55] with microsporidia, cryptomycota and aphelidea grouping from [214], and filled out from the Open Tree of Life, version 3. Some other branches from [402], with remaining phyla filled out from the Open Tree of Life, version 4. For discussion of the rendezvous date, see page 517 and [321, 329, 402].

RENDEZVOUS 36, 37 & 38 Based on [50], with corrections from [106], and rooted at the solid arrow on the diagram at page 574. Groups have been filled out using the Open Tree of Life. Also see [4, fig. 1], which gives the same basic arrangement.

RENDEZVOUS 39 As discussed on page 625, also see [237, 538]. Phylogenies for each branch filled out from the Open Tree of Life.

RENDEZVOUS 40 It is debatable whether relationships at this level can be represented as a simply branching tree, since different bacterial 'species' swap genes regularly. Nevertheless, the groupings shown here are based on a study of potentially unswapped ribosomal genes [237], with the six branches of the fractal corresponding to (clockwise from the upper left): 1) a collection of a few obscure phyla such as Deferribacteres and Thermodesulfobacteria; 2) Archaea and eukaryotes; 3) most proteobacteria, including *E. coli*, mitochondria, and *Rhizobium* (one of our tale-tellers), along with other bacteria such as the Chlamydia group and the green sulphur bacteria; 4) various heat-loving groups such as Aquificae and the Hadobacteria (which includes another tale-teller, *Thermus aquaticus*); 5) Cyanobacteria together with the Actinobacteria (e.g. *Streptomyces*); 6) Fusobacteria and the large group known as 'firmicutes' which includes *Lactobacillus* (in yoghurt), *Bacillus anthracis* (anthrax) and *Clostridium botulinum* (which causes botulism, and from which is derived Botulinum toxin or Botox). Phylogenetic detail down to the species level has been filled out from the Open Tree of Life, but there are an unknown number of additional species, families, orders, and phyla which have not been represented here.

CONCESTOR RECONSTRUCTIONS
Concestor reconstructions by Malcolm Godwin.

Reconstructions are intended to give a general impression of the probable appearance and habitat of each concestor, based on current scientific knowledge. Non-skeletal features (e.g. colour of fur or skin) are inevitably a matter of considerable conjecture. Henry Bennett-Clark, Tom Cavalier-Smith, Hugh Dickinson, William Hawthorne, Peter Holland, Tom Kemp, Anna Nekaris, Marcello Ruta, Mark Sutton, and Keith Thomson provided various advice for the reconstructions. However, they bear no responsibility for the final pictures: any errors in interpretation are solely my responsibility.

Concestor 3 Large arboreal quadrupedal ape [25], which probably lived in Asia [403]. The face protrudes less than in orang utans, with rounder, more widely-spaced orbits (inferred from the Miocene ape *Ankarapithecus*). Forelimbs are suspensory, although less so than in orang utans; locomotion similar to the proboscis monkey *Nasalis*. Note also the brow ridges, prominent glabella, relatively high degree of encephalisation, predominantly fruit-based diet, and (relative to gibbons and Old World monkeys) the enlarged mammary glands and more bowed radius bone [168].

Concestor 8 Based on fossil adapids and omomyids. Likely to have been 1–4 kg, and nocturnal or (more likely) cathemeral (active during both day and night). Note forward-pointing eyes [187] for locating fruit and/or insects, short whiskers, strepsirrhine-type nose, and grasping hands and feet with nails rather than claws [35] for climbing among small terminal branches.

Concestor 16 Loosely based on the Upper Carboniferous amniote *Hylonomus* [57]. Note the robust skull with peg-like teeth for catching insects, variation in scale morphology across the body, relative toe lengths, eardrums and amniotic eggs.

Concestor 17 Based partly on the Lower Carboniferous tetrapod *Balanerpeton* [290]. Note the large protruding eyes, eardrums, ridged muscle blocks and head:body:tail ratio of about 1:3:3. Although some fossil amphibians had fewer digits on some limbs, given the apparent difficulty of regaining digits it is most likely that the concestor itself had five. Skin colouration is duller than for many living amphibians, because warning colouration against terrestrial predators was probably unnecessary.

Concestor 18 Informed by the Lower Devonian rhipidistian *Styloichthys*. Note the fin lobation, the headshield, the lateral line, and the heterocercal tail.

Concestor 23 Similar to lancelets, but notochord does not reach the rostrum, and specialised wheel organ absent. Note the pigment spot eye, gill bars, notochord, myomeres (V-shaped muscle blocks) and atrium (enclosed space below the main body).

Concestor 26 A bilaterally symmetrical worm, with a head end and through-flow gut. Note the eyes, the oral feeding appendages, a body formed by serial repetition (but not true segmentation) and some degree of body ornamentation.

Concestor 31 Thought to have been a hollow ball consisting of outward-pointing choanocyte cells [366] (similar to a sponge embryo). Cilia used for locomotion and for wafting food particles into choanocyte 'collar'. Note also cellular specialisation: sexual reproduction is via egg cells and free-swimming sperm. Concestor reconstructed with a pelagic lifestyle, similar to sponge embryos.

Concestor 38 Typical single-celled eukaryote, hence with a pervasive micro-tubular cytoskeleton, cilium (eukaryotic 'flagella') associated with a centriole (basal body) acting as a microtubule organising centre, a nucleus with pore structure surrounded by perforated sheets of rough ER which graduate into the cytosol, and a grainy appearance caused by tiny ribosomes. Note also the mitochondria with tubular cristae, small numbers of peroxisomes and other cellular vesicles, and movement via a combination of cilium and short pseudopods. Concestor depicted engulfing a food particle (note localised cytoskeleton build-up).

BIBLIOGRAPHY

Numbers in square brackets are used in the captions, notes and index for reference.

[1] Achatz, J.G., Chiodin, M., Salvenmoser, W., *et al.* (2013) The Acoela: on their kind and kinships, especially with nemertodermatids and xenoturbellids (Bilateria *incertae sedis*). *Organisms Diversity & Evolution* **13**: 267–286.

[2] Adams, D. (1987) *Dirk Gently's Holistic Detective Agency*. William Heinemann, London.

[3] Adams, D. & Carwardine, M. (1991) *Last Chance to See*. Pan Books, London, 2nd edn.

[4] Adl, S.M., Simpson, A.G.B., Lane, C.E., *et al.* (2012) The revised classification of eukaryotes. *Journal of Eukaryotic Microbiology* **59**: 429–514.

[5] Alexander, R.D., Hoogland, J.L., Howard, R.D., *et al.* (1979) Sexual dimorphisms and breeding systems in pinnipeds, ungulates, primates, and humans. In *Evolutionary Biology and Human Social Behavior: An Anthropological Perspective* (Chagnon, N.A. & Irons, W., eds.), pp. 402–435, Duxbury Press, North Scituate, Mass.

[6] Ali, J.R. & Huber, M. (2010) Mammalian biodiversity on Madagascar controlled by ocean currents. *Nature* **463**: 653–656.

[7] Allentoft, M.E., Sikora, M., Sjögren, K.-G., *et al.* (2015) Population genomics of Bronze Age Eurasia. *Nature* **522**: 167–172.

[8] Amemiya, C.T., Alföldi, J., Lee, A.P., *et al.* (2013) The African coelacanth genome provides insights into tetrapod evolution. *Nature* **496**: 311–316.

[9] Amson, E., de Muizon, C., Laurin, M., *et al.* (2014) Gradual adaptation of bone structure to aquatic lifestyle in extinct sloths from Peru. *Proceedings of the Royal Society B: Biological Sciences* **281**: 20140192–20140192.

[10] Antcliffe, J.B., Callow, R.H.T. & Brasier, M.D. (2014) Giving the early fossil record of sponges a squeeze: The early fossil record of sponges. *Biological Reviews* **89**: 972–1004.

[11] *Arabian Nights, The* (1885) (Burton, R.F., trans.). The Kamashastra Society, Benares.

[12] Arnaud, E., Halverson, G.P. & Shields-Zhou, G. (eds.) (2011) *The geological record of Neoproterozoic glaciations*. Geological Society memoir 36. Geological Society, London.

[13] Arrese, C.A., Hart, N.S., Thomas, N., *et al.* (2002) Trichromacy in Australian marsupials. *Current Biology* **12:** 657–660.

[14] Arsuaga, J. L., Martinez, I., Arnold, L. J., *et al.* (2014) Neandertal roots: Cranial and chronological evidence from Sima de los Huesos. *Science* **344**: 1358–1363.

[15] Aschliman, N. C., Nishida, M., Miya, M., *et al.* (2012) Body plan convergence in the evolution of skates and rays (Chondrichthyes: Batoidea). *Molecular Phylogenetics and Evolution.* **63**: 28–42.

[16] Bada, J. L. & Lazcano, A. (2003) Prebiotic soup – revisiting the Miller experiment. *Science* **300**: 745–746.

[17] Bakker, R. (1986) *The Dinosaur Heresies: A Revolutionary View of Dinosaurs.* Longman Scientific and Technical, Harlow.

[18] Balbus, S. A. (2014) Dynamical, biological and anthropic consequences of equal lunar and solar angular radii. *Proceedings of the Royal Society A: Mathematical, Physical and Engineering Sciences* **470**: 20140263.

[19] Baldwin, J. M. (1896) A new factor in evolution. *American Naturalist* **30**: 441–451.

[20] Barlow, G. W. (2002) *The Cichlid Fishes: Nature's Grand Experiment in Evolution.* Perseus Publishing, Cambridge, Mass.

[21] Bateson, P. P. G. (1976) Specificity and the origins of behavior. In *Advances in the Study of Behavior* (Rosenblatt, J., Hinde, R. A., & Beer, C., eds.), vol. 6, pp. 1–20, Academic Press, New York.

[22] Bateson, W. (1894) *Materials for the Study of Variation Treated with Especial Regard to Discontinuity in the Origin of Species.* Macmillan and Co, London.

[23] Bauer, M. & von Halversen, O. (1987) Separate localization of sound recognizing and sound producing neural mechanisms in a grasshopper. *Journal of Comparative Physiology A* **161**: 95–101.

[24] Beerling, D. (2008) *The Emerald Planet: How Plants Changed Earth's History.* Oxford: Oxford Univ. Press.

[25] Begun, D. R. (1999) Hominid family values: Morphological and molecular data on relations among the great apes and humans. In *The Mentalities of Gorillas and Orangutans* (Parker, S. T., Mitchell, R. W., & Miles, H. L., eds.), chap. 1, pp. 3–42, Cambridge University Press, Cambridge.

[26] Bell, G. (1982) *The Masterpiece of Nature: The Evolution and Genetics of Sexuality.* Croom Helm, London.

[27] Belloc, H. (1999) *Complete Verse.* Random House Children's Books, London.

[28] Benton, M. J., Donoghue, P. C., Asher, R. J., *et al.* (2015) Constraints on the timescale of animal evolutionary history. *Palaeontologia Electronica* **18**: 1–106.

[29] Bergmann, P. J. & Irschick, D. J. (2012) Vertebral evolution and the diversification of squamate reptiles. *Evolution* **66**: 1044–1058.

[30] Berna, L. & Alvarez-Valin, F. (2014) Evolutionary genomics of fast evolving tunicates. *Genome Biology and Evolution* **6**: 1724–1738.

[31] Betzig, L. (1995) Medieval monogamy. *Journal of Family History* **20**: 181–216.

[32] Bininda-Emonds, O. R. P., Cardillo, M., Jones, K. E., *et al.* (2007) The delayed rise of present-day mammals. *Nature* **446**: 507–512.

[33] Blackmore, S. (1999) *The Meme Machine.* Oxford University Press, Oxford.

[34] Blair, W. F. (1955) Mating call and stage of speciation in the *Microhyla olivacea – M. carolinensis* complex. *Evolution* **9**: 469–480.

[35] Bloch, J. I. & Boyer, D. M. (2002) Grasping primate origins. *Science* **298**: 1606–1610.

[36] Blum, M. G. B. & Jakobsson, M. (2011) Deep divergences of human gene trees and models of human origins. *Molecular Biology and Evolution* **28**: 889–898.

[37] Bond, M., Tejedor, M. F., Campbell, K. E., *et al.* (2015) Eocene primates of South America and the African origins of New World monkeys. *Nature* **520**: 538–541.

[38] Bonner, J. T. (1993) *Life Cycles: Reflections of an Evolutionary Biologist.* Princeton University Press, Princeton, N.J.

[39] Bourlat, S. J., Juliusdottir, T., Lowe, C. J., *et al.* (2006) Deuterostome phylogeny reveals monophyletic chordates and the new phylum Xeno-turbellida. *Nature* **444**: 85–88.

[40] Bourlat, S. J., Nielsen, C., Lockyer, A. E., *et al.* (2003) *Xenoturbella* is a deuterostome that eats molluscs. *Nature* **424**: 925–928.

[41] Briggs, D., Erwin, D., & Collier, F. (1994) *The Fossils of the Burgess Shale.* Smithsonian Institution Press, Washington, D.C.

[42] Briggs, D. E. G. & Fortey, R. A. (2005) Wonderful strife – systematics, stem groups and the phylogenetic signal of the Cambrian radiation. *Paleobiology* **31**: 94–112.

[43] Bromham, L. & Penny, D. (2003) The modern molecular clock. *Nature Reviews Genetics* **4**: 216–224.

[44] Bromham, L., Woolfit, M., Lee, M. S. Y., & Rambaut, A. (2002) Testing the relationship between morphological and molecular rates of change along phylogenies. *Evolution* **56**: 1921–1930.

[45] Broughton, R. E., Betancur-R., R., Li, C., *et al.* (2013) Multi-locus phylogenetic analysis reveals the pattern and tempo of bony fish evolution. *PLoS Currents.*

[46] Brown, C. T., Hug, L. A., Thomas, B. C., *et al.* (2015) Unusual biology across a group comprising more than 15% of domain Bacteria. *Nature* **523**: 208–211.

[47] Brunet, M., Guy, F., Pilbeam, D., *et al.* (2002) A new hominid from the Upper Miocene of Chad, central Africa. *Nature* **418**: 145–151.

[48] Buchsbaum, R. (1987) *Animals Without Backbones.* University of Chicago Press, Chicago, 3rd edn.

[49] Budd, G.E. (2009) The earliest fossil record of animals and its significance. In *Animal Evolution: Genomes, Fossils, and Trees* (Telford, M.J. &
Littlewood, D.T.J., eds.), Oxford University Press, Oxford.

[50] Burki, F. (2014) The eukaryotic tree of life from a global phylogenomic
perspective. *Cold Spring Harbor Perspectives in Biology* 6: a016147–
a016147.

[51] Burt, A. & Trivers, R. (2006) *Genes in Conflict: The Biology of Selfish
Genetic Elements.* Cambridge, MA: Belknap Press of Harvard University
Press.

[52] Butterfield, N.J. (2000) *Bangiomorpha pubescens* n. gen., n. sp.: implications for the evolution of sex, multicellularity, and the Mesoproter-
ozoic/Neoproterozoic radiation of eukaryotes. *Paleobiology* 26: 386–404.

[53] Butterfield, N.J. (2001) Paleobiology of the late Mesoproterozoic (ca.
1200 Ma) Hunting Formation, Somerset Island, arctic Canada. *Precambrian Research* 111: 235–256.

[54] Cairns-Smith, A.G. (1985) *Seven Clues to the Origin of Life.* Cambridge
University Press, Cambridge.

[55] Capella-Gutiérrez, S., Marcet-Houben, M. & Gabaldón, T. (2012)
Phylogenomics supports microsporidia as the earliest diverging clade
of sequenced fungi. *BMC Biology* 10: 47.

[56] Carbone, L., Alan Harris, R., Gnerre, S., *et al.* (2014) Gibbon genome
and the fast karyotype evolution of small apes. *Nature* 513: 195–201.

[57] Carroll, R.L. (1988) *Vertebrate Paleontology and Evolution.* W.H. Freeman, New York.

[58] Casane, D. & Laurenti, P. (2013) Why coelacanths are not 'living
fossils': A review of molecular and morphological data. *BioEssays* 35:
332–338.

[59] Castellano, S., Parra, G., Sanchez-Quinto, F.A., *et al.* (2014) Patterns of
coding variation in the complete exomes of three Neandertals. *Proceedings of the National Academy of Sciences* 111: 6666–6671.

[60] Catania, K.C. & Kaas, J.H. (1997) Somatosensory fovea in the star-
nosed mole: Behavioral use of the star in relation to innervation
patterns and cortical representation. *Journal of Comparative Neurology*
387: 215–233.

[61] Cavalier-Smith, T. (1991) Intron phylogeny: A new hypothesis. *Trends
in Genetics* 7: 145–148.

[62] Cavalier-Smith, T. (2002) The neomuran origin of archaebacteria, the
negibacterial root of the universal tree and bacterial megaclassification.
International Journal of Systematic and Evolutionary Microbiology 52: 7–76.

[63] Cavalier-Smith, T. & Chao, E.E. (2010) Phylogeny and evolution of
Apusomonadida (Protozoa: Apusozoa): New genera and species. *Protist* 161: 549–576.

[64] Censky, E.J., Hodge, K., & Dudley, J. (1998) Overwater dispersal of
lizards due to hurricanes. *Nature* 395: 556.

[65] Chang, J. T. (1999) Recent common ancestors of all present-day individuals. *Advances in Applied Probability* **31:** 1002–1026.

[66] Chaucer, G. (2000) *Chaucer: The General Prologue on CD-ROM* (Solopova, E., ed.). Cambridge University Press, Cambridge.

[67] Chiari, Y., Cahais, V., Galtier, N., *et al.* (2012) Phylogenomic analyses support the position of turtles as the sister group of birds and crocodiles (Archosauria). *Bmc Biology* **10:** 65.

[68] Clack, J. (2002) *Gaining Ground: The Origin and Evolution of Tetrapods.* Indiana University Press, Bloomington.

[69] Clarke, R. J. (1998) First ever discovery of a well-preserved skull and associated skeleton of *Australopithecus. South African Journal of Science* **94:** 460–463.

[70] Cleveland, L. R. & Grimstone, A. V. (1964) The fine structure of the flagellate *Mixotricha paradoxa* and its associated micro-organisms. *Proceedings of the Royal Society of London: Series B* **159:** 668–686.

[71] Collins, A. G. (2009) Recent insights into cnidarian phylogeny. *Smithsonian Contributions to Marine Sciences* **38:** 139–149.

[72] Conway-Morris, S. (1998) *The Crucible of Creation: The Burgess Shale and the Rise of Animals.* Oxford University Press, Oxford.

[73] Conway-Morris, S. (2003) *Life's Solution: Inevitable Humans in a Lonely Universe.* Cambridge University Press, Cambridge.

[74] Cooper, A. & Fortey, R. (1998) Evolutionary explosions and the phylogenetic fuse. *Trends in Ecology and Evolution* **13:** 151–156.

[75] Cooper, A. & Stringer, C. B. (2013) Did the Denisovans cross Wallace's line? *Science* **342:** 321–323.

[76] Cott, H. B. (1940) *Adaptive Coloration in Animals.* Methuen, London.

[77] Crick, F. H. C. (1981) *Life Itself: Its Origin and Nature.* Macdonald, London.

[78] Criswell, K. E. (2015) The comparative osteology and phylogenetic relationships of African and South American lungfishes (*Sarcoptergii: Dipnoi*). *Zoological Journal of the Linnean Society* **174:** 801–858.

[79] Crockford, S. (2002) *Dog Evolution: A Role for Thyroid Hormone Physiology in Domestication Changes.* Johns Hopkins University Press, Baltimore.

[80] Cronin, H. (1991) *The Ant and the Peacock: Altruism and Sexual Selection from Darwin to Today.* Cambridge University Press, Cambridge.

[81] Csányi, V. (2006) *If Dogs Could Talk: Exploring the Canine Mind.* The History Press.

[82] Daeschler, E. B., Shubin, N. H. & Jenkins, F. A. (2006) A Devonian tetrapod-like fish and the evolution of the tetrapod body plan. *Nature* **440:** 757–763.

[83] Dalen, L., Orlando, L., Shapiro, B., *et al.* (2012) Partial genetic turnover in Neanderthals: continuity in the east and population replacement in the west. *Molecular Biology and Evolution* **29:** 1893–1897.

[84]   Darwin, C. (1860/1859) *On The Origin of Species by Means of Natural Selection*. John Murray, London.

[85]   Darwin, C. (1987/1842) *The Geology of the Voyage of HMS Beagle: The Structure and Distribution of Coral Reefs*. New York University Press, New York.

[86]   Darwin, C. (2002/1839) *The Voyage of the Beagle*. Dover Publications, New York.

[87]   Darwin, C. (2003/1871) *The Descent of Man*. Gibson Square Books, London.

[88]   Darwin, F. (ed.) (1888) *The Life And Letters of Charles Darwin*. John Murray, London.

[89]   Daubin, V., Gouy, M., & Perrière, G. (2002) A phylogenomic approach to bacterial phylogeny: Evidence for a core of genes sharing common history. *Genome Research* **12:** 1080–1090.

[90]   Davies, P. (1998) *The Fifth Miracle: The Search for the Origin of Life*. Allen Lane, The Penguin Press, London.

[91]   Dawkins, R. (1982) *The Extended Phenotype*. W. H. Freeman, Oxford.

[92]   Dawkins, R. (1986) *The Blind Watchmaker*. Longman, London.

[93]   Dawkins, R. (1989) The evolution of evolvability. In *Artificial Life* (Langton, C., ed.), pp. 201–220, Addison-Wesley, New York.

[94]   Dawkins, R. (1989) *The Selfish Gene*. Oxford University Press, Oxford, 2nd edn.

[95]   Dawkins, R. (1995) *River Out of Eden*. Weidenfeld & Nicolson, London.

[96]   Dawkins, R. (1996) *Climbing Mount Improbable*. Viking, London.

[97]   Dawkins, R. (1998) *Unweaving the Rainbow*. Penguin, London.

[98]   Dawkins, R. (2003) *A Devil's Chaplain*. Weidenfeld & Nicolson, London.

[99]   Dawkins, R. & Krebs, J. R. (1979) Arms races between and within species. *Proceedings of the Royal Society of London: Series B* **205:** 489–511.

[100] de Morgan, A. (2003/1866) *A Budget of Paradoxes*. The Thoemmes Library, Poole, Dorset.

[101] de Waal, F. (1995) Bonobo sex and society. *Scientific American* **272** (March): 82–88.

[102] de Waal, F. (1997) *Bonobo: The Forgotten Ape*. University of California Press, Berkeley.

[103] Delsuc, F., Brinkmann, H., Chourrout, D., *et al.* (2006) Tunicates and not cephalochordates are the closest living relatives of vertebrates. *Nature* **439:** 965–968.

[104] Dennett, D. (1991) *Consciousness Explained*. Little, Brown, Boston.

[105] Dennett, D. (1995) *Darwin's Dangerous Idea: Evolution and the Meaning of Life*. Simon & Schuster, New York.

[106] Derelle, R., Torruella, G., Klimeš, V., *et al.* (2015) Bacterial proteins pinpoint a single eukaryotic root. *Proceedings of the National Academy of Sciences of the USA* **112:** E693–E699.

[107] Deutsch, D. (1997) *The Fabric of Reality*. Allen Lane, The Penguin Press, London.

[108] Diamond, J. (1991) *The Rise and Fall of the Third Chimpanzee*. Radius, London.

[109] Dias, B.G. & Ressler, K.J. (2013) Parental olfactory experience influences behavior and neural structure in subsequent generations. *Nature Neuroscience* **17**: 89–96.

[110] Dixon, D. (1981) *After Man: A Zoology of the Future*. Granada, London.

[111] Doolittle, W.F. & Sapienza, C. (1980) Selfish genes, the phenotype paradigm and genome evolution. *Nature* **284**: 601–603.

[112] dos Reis, M., Donoghue, P.C.J. & Yang, Z. (2014) Neither phylogenomic nor palaeontological data support a Palaeogene origin of placental mammals. *Biology Letters* **10**: 20131003.

[113] dos Reis, M., Inoue, J., Hasegawa, M., *et al.* (2012) Phylogenomic datasets provide both precision and accuracy in estimating the timescale of placental mammal phylogeny. *Proceedings of the Royal Society B: Biological Sciences* **279**: 3491–3500.

[114] dos Reis, M., Thawornwattana, Y., Angelis, K., *et al.* (2015) Uncertainty in the timing of origin of animals and the limits of precision in molecular timescales. *Current Biology* **25**: 2939–2950.

[115] Douady, C.J., Catzeflis, F., Kao, D.J., *et al.* (2002) Molecular evidence for the monophyly of Tenrecidae (Mammalia) and the timing of the colonization of Madagascar by malagasy tenrecs. *Molecular Phylogenetics and Evolution* **22**: 357–363.

[116] Drayton, M. (1931–1941) *The Works of Michael Drayton*. Blackwell, Oxford.

[117] Dudley, J.W. & Lambert, R.J. (1992) Ninety generations of selection for oil and protein in maize. *Maydica* **37**: 96–119.

[118] Dulai, K.S., von Dornum, M., Mollon, J.D., & Hunt, D.M. (1999) The evolution of trichromatic color vision by opsin gene duplication in New World and Old World primates. *Genome Research* **9**: 629–638.

[119] Dunn, C.W., Hejnol, A., Matus, D.Q., *et al.* (2008) Broad phylogenomic sampling improves resolution of the animal tree of life. *Nature* **452**: 745–749.

[120] Durham, W.H. (1991) *Coevolution: Genes, Culture and Human diversity*. Stanford University Press, Stanford.

[121] Dyson, F.J. (1999) *Origins of Life*. Cambridge University Press, Cambridge, 2nd ed.

[122] Eddy, S.R. (2012) The C-value paradox, junk DNA and ENCODE. *Current Biology* **22**: R898–R899.

[123] Edgecombe, G.D., Giribet, G., Dunn, C.W., *et al.* (2011) Higher-level metazoan relationships: Recent progress and remaining questions. *Organisms Diversity & Evolution* **11**: 151–172.

[124] Edwards, A.W.F. (2003) Human genetic diversity: Lewontin's fallacy. *BioEssays* **25**: 798–801.

[125] Eigen, M. (1992) *Steps Towards Life: A Perspective on Evolution.* Oxford University Press, Oxford.

[126] Eitel, M., Osigus, H.-J., DeSalle, R., & Schierwater, B. (2013) Global diversity of the Placozoa. *PLoS ONE* **8**: e57131.

[127] Erdmann, M. (1999) An account of the first living coelacanth known to scientists from Indonesian waters. *Environmental Biology of Fishes* **54**: 439–443.

[128] Ezkurdia, I., Juan, D., Rodriguez, J.M., *et al.* (2014) Multiple evidence strands suggest that there may be as few as 19 000 human protein-coding genes. *Human Molecular Genetics* **23**: 5866–5878.

[129] Fabre, P.-H., Hautier, L., Dimitrov, D., *et al.* (2012) A glimpse on the pattern of rodent diversification: a phylogenetic approach. *BMC Evolutionary Biology* **12**: 88.

[130] Farina, R.A., Tambusso, P.S., Varela, L., *et al.* (2013) Arroyo del Vizcaino, Uruguay: a fossil-rich 30-ka-old megafaunal locality with cut-marked bones. *Proceedings of the Royal Society B: Biological Sciences* **281**: 20132211–20132211.

[131] Felsenstein, J. (2004) *Inferring phylogenies.* Sunderland, Mass: Sinauer Associates.

[132] Fernholm, B., Norén, M., Kullander, S.O., *et al.* (2013) Hagfish phylogeny and taxonomy, with description of the new genus *Rubicundus* (Craniata, Myxinidae). *Journal of Zoological Systematics and Evolutionary Research* **51**: 296–307.

[133] Ferrier, D.E.K. & Holland, P.W.H. (2001) Ancient origin of the Hox gene cluster. *Nature Reviews Genetics* **2**: 33–38.

[134] Ferrier, D.E.K., Minguillón, C., Holland, P.W.H., & Garcia-Fernàndez, J. (2000) The amphioxus Hox cluster: Deuterostome posterior flexibility and Hox14. *Evolution and Development* **2**: 284–293.

[135] Fisher, R.A. (1999/1930) *The Genetical Theory of Natural Selection: A Complete Variorum Edition.* Oxford University Press, Oxford.

[136] Fisher, S.E. & Scharff, C. (2009) FOXP2 as a molecular window into speech and language. *Trends in Genetics* **25**: 166–177.

[137] Fogle, B. (1993) *101 Questions Your Dog Would Ask Its Vet.* Michael Joseph, London.

[138] Fortey, R. (1997) *Life: An Unauthorised Biography: A Natural History of the First Four Thousand Million Years of Life on Earth.* HarperCollins, London.

[139] Fortey, R. (2012) *Survivors: The Animals and Plants that Time has Left Behind.* HarperPress, London.

[140] Freedman, A.H., Gronau, I., Schweizer, R.M., *et al.* (2014) Genome sequencing highlights the dynamic early history of dogs. *PLoS Genetics* **10**: e1004016.

[141]   Fu, Q., Hajdinjak, M., Moldovan, O. T., *et al.* (2015) An early modern human from Romania with a recent Neanderthal ancestor. *Nature* **524**: 216–219.

[142]   Fu, Q., Li, H., Moorjani, P., *et al.* (2014) Genome sequence of a 45,000-year-old modern human from western Siberia. *Nature* **514**: 445–449.

[143]   Fu, Q., Mittnik, A., Johnson, P. L. F., *et al.* (2013) A revised timescale for human evolution based on ancient mitochondrial genomes. *Current Biology* **23**: 553–559.

[144]   Gallant, J. R., Traeger, L. L., Volkening, J. D., *et al.* (2014) Genomic basis for the convergent evolution of electric organs. *Science* **344**: 1522–1525.

[145]   Gardner, A. & Conlon, J. P. (2013) Cosmological natural selection and the purpose of the universe. *Complexity* **18**: 48–56.

[146]   Garstang, W. (1951) *Larval Forms and Other Zoological Verses by the late Walter Garstang* (Hardy, A. C., ed.). Blackwell, Oxford.

[147]   Gavelis, G. S., Hayakawa, S., White III, R. A., *et al* (2015) Eye-like ocelloids are built from different endosymbiotically acquired components. *Nature* **523**: 204–207.

[148]   Geissmann, T. (2002) Taxonomy and evolution of gibbons. *Evolutionary Anthropology* **11, Supplement 1**: 28–31.

[149]   Gensel, P. G. (2008) The earliest land plants. *Annual Review of Ecology, Evolution, and Systematics* **39**: 459–477.

[150]   Georgy, S. T., Widdicombe, J. G., & Young, V. (2002) The pyrophysiology and sexuality of dragons. *Respiratory Physiology & Neurobiology* **133**: 3–10.

[151]   Giribet, G. (2008) Assembling the lophotrochozoan (=spiralian) tree of life. *Philosophical Transactions of the Royal Society B: Biological Sciences* **363**: 1513–1522.

[152]   Gold, T. (1992) The deep, hot biosphere. *Proceedings of the National Academy of Sciences of the USA* **89**: 6045–6049.

[153]   Goodman, M. (1985) Rates of molecular evolution: The hominoid slowdown. *BioEssays* **3**: 9–14.

[154]   Gould, S. J. (1977) *Ontogeny and Phylogeny*. The Belknap Press of Harvard University Press, Cambridge, Mass.

[155]   Gould, S. J. (1985) *The Flamingo's Smile: Reflections in Natural History*. Norton, New York.

[156]   Gould, S. J. (1989) *Wonderful Life: The Burgess Shale and the Nature of History*. Hutchinson Radius, London.

[157]   Gould, S. J. & Calloway, C. B. (1980) Clams and brachiopods: Ships that pass in the night. *Paleobiology* **6**: 383–396.

[158]   Grafen, A. (1990) Sexual selection unhandicapped by the Fisher process. *Journal of Theoretical Biology* **144**: 473–516.

[159]   Granger, D. E., Gibbon, R. J., Kuman K., *et al.* (2015) New cosmogenic burial ages for Sterkfontein Member 2 Australopithecus and Member 5 Oldowan. *Nature* **522**: 85–88.

[160] Grant, P. R. (1999/1986) *Ecology and Evolution of Darwin's Finches.* Princeton University Press, Princeton, N. J., revised ed.

[161] Grant, P. R. & Grant, B. R. (2014) *40 Years of Evolution: Darwin's Finches on Daphne Major Island.* Princeton University Press, Princeton, N.J.

[162] Graur, D. & Martin, W. (2004) Reading the entrails of chickens: Molecular timescales of evolution and the illusion of precision. *Trends in Genetics* **20**: 80–86.

[163] Green, R. E., Krause, J., Briggs, A. W., *et al.* (2010) A draft sequence of the Neandertal genome. *Science* **328**: 710–722.

[164] Greenwalt, D. E., Goreva, Y. S., Siljestrom, S. M., *et al.* (2013) Hemoglobin-derived porphyrins preserved in a Middle Eocene blood-engorged mosquito. *Proceedings of the National Academy of Sciences* **110**: 18496–18500.

[165] Gribbin, J. & Cherfas, J. (1982) *The Monkey Puzzle.* The Bodley Head, London.

[166] Gribbin, J. & Cherfas, J. (2001) *The First Chimpanzee: In Search of Human Origins.* Penguin, London.

[167] Gross, J. B. (2012) The complex origin of *Astyanax* cavefish. *BMC evolutionary biology* **12**: 105.

[168] Groves, C. P. (1986) Systematics of the great apes. In *Systematics, Evolution, and Anatomy* (Swindler, D. R. & Erwin, J., eds.), *Comparative Primate Biology*, vol. 1, pp. 186–217, Alan R. Liss, New York.

[169] Hadzi, J. (1963) *The Evolution of the Metazoa.* Pergamon Press, Oxford.

[170] Haeckel, E. (1866) *Generelle Morphologie der Organismen.* Georg Reimer, Berlin.

[171] Haeckel, E. (1899–1904) *Kunstformen der Natur.*

[172] Haig, D. (1993) Genetic conflicts in human pregnancy. *The Quarterly Review of Biology* **68**: 495–532.

[173] Haldane, J. B. S. (1952) Introducing Douglas Spalding. *British Journal for Animal Behaviour* **2**: 1.

[174] Haldane, J. B. S. (1985) *On Being the Right Size and Other Essays* (Maynard Smith, J., ed.). Oxford University Press, Oxford.

[175] Halder, G., Callaerts, P., & Gehring, W. J. (1995) Induction of ectopic eyes by targeted expression of the eyeless gene in *Drosophila*. *Science* **267**: 1788–1792.

[176] Hallam, A. & Wignall, P. B. (1997) *Mass Extinctions and their Aftermath.* Oxford University Press, Oxford.

[177] Hamilton, W. D. (2001) *Narrow Roads of Gene Land*, vol. 2. Oxford University Press, Oxford.

[178] Hamilton, W. D. (2006) *Narrow Roads of Gene Land* (Ridley, M., ed.), vol. 3. Oxford University Press, Oxford.

[179] Hamrick, M. W. (2001) Primate origins: Evolutionary change in digital ray patterning and segmentation. *Journal of Human Evolution* **40**: 339–351.

[180] Harcourt, A. H., Harvey, P. H., Larson, S. G., & Short, R. V. (1981) Testis weight, body weight and breeding system in primates. *Nature* **293**: 55–57.

[181] Hardy, A. (1965) *The Living Stream*. Collins, London.

[182] Hardy, A. C. (1954) The escape from specialization. In *Evolution as a Process* (Huxley, J., Hardy, A. C., & Ford, E. B., eds.), Allen and Unwin, London, 1st edn.

[183] Harmand, S., Lewis, J. E., Fiebel, C. S., *et al.* (2015) 3.3-million-year-old stone tools from Lomekwi 3, West Turkana, Kenya. *Nature* **521**: 310–315.

[184] Hartig, G., Churakov, G., Warren, W. C., *et al.* (2013) Retrophylogenomics place tarsiers on the evolutionary branch of anthropoids. *Scientific Reports* **3**.

[185] Harvey, P. H. & Pagel, M. D. (1991) *The Comparative Method in Evolutionary Biology*. Oxford University Press, Oxford.

[186] Hay, J. M., Subramanian, S., Millar, C. D., *et al.* (2008) Rapid molecular evolution in a living fossil. *Trends in Genetics* **24**: 106–109.

[187] Heesy, C. P. & Ross, C. F. (2001) Evolution of activity patterns and chromatic vision in primates: Morphometrics, genetics and cladistics. *Journal of Human Evolution* **40**: 111–149.

[188] Heimberg, A. M., Cowper-Sal·lari, R., Semon, M., *et al.* (2010) microRNAs reveal the interrelationships of hagfish, lampreys, and gnathostomes and the nature of the ancestral vertebrate. *Proceedings of the National Academy of Sciences* **107**: 19379–19383.

[189] Heinicke, M. P., Sander, J. M. & Hedges, S. B. (2009) Lungfishes (Dipnoi). In *The Timetree of Life*. Hedges, S. B. & Kumar, S. (Eds). . Oxford?; New York: Oxford University Press.

[190] Hejnol, A., Obst, M., Stamatakis, A., *et al.* (2009) Assessing the root of bilaterian animals with scalable phylogenomic methods. *Proceedings of the Royal Society B: Biological Sciences* **276**: 4261–4270.

[191] Higham, T., Douka, K., Wood, R., *et al.* (2014) The timing and spatiotemporal patterning of Neanderthal disappearance. *Nature* **512**: 306–309.

[192] Hinchliff, C. E., Smith, S. A., Allman, J. F., *et al.* (2015) Synthesis of phylogeny and taxonomy into a comprehensive tree of life. *Proceedings of the National Academy of Sciences*.

[193] Hobolth, A., Dutheil, J. Y., Hawks, J., *et al.* (2011) Incomplete lineage sorting patterns among human, chimpanzee, and orangutan suggest recent orangutan speciation and widespread selection. *Genome Research* **21**: 349–356.

[194] Holland, N. D. (2003) Early central nervous system evolution: An era of skin brains? *Nature Reviews Neuroscience* **4**: 617–627.

[195] Hollmann, J., Myburgh, S., Van der Schijff, M., *et al.* (1995) Aardvark and cucumber: A remarkable relationship. *Veld and Flora* 108–109.

[196] Home, E. (1802) A description of the anatomy of the *Ornithorhynchus paradoxus*. *Philosophical Transactions of the Royal Society of London* **92**: 67–84.

[197]   Hou, X.-G., Aldridge, R. J., Bergstrom, J., *et al.* (2004) *The Cambrian Fossils of Chengjiang, China: The Flowering of Early Animal Life*. Blackwell Science, Oxford.

[198]   Huerta-Sánchez, E., Jin, X., Asan, *et al.* (2014) Altitude adaptation in Tibetans caused by introgression of Denisovan-like DNA. *Nature* **512**: 194–197.

[199]   Hume, D. (1957/1757) *The Natural History of Religion* (Root, H. E., ed.). Stanford University Press, Stanford.

[200]   Huxley, A. (1939) *After Many a Summer*. Chatto and Windus, London.

[201]   Huxley, T. H. (2001/1836) *Man's Place in Nature*. Random House USA, New York.

[202]   Ibarra-Laclette, E., Lyons, E., Hernández-Guzmán, G., *et al.* (2013) Architecture and evolution of a minute plant genome. *Nature* **498**: 94–98.

[203]   Inoue, J. G., Miya, M., Lam, K., *et al.* (2010) Evolutionary origin and phylogeny of the modern holocephalans (chondrichthyes: chimaeriformes): A mitogenomic perspective. *Molecular Biology and Evolution* **27**: 2576–2586.

[204]   Isaac, N. J. B., Redding, D. W., Meredith, H. M., *et al.* (2012) Phylogenetically-informed priorities for amphibian conservation. *PLoS ONE* **7**: e43912.

[205]   Jaffe, A. L., Slater, G. J. & Alfaro, M. E. (2011) The evolution of island gigantism and body size variation in tortoises and turtles. *Biology Letters* **7**: 558–561.

[206]   Janečka, J. E., Helgen, K. M., Lim, N.T.-L., *et al.* (2008) Evidence for multiple species of Sunda colugo. *Current Biology* **18**: R1001–R1002.

[207]   Janečka, J. E., Miller, W., Pringle, T. H., *et al.* (2007) Molecular and genomic data identify the closest living relative of primates. *Science* **318**: 792–794.

[208]   Jensen, S., Droser, M. L. & Gehling, J. G. (2005) Trace fossil preservation and the early evolution of animals. *Palaeogeography, Palaeoclimatology, Palaeoecology* **220**: 19–29.

[209]   Jerison, H. J. (1973) *Evolution of the Brain and Intelligence*. Academic Press, New York.

[210]   Jimenez-Guri, E., Philippe, H., Okamura, B., *et al.* (2007) *Buddenbrockia* is a cnidarian worm. *Science* **317**: 116–118.

[211]   Johanson, D. C. & Edey, M. A. (1981) *Lucy: The Beginnings of Humankind*. Grenada, London.

[212]   Jones, S. (1993) *The Language of the Genes: Biology, History, and the Evolutionary Future*. HarperCollins, London.

[213]   Judson, O. (2002) *Dr. Tatiana's Sex Advice to all Creation*. Metropolitan Books, New York.

[214]   Karpov, S. A., Mamkaeva, M. A., Aleoshin, V. V., *et al.* (2014) Morphology, phylogeny, and ecology of the aphelids (Aphelidea, Opisthokonta) and

proposal for the new superphylum Opisthosporidia. *Frontiers in Microbiology* 5.

[215] Katzourakis, A. & Gifford, R. J. (2010) Endogenous viral elements in animal genomes. *PLoS Genetics* **6**: e1001191.

[216] Kauffman, S. A. (1985) Self-organization, selective adaptation, and its limits. *In Evolution at a Crossroads* (Depew, D. J. & Weber, B. H., eds.), pp. 169–207, MIT Press, Cambridge, Mass.

[217] Keeling, P. J. (2013) The number, speed, and impact of plastid endosymbioses in eukaryotic evolution. *Annual Review of Plant Biology* **64**: 583–607.

[218] Kemp, T. S. (1982) The reptiles that became mammals. *New Scientist* **93**: 581–584.

[219] Kemp, T. S. (2005) *The Origin and Evolution of Mammals.* Oxford University Press, Oxford.

[220] Kimura, M. (1994) *Population Genetics, Molecular Evolution and the Neutral Theory* (Takahata, N., ed.). University of Chicago Press, Chicago.

[221] King, H. M., Shubin, N. H., Coates, M. I., *et al.* (2011) Behavioral evidence for the evolution of walking and bounding before terrestriality in sarcopterygian fishes. *Proceedings of the National Academy of Sciences* **108**: 21146–21151.

[222] Kingdon, J. (1990) *Island Africa.* Collins, London.

[223] Kingdon, J. (2003) *Lowly Origin: Where, When and Why our Ancestors First Stood Up.* Princeton University Press, Princeton/Oxford.

[224] Kingsley, C. (1995/1863) *The Water Babies.* Puffin, London.

[225] Kipling, R. (1995/1906) *Puck of Pook's Hill.* Penguin, London.

[226] Kirschner, M. & Gerhart, J. (1998) Evolvability. *Proceedings of the National Academy of Sciences of the USA* **95**: 8420–8427.

[227] Kittler, R., Kayser, M., & Stoneking, M. (2003) Molecular evolution of *Pediculus humanus* and the origin of clothing. *Current Biology* **13**: 1414–1417.

[228] Kivell, T. L. & Schmitt, D. (2009) Independent evolution of knuckle-walking in African apes shows that humans did not evolve from a knuckle-walking ancestor. *Proceedings of the National Academy of Sciences* **106**: 14241–14246.

[229] Klein, R. G. (1999) *The Human Career: Human Biological and Cultural Origins.* Chicago University Press, Chicago/London, 2nd ed.

[230] Kong, A., Frigge, M. L., Masson, G., *et al.* (2012) Rate of de novo mutations and the importance of father's age to disease risk. *Nature* **488**: 471–475.

[231] Krings, M., Stone, A., Schmitz, R. W., *et al.* (1997) Neanderthal DNA sequences and the origin of modern humans. *Cell* **90**: 19–30.

[232] Kruuk, H. (2003) *Niko's Nature.* Oxford University Press, Oxford.

[233] Kuraku, S. (2008) Insights into cyclostome phylogenomics: Pre-2R or post-2R. *Zool. Sci.* **25**: 960–968.

[234] Lack, D. (1947) *Darwin's Finches*. Cambridge University Press, Cambridge.

[235] Lambourn, W. A. (1930) The remarkable adaptation by which a dipterous pupa (Tabanidae) is preserved from the dangers of fissures in drying mud. *Proceedings of the Royal Society of London: Series B* **106**: 83–87.

[236] Lamichhaney, S., Berglund, J., Almén, M. S., *et al.* (2015) Evolution of Darwin's finches and their beaks revealed by genome sequencing. *Nature* **518**: 371–375.

[237] Lasek-Nesselquist, E. & Gogarten, J. P. (2013) The effects of model choice and mitigating bias on the ribosomal tree of life. *Molecular Phylogenetics and Evolution* **69**: 17–38.

[238] Laskey, R. A. & Gurdon, J. B. (1970) Genetic content of adult somatic cells tested by nuclear transplantation from cultured cells. *Nature* **228**: 1332–1334.

[239] Leakey, M. (1987) The hominid footprints: Introduction. In *Laetoli: A Pliocene Site in Northern Tanzania* (Leakey, M. D. & Harris, J. M., eds.), pp. 490–496, Clarendon Press, Oxford.

[240] Leakey, M., Feibel, C., McDougall, I., & Walker, A. (1995) New four-million-year-old hominid species from Kanapoi and Allia Bay, Kenya. *Nature* **376**: 565–571.

[241] Leakey, R. (1994) *The Origin of Humankind*. Basic Books, New York.

[242] Leakey, R. & Lewin, R. (1992) *Origins Reconsidered: In Search of What Makes us Human*. Little, Brown, London.

[243] Leakey, R. & Lewin, R. (1996) *The Sixth Extinction: Biodiversity and its Survival*. Weidenfeld & Nicolson, London.

[244] Legg, D. A., Sutton, M. D. & Edgecombe, G. D. (2013) Arthropod fossil data increase congruence of morphological and molecular phylogenies. *Nature Communications* **4**.

[245] Lewis-Williams, D. (2002) *The Mind in the Cave*. Thames and Hudson, London.

[246] Lewontin, R. C. (1972) The apportionment of human diversity. *Evolutionary Biology* **6**: 381–398.

[247] Li, H. & Durbin, R. (2011) Inference of human population history from individual whole-genome sequences. *Nature* **475**: 493–496.

[248] Liang, D., Shen, X. X. & Zhang, P. (2013) One thousand two hundred ninety nuclear genes from a genome-wide survey support lungfishes as the sister group of tetrapods. *Molecular Biology and Evolution* **30**: 1803–1807.

[249] Liem, K. F. (1973) Evolutionary strategies and morphological innovations: cichlid pharyngeal jaws. *Systematic Zoology* **22**: 425–441.

[250] Lindblad-Toh, K., Wade, C. M., Mikkelsen, T. S., *et al.* (2005) Genome sequence, comparative analysis and haplotype structure of the domestic dog. *Nature* **438**: 803–819.

[251] Liu, Y., Cotton, J.A., Shen, B., *et al.* (2010) Convergent sequence evolution between echolocating bats and dolphins. *Current Biology* **20**: R53–R54.

[252] Long, J.A., Trinajstic, K. & Johanson, Z. (2009) Devonian arthrodire embryos and the origin of internal fertilization in vertebrates. *Nature* **457**: 1124–1127.

[253] Lordkipanidze, D., Ponce de Leon, M.S., Margvelashvili, A., *et al.* (2013) A complete skull from Dmanisi, Georgia, and the evolutionary biology of early *Homo. Science* **342**: 326–331.

[254] Lorenz, K. (2002) *Man Meets Dog.* Routledge Classics, Routledge, London.

[255] Lovejoy, C.O. (1981) The origin of man. *Science* **211**: 341–350.

[256] Lovejoy, C.O. (2009) Reexamining human origins in light of *Ardipithecus ramidus. Science* **326**: 74–74, 74e1–74e8.

[257] Ludeman, D.A., Farrar, N., Riesgo, A., *et al.* (2014) Evolutionary origins of sensation in metazoans: functional evidence for a new sensory organ in sponges. *BMC Evolutionary Biology* **14**: 3.

[258] Luo, Z.-X., Cifelli, R.L., & Kielan-Jaworowska, Z. (2001) Dual origin of tribosphenic mammals. *Nature* **409**: 53–57.

[259] Luo, Z.-X., Ji, Q., Wible, J.R., *et al.* (2003) An Early Cretaceous tribosphenic mammal and metatherian evolution. *Science* **302**: 1934–1940.

[260] Luo, Z.-X., Yuan, C.-X., Meng, Q.-J., *et al.* (2011) A Jurassic eutherian mammal and divergence of marsupials and placentals. *Nature* **476**: 442–445.

[261] Macaulay, V. (2005) Single, rapid coastal settlement of Asia revealed by analysis of complete mitochondrial genomes. *Science* **308**: 1034–1036.

[262] Magiorkinis, G., Blanco-Melo, D. & Belshaw, R. (2015) The decline of human endogenous retroviruses: extinction and survival. *Retrovirology* **12**: 8.

[263] Mandelbrot, B.B. (1982) *The Fractal Geometry of Nature.* San Francisco: W.H. Freeman.

[264] Manger, P.R. & Pettigrew, J.D. (1995) Electroreception and feeding behaviour of the platypus (*Ornithorhychus anatinus:* Monotrema: Mammalia). *Philosophical Transactions of the Royal Society of London: Biological Sciences* **347**: 359–381.

[265] Margulis, L. (1981) *Symbiosis in Cell Evolution.* W.H. Freeman, San Francisco.

[266] Maricic, T., Gunther, V., Georgiev, O., *et al.* (2013) A recent evolutionary change affects a regulatory element in the human FOXP2 gene. *Molecular Biology and Evolution* **30**: 844–852.

[267] Mark Welch, D. & Meselson, M. (2000) Evidence for the evolution of bdelloid rotifers without sexual reproduction or genetic exchange. *Science* **288**: 1211–1219.

[268] Martin, R. D. (1981) Relative brain size and basal metabolic rate in terrestrial vertebrates. *Nature* **293:** 57–60.

[269] Martin, W. & Koonin, E. V. (2006) Introns and the origin of nucleus–cytosol compartmentalization. *Nature* **440:** 41–45.

[270] Mash, R. (2003/1983) *How to Keep Dinosaurs.* Weidenfeld & Nicholson, London.

[271] Mathieson, I., Lazaridis, I., Rohland, N., *et al.* (2015) Eight thousand years of natural selection in Europe. *bioRxiv.*

[272] Maynard Smith, J. (1978) *The Evolution of Sex.* Cambridge University Press, Cambridge.

[273] Maynard Smith, J. (1986) Evolution – contemplating life without sex. *Nature* **324:** 300–301.

[274] Maynard Smith, J. & Szathmàry, E. (1995) *The Major Transitions in Evolution.* Oxford University Press, Oxford.

[275] Mayr, E. (1985/1982) *The Growth of Biological Thought.* Harvard University Press, Cambridge, Mass.

[276] McBrearty, S. & Jablonski, N. G. (2005) First fossil chimpanzee. *Nature* **437:** 105–108.

[277] McDougall, I., Brown, F. H. & Fleagle, J. G. (2005) Stratigraphic placement and age of modern humans from Kibish, Ethiopia. *Nature* **433:** 733–736.

[278] McPherron, S. P., Alemseged, Z., Marean, C. W., *et al.* (2010) Evidence for stone-tool assisted consumption of animal tissues before 3.39 million years ago at Dikka, Ethiopia. *Nature* **466:** 857–860.

[279] Mendez, F. L., Krahn, T., Schrack, B., *et al.* (2013) An African American paternal lineage adds an extremely ancient root to the human Y chromosome phylogenetic tree. *The American Journal of Human Genetics* **92:** 454–459.

[280] Mendez, F. L., Veeramah, K. R., Thomas, M. G., *et al.* (2015) Reply to 'The "extremely ancient" chromosome that isn't' by Elhaik *et al. European Journal of Human Genetics* **23:** 564–567.

[281] Menon, L. R., McIlroy, D. & Brasier, M. D. (2013) Evidence for Cnidaria-like behavior in ca. 560 Ma Ediacaran *Aspidella. Geology* **41:** 895–898.

[282] Menotti-Raymond, M. & O'Brien, S. J. (1993) Dating the genetic bottleneck of the African cheetah. *Proceedings of the National Academy of Sciences of the USA* **90:** 3172–3176.

[283] Meyer, M., Fu, Q., Aximu-Petri, A., *et al.* (2013) A mitochondrial genome sequence of a hominin from Sima de los Huesos. *Nature* **505:** 403–406.

[284] Meyer, M., Kircher, M., Gansauge, M.-T., *et al.* (2012) A high-coverage genome sequence from an archaic Denisovan individual. *Science* **338:** 222–226.

[285] Milius, S. (2000) Bdelloids: No sex for over 40 million years. *Science News* **157:** 326.

[286] Miller, G. (2000) *The Mating Mind: How Sexual Choice Shaped the Evolution of Human Nature.* Heinemann, London.

[287] Miller, K.R. (1999) *Finding Darwin's God: A Scientist's Search for Common Ground Between God and Evolution.* Cliff Street Books (HarperCollins), New York.

[288] Miller, K.R. (2004) The flagellum unspun: the collapse of 'irreducible complexity'. In *Debating Design: From Darwin to DNA* (Ruse, M. & Dembski, W., eds.), Cambridge University Press, Cambridge.

[289] Mills, D.R., Peterson, R.L., & Spiegelman, S. (1967) An extracellular Darwinian experiment with a self-duplicating nucleic acid molecule. *Proceedings of the National Academy of Sciences of the USA* **58:** 217–224.

[290] Milner, A.R. & Sequeira, S.E.K. (1994) The temnospondyl amphibians from the Viséan of East Kirkton. *Transactions of the Royal Society of Edinburgh, Earth Sciences* **84:** 331–361.

[291] Mitchell, K.J., Llamas, B., Soubrier, J., *et al.* (2014) Ancient DNA reveals elephant birds and kiwi are sister taxa and clarifies ratite bird evolution. *Science* **344:** 898–900.

[292] Mitchell, K.J., Pratt, R.C., Watson, L.N., *et al.* (2014) Molecular phylogeny, biogeography, and habitat preference evolution of marsupials. *Molecular Biology and Evolution* **31:** 2322–2330.

[293] Mollon, J.D., Bowmaker, J.K., & Jacobs, G.H. (1984) Variations of colour vision in a New World primate can be explained by polymorphism of retinal photopigments. *Proceedings of the Royal Society of London: Series B* **222:** 373–399.

[294] Monod, J. (1972) *Chance and Necessity: Essay on the Natural Philosophy of Modern Biology.* Collins, London.

[295] Morgan, C.C., Foster, P.G., Webb, A.E., *et al.* (2013) Heterogeneous models place the root of the placental mammal phylogeny. *Molecular Biology and Evolution* **30:** 2145–2156.

[296] Morgan, E. (1997) *The Aquatic Ape Hypothesis.* Souvenir Press, London.

[297] Moroz, L.L., Kocot, K.M., Citarella, M.R., *et al.* (2014) The ctenophore genome and the evolutionary origins of neural systems. *Nature* **510:** 109–114.

[298] Müller, J. & Reisz, R.R. (2005) Four well-constrained calibration points from the vertebrate fossil record for molecular clock estimates. *BioEssays* **27:** 1069–1075.

[299] Murdock, G.P. (1967) *Ethnographic Atlas.* University of Pittsburgh Press, Pittsburgh.

[300] Narkiewicz, M., Grabowski, J., Narkiewicz, K., *et al.* (2015) Palaeoenvironments of the Eifelian dolomites with earliest tetrapod trackways (Holy Cross Mountains, Poland). *Palaeogeography, Palaeoclimatology, Palaeoecology* **420:** 173–192.

[301] Nesnidal, M.P., Helmkampf, M., Bruchhaus, I., *et al.* (2013) Agent of whirling disease meets orphan worm: Phylogenomic analyses firmly place Myxozoa in Cnidaria. *PLoS ONE* 8: e54576.

[302] Nesse, R.M. & Williams, G.C. (1994) *The Science of Darwinian Medicine.* Orion, London.

[303] Ni, X., Gebo, D.L., Dagosto, M., et al. (2013) The oldest known primate skeleton and early haplorhine evolution. Nature 498: 60–64.

[304] Nie, W., Fu, B., O'Brien, P.C., et al. (2008) Flying lemurs – The 'flying tree shrews'? Molecular cytogenetic evidence for a Scandentia-Dermoptera sister clade. BMC Biology 6: 18.

[305] Niedźwiedzki, G., Szrek, P., Narkiewicz, K., *et al.* (2010) Tetrapod trackways from the early Middle Devonian period of Poland. *Nature* 463: 43–48.

[306] Nilsson, M.A., Churakov, G., Sommer, M., *et al.* (2010) Tracking marsupial evolution using archaic genomic retroposon insertions. *PLoS Biology* 8: e1000436.

[307] Nishihara, H., Maruyama, S. & Okada, N. (2009) Retroposon analysis and recent geological data suggest near-simultaneous divergence of the three superorders of mammals. *Proceedings of the National Academy of Sciences* 106: 5235–5240.

[308] Norman, D. (1991) *Dinosaur!* Boxtree, London.

[309] Oaks, J.R. (2011) A time-calibrated species tree of crocodylia reveals a recent radiation of the true crocodiles. *Evolution* 65: 3285–3297.

[310] Ohta, T. (1992) The nearly neutral theory of molecular evolution. *Annual Review of Ecology and Systematics* 23: 263–286.

[311] O'Leary, M.A., Bloch, J.I., Flynn, J.J., *et al.* (2013) The placental mammal ancestor and the post-K-Pg radiation of placentals. *Science* 339: 662–667.

[312] Oparin, A.I. (1938) *The Origin of Life.* Macmillan, New York.

[313] Orgel, L.E. (1998) The origin of life – a review of facts and speculations. *Trends in Biochemical Sciences* 23: 491–495.

[314] Orgel, L.E. & Crick, F.H.C. (1980) Selfish DNA: the ultimate parasite. *Nature* 284: 604–607.

[315] Orlando, L. & Cooper, A. (2014) Using ancient DNA to understand evolutionary and ecological processes. *Annual Review of Ecology, Evolution, and Systematics* 45: 573–598.

[316] Orlando, L., Ginolhac, A., Zhang, G., *et al.* (2013) Recalibrating *Equus* evolution using the genome sequence of an early Middle Pleistocene horse. *Nature* 499: 74–78.

[317] Ota, K.G., Fujimoto, S., Oisi, Y., *et al.* (2011) Identification of vertebra-like elements and their possible differentiation from sclerotomes in the hagfish. *Nature Communications* 2: 373.

[318] Pääbo, S. (2014) *Neanderthal Man: In Search of Lost Genomes.* New York, Basic Books.

[319]   Pagel, M. & Bodmer, W. (2003) A naked ape would have fewer parasites. *Proceedings of the Royal Society of London: Biological Sciences (Suppl.)* **270**: S117–S119.

[320]   Panchen, A. L. (2001) Étienne Geoffroy St.-Hilaire: Father of 'evo-devo'? *Evolution and Development* **3**: 41–46.

[321]   Parfrey, L. W., Lahr, D. J. G., Knoll, A. H., *et al.* (2011) Estimating the timing of early eukaryotic diversification with multigene molecular clocks. *Proceedings of the National Academy of Sciences* **108**: 13624–13629.

[322]   Parker, A. (2003) *In the Blink of an Eye: The Cause of the Most Dramatic Event in the History of Life.* Free Press, London.

[323]   Parker, J., Tsagkogeorga, G., Cotton, J. A., *et al.* (2013) Genome-wide signatures of convergent evolution in echolocating mammals. *Nature* **502**: 228–231.

[324]   Partridge, T. C., Granger, D. E., Caffee, M. W., & Clarke, R. J. (2003) Lower Pliocene hominid remains from Sterkfontein. *Science* **300**: 607–612.

[325]   Patel, B. H., Percivalle, C., Ritson, D. J., *et al.* (2015) Common origins of RNA, protein and lipid precursors in a cyanosulfidic protometabolism. *Nature Chemistry* **7**: 301–307.

[326]   Penfield, W. & Rasmussen, T. (1950) *The Cerebral Cortex of Man: A Clinical Study of Localization of Function.* Macmillan, New York.

[327]   Peng, Z., Ludwig, A., Wang, D., *et al.* (2007) Age and biogeography of major clades in sturgeons and paddlefishes (Pisces: Acipenseriformes). *Molecular Phylogenetics and Evolution.* **42**: 854–862.

[328]   Perdeck, A. C. (1957) The isolating value of specific song patterns in two sibling species of grasshoppers. *Behaviour* **12**: 1–75.

[329]   Peterson, K. J., Cotton, J. A., Gehling, J. G., *et al.* (2008) The Ediacaran emergence of bilaterians: congruence between the genetic and the geological fossil records. *Philosophical Transactions of the Royal Society B: Biological Sciences* **363**: 1435–1443.

[330]   Pettigrew, J. D., Manger, P. R., & Fine, S. L. B. (1998) The sensory world of the platypus. *Philosophical Transactions of the Royal Society of London: Biological Sciences* **353**: 1199–1210.

[331]   Pierce, S. E., Clack, J. A. & Hutchinson, J. R. (2012) Three-dimensional limb joint mobility in the early tetrapod *Ichthyostega*. *Nature* **486**: 523–526.

[332]   Pinker, S. (1994) *The Language Instinct: The New Science of Language and Mind.* Allen Lane, The Penguin Press, London.

[333]   Pinker, S. (1997) *How the Mind Works.* Norton, New York.

[334]   Pisani, D., Feuda, R., Peterson, K. J., *et al.* (2012) Resolving phylogenetic signal from noise when divergence is rapid: A new look at the old problem of echinoderm class relationships. *Molecular Phylogenetics and Evolution* **62**: 27–34.

[335]   Podar, M., Haddock, S. H., Sogin, M. L., *et al.* (2001) A molecular phylogenetic framework for the phylum Ctenophora using 18S rRNA genes. *Molecular Phylogenetics and Evolution.* **21**: 218–230.

[336] Poulakakis, N. & Stamatakis, A. (2010) Recapitulating the evolution of Afrotheria: 57 genes and rare genomic changes (RGCs) consolidate their history. *Systematics and Biodiversity* **8**: 395–408.

[337] Poux, C., Madsen, O., Marquard, E., *et al.* (2005) Asynchronous colonization of Madagascar by the four endemic clades of primates, tenrecs, carnivores, and rodents as inferred from nuclear genes. *Systematic Biology* **54**: 719–730.

[338] Prüfer, K., Munch, K., Hellmann, I., *et al.* (2012) The bonobo genome compared with the chimpanzee and human genomes. *Nature* **486**: 527–531.

[339] Prüfer, K., Racimo, F., Patterson, N., *et al.* (2013) The complete genome sequence of a Neanderthal from the Altai Mountains. *Nature* **505**: 43–49.

[340] Pullman, P. (2001) *His Dark Materials Trilogy.* Scholastic Press, London.

[341] Pyron, R. A. (2011) Divergence time estimation using fossils as terminal taxa and the origins of Lissamphibia. *Systematic Biology* **60**: 466–481.

[342] Pyron, R. A., Kandambi, H. K. D., Hendry, C. R., *et al.* (2013) Genus-level phylogeny of snakes reveals the origins of species richness in Sri Lanka. *Molecular Phylogenetics and Evolution.* **66**: 969–978.

[343] Ralph, P. & Coop, G. (2013) The geography of recent genetic ancestry across Europe. *PLoS Biology* **11**: e1001555.

[344] Ramsköld, L. (1992) The second leg row of *Hallucigenia* discovered. *Lethaia* **25**: 221–224.

[345] Rands, C. M., Meader, S., Ponting, C. P., *et al.* (2014) 8.2% of the human genome is constrained: Variation in rates of turnover across functional element classes in the human lineage. *PLoS Genetics* **10**: e1004525.

[346] Reader, J. (1988) *Man on Earth.* Collins, London.

[347] Reader, J. (1998) *Africa: A Biography of the Continent.* Penguin, London.

[348] Rebek, J. (1994) Synthetic self-replicating molecules. *Scientific American* **271** (July): 48–55.

[349] Rees, M. (1999) *Just Six Numbers.* Science Masters, Weidenfeld & Nicolson, London.

[350] Reich, D., Green, R. E., Kircher, M., *et al.* (2010) Genetic history of an archaic hominin group from Denisova Cave in Siberia. *Nature* **468**: 1053–1060.

[351] Reich, D., Patterson, N., Kircher, M., *et al.* (2011) Denisova admixture and the first modern human dispersals into Southeast Asia and Oceania. *The American Journal of Human Genetics* **89**: 516–528.

[352] Richardson, M. K. & Keuck, G. (2002) Haeckel's ABC of evolution and development. *Biological Reviews* **77**: 495–528.

[353] Ridley, Mark (1983) *The Explanation of Organic Diversity – The Comparative Method and Adaptations for Mating.* Clarendon Press/Oxford University Press, Oxford.

[354] Ridley, Mark (1986) Embryology and classical zoology in Great Britain. In *A History of Embryology: The Eighth Symposium of the British Society for Developmental Biology* (Horder, T. J., Witkowski, J., & Wylie, C. C., eds.), pp. 35–67, Cambridge University Press, Cambridge.

[355] Ridley, Mark (2000) *Mendel's Demon: Gene Justice and the Complexity of Life.* Weidenfeld & Nicolson, London.

[356] Ridley, Matt (1993) *The Red Queen: Sex and the Evolution of Human Nature.* Viking, London.

[357] Ridley, Matt (2003) *Nature Via Nurture: Genes, Experience and What Makes Us Human.* Fourth Estate, London.

[358] Rinke, C., Schwientek, P., Sczyrba, A., *et al.* (2013) Insights into the phylogeny and coding potential of microbial dark matter. *Nature* **499**: 431–437.

[359] Rogers, A. R., Iltis, D. & Wooding, S. (2004) Genetic variation at the MC1R locus and the time since loss of human body hair. *Current Anthropology* **45**: 105–108.

[360] Rohde, D. L. T., Olson, S. & Chang, J. T. (2004) Modelling the recent common ancestry of all living humans. *Nature* **431**: 562–566.

[361] Rokas, A. & Holland, P. W. H. (2000) Rare genomic changes as a tool for phylogenetics. *Trends in Ecology and Evolution* **15**: 454–459.

[362] Romiguier, J., Ranwez, V., Delsuc, F., *et al.* (2013) Less is more in mammalian phylogenomics: AT-rich genes minimize tree conflicts and unravel the root of placental mammals. *Molecular Biology and Evolution* **30**: 2134–2144.

[363] Rooney, A. D., Strauss, J. V., Brandon, A. D., *et al.* (2015) A Cryogenian chronology: Two long-lasting synchronous Neoproterozoic glaciations. *Geology* **43**: 459–462.

[364] Rosindell, J. & Harmon, L. J. (2012) OneZoom: A fractal explorer for the tree of life. *PLoS Biology* **10**: e1001406.

[365] Ruiz-Trillo, I. & Paps, J. (2015) Acoelomorpha: Earliest branching bilaterans or deuterostomes? *Organisms Diversity & Evolution.*

[366] Ruppert, E. E. & Barnes, R. D. (1994) *Invertebrate Zoology.* Saunders College Publishing, Fort Worth, 6th ed.

[367] Sacks, O. (1996) *The Island of the Colour-blind and Cycad Island.* Picador, London.

[368] Saffhill, R., Schneider-Bernloer, H., Orgel, L. E., & Spiegelman, S. (1970) *In vitro* selection of bacteriophage Q ribonucleic acid variants resistant to ethidium bromide. *Journal of Molecular Biology* **51**: 531–539.

[369] Sankararaman, S., Mallick, S., Dannemann, M., *et al.* (2014) The genomic landscape of Neanderthal ancestry in present-day humans. *Nature* **507**: 354–357.

[370] Sansom, I. J., Davies, N. S., Coates, M. I., *et al.* (2012) Chondrichthyan-like scales from the Middle Ordovician of Australia. *Palaeontology* **55**: 243–247.

[371] Sarich, V.M. & Wilson, A.C. (1967) Immunological time scale for hominid evolution. *Science* **158**: 1200–1203.

[372] Scally, A. & Durbin, R. (2012) Revising the human mutation rate: implications for understanding human evolution. *Nature Reviews Genetics* **13**: 745–753.

[373] Scally, A., Dutheil, J.Y., Hillier, L.W., *et al.* (2012) Insights into hominid evolution from the gorilla genome sequence. *Nature* **483**: 169–175.

[374] Schiffels, S. & Durbin, R. (2014) Inferring human population size and separation history from multiple genome sequences. *Nature Genetics* **46**: 919–925.

[375] Schluter, D. (2000) *The Ecology of Adaptive Radiation*. Oxford University Press, Oxford.

[376] Schreiweis, C., Bornschein, U., BurguiÈre, E., *et al.* (2014) Humanized Foxp2 accelerates learning by enhancing transitions from declarative to procedural performance. *Proceedings of the National Academy of Sciences* **111**: 14253–14258.

[377] Schweitzer, M.H., Zheng, W., Organ, C.L., *et al.* (2009) Biomolecular characterization and protein sequences of the campanian hadrosaur *B. canadensis*. *Science* **324**: 626–631.

[378] Scotese, C.R. (2001) *Atlas of Earth History*, vol. 1, Palaeography. PALEOMAP project, Arlington, Texas.

[379] Seehausen, O. & van Alphen, J.J.M. (1998) The effect of male coloration on female mate choice in closely related Lake Victoria cichlids (*Haplochromis nyererei* complex). *Behavioral Ecology and Sociobiology* **42**: 1–8.

[380] Segurel, L., Thompson, E.E., Flutre, T., *et al.* (2012) The ABO blood group is a trans-species polymorphism in primates. *Proceedings of the National Academy of Sciences* **109**: 18493–18498.

[381] Senut, B., Pickford, M., Gommery, D., *et al.* (2001) First hominid from the Miocene (Lukeino Formation, Kenya). *Comptes Rendus de l'Academie des Sciences, Series IIA – Earth and Planetary Science* **332**: 137–144.

[382] Sepkoski, J.J. (1996) Patterns of Phanerozoic extinction: A perspective from global databases. In *Global Events and Event Stratigraphy in the Phanerozoic* (Walliser, O.H., ed.), pp. 35–51, Springer-Verlag, Berlin.

[383] Seton, M., Müller, R.D., Zahirovic, S., *et al.* (2012) Global continental and ocean basin reconstructions since 200Ma. *Earth-Science Reviews* **113**: 212–270.

[384] Shalchian-Tabrizi, K., Minge, M.A., Espelund, M., *et al.* (2008) Multigene phylogeny of choanozoa and the origin of animals. *PLoS ONE* **3**: e2098.

[385] Shapiro, B., Sibthorpe, D., Rambaut, A., *et al.* (2002) Flight of the dodo. *Science* **295**: 1683.

[386] Sheets-Johnstone, M. (1990) *The Roots of Thinking*. Temple University Press, Philadelphia.

[387] Shimeld, S. M. & Donoghue, P. C. J. (2012) Evolutionary crossroads in developmental biology: cyclostomes (lamprey and hagfish). *Development* **139**: 2091–2099.

[388] Shu, D.-G., Luo, H.-L., Conway-Morris, S., *et al.* (1999) Lower Cambrian vertebrates from south China. *Nature* **402**: 42–46.

[389] Simpson, G. G. (1980) *Splendid Isolation: The Curious History of South American Mammals.* Yale University Press, New Haven.

[390] Sistiaga, A., Mallol, C., Galván, B., *et al.* (2014) The Neanderthal meal: a new perspective using faecal biomarkers. *PLoS ONE* **9**: e101045.

[391] Slack, J. M. W., Holland, P. W. H., & Graham, C. F. (1993) The zootype and the phylotypic stage. *Nature* **361**: 490–492.

[392] Smith, C. L., Varoqueaux, F., Kittelmann, M., *et al.* (2014) Novel cell types, neurosecretory cells, and body plan of the early-diverging metazoan *Trichoplax adhaerens.* *Current Biology* **24**: 1565–1572.

[393] Smith, D. C. (1979) From extracellular to intracellular: The establishment of a symbiosis. In *The Cell as a Habitat* (Richmond, M. H. & Smith, D. C., eds.), Royal Society of London, London.

[394] Smith, M. R. & Caron, J.-B. (2015) Hallucigenia's head and the pharyngeal armature of early ecdysozoans. *Nature* **523**: 75–78.

[395] Smolin, L. (1997) *The Life of the Cosmos.* Weidenfeld & Nicolson, London.

[396] Southwood, T. R. E. (2003) *The Story of Life.* Oxford University Press, Oxford.

[397] Spang, A., Saw, J. H., Jørgensen, S. L., *et al.* (2015) Complex archaea that bridge the gap between prokaryotes and eukaryotes. *Nature* **521**: 173–179.

[398] Sperling, E. A., Peterson, K. J. & Pisani, D. (2009) Phylogenetic-signal dissection of nuclear housekeeping genes supports the paraphyly of sponges and the monophyly of eumetazoa. *Molecular Biology and Evolution* **26**: 2261–2274.

[399] Sperling, E. A., Robinson, J. M., Pisani, D., *et al.* (2010) Where's the glass? Biomarkers, molecular clocks, and microRNAs suggest a 200-Myr missing Precambrian fossil record of siliceous sponge spicules: Sponge biomarkers, molecular clocks and microRNAs. *Geobiology* **8**: 24–36.

[400] Springer, M. S., Meredith, R. W., Gatesy, J., *et al.* (2012) Macroevolutionary dynamics and historical biogeography of primate diversification inferred from a species supermatrix. *PLoS ONE* **7**: e49521.

[401] Srivastava, M., Begovic, E., Chapman, J., *et al.* (2008) The *Trichoplax* genome and the nature of placozoans. *Nature* **454**: 955–960.

[402] Stajich, J. E., Berbee, M. L., Blackwell, M., *et al.* (2009) The Fungi. *Current Biology* **19**: R840–R845.

[403] Stewart, C. B. & Disotell, T. R. (1998) Primate evolution – in and out of Africa. *Current Biology* **8**: R582–R588.

[404] Storz, J. F., Opazo, J. C. & Hoffmann, F. G. (2013) Gene duplication, genome duplication, and the functional diversification of vertebrate globins. *Molecular Phylogenetics and Evolution* **66**: 469–478.

[405] Suga, H., Chen, Z., de Mendoza, A., *et al.* (2013) The *Capsaspora* genome reveals a complex unicellular prehistory of animals. *Nature Communications* **4**: 131.

[406] Sumper, M. & Luce, R. (1975) Evidence for *de novo* production of self-replicating and environmentally adapted RNA structures by bacteriophage Qβ replicase. *Proceedings of the National Academy of Sciences of the USA* **72**: 162–166.

[407] Sutherland, J. L. (1933) Protozoa from Australian termites. *Quarterly Journal of Microscopic Science* **76**: 145–173.

[408] Suzuki, D., Brandley, M. C. & Tokita, M. (2010) The mitochondrial phylogeny of an ancient lineage of ray-finned fishes (Polypteridae) with implications for the evolution of body elongation, pelvic fin loss, and craniofacial morphology in Osteichthyes. *BMC Evolutionary Biology* **10**: 209.

[409] Swift, J. (1733) *Poetry, A Rhapsody.*

[410] Syed, T. & Schierwater, B. (2002) *Trichoplax adhaerens:* discovered as a missing link, forgotten as a hydrozoan, rediscovered as a key to metazoan evolution. *Vie et Milieu* **52**: 177–187.

[411] Takezaki, N. (2004) The phylogenetic relationship of tetrapod, coelacanth, and lungfish revealed by the sequences of forty-four nuclear genes. *Molecular Biology and Evolution* **21**: 1512–1524.

[412] Tamm, S. L. (1982) Flagellated endosymbiotic bacteria propel a eukaryotic cell. *Journal of Cell Biology* **94**: 697–709.

[413] Taylor, C. R. & Rowntree, V. J. (1973) Running on two or four legs: which consumes more energy? *Science* **179**: 186–187.

[414] The Chimpanzee Sequencing and Analysis Consortium. (2005) Initial sequence of the chimpanzee genome and comparison with the human genome. *Nature* **437**: 69–87.

[415] Thomson, K. S. (1991) *Living Fossil: The Story of the Coelacanth.* Hutchinson Radius, London.

[416] Toups, M. A., Kitchen, A., Light, J. E., *et al.* (2011) Origin of clothing lice indicates early clothing use by anatomically modern humans in Africa. *Molecular Biology and Evolution* **28**: 29–32.

[417] Trivers, R. L. (1972) Parental investment and sexual selection. In *Sexual Selection and the Descent of Man* (Campbell, B., ed.), pp. 136–179, Aldine, Chicago.

[418] Trut, L. N. (1999) Early canid domestication: The farm-fox experiment. *American Scientist* **87**: 160–169.

[419] Tsagkogeorga, G., Turon, X., Hopcroft, R. R., *et al.* (2009) An updated 18S rRNA phylogeny of tunicates based on mixture and secondary structure models. *BMC Evolutionary Biology* **9**: 187.

[420] Tschopp, E., Mateus, O. & Benson, R.B.J. (2015) A specimen-level
       phylogenetic analysis and taxonomic revision of Diplodocidae (Dino-
       sauria, Sauropoda). *PeerJ* **3**: e857.

[421] Tudge, C. (1998) *Neanderthals, Bandits and Farmers: How Agriculture
       Really Began*. Weidenfeld & Nicolson, London.

[422] Tudge, C. (2000) *The Variety of Life*. Oxford University Press, Oxford.

[423] Vaidya, N., Manapat, M.L., Chen, I.A., *et al.* (2012) Spontaneous
       network formation among cooperative RNA replicators. *Nature* **491**:
       72–77.

[424] van Iten, H., Marques, A.C., Leme, J. de M., *et al.* (2014) Origin and
       early diversification of the phylum Cnidaria Verrill: major develop-
       ments in the analysis of the taxon's Proterozoic-Cambrian history.
       Palaeontology **57**: 677–690.

[425] van Schaik, C.P., Ancrenaz, M., Borgen, G., *et al.* (2003) Orangutan
       cultures and the evolution of material culture. *Science* **299**: 102–105.

[426] Varricchio, D.J., Martin, A.J. & Katsura, Y. (2007) First trace and body
       fossil evidence of a burrowing, denning dinosaur. *Proceedings of the
       Royal Society B: Biological Sciences* **274**: 1361–1368.

[427] Verheyen, E., Salzburger, W., Snoeks, J., & Meyer, A. (2003) Origin of
       the superflock of cichlid fishes from Lake Victoria, East Africa. *Science*
       300: 325–329.

[428] Vernot, B. & Akey, J.M. (2014) Resurrecting surviving Neanderthal
       lineages from modern human genomes. *Science* **343**: 1017–1021.

[429] Villmoare, B., Kimbel, W.H., Seyoum, C., *et al.* (2015) Early Homo at
       2.8 Ma from Ledi-Geraru, Afar, Ethiopia. *Science* **347**: 1352–1355.

[430] Vine, F.J. & Matthews, D.H. (1963) Magnetic anomalies over oceanic
       ridges. *Nature* **199**: 947–949.

[431] Wacey, D., Kilburn, M.R., Saunders, M., *et al.* (2011) Microfossils of
       sulphur-metabolizing cells in 3.4-billion-year-old rocks of Western
       Australia. *Nature Geoscience* **4**: 698–702.

[432] Wake, D.B. (1997) Incipient species formation in salamanders of the
       Ensatina complex. *Proceedings of the National Academy of Sciences of the
       USA* **94**: 7761–7767.

[433] Ward, C.V., Walker, A., & Teaford, M.F. (1991) *Proconsul* did not have
       a tail. *Journal of Human Evolution* **21**: 215–220.

[434] Weinberg, S. (1993) *Dreams of a Final Theory*. Hutchinson Radius,
       London.

[435] Weiner, J. (1994) *The Beak of the Finch*. Jonathan Cape, London.

[436] Welker, F., Collins, M.J., Thomas, J.A., *et al.* (2015) Ancient proteins
       resolve the evolutionary history of Darwin's South American ungulates.
       Nature **522**: 81–84.

[437] Wesenberg-Lund, C. (1930) Contributions to the biology of the Roti-
       fera. Part II. The periodicity and sexual periods. *Det Kongelige Danske
       Videnskabers Selskabs Skrifter* **9, II**: 1–230.

[438] West, G. B., Brown, J. H., & Enquist, B. J. (2000) The origin of universal scaling laws in biology. In *Scaling in Biology* (Brown, J. H. & West, G. B., eds.), Oxford University Press, Oxford.

[439] West-Eberhard, M. J. (2003) *Developmental Plasticity and Evolution.* Oxford University Press, New York.

[440] Wheeler, J. A. (1990) Information, physics, quantum: The search for links. In *Complexity, Entropy, and the Physics of Information* (Zurek, W. H., ed.), pp. 3–28, Addison-Wesley, New York.

[441] White, T. D., Asfaw, B., DeGusta, D., *et al.* (2003) Pleistocene *Homo sapiens* from Middle Awash, Ethiopia. *Nature* **423:** 742–747.

[442] White, T. D., Lovejoy, C. O., Asfaw, B., *et al.* (2015) Neither chimpanzee nor human, *Ardipithecus* reveals the surprising ancestry of both. *Proceedings of the National Academy of Sciences* **112:** 4877–4884.

[443] Whiten, A., Goodall, J., McGrew, W. C., *et al.* (1999) Cultures in chimpanzees. *Nature* **399:** 682–685.

[444] Williams, G. C. (1975) *Sex and Evolution.* Princeton University Press, Princeton, N.J.

[445] Williams, G. C. (1992) *Natural Selection: Domains, Levels and Challenges.* Oxford University Press, Oxford.

[446] Williams, T. A., Foster, P. G., Cox, C. J., *et al.* (2013) An archaeal origin of eukaryotes supports only two primary domains of life. *Nature* **504:** 231–236.

[447] Wilson, E. O. (1992) *The Diversity of Life.* Harvard University Press, Cambridge, Mass.

[448] Wilson, H. V. (1907) On some phenomena of coalescence and regeneration in sponges. *Journal of Experimental Zoology* **5:** 245–258.

[449] Woese, C. R., Kandler, O., & Wheelis, M. L. (1990) Towards a natural system of organisms: Proposal for the domains Archaea, Bacteria, and Eucarya. *Proceedings of the National Academy of Sciences of the USA* **87:** 4576–4579.

[450] Wolpert, L. (1991) *The Triumph of the Embryo.* Oxford University Press, Oxford.

[451] Wolpert, L., Tickle, C., Arias, A. M., *et al.* (2015) *Principles of Development.* Oxford University Press, Oxford, 5th edn.

[452] Wray, G. A., Levinton, J. S., & Shapiro, L. H. (1996) Molecular evidence for deep Precambrian divergences among metazoan phyla. *Science* **274:** 568–573.

[453] Xiao, S. H., Yuan, X. L., & Knoll, A. H. (2000) Eumetazoan fossils in terminal Proterozoic phosphorites? *Proceedings of the National Academy of Sciences of the USA* **97:** 13684–13689.

[454] Yeats, W. B. (1984) *The Poems* (Finneran, R. J., ed.). Macmillan, London.

[455] Yin, Z., Zhu, M., Davidson, E. H., *et al.* (2015) Sponge grade body fossil with cellular resolution dating 60 Myr before the Cambrian. *Proceedings of the National Academy of Sciences* **112:** E1453–E1460.

[456] Zahavi, A. & Zahavi, A. (1997) *The Handicap Principle*. Oxford University Press, Oxford.

[457] Zhu, M. & Yu, X. (2002) A primitive fish close to the common ancestor of tetrapods and lungfish. *Nature* **418:** 767–770.

[458] Zintzen, V., Roberts, C.D., Anderson, M.J., *et al.* (2011) Hagfish predatory behaviour and slime defence mechanism. *Scientific Reports* **1:** 131.

[459] Zuckerkandl, E. & Pauling, L. (1965) Evolutionary divergence and convergence in proteins. In *Evolving Genes and Proteins* (Bryson, V. & Vogels, H.J., eds.), pp. 97–166, Academic Press, New York.

ILLUSTRATION CREDITS

Multiple images in one plate are credited clockwise from top left.

22 CONCESTOR 18, © Moonrunner Design.

23 TETRAPOD FOOTPRINTS, Per Ahlberg, Uppsala University.

24 COELACANTH, Peter Scoones/Science Photo Library.

25 DEEP SEA SNIPE EEL, Peter Parks/imagequestmarine.com; DEEP SEA BLACK SWALLOWER, Peter Herring/imagequestmarine.com; PLAICE, © Linda Pitkin/NHPA; OCEAN SUNFISH, Masa Ushioda/imagequest marine.com; LEAFY SEA DRAGON, Paul Zahl/Science Photo Library; DEEP-SEA GULPER, © Norbert Wu/Minden Pictures.

26 UNROOTED HAPLOTYPE NETWORK OF THE HAPLOCHROMINE SUPERFLOCK from Verheyen *et al.* Reprinted with permission from [427], fig 3c. Copyright 2003 AAAS.

27 CICHILIDS, *PUNDAMILIA PUNDAMILIA* AND *PUNDAMILIA NYER-EREI*, Kevin Bauman, www.african-cichlid.com.

28 ELEPHANT SHARK OR PLOUGH-NOSED CHIMAERA, © ANT Photo Library/NHPA.

29 GREAT HAMMERHEAD SHARK, James, D. Watt/imagequestmarine.com; SMALLTOOTH SAWFISH, © Yves Lanceau/NHPA.

30 SEA SQUIRTS, © B. Jones & M. Shimlock/NHPA.

31 CUSHION SEA STAR, Bob Peterson.

32 CONCESTOR 23, © Moonrunner Design.

33 CONCESTOR 26, © Moonrunner Design.

34 LEAF CUTTER ANTS, © Michael & Patricia Fogden/Minden Pictures.

35 CONDOLEEZZA RICE, COLIN POWELL, GEORGE W. BUSH AND DONALD H. RUMSFELD, Associated Press.

36 DROSOPHILA MELANOGASTER from P. A. Lawrence, *The Making of a Fly*, Blackwell Science (1992), plate 5.1.

37 LIGHT MICROGRAPH OF A ROTIFER, John Walsh/Science Photo Library.

38 VELVET WORM, Dr Morley Read/Science Photo Library.

39 *DICKINSONIA COSTATA*, all rights reserved by J. G. Gehling, South Australian Museum.

40 *PORPITA PORPITA*, Peter Parks/imagequestmarine.com.

41 MASS OF MASTIGIAS JELLYFISH AT SURFACE PALAU, WESTERN PACIFIC ISLANDS, Michael Pitts/naturepl.com.

42 AERIAL VIEW OF HERON ISLAND (CORAL ISLAND), GREAT BARRIER REEF MARINE PARK, © Gerry Ellis/Minden Pictures.

43 CLEANER WRASSE CLEANING ROSY GOATFISH, Georgette Douwma/naturepl.com.

44 VENUS'S GIRDLE, A TYPE OF COMB JELLY, Sinclair Stammers/Science Photo Library.

45 COMB JELLY, Alexander Semenov/Science Photo Library

46 CONCESTOR 31, © Moonrunner Design.

47 GLASS SPONGE, © The Natural History Museum, London.

48 STINKHORN MUSHROOMS, Vaughan Fleming/Science Photo Library.

49 CONCESTOR 38, © Moonrunner Design.

50 SEQUOIA TREE, SEQUOIA NATIONAL PARK, Tony Craddock/Science Photo Library.

51 TRANSFER RNA MOLECULE, COMPUTER ARTWORK, Alfred Pasieka/ Science Photo Library.

52 SIDE VIEW OF *NAUTILUS POMPILIUS* SHOWING 'PIN-HOLE' EYE, © The Natural History Museum, London; COLOURED SCANNING ELECTRON MICROGRAPH OF TRILOBITE EYE FOSSIL, VVG/Science Photo Library; SCANNING ELECTRON MICROSCOPE IMAGE SHOWING THE COMPOUND EYE OF A SIMULIAN BLACKFLY, © The Natural History Museum, London; STOPLIGHT PARROTFISH EYE, © B. Jones & M. Shimlock/NHPA; GREAT HORNED OWL EYE, Simon Fraser/ Science Photo Library.

Page

15 Published with kind permission of the International Commission on Stratigraphy.

55 VICTORIA FAMILY TREE by Yan Wong.

68 RICHARD DAWKINS GENOME DATA by Yan Wong.

69 MOST RECENT COALESCENCE OF NINE INDIGENOUS HUMANS by Stephan Schiffels, from [374].

95 SCATTER GRAPH OF LOG BRAIN WEIGHT AGAINST LOG BODY WEIGHT, from R. D. Martin *et al.* [268], fig 1. Reproduced with kind permission of Nature Publishing Group.

101 EQ OR 'BRAININESS' INDEX by Yan Wong.

102 COMPARISON OF SKULLS by Yan Wong

108 SKELETAL RECONSTRUCTION OF *ARDIPITHECUS RAMIDUS*, © J. H. Matternes.

141 HISTORICAL ENGRAVING OF APELIKE HUMANS, Science Photo Library.

147 A SYNTHETIC HYPOTHESIS OF CATARRHINE PRIMATE EVOLUTION. Reprinted from C. B. Stewart & T. R. Disotell [403], fig. 2. Copyright 1998, with permission from Elsevier.

155, 156 UNROOTED TREES by Yan Wong; ROOTED CLADOGRAMS by Yan Wong.

161, 162, 165 CHAUCER DIAGRAMS by Yan Wong based on data from [66].

166 'NONCOMMUNICATORY' DATA DIAGRAM from T. Geissmann [148], fig. 1.6. Reprinted by permission of Wiley-Liss, Inc., a subsidiary of John Wiley & Sons, Inc.

167 UNROOTED MAXIMUM LIKELIHOOD TREE. Adapted from L. Carbone *et al.* [56].

169 GENETIC RELATIONSHIPS BETWEEN GIBBONS. Adapted from L. Carbone *et al.* [56].

193 TARSIER SKELETON by Stephen D. Nash, from Primate Adaptation and Evolution by John G. Fleagle, 2nd ed., Academic Press, San Diego (1999), Fig. 4.25.

239 TREE, from E. Haeckel [170].

243 GRAPH based on data from R. D. Alexander *et al.* [5].

247 TESTIS MASS AGAINST BODY MASS from P. H. Harvey & M. D. Pagel [185], fig. 1.3.

249 GLYPTODONT RECONSTRUCTION, Science Photo Library.

262 SEA FLOOR SPREADING DIAGRAM redrawn by Ken Wilson.

266 BREAKUP OF GONDWANA, adapted from data from [383] and www.gplates.org.

282 HENKELOTHERIUM by Elke Groëning, from B. Krebs, 'Das Skelt von Henkelotherium guimarotae gen. et. sp. nov (*Eupantotheria, Mammalia*) aus dem Oberen Jura von Portugal', *Berliner Geowiss* (1991), Abh. A 133: 1–110.

285 PENFIELD BRAIN DIAGRAM from [326]; PLATYPUNCULUS from J. D. Pettigrew, P. R. Manger & S. L. B. Fine, [330], fig. 7a.

287 PLATYPUS ISO-SENSITIVITY LINES from P. R. Manger & J. D. Pettigrew [264], fig. 13e.

292 STAR-NOSED MOLE, Rod Planck/Science Photo Library.

294 MOLUNCULUS from K. C. Catania & J. H. Kaas [60]. Reprinted by permission of Wiley-Liss, Inc., a subsidiary of John Wiley & Sons, Inc.

297 EXTINCTION RATES DIAGRAM from J. J. Sepkoski [382].

300 MAMMAL-LIKE REPTILES DIAGRAM from T. Kemp [218].

301 GORGONOPSID SKULL by Tom Kemp.

311 MAIZE OIL GRAPH from M. Ridley, Evolution, 3rd Edn. Blackwell (2003), fig. 9.7. Modified from J. W. Dudley & R. L. Lambert [117].

334 RICHARD OWEN, BRITISH ZOOLOGIST, WITH A GIANT MOA, George Bernard/Science Photo Library.

336 RATITE EVOLUTIONARY TREE adapted from Mitchell *et al.* [291].

350 BELL CURVE DIAGRAM and DISTRIBUTION DIAGRAM by Richard Dawkins.

378 RATES OF PROTEIN EVOLUTION IN VERTEBRATES, adapted from D. Liang [248].

399 TWO WAYS OF BEING A FLATFISH, by L. Ward, from R. Dawkins [96], fig. 4.7.

438 BRACHIOPOD, © Natural History Museum, London.

447 UPSIDE-DOWN CATFISH, © Slip Nicklin/Minden Pictures; APOLLO 11 IMAGE OF MOON CRATERS, NASA/DvR/Science Photo Library.

455 COLIN POWELL AND DANIEL ARAP MOI, Associated Press.

493 *THAUMATOXENA ANDREINII SILVESTRI*, from R. H. L. Disney and D. H. Kistner, 'Revision of the termitophilous Thaumatoxeninae (Diptera: Phoridae)', *Journal of Natural History* (1992) 26: 953–991; megaselia scalaris (Loew) by Arthur Smith, © Natural History Museum, London.

495 *HALLUCIGENIA* RECONSTRUCTION reprinted by kind permission of Simon Conway Morris, University of Cambridge. From [72], fig. 19a by permission of Oxford University Press.

496 *HALLUCIGENIA* RECONSTRUCTION, by kind permission of David Legg, Oxford University Museum of Natural History.

497 ANIMALOCARIS RECONSTRUCTION, by kind permission of Dr Derek J. Siveter, Oxford University Museum of Natural History. From [197].

498 HALKIERIA reprinted by kind permission of Simon Conway-Morris, University of Cambridge. From S. Conway-Morris and J. S. Peel, 'Articulated Halkieriids from the Lower Cambrian of North Greenland and Their Role in Early Protostome Evolution', Phil. Trans. Roy. Soc. Lond. (1995), B 347: 305–358, figs. 49a, b, c.

499 MYLLOKUNMINGIA SPECIMEN AND DIAGRAM from D-G Shu *et al.* [388], figs. 2a and 3. Reproduced with kind permission of Nature Publishing Group.

524 CNIDARIAN HARPOON, courtesy of BIODIDAC (biodidac.bio.uottawa.ca).

549 DIAGRAM OF PORTION OF SPONGE WALL from R. C. Brusca & G. J. Brusca, Invertebrates, 2nd ed., Sinauer Associates, Inc. (2003). Drawn from original by Tim Brown.

551 CHOANOFLAGELLATE from Haeckel, *Art Forms in Nature*, 1904

555 *PROTEROSPONGIA* from B. S. C. Leadbeater, 'Life-history and ultra-structure of a new marine species of *Proterospongia*', Jour. Mar. Biol. Assoc. UK (1983), 63:135–60.

574 TREE OF EURYKAROTIC LIFE, adapted with permission from F. Burki [50].

580 ARTWORK OF RADIOLARIANS (from Ernst Haeckel), Scott Camazine/ Science Photo Library.

581 POSTCARD FROM THE EXPOSITION UNIVERSELLE, 1900, from akg-images.

585 KLIEBER'S LAW SCATTER PLOT from West, Brown & Enquist [438], fig. 2. From diagram in Hemmingsen, 'Energy meta as related to body size', Rep. Steno. Mem. Hosp., Copenhagen: 1–110 (1960). Drawn from original by Tim Brown.

587 CROSS-SECTION OF CAULIFLOWER by Yan Wong.

613 DIAGRAM OF MIXOTRICHA from L. R. Cleveland & A. V. Grimstone [70].

631 BACTERIAL FLAGELLAR MOTOR from D. Voet & J. G. Voet, Biochemistry, John Wiley & Sons, Inc. (1994), p. 1259. Drawn from original by Tim Brown.

638 MODERN VIEW OF THE TREE OF LIFE by Yan Wong, from data E. Lasek-Nesselquist and J. P. Gogarten [237].

670 AWFUL CHANGES © Department of Geology, National Museum of Wales.

672 EYE LANDSCAPE drawn by L. Ward after original by M. F. Land, from R. Dawkins [96].

673 TEM OF THE OCELLOID OF *ERYTHROPSIDINIUM*, from S. Gavelis *et al.* [147], fig. 1b. Reproduced with kind permission of Nature Publishing Group.

Silhouettes for the fractal diagrams are from phylopic.org, with thanks to Mike Keesey for the website and many of the images. All silhouettes are public domain apart from the following, which are all distributed under the Creative Commons Attribution 3.0 unported license.

COMMON CHIMPANZEE: Tony Hisgett, vectorised by T. Michael Keesey.

SUMATRAN ORANG UTAN, NEMATODE, AMOEBA: Gareth Monger.

CAPUCHIN MONKEY, RABBIT, SLOTH, KANGAROO, MONITO DEL MONTE, TASMANIAN DEVIL, MARSUPIAL MOLE, SILKY SHREW OPOSSUM, VIRGINIA OPOSSUM, PLATYPUS, TAILED FROG: Sarah Werning.

PIGMY HIPPO: Marek Velechovský, vectorised by T. Michael Keesey.

PRZEWALSKII'S HORSE: Mercedes Yrayzoz, vectorised by T. Michael Keesey.

TORTOISE: Andrew A. Farke.

OSTRICH, SALAMANDERFISH: Darren Naish, vectorised by T. Michael Keesey.

ROBIN: Rebecca Groom.

HAGFISH: J. H. Richard, vectorised by Roberto Díaz Sibaja.

LAMPREY: Original drawing by unknown artist, vectorised by Roberto Díaz Sibaja.

STARFISH: Flickr user Wildcat Dunny, vectorised by Mali'o Kodis.

HAIRWORM: Eduard Solà Vázquez, vectorised by Yan Wong.

BEETLE: Birgit Lang.

NEMATODEMATID: I. Meyer-Wachsmuth, M. Curini Galletti, U. Jondelius, vectorised by Yan Wong.

JELLYFISH: Flickr user Ching, vectorised by Mali'o Kodis.

SPONGE: Derek Keats, vectorised by Mali'o Kodis.

INDEX

Key:
    208   image or diagram
    208n   footnote
    [**208**]   bibliography entry
    (plate 1)   coloured plates